REVIEWS in MINERALOGY
Volume 14

MICROSCOPI
MACROSC

Atomic Environments to Mineral Thermodynamics

S. W. KIEFFER & A. NAVROTSKY, Editors

The AUTHORS:

CHARLES W. BURNHAM
Dept. of Geological Sciences
Harvard University
Cambridge, Massachusetts 02138

ROGER G. BURNS
Dept. of Earth, Atmospheric, and Planetary Sciences
Massachusetts Institute of Technology
Cambridge, Massachusetts 02139

MICHAEL A. CARPENTER
Dept. of Earth Sciences
University of Cambridge
Cambridge CB2 3EQ, England

SUBRATA GHOSE
Dept. of Geological Sciences
University of Washington
Seattle, Washington 98195

ROBERT M. HAZEN
Geophysical Laboratory
2801 Upton Street NW
Washington, D.C. 20008

RAYMOND JEANLOZ
Dept. of Geology & Geophysics
University of California
Berkeley, California 94720

SUSAN WERNER KIEFFER
United States Geological Survey
Flagstaff, Arizona 86001

J. DESMOND C. McCONNELL
Dept. of Earth Sciences
University of Cambridge
Cambridge CB2 3EQ, England
[Also: Schlumberger Cambridge Research P.O. Box 153, Cambridge CB2 3BE]

PAUL McMILLAN
Dept. of Chemistry
Arizona State University
Tempe, Arizona 85287

ALEXANDRA NAVROTSKY
Dept. of Chemistry
Arizona State University
Tempe, Arizona 85287
[From 9/85: Dept. of Geological & Geophysical Sciences, Princeton University, Princeton, New Jersey 08544]

SERIES EDITOR: Paul H. Ribbe
Department of Geological Sciences
Virginia Polytechnic Institute & State University
Blacksburg, Virginia 24061

MINERALOGICAL SOCIETY OF AMERICA

PRINTED BY

BookCrafters, Inc.
Chelsea, Michigan 48118

REVIEWS in MINERALOGY

(Formerly: SHORT COURSE NOTES)

ISSN 0275-0279

Volume 14: MICROSCOPIC TO MACROSCOPIC
Atomic Environments to Mineral Thermodynamics

ISBN 0-939950-18-9

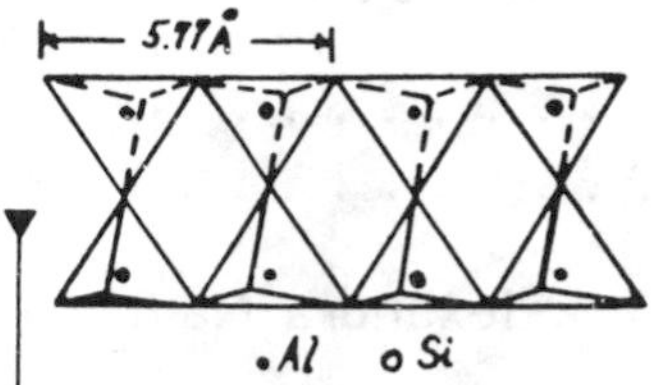

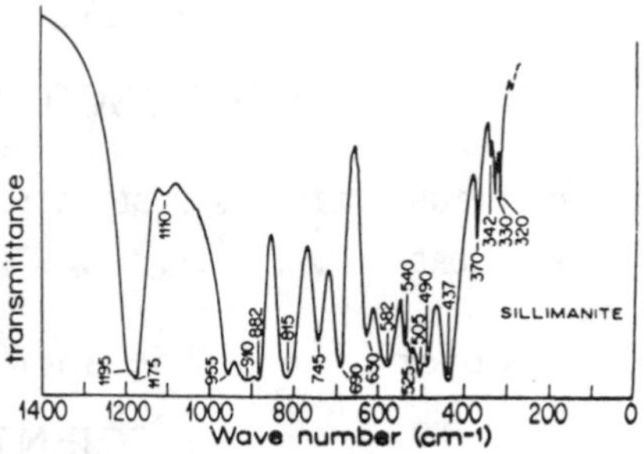

MICROSCOPIC to MACROSCOPIC

Atomic Environments to Mineral Thermodynamics

Mineralogical Society of America Short Course
May 24-26, 1985

FOREWORD

As Volume 14 in the REVIEWS in MINERALOGY series, "Microscopic to Macroscopic" represents an attempt to answer these questions: "What minerals exist under given constraints of pressure, temperature, and composition, and why?"

The upper and right margins of this page display a modified version of the logo of the short course that provided the impetus for the publication of this volume. Organized by Sue Kieffer and Alex Navrotsky, the course was presented by the ten authors of this book on the campus of Washington College in Chestertown, Maryland. This was the second of MSA's short courses to be given in conjunction with meetings of the American Geophysical Union.

Editorially, this book is the first one in the REVIEWS series to be produced entirely from manuscripts prepared on word processors. Consequently, there in some unevenness in style, but the benefits of this method of preparation of camera-ready copy far outweigh the disadvantages. In the near future, technical advances should greatly improve our appearance, if not our content.

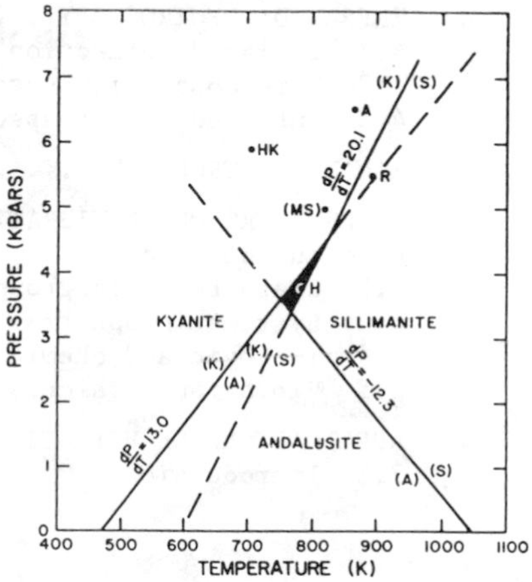

I thank Margie Strickler for her pleasant and untiring spirit of cooperation in typing and editing several of the chapters and Barbara Minich for her assistance with details of the printing arrangements.

Paul H. Ribbe
Series Editor
Blacksburg, VA
April 2, 1985

TABLE OF CONTENTS

MICROSCOPIC TO MACROSCOPIC

Chapter 3. Susan Werner Kieffer

HEAT CAPACITY and ENTROPY: SYSTEMATIC RELATIONS to LATTICE VIBRATIONS

Chapter 4. Subrata Ghose

LATTICE DYNAMICS, PHASE TRANSITIONS and SOFT MODES

Page

Chapter 5. J. Desmond C. McConnell

SYMMETRY ASPECTS of ORDER-DISORDER and the APPLICATION of LANDAU THEORY

Chapter 6. Michael A. Carpenter

ORDER-DISORDER TRANSFORMATIONS in MINERAL SOLID SOLUTIONS

Chapter 7. Alexandra Navrotsky

CRYSTAL CHEMICAL CONSTRAINTS on the THERMOCHEMISTRY of MINERALS

Chapter 8. Roger G. Burns

THERMODYNAMIC DATA from CRYSTAL FIELD SPECTRA

Chapter 10. Charles W. Burnham

MINERAL STRUCTURE ENERGETICS and MODELING USING the IONIC APPROACH

Chapter I. Susan Werner Kieffer & Alexandra Navrotsky

SCIENTIFIC PERSPECTIVE

1. PURPOSE OF THE COURSE AND THE VOLUME

The purpose of this short course is to examine the relations among the microscopic structure of minerals and their macroscopic thermodynamic properties. Understanding the micro-to-macro relations provides a rigorous theoretical foundation for formulation of energy relations. With such a foundation, measured parameters can be understood, and extrapolation and prediction of thermodynamic properties beyond the range of measurement can be done with more confidence than if only empirical relations are used.

Mineral systems are sufficiently complex in structure and properties that a balance must be sought between rigorous complexity and useless simplicity. Eventually, even the most rigorous thermodynamic analysis requires simplifying assumptions in order to be tractable for complex minerals, and a firm foundation in the microscopic fundamentals should underlie those assumptions.

The most fundamental questions of mineral physics and chemistry are "What minerals exist under given constraints of pressure, temperature, and composition, and why?" The macroscopic thermodynamic parameter defining mineral stability at a given pressure and temperature is the Gibbs free energy. The purpose of this course is to consider the microscopic factors that influence the free energy of minerals: atomic environments, bonding, and crystal structure. These factors influence the structural energy and the detailed nature of the lattice vibrations which are an important source of entropy and enthalpy at temperatures greater than 0 K. The same factors determine the relative energy of different phases, and thereby, the relative stability of different minerals. Configurational entropy terms arising from disorder also contribute to the energy and entropy. In transition metal compounds there are additional energy and entropy terms arising from the electronic configurations, leading to additional stabilizations, magnetic ordering, and, incidentally, color.

2. THEORETICAL FOUNDATIONS

A first-principles approach to the problems posed requires the tools of quantum mechanics and statistical mechanics, including the mathematics inherent in these disciplines. These cannot be developed in this short course but we recommend a background in the following subjects and suggest a few references:

<u>Quantum Mechanics</u>

Moelwyn-Hughes, E.A. (1961) Physical Chemistry, 2nd Ed. Pergamon Press.

Moore, W.J. (1972) Physical Chemistry, 4th Ed. Prentice Hall.

<u>Statistical Mechanics</u>

Davidson, N. (1962) Statistical Mechanics. McGraw-Hill, New York.

Feynman, R.P., Leighton, T.B. and Sands, M. (1963) The Feynman Lectures on Physics, v. 1. Addison-Wesley, Reading, Massachusetts.

Physical Chemistry texts listed above.

Solid State Physics

Brüesch, P. (1982) Phonons: Theory and Experiments I.. Springer-Verlag, Berlin.

Kittel, C. (1968) Introduction to Solid State Physics, 3rd ed. John Wiley, New York.

Ziman, J.M. (1972) Principles of the Theory of Solids. Cambridge, London.

Lattice Vibrational Theory

Birman, J.L. (1984) Theory of Crystal Space Groups and Lattice Dynamics. Springer-Verlag, New York.

Born, M. and Huang, K. (1954) Dynamical Theory of Crystal Lattices. Oxford, London.

Brillouin, L. (1946) Wave Propagation in Periodic Structures. McGraw-Hill, New York (republished by Dover in 1953).

Group theory

Burns, G. and Glazer, A.M. (1978) Space Groups for Solid State Scientists. Academy Press, New York.

Cotton, F.A. (1971) Chemical Applications of Group Theory. John Wiley, New York.

Decius, J.C., and Hexter, R.M. (1977) Molecular Vibrations in Crystals. McGraw-Hill, New York.

Mathematical background

Boas, M.Z. (1966) Mathematical Methods in the Physical Sciences. John Wiley and Sons, New York.

The atoms in a solid, according to the concepts of quantum mechanics, follow the Schroedinger equation:

$$H\Psi = E\Psi, \tag{1}$$

in which the term H, the Hamiltonian, is a quantum mechanical expression for the kinetic and potential energy of all the particles in the system, E is the energy, and Ψ is a wave function of the system. The correct set of wave functions are those for which the energy is an eigenvalue of the Hamiltonian, i.e. the energy is a real and scalar quantity. The main and originally revolutionary finding of quantum mechanics is that not all energy values occur as eigenvalues; rather, only a certain discrete set of energy levels are allowed. The computational difficulty of quantum chemistry arises not because the Schroedinger equation is intrinsically mysterious but because the Hamiltonian is generally very complex. It must contain terms which describe interactions among all the nuclei and electrons in the system. For atoms and isolated small molecules, the formalisms of atomic and molecular orbital theory (used largely by chemists), taken at different levels of computational complexity, are the common approach. For crystalline solids, the approach used by solid state physicists generally exploits the periodicity of the lattice to simplify the Hamiltonian, leading to various formulations ranging from the nearly-free-electron gas to pseudopotentials. Thus the reader is often faced with several superficially rather different descriptions of the same problem. For further details, see the references given above.

The revolutionary impact of quantum mechanics on the thermodynamics of solids arises from the quantization of energy levels. At high temperatures, when the spacing of energy levels is small compared to the available thermal energy, solids approach the classical behavior predicted by kinetic theory. In particular, the harmonic contribution to the heat capacity approaches 3nR per mole, where n moles of atoms per gram-formula-unit of solid each have freedom to vibrate in each of three independent directions; R is the gas constant. At low temperatures, when the thermal energy is no longer adequate to populate energy levels randomly, the heat capacity diminishes, and both C_P and C_V approach zero as $T \to 0$ K. The quantitative description of this behavior is one of the major subjects dealt with in this short course.

Quantitatively, the link between quantum mechanics and bulk thermodynamic properties is statistical mechanics. Once more, the difficulty of this subject arises not from its basic concepts but from the mathematical complexity of dealing with many-body interactions and complex Hamiltonians. Two concepts pervade this complexity which, in themselves, can be described simply and physically. These are the Boltzmann factor and the partition function.

Consider a system with energy levels ε_0, ε_1, ε_2 ...(with ε_0 the lowest energy as ground state). If the components are distinguishable, the ratio of particles in an excited state (N_i particles having energy $\Delta\varepsilon_i = \varepsilon_i - \varepsilon_0$) to those in the ground state (N_0 particles with $\Delta\varepsilon = 0$) is given by the Boltzmann factor

$$\frac{N_i}{N_0} = \exp(-\Delta\varepsilon_i/kT) = \exp(-\Delta E_i/RT), \qquad (2)$$

where $\Delta\varepsilon_i$ is the energy difference per molecule, k the Boltzmann constant, and ΔE_i is the energy difference per mole.

The partition function, or sum over states (german "Zustandsumme"), is given by

$$q = \sum_{i=0}^{\infty} \exp(-\Delta\varepsilon_i/kT). \qquad (3)$$

The total energy is given by

$$\varepsilon = \sum_{i=0}^{\infty} N_i \varepsilon_i . \qquad (4)$$

Thus if one knows the energy level spacings and the temperature, one can calculate the Boltzmann factors, the partition function, and the total energy. The temperature dependence of the energy gives the heat capacity, C_V, and the integral of $(C_V/T)/dT$ gives the harmonic entropy. Thus the partition function lets one calculate macroscopic thermodynamic properties from microscopic energy level spacings.

If there is only one energy level having the lowest energy (nondegenerate ground state) then $q \to 1$ as $T \to 0$. In that case the entropy also is zero at $T = 0$. A degenerate ground state arises when several (n) physically distinguishible configurations all have the same lowest energy. Then the system has a zero point entropy of R ln(n). Substitutional disorder in a solid solution is one common example of such a situation.

An example applying the above ideas is the quantum harmonic oscillator described by McMillan (this volume). Its energy levels are quantized and given by

$$\varepsilon_i = (i + 1/2)h\nu = (i + 1/2)\hbar\omega. \tag{5}$$

where ω is the angular frequency and $\nu = \omega/2\pi$. The vibrational partition function is

$$\sum_{i=0}^{\infty} \exp\left(-(i + 1/2)h\nu/kT\right) = \frac{\exp\left(-h\nu/2kT\right)}{\left(1 - \exp(-h\nu/kT)\right)}. \tag{6}$$

Alternatively to this description in "oscillator language", the quantum harmonic oscillator can be described in "phonon language". In the phonon description, the analogy to the quantum theory of the electromagnetic field (photons) is used. In that theory, it is said that there are n_i indistinguishable particles, called phonons, in the system. The various n_i's are called the occupation numbers. The particle formulation is especially convenient when interactions between phonons, photons, electrons, or neutrons are being discussed. Phonons obey Bose-Einstein statistics, and at temperature T, the mean occupation number of phonons in the state i is given by the Bose-Einstein distribution:

$$n_i(T) = \left(\exp(h\nu/kT) - 1\right)^{-1}. \tag{7}$$

At low temperatures, only low-frequency phonons are excited to any appreciable extent, and $n_i \sim \exp(h\nu/kT)$; at high temperatures, more phonons are excited, and $n_i \sim kT/h\nu$. Although the statistical counting process is different in the oscillator and phonon treatments, the thermodynamic results are the same. The reader will need to grasp these concepts to really work with the ideas presented in this course.

If one considers different chemical entities (e.g. reactants and products in a chemical reaction, polymorphs in a phase transition) then each phase will have its own Hamiltonian, energy levels, partition functions, and thermodynamic properties. To compare the stabilities of different phase assemblages, one must know not only the vibrational and configurational thermodynamic properties within that phase but the difference in ground state energies between the different assemblages. These ground state energies are often calculated relative to isolated atoms or ions, and the differences between them are often several orders of magnitude smaller than the total stabilization energy of the condensed phase relative to isolated species. Proper consideration of these energy differences requires detailed understanding of chemical bonding within each phase. Thus phase stability (the major petrologic and mineralogic interest) is governed by a fine balance of microscopic factors involving both static atomic environments and lattice vibrations.

3. SPECTRA

The window into the microscopic world is the interaction of radiation with atomic matter. The parts of the acoustic and electromagnetic spectrum are shown in Figure 1, along with useful conversions between different units of frequency and wavelength (see also Burns, this volume, Fig. 2). The reader should be prepared for a dizzying journey in scale -- from the world of sound waves (approaching infinite wavelength compared to the scale of a unit cell) through the worlds of infrared, visible, ultraviolet and x-ray wavelengths.

Sound, as popularly conceived, is the phenomenon of propagation of waves in a solid, liquid, or gas, although, as discussed above, sound waves also have a particle-like nature, the particles being called phonons. In solids, the sound waves have transverse and longitudinal components. Sound waves, in the sense important to thermodynamics, are not limited to the audible range of the human ear (16 to 16,000 cycles per second), but extend to much lower frequency and very long wavelength. The speed of sound of the longitudinal and transverse waves in minerals varies with composition and structure, but is typically a few kilometers per second.

Radiant energy is emitted over a tremendous range of frequencies, referred to as the electromagnetic spectrum, shown schematically in Figure 1. The main difference between the various forms is in the way in which they are excited and manifested, and in the frequencies at which they occur. Electromagnetic radiation travels through free space at 3×10^{10} cm/sec. Under the proper conditions, all electromagnetic radiation fields can exhibit the wave-like properties of refraction, interference, and diffraction, these being the basis for a wide range of experimental techniques involving the use of radiation to probe the interatomic nature of solids. Other phenomena -- such as radiation from "hot" bodies, the production of x-rays, and the photoelectric effect -- cannot be explained by the behavior of waves, but can be explained by considering the radiation to consist of a stream of small particles called photons.

The range of magnitudes of wavelengths of radiation is so great that the electromagnetic spectrum is conveniently subdivided according to the ways in which the various types of radiation are produced and used. Unfortunately, it is traditional to change units of measurement in discussions of different parts of the spectrum, e.g., wavelengths corresponding to radio frequencies are usually expressed in meters; visible light in centimeters, angstroms, or micrometers (microns); and x-rays or gamma rays in angstroms or millimicrons. The reader might find the following conversion factors helpful:

$$10{,}000\ \mathrm{cm}^{-1} \triangleq 1000\ \mathrm{nm} \triangleq 1\ \mu\mathrm{m} \triangleq 10^{-6}\ \mathrm{m}$$

$$10{,}000\ \mathrm{cm}^{-1} \triangleq 1.99 \times 10^{-19}\ \mathrm{J} \triangleq 4.75 \times 10^{-20}\ \mathrm{cal} \triangleq 1.24\ \mathrm{eV}$$

$$1\ \mathrm{eV} \triangleq 8066\ \mathrm{cm}^{-1}$$

$$1\ \mathrm{cm}^{-1} \triangleq 1.438\ \mathrm{K}$$

$$1\ \mathrm{K} \triangleq 0.695\ \mathrm{cm}^{-1}$$

$$10{,}000\ \mathrm{cm}^{-1} = 3 \times 10^{14}\ \mathrm{sec}^{-1}(\nu) = 1.88 \times 10^{15}\ \mathrm{cps}\ (\omega)$$

($\triangleq$ means "corresponds to".)

FIGURE 1. THE ELECTROMAGNETIC SPECTRUM

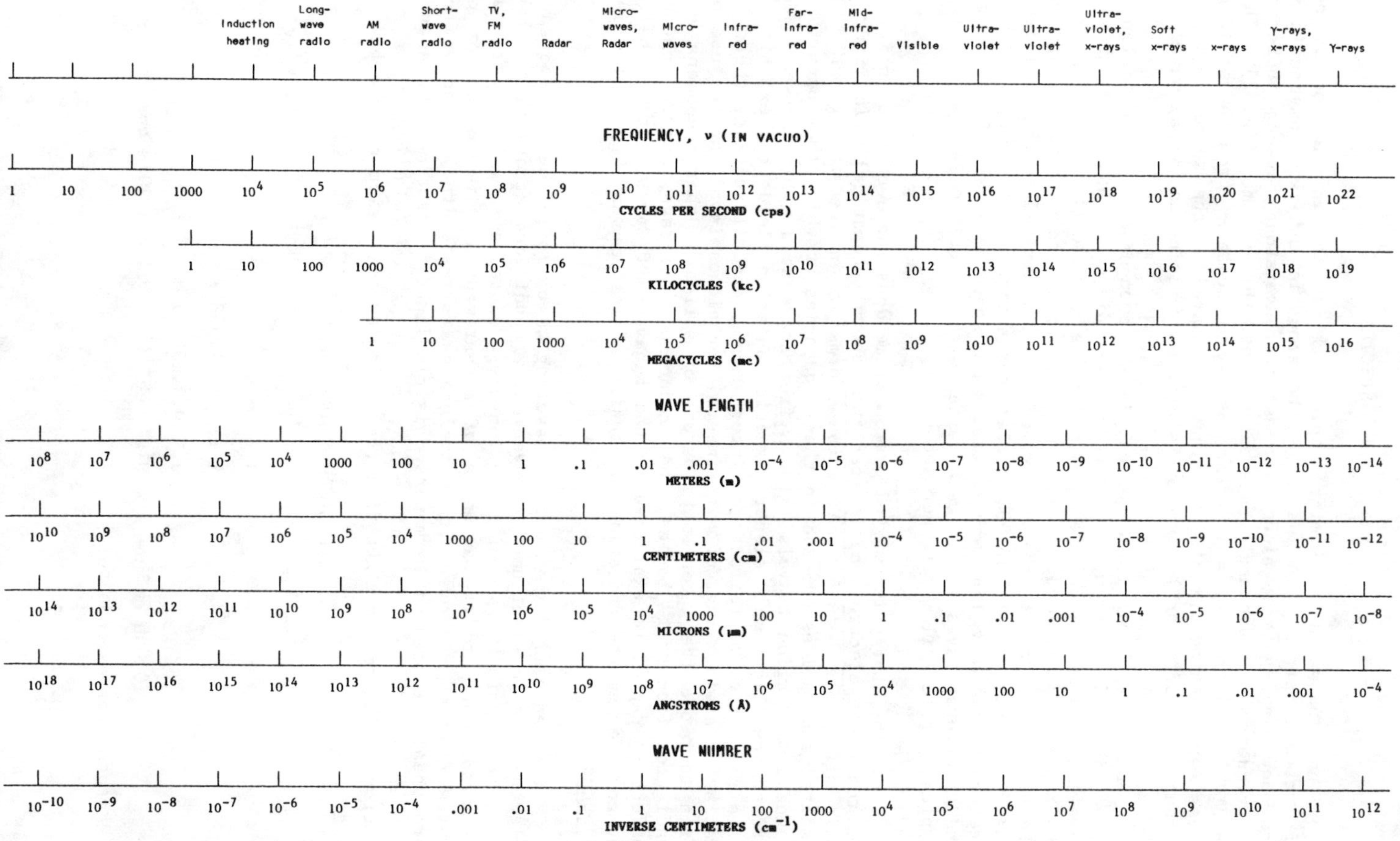
Induction heating
Long-wave radio
AM radio
Short-wave radio
TV, FM radio
Radar
Micro-waves, Radar
Micro-waves
Infra-red
Far-Infra-red
Mid-Infra-red
Visible
Ultra-violet
Ultra-violet
Ultra-violet, x-rays
Soft x-rays
x-rays
γ-rays, x-rays
γ-rays
FREQUENCY, ν (IN VACUO)
1 10 100 1000 10^4 10^5 10^6 10^7 10^8 10^9 10^{10} 10^{11} 10^{12} 10^{13} 10^{14} 10^{15} 10^{16} 10^{17} 10^{18} 10^{19} 10^{20} 10^{21} 10^{22}
CYCLES PER SECOND (cps)
1 10 100 1000 10^4 10^5 10^6 10^7 10^8 10^9 10^{10} 10^{11} 10^{12} 10^{13} 10^{14} 10^{15} 10^{16} 10^{17} 10^{18} 10^{19}
KILOCYCLES (kc)
1 10 100 1000 10^4 10^5 10^6 10^7 10^8 10^9 10^{10} 10^{11} 10^{12} 10^{13} 10^{14} 10^{15} 10^{16}
MEGACYCLES (mc)
WAVE LENGTH
10^8 10^7 10^6 10^5 10^4 1000 100 10 1 .1 .01 .001 10^{-4} 10^{-5} 10^{-6} 10^{-7} 10^{-8} 10^{-9} 10^{-10} 10^{-11} 10^{-12} 10^{-13} 10^{-14}
METERS (m)
10^{10} 10^9 10^8 10^7 10^6 10^5 10^4 1000 100 10 1 .1 .01 .001 10^{-4} 10^{-5} 10^{-6} 10^{-7} 10^{-8} 10^{-9} 10^{-10} 10^{-11} 10^{-12}
CENTIMETERS (cm)
10^{14} 10^{13} 10^{12} 10^{11} 10^{10} 10^9 10^8 10^7 10^6 10^5 10^4 1000 100 10 1 .1 .01 .001 10^{-4} 10^{-5} 10^{-6} 10^{-7} 10^{-8}
MICRONS (μm)
10^{18} 10^{17} 10^{16} 10^{15} 10^{14} 10^{13} 10^{12} 10^{11} 10^{10} 10^9 10^8 10^7 10^6 10^5 10^4 1000 100 10 1 .1 .01 .001 10^{-4}
ANGSTROMS (Å)
WAVE NUMBER
10^{-10} 10^{-9} 10^{-8} 10^{-7} 10^{-6} 10^{-5} 10^{-4} .001 .01 .1 1 10 100 1000 10^4 10^5 10^6 10^7 10^8 10^9 10^{10} 10^{11} 10^{12}
INVERSE CENTIMETERS (cm^{-1})

Other useful values are: h (Planck's constant) = 6.6255×10^{-34} Js; k (Boltzmann's constant) = 1.3804×10^{-23} JK^{-1}; N (Avogadro's number of particles per mole) = 6.0248×10^{23} mol^{-1}. Some of the conversions are shown in Figure 1; others are given in the article by Burns. Although we have asked our authors to use SI (System International or, sometimes, System Inconvenient) wherever possible without distorting the clarity of their work, it is simply a fact of science in the 20th century that there is no single system of units convenient and intuitive to all researchers.

Different experimental techniques provide information about a mineral on different scales. One way to summarize this is to say that the characteristic wavelengths, λ, and frequencies, ω (the circular frequency), or ν (the frequency) are different. (Remember that $\omega = 2\pi\nu$; $\lambda = v/\nu$, where v is the appropriate speed of wave propagation; and $K = 2\pi/\lambda$, where K is the wave vector, a vector of length $2\pi/\lambda$ pointing in the direction of propagation of the wave.)

Diffraction experiments give information about the periodicity of mineral structures on the atomic scale; absorption and emission experiments give information about the energy levels of the atomic or molecular constituents. A powerful tool for investigating phonon behavior is inelastic neutron scattering, because, in contrast to x-rays, neutrons have energies of the same order of magnitude as phonons. The neutron beam is diffracted by the crystal, and there is elastic scattering according to Bragg's law. The neutron beam also exchanges energies with the phonons, and consequently, by measuring the change in direction and energy of scattered neutrons, it is possible to analyze the phonon frequencies as a function of wave vector. In this course, experimental information from many different techniques is brought to bear on the subject of mineral lattice dynamics and energetics.

Different experimental techniques provide information about different energy levels in a crystal. If a molecule with rotational, vibrational, and electronic energy levels is placed in an electromagnetic field, a transfer of energy from the field to the molecule (or the reverse) can only occur if

$$\Delta E = h\nu, \tag{8}$$

where ΔE is the difference in energy between two quantum states. The energy differences between rotational levels are typically small, so the frequency of light emitted or absorbed in such a transition is small -- usually between 1 cm^{-1} (10^4 μm) and 10^2 cm^{-1} (10^2 μm). This corresponds to the microwave to far-infrared portion of the spectrum. The energy differences between vibrational levels are greater, and the transitions therefore occur at higher frequencies -- typically between 10^2 cm^{-1} (10^2 μm) and 10^4 cm^{-1} (1 μm). This corresponds to the far-infrared to mid-infrared portion of the spectrum. Electronic energy levels are widely spaced, and electronic spectra are observed at correspondingly higher frequencies -- between 10^4 cm^{-1} (1 μm and 10^5 cm^{-1} (10^{-1} μm). This corresponds to the near-infrared, visible and ultraviolet portion of the spectrum. Overlap of these ranges can, of course, occur, even for simple diatomic molecules. If the molecule (or ion) is situated in a crystalline lattice, the frequencies it would have as an isolated molecule are perturbed by the change in symmetry and by intermolecular interactions, and extra lattice modes are introduced. Nevertheless, the order of magnitude of the rotational (librational), vibrational, and electronic energies is preserved. The full range of energy levels is covered by the authors of the papers in this volume.

4. ORGANIZATION OF THIS VOLUME

We begin our course with a discussion of the techniques used, and information obtained, about atomic environments and lattice vibrations from vibrational spectroscopy (McMillan). This paper introduces concepts and terminology common to many of the contributions. Kieffer then discusses interpretations of vibrational spectra, their "integration" (literally) to give the harmonic contribution to entropy and energy, and systematic trends in vibrational spectra and thermodynamic properties.

In the third paper, Ghose shows how inelastic neutron scattering data provide information about vibrations at all wave vectors throughout the Brillouin zone, information that supplements the infrared and Raman techniques discussed by McMillan. Such data give information about anharmonic effects, and Ghose discusses the phase transitions associated with "soft" vibrational modes. He also introduces some constraints imposed on phase transitions by the symmetry of the phases involved as formalized in "Landau theory". McConnell takes symmetry arguments further in delineating possible phase transitions in minerals, and Carpenter combines these theoretical constraints with experimental observations on aluminum-silicon order-disorder in framework silicates.

The role of atomic environments in determining energetics is next described by Navrotsky, who discusses thermochemical systematics of heats and free energies of formation and phase transitions. The roles of coordination environments and order-disorder are stressed. Burns supplements this discussion with a paper on electronic energy levels and their influence on energetics for transition metal compounds.

Hazen summarizes the systematic variations of the geometry of the coordination polyhedra with pressure, temperature, and composition. Burnham then discusses computational techniques which permit the calculation of crystal sructure geometries and energetics using optimum interatomic distances (DLS methods) and the minimization of cohesive energy (lattice or structural energy). These methods relate local atomic environments to long-range crystal structure and relative energetic stability. Finally, Jeanloz explores some general trends in the thermodynamics of phase transitions at high pressure and temperature.

An innovation in this volume is a set of worked problems contributed by each author to amplify details, methods, and approaches. We hope that these problems will help students overcome the conceptual and computational barriers that seem to make thermodynamics formidable -- it´s really nothing but fun!

ACKNOWLEDGMENTS

We much appreciate the expert guidance and help of Paul Ribbe as series editor; during the many months of preparation of manuscripts we came to know that every question was going to be answered with a cheerful "No problem!". Charlie Prewitt provided thoughtful and constructive advice during the early stages of planning for the Short Course. Barbara Minich expertly handled the logistics of the Short Course.

Chapter 2. Paul McMillan

VIBRATIONAL SPECTROSCOPY in the MINERAL SCIENCES

1. INTRODUCTION

The atomic vibrations of a molecular system are governed by the interatomic potential determined by the type and relative arrangement of its constituent atoms. At the same time, many of the bulk thermodynamic and material properties of the system are related directly or indirectly to its vibrational nature, hence a study of vibrational properties can form a true link between the microscopic and macroscopic characteristics of the material. The purpose of the present chapter is to introduce some of the basic concepts and terminology of vibrational theory and spectroscopy, describe certain aspects of experimental infrared and Raman spectroscopy, and discuss some of their applications to problems in geochemistry and mineralogy.

The following discussion is intended to provide a coherent description of the vibrations of both small molecules and crystalline systems. In each case, the standard terminology has been adhered to, except when this might have led to ambiguity. In general, references to standard models and derivations have been omitted. The interested reader is referred to one of the following texts and classic works in the field for more detailed treatment and original citations. The vibrations of small molecules are discussed in great detail by Herzberg (1945, 1950) and by Wilson et al. (1955), while Born and Huang (1954) remains one of the definitive treatments of the dynamics of crystalline lattices. Most of the fundamental concepts of molecular vibrational theory are described in physical chemistry texts, for example Moore (1972), Berry et al. (1980), or Atkins (1982), while the dynamics of crystals are developed in texts in solid state physics such as Kittel (1976) or Ashcroft and Mermin (1976). A number of recent books give an introduction to the theory and practice of solid state vibrational spectroscopy; these include Sherwood (1972), Turrell (1972), Cochran (1973), and Decius and Hexter (1977). Banwell (1972) provides a useful introduction to molecular spectroscopy including infrared absorption and Raman scattering. Hadni (1967) gives a comprehensive account of infrared spectroscopy, while Long (1977) gives an excellent discussion of Raman scattering and related spectroscopies. The compilations by Ross (1972) and Nakamoto (1978) form useful sources of vibrational data, while the books by Lazarev (1972), Farmer (1974) and Karr (1975) present many aspects of the application of vibrational spectroscopy in the mineral sciences.

2. SIMPLE MODELS AND CONCEPTS

2.1 The simple harmonic oscillator.

The simplest model for vibration of a diatomic molecule is provided by the classical harmonic oscillator. Two masses m_1 and m_2 are connected by an ideal spring. At rest, the masses have an equilibrium separation r_o: on extension or compression ($r_o \pm \Delta r$) the masses are subject to a restoring force proportional to the displacement

$$F = -k\Delta r \tag{1}$$

This is known as Hooke's law, and the proportionality constant k is the force constant of the spring, expressing its resistance to compression or extension. The potential energy function for the harmonic oscillator is then

$$V = \frac{1}{2} k(\Delta r)^2 \quad . \tag{2}$$

The force constant k is the curvature of the function V(r); $k = d^2V/d(\Delta r)^2$ (Fig. 1). Most diatomic molecules have force constants in the range $10^2 - 10^3\,Nm^{-1}$ ($10^7 - 10^8$ dynes/cm) (Huber and Herzberg, 1979).

It is often taken as a general rule that molecules with stronger bonds have higher force constants, and this is commonly the case. There are however a few exceptions; for example HI has a potential well depth (D_e: see Section 2.3) of 308 $kJmol^{-1}$ and a force constant k = 314 Nm^{-1}, while Cl_2 with a weaker bond (D_e = 242 $kJmol^{-1}$) has k = 333 Nm^{-1} (Herzberg, 1950; Huber and Herzberg, 1979). A common expression for estimating force constants for diatomic molecules which is often extended to atom pairs in polyatomic systems is known as Badger's rule (Badger, 1934, 1935; Herzberg, 1950, p. 457),

$$k = \frac{186}{(r_o - d_{ij})^3}\ Nm^{-1}, \tag{3}$$

where r_o is the equilibrium bond length (tabulated as r_e for real, anharmonic molecules: Section 2.3) in Angstroms, and d_{ij} is a constant depending on atoms i and j in the atom pair. Values for d_{ij} are tabulated in Table 1, along with some calculated and observed values of k using Equation 3. The agreement is generally good, except for some anomalous cases such as MgO (Table 1).

The Newtonian equation of motion for the harmonic oscillator is given by equating the expressions for force:

$$\mu \frac{d^2 \Delta r}{dt^2} = -k\Delta r \ . \tag{4}$$

Here, μ is the reduced mass,

$$\mu = \frac{m_1 m_2}{m_1 + m_2} \quad , \tag{5}$$

with typical values of 10^{-26} to 10^{-27} kg. One solution to the differential Equation 4 is given by

$$r = A \sin\left(t\sqrt{k/\mu} + \phi\right) \quad . \tag{6}$$

There are two integration constants A and ϕ. ϕ is a phase angle given by the initial value of Δr and usually set to zero; A is the amplitude

Table 1. Estimating force constants using Badger's rule.

(a) Badger's rule parameters d_{ij} (Berry et al., 1980, p. 268)

	Atom j			
Atom i	H	Row 1	Row 2	Row3
H	0.025	0.335	0.585	0.6550
Row 1	0.335	0.680	0.900	
Row 2	0.585	0.900	1.180	

(b) Comparison of calculated and observed force constants for diatomic molecules (data from Huber and Herzberg, 1979).

Molecule	r_e(Å)	k_e(Nm^{-1})	k(Badger's rule; Nm^{-1})
H_2	0.7414	580	506
O_2	1.2075	1186	1271
MgO	1.749	351	152*
SiO	1.510	933	819
OH	0.970	787	726

(c) Estimates of stretching forces for atom pairs in silicates

Bond	r(Å)	k(Nm^{-1})	Range used in calculations
Si^{IV}-O	1.6	498	300-700 [a]
Al^{IV}-O	1.8	255	134 [b]
Al^{VI}-O	2.0	140	71-147 [c]
Mg^{VI}-O	2.1	108	30-90 [a]

[a]McMillan (1984). [b]Iishi et al. (1979). [c]Iishi (1978c).

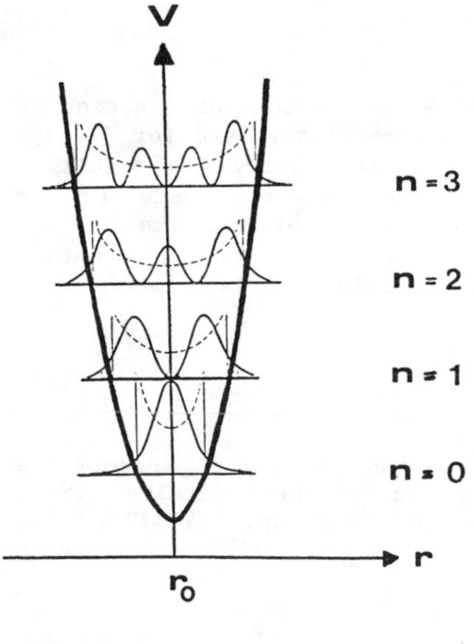

Figure 1. The harmonic oscillator: energy levels and wave functions. The potential is bounded by the curve V = 1/2 $k(\Delta r)^2$ (heavy solid line). The quantum mechanical probability density functions $|\psi_n|^2$ are shown as light solid lines for each energy level, while the corresponding clasical probabilities are shown as dashed lines.

of the oscillation and is defined by the energy of the system. The frequency of the oscillation is defined as the number of cycles in unit time,

$$\nu_o = \frac{1}{2\pi}\sqrt{k/\mu} \ , \qquad (7)$$

with ν_o expressed in Hz, or s^{-1}. Assuming $k = 10^2 - 10^3$ Nm^{-1} and $\mu = 10^{-27} - 10^{-26}$ kg, typical diatomic molecules have frequencies around 10^{12}-10^{14} s^{-1}, corresponding to the infrared region of the electromagnetic spectrum and suggesting that infrared radiation should resonate with molecular vibrations (see introduction to this volume).

2.2 The quantized harmonic oscillator

The classical harmonic oscillator model does not rationalize the interaction of molecular vibrations with light, nor does it suggest why a discrete line spectrum rather than a continuous exchange of energy is observed. To understand these, the quantum mechanical oscillator must be considered.

Schrödinger's wave equation may be expressed by

$$\hat{H}\psi = E\psi \ , \qquad (8)$$

where ψ is the wave function describing the phenomenon of interest (in this case, the vibrational motion), E is the energy of the system, and $\hat{H}$ is the Hamiltonian operator

$$\hat{H} = \frac{h^2}{8\pi^2 m} \nabla^2 + V \quad . \qquad (9)$$

The Laplacian operator ∇^2 is defined as

$$\nabla^2 = \frac{\partial^2}{\partial x^2} + \frac{\partial^2}{\partial y^2} + \ldots . \qquad (10)$$

over all space coordinates x,y,..., while V is a potential function in these coordinates. The wave Equation 8 is generally solved for ψ and E by choosing an appropriate expression for V and solving the resulting differential equation. For the harmonic oscillator, there is one coordinate Δr, and the potential function is given by Equation 2. The mass m in Equation 9 must be replaced by the reduced mass μ (Equation 5). This gives the wave equation for the harmonic oscillator as

$$\frac{d^2\psi}{dr^2} + \frac{4\pi^2\mu}{h}(2E - k(\Delta r)^2)\psi = 0 \quad . \qquad (11)$$

This may be solved exactly for ψ and E using appropriate mathematical techniques, for example via factorization (Atkins, 1980, pp. 57-65), or a power series method (Moore, 1972, pp. 619-622). The solution forms a set of vibrational eigenfunctions ψ_n and their associated energies E_n. The subscripts n are integers corresponding to the vibrational quantum numbers. These arise naturally from indices in the power series solution, or as orders of differentiation in the factorization method. The vibrational energy levels are found to be

$$E_n = \left(n + \frac{1}{2}\right) \frac{h}{2\pi} \sqrt{k/\mu}$$

$$= \left(n + \frac{1}{2}\right) h\nu_0 \quad , \tag{12}$$

where ν_0 is the classical harmonic oscillator frequency (Equation 7). These form a set of equidistant energy levels, spaced by $\Delta E = h\nu_0$, bounded by the harmonic oscillator potential curve (Fig. 1). The lowest, or ground state level with n = 0 has an energy $E_0 = 1/2\ h\nu_0$, known as the zero-point energy. This has the consequence that all molecular systems have vibrational energy even at 0°K, contributing to their internal energy (see Herzberg, 1945, p. 501-526).

The corresponding vibrational eigenfunctions are given by

$$\psi_n = C_n H_n(y) e^{-y^2/\Delta r} \quad , \tag{13}$$

where $y = \left(\frac{4\pi^2 \mu k}{h^2}\right)^{1/4} \Delta r$, and C_n is a normalization factor, $C_n = (2^n n! \sqrt{\pi})^{-1/2}$. The functions $H_n(y)$ are the Hermite polynomials, for example, $H_0(y) = 1$; $H_1(y) = 2y$; $H_2(y) = 4y^2 - 2$; $H_3(y) = 8y^3 - 12y$ . . . These wave functions are usually interpreted in terms of the probability density functions $|\psi_n|^2$, drawn for the first few vibrational levels in Figure 1. Also shown for comparison at each level is the classical probability function. Classically, the most probable values of r occur when the oscillator is moving most slowly, i.e. at the turning points at the extrema of extension and compression of the spring. In contrast, the maximum quantum mechanical probability in the ground state is at $r = r_0$ (Fig. 1). The number of nodes in the wavefunction increases with higher quantum numbers, to approach the classical probability function for highly excited vibrational states. The relative population of these vibrational states as a function of temperature is governed by the Maxwell-Boltzmann distribution,

$$\frac{N_n}{N_0} = e^{\frac{-(E_n - E_0)}{kT}} \quad , \tag{14}$$

where N_n/N_0 is the population of the nth state relative to the ground state, k is the Boltzmann constant, and T is the absolute temperature (Kieffer, this volume).

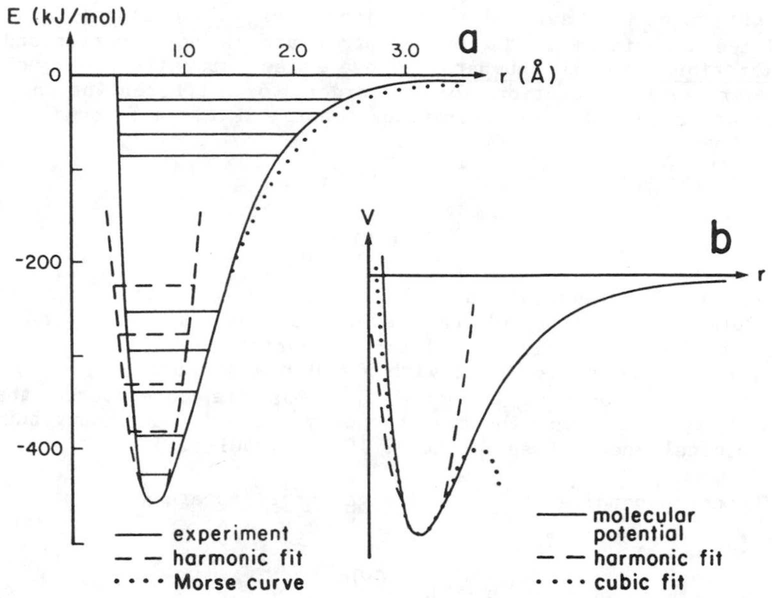

Figure 2. (a) The experimental potential curve for H_2, redrawn from Berry et al. (1980, p. 259) (full curve). The dotted curve represents a Morse fit to the data (Herzberg, 1950, p. 99), and the dashed curve shows a harmonic oscillator function with the force constant taken at $r = r_e$ (0.7414 Å). The first five energy levels are drawn for the harmonic (dashed lines) and the observed (full lines) potential functions. (b) Comparison of a typical molecular potential function (full curve), a harmonic fit (dashed curved), and a cubic anharmonic fit (dotted curve).

2.3 The anharmonic oscillator

Although the harmonic oscillator is a useful model for beginning to describe molecular vibrations, it is known that true interatomic potentials are not harmonic. The repulsive forces resisting bond compression vary more rapidly with distance than the attractive forces resisting separation. The two effects combine to give a minimum in energy at the equilibrium bond distance r_e, but the potential rises steeply at small r while it flattens off to zero energy at large r (Fig. 2). The analytic form of such potentials may be measured for some molecules (by analysis of vibration-rotation structure in electronic spectra (Herzberg, 1945), from molecular beam scattering experiments (Fluendy and Lawley, 1973) or may be accurately calculated via high level ab initio calculations (Szabo and Ostlund, 1982). However, these potentials are usually approximated by an empirical potential function, for example the Morse potential (Fig. 2)

$$V(r) = D_e(1 - e^{-ar})^2, \tag{15}$$

where a is a constant for the system, $\sim \pi\nu_0\sqrt{2\mu/D_e}$, and D_e is the depth of the potential well at the minimum. This is distinguished from the spectroscopically measured bond dissociation energy D_o, corresponding to the height of the well above the zero point level (Figure 2).

The effects of anharmonicity can also be approximated by adding higher order terms in r to the harmonic potential function:

$$V = \frac{1}{2} kr^2 - \frac{1}{3} gr^3 + ... \tag{16}$$

This anharmonic function has energy levels

$$E_n = h\nu_e[(n + \frac{1}{2}) - x_e(n + \frac{1}{2})^2 + y_e(n + \frac{1}{2})^3 - ...] \tag{17}$$

where ν_e, x_e, y_e (higher terms are also often given in tabulations) are known as the anharmonicity constants for the molecule, and are measured from series of vibrational lines (Herzberg, 1950, p. 53-55, 92-93). The subscripts "e" (i.e., x_e, y_e, ν_e, r_e) refer to quantities measured with respect to the bottom of the energy well. Due to anharmonicity, the bond length in the ground vibrational state (r_o) is not the same as the distance corresponding to the minimum, r_e (in general, $r_e < r_o$). Further, the curvature of the potential energy function at the minimum (k_e) gives a harmonic frequency ν_e which is generally greater than the observed fundamental frequency ν_o (there are only a few exceptions: see Huber and Herzberg, 1979). The quantity ν_e (or its equivalent in wavenumbers: see Section 2.4) is often tabulated as the "frequency corrected for anharmonicity" (e.g., Nakamoto, 1978, p. 105-111). These corrected frequencies ν_e must be used for realistic comparisons of molecular force constants, but do not correspond to the fundamental frequencies observed in infrared or Raman spectra. The observed frequencies can be calculated from ν_e via

$$\nu_o = \nu_e(1 - 2x_e + \frac{26}{8} y_e \; . \; . \; .) \quad , \tag{18}$$

but x_e and y_e must also be known. Finally, the zero-point energy is given by

$$E_o = h\nu_e(\frac{1}{2} - \frac{x_e}{4} + \frac{y_e}{8} + \; . \; . \; .) \quad . \tag{19}$$

In general, x_e and y_e are positive, with $1 \gg x_e \gg y_e$. The zero-point level is lower in energy for the anharmonic than for the corresponding harmonic oscillator. From Equation 17, the spacing between adjacent levels is also smaller, and this spacing decreases with increasing vibrational quantum number (Fig. 2). At the same time, the equilibrium bond length r_o increases for higher vibrational levels, and in general, anharmonic effects are responsible for the thermal expansion of materials.

2.4 Interaction with light

The final topic to approach in this section is the interaction of light with molecular vibrations. Electromagnetic radiation is simply described as mutually perpendicular electric and magnetic fields oscillating in phase in free space and travelling with constant velocity $c = 2.9977 \times 10^8 \ ms^{-1}$ (see Introduction to this volume). The energy of the radiation is characterized by the frequency (ν) or wavelength (λ) of the wave

$$E = h\nu = \frac{hc}{\lambda} \quad . \tag{20}$$

In vibrational spectroscopy, it is also common to express the energy in terms of wavenumbers defined by $\bar{\nu} = 1/\lambda$. The common units used are wavenumbers in cm^{-1} (Raman spectroscopy), wavelength in μm or nm (infrared spectroscopy), or frequency in THz ($10^{12}s^{-1}$: inelastic neutron spectroscopy). These may be simply inter-converted using Equation 20 and values for the physical constants c and h (Table 2).

The propagation of the electromagnetic wave is described by defining a <u>wave vector</u> $\vec{k}$ in the direction of propagation, of magnitude

$$k = \frac{1}{\lambda} \quad . \tag{21}$$

(This should not be confused with the <u>wavenumber</u> $\bar{\nu}$ defined above, used to express an <u>energy</u> term). The relation between energy and wave vector may be written as the <u>dispersion relation</u>

$$E = hck \quad . \tag{22}$$

The dispersion relation E(k) of light then forms a straight line passing through the origin, with constant slope hc.

<u>Infrared absorption</u>. It was noted earlier that the mechanical frequency associated with molecular vibrations corresponded with that of the infrared region of the electromagnetic spectrum. Infrared absorption occurs when the oscillating electric field of the light resonates with a fluctuating electric dipole caused by the vibration, when the energy of the radiation matches that of the vibration.

If two centers of equal and opposite charge $\pm Q$ are separated by a distance r, these constitute an electric dipole with a <u>dipole moment</u> of magnitude

$$\mu = Qr \quad . \tag{23}$$

Such a dipole moment may be defined between any two atoms of differing electronegativity. If the atoms are at rest with $r = r_o$, Equation 23 defines the equilibrium dipole moment μ_o. For a polyatomic molecule, each atom pair defines a dipole moment, and these <u>bond dipole moments</u> may be summed vectorially to give a net molecular dipole moment. The relative motion of the nuclei during a vibrational mode may cause a change in the net dipole moment.

$$\tilde{\mu}(r) = \tilde{\mu}_o + \left(\frac{d\mu}{dr}\right)\Delta r \quad . \tag{24}$$

Since the vibrational motion is oscillatory, this gives rise to a fluctuating electric field which may interact with the electric part of the electromagnetic radiation. By analogy with Equation 23, the derivative in 24 defines an <u>effective charge</u> Q^*, whose magnitude is related to the strength of the infrared absorption as described below.

Table 2. Units and conversion factors.

Standard Physical Constants

Speed of light in vacuum	$c = 2.997925 \times 10^{8}\ \mathrm{ms^{-1}}$
Planck constant	$h = 6.6262 \times 10^{-34}$ Js
Electronic charge	$e = 1.60210 \times 10^{-19}$ C
Boltzmann constant	$k = 1.38062 \times 10^{-23}\ \mathrm{JK^{-1}}$
Proton rest mass	$m_p = 1.672614 \times 10^{-27}$ kg
Avogadro's number	$N_A = 6.022169 \times 10^{23}\ \mathrm{mol^{-1}}$

Common Energy Units

Unit	$\mathrm{cm^{-1}}$	kJ/mol	eV
1 $\mathrm{cm^{-1}}$	1	0.01196	0.000124
1 kJ/mol	83.59	1	0.0104
1 eV	8065.6	96.49	1

In general, the probability of a transition between vibrational states n and n' is proportional to the square of the transition moment $\tilde{M}_{n'n}$, given by

$$\tilde{M}_{n'n} = \int_0^\infty \psi^*_{n'}\ \tilde{\mu}(r)\ \psi_n dr$$

$$= \tilde{\mu}_o \int_o^\infty \psi^*_{n'} \psi_n dr + \int_o^\infty \psi^*_{n'} Q^* \Delta r \psi_n dr \ . \qquad (25)$$

The first term in $\int \psi^*_{n'} \psi_n dr$ is defined to be zero since the wave functions $\psi_{n'}$ and ψ_n are defined as orthogonal, hence vibrations are only infrared active when $Q^* \neq o$. Diatomic molecules where both atoms are the same (termed homonuclear diatomics) have no permanent dipole moment, and this does not change during the vibration, hence $Q^* = o$ and these molecules do not have an infrared absorption spectrum. (In fact, there generally is a weak absorption due to multipolar effects not considered in Eqn. 23.) Special circumstances may also allow an infrared spectrum to appear. For example, Freund and co-workers (e.g., Martens and Freund, 1976; Freund, 1981; Freund and Wengeler, 1981; Freund et al., 1983) have recently observed infrared absorption due to H_2 trapped in crystalline MgO, CaO and perhaps Mg_2SiO_4, indicating that the site is not centrosymmetric or that some polarization of the H_2 molecule has occurred. Heteronuclear diatomic molecules all have infrared absorption spectra with the absorption intensity proportional to the magnitude of the change in dipole moment during the vibration. Polyatomic molecules with no net dipole moment may have some vibrations infrared-active if these cause a change in dipole moment.

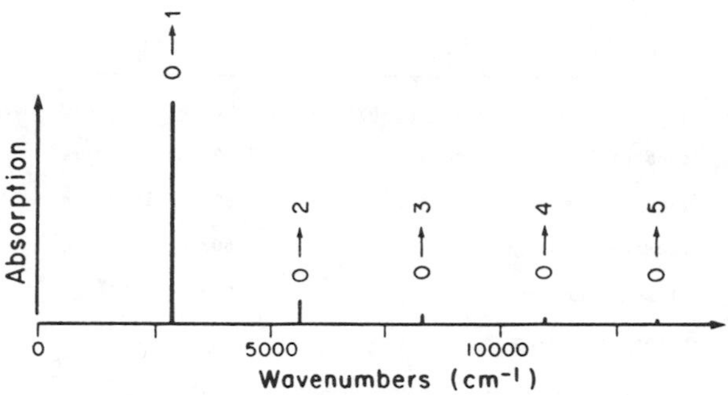

Figure 3. Schematic infrared absorption spectrum for HCl gas (Herzberg, 1950, p 54). The numbers above the peaks refer to the transitions from initial to final states. The 0 → 1 transition is the fundamental at 2885.9 cm^{-1}; the intensities for higher transitions fall off five times faster than shown here.

If the harmonic model is assumed, the transition moment is only non-zero for transitions $\Delta n = \pm 1$ due to the properties of the Hermite polynomials. This is known as the <u>selection rule</u> for infrared absorption. This restriction is lifted for the anharmonic oscillator, but transitions $\Delta n = \pm 2, \pm 3$ etc. remain much weaker than $\Delta n = \pm 1$. Since the intensity of a particular absorption is also proportional to the population of the initial state, and molecules are most commonly in the ground state at standard temperatures, the fundamental transition $E_0 \rightarrow E_1$ generally gives rise to a strong absorption, with a much weaker series of overtones due to $E_0 \rightarrow E_2$, $E_0 \rightarrow E_3$, etc. (Fig. 3) (Transitions from higher states may occur, especially at high temperature, and these are known as <u>hot bands</u>.)

<u>Raman scattering</u>. When visible light is passed through a sample, some of the light is scattered in all directions. When the energy of this scattered light is analyzed, a small fraction is generally found to have gained or lost energy compared with the incident beam. This corresponds to an inelastic scattering process, and is known as the <u>Raman</u> effect. The change in the energy of the light beam is termed the <u>Raman shift</u>, generally measured in wavenumbers (cm^{-1}) relative to the incident beam energy. Raman scattering with a <u>decrease</u> in incident energy is known as a <u>Stokes shift</u>, and with an <u>increase</u> in energy, <u>anti-Stokes</u>. The greater part of the scattered light is scattered elastically. This is known as <u>Rayleigh scattering</u>. In general, around 10^{-3} of the incident beam is scattered elastically, while only around 10^{-6} - 10^{-7} of the incident intensity gives rise to Raman scattering, hence the Raman effect is a very weak one. For this reason, high intensity light sources such as lasers are normally used to excite the sample.

The Stokes and anti-Stokes Raman spectra for α-quartz are shown in Figure 4. The incident beam was the blue line of an argon ion gas laser, with λ = 487.986 nm corresponding to an incident energy of 20,492 cm^{-1}. An intense line is observed on the Stokes side at 20,027 cm^{-1}, corresponding to a Raman shift of 465 cm^{-1}. A corresponding feature appears in the weaker anti-Stokes spectrum at 20,957 cm^{-1}, or -465 cm^{-1} Raman shift. These energy differences correspond to the energies associated with vibrational transitions, leading to the following classical description of the process.

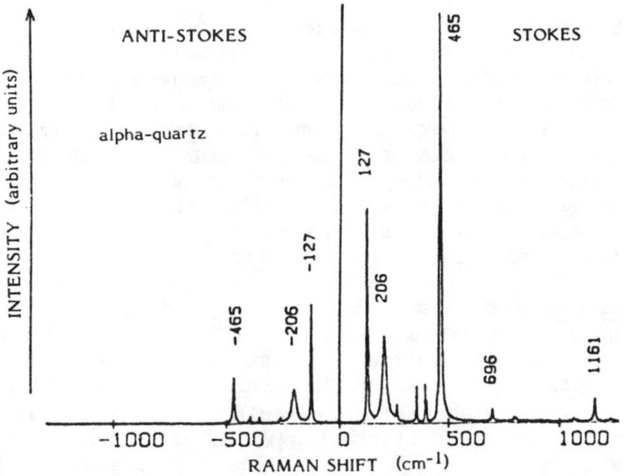

Figure 4. Stokes and anti-Stokes Raman spectra for α-quartz.

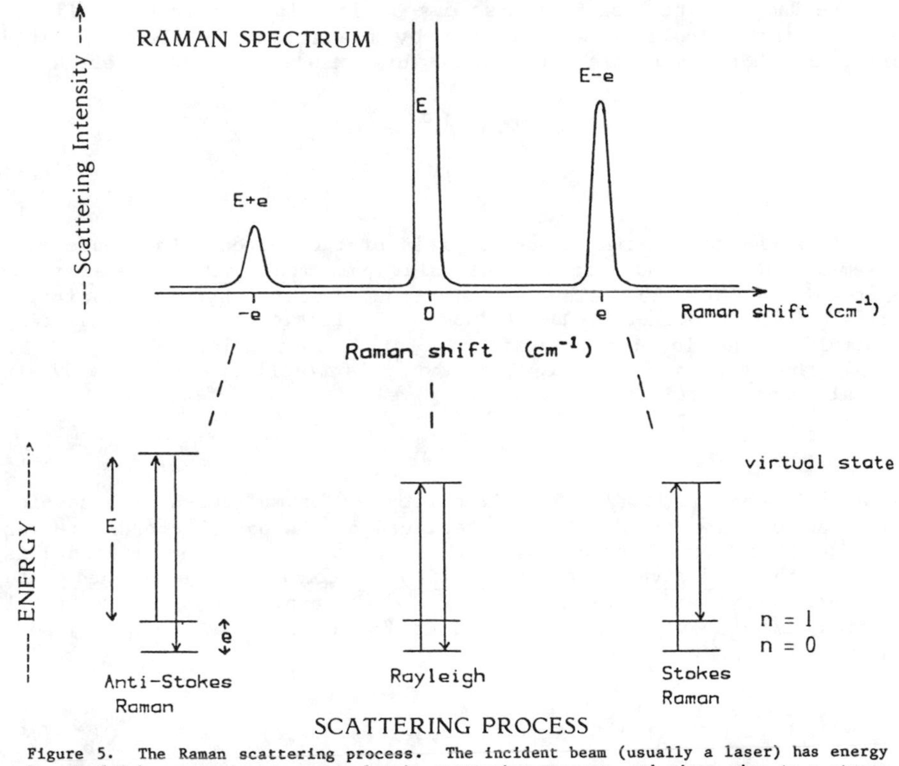

Figure 5. The Raman scattering process. The incident beam (usually a laser) has energy E. Rayleigh scattering is an inelastic scattering process, and gives rise to a strong central peak at the incident frequency (zero Raman shift). Raman scattering is accompanied by a change in vibrational energy level to give peaks at E - e (Stokes shift) or E + e (anti-Stokes shift), where e is the vibrational level separation. At normal temperatures, the ground state is more populated than higher vibrational states, hence the Stokes spectrum is more intense than the anti-Stokes.

The visible radiation used to excite the molecule is much higher in energy than vibrational motions, but approaches that of electronic transitions (see R. Burns, this volume). In normal Raman scattering, the incident beam is not absorbed by the sample. When absorption does occur, the process is known as resonance Raman scattering, briefly described later. The incident beam then instantaneously deforms the electron distribution around the molecule during the photon-molecule collision. The heavier nuclei move to follow the new, deformed electron distribution. In most cases, the photon simply leaves, the nuclei relax to their original positions, and the collision is elastic. However, if the nuclear motions correspond to their displacements during a vibrational normal mode (see Section 3) and the perturbation is sufficient, the vibrational motion is stimulated and the energy of the vibrational transition is subtracted from the incident photon to give Stokes Raman scattering (Fig. 5). The reverse process will give rise to anti-Stokes scattering, where the molecule is initially in an excited vibrational state, and the interaction with light induces a transition to a lower (usually the ground) state. These processes may be depicted in a schematic energy level diagram (Fig. 5).

The Raman scattering mechanism may be described in more detail via the transient dipole moment induced by the radiation electric field during the photon-molecule collision. This induced dipole is defined by

$$\begin{aligned}\tilde{\mu}_{ind} &= \alpha\tilde{E} \\ &= \alpha E_o \cos 2\pi\nu t,\end{aligned} \qquad (26)$$

where $\tilde{E}$ is the oscillating electric field of the photon with frequency ν and amplitude $\tilde{E}_o$, and α is the molecular polarizability expressing the deformability of the molecular electron density by the radiation field. Since in general, the induced polarization is not necessarily parallel to the incident electric vector $\tilde{E}$, this polarizability is a second rank tensor (see below). The polarizability itself will in general change during a vibration, to give

$$\alpha = \alpha_o + \left(\frac{d\alpha}{dr}\right)\Delta r \qquad (27)$$

where α_o is the equilibrium polarizability. (For molecules more complex than diatomic, the coordinate r is replaced by the normal coordinate for the vibration: see Sections 3.1 and 3.4). The term in $\Delta r(d\alpha/dr)$ is termed the derived polarizability, commonly shortened to "polarizability" and given the symbol "α" in replacement of the total polarizability above. From expressions 26 and 27, the induced dipole moment is given by

$$\tilde{\mu}_{ind} = 2\alpha_o\tilde{E}_o \cos 2\pi\nu_o t + E_o(\Delta r)\left(\frac{d\alpha}{dr}\right)\left[\cos 2\pi(\nu_o + \nu')t + \cos 2\pi\,(\nu_o - \nu')t\right], \qquad (28)$$

where ν' is the vibrational frequency. The equilibrium polarizability α_o contributes to Rayleigh scattering at the incident light frequency ν_o, while the derived polarizability is associated with Raman scattering at the shifted frequencies $\nu_o \mp \nu'$.

The total instantaneous dipole moment is given by

$$\tilde{\mu} = \tilde{\mu}_o + \tilde{\mu}_{ind} , \tag{29}$$

with $\tilde{\mu}_o$ the permanent molecular dipole moment. For a transition between levels n and n', the transition moment is given as before as

$$\tilde{M}_{n'n} = \int_o^\infty \psi_{n'}^* \tilde{\mu}(r) \psi_n dr \quad . \tag{30}$$

The transition moment for Raman scattering is obtained by substituting for $\mu(r)$ from Equations 26 and 29

$$\begin{aligned} \tilde{M}_{n'n} &= \tilde{E} \int_o^\infty \psi_n^* \Delta r(\frac{d\alpha}{dr})\psi_n \, dr \\ &= \tilde{E} \int_o^\infty \psi_{n'}^* \alpha\psi_n \, dr , \end{aligned} \tag{31}$$

where α is now the derived polarizability tensor expressing the change in molecular polarizability during the vibration. This shows that a vibration is only Raman active when one or more elements of α are non-zero. Like the total polarizability, the derived polarizability is a second rank tensor with elements

$$\alpha = \begin{bmatrix} \alpha_{xx} & \alpha_{yx} & \alpha_{zx} \\ \alpha_{xy} & \alpha_{yy} & \alpha_{zy} \\ \alpha_{xz} & \alpha_{yz} & \alpha_{zz} \end{bmatrix} \quad . \tag{32}$$

This is useful in polarized Raman scattering studies (Section 5). The molecular polarizability is often expressed schematically as a <u>polarizability ellipsoid</u> (Fig. 6). However, it is generally more difficult to imagine relative changes in molecular polarizability during vibrations than it is dipole moment changes, hence relative Raman activities and intensities are usually more difficult to visualize phenomenologically than infrared intensities and band activities. For polyatomic molecules, this is most simply done using the molecular symmetry (Section 3). Finally, through substitution of the appropriate vibrational wave functions into the above expressions, the selection rule for Raman scattering is found to be the same as for infrared absorption: $\Delta n = \pm 1$ for the harmonic oscillator, and $\Delta n = \pm 2, \pm 3$ etc. weakly allowed for anharmonic molecules.

3. POLYNUCLEAR SYSTEMS

3.1 Equations of motion and normal modes.

The equations of motion for a molecular system containing N atoms

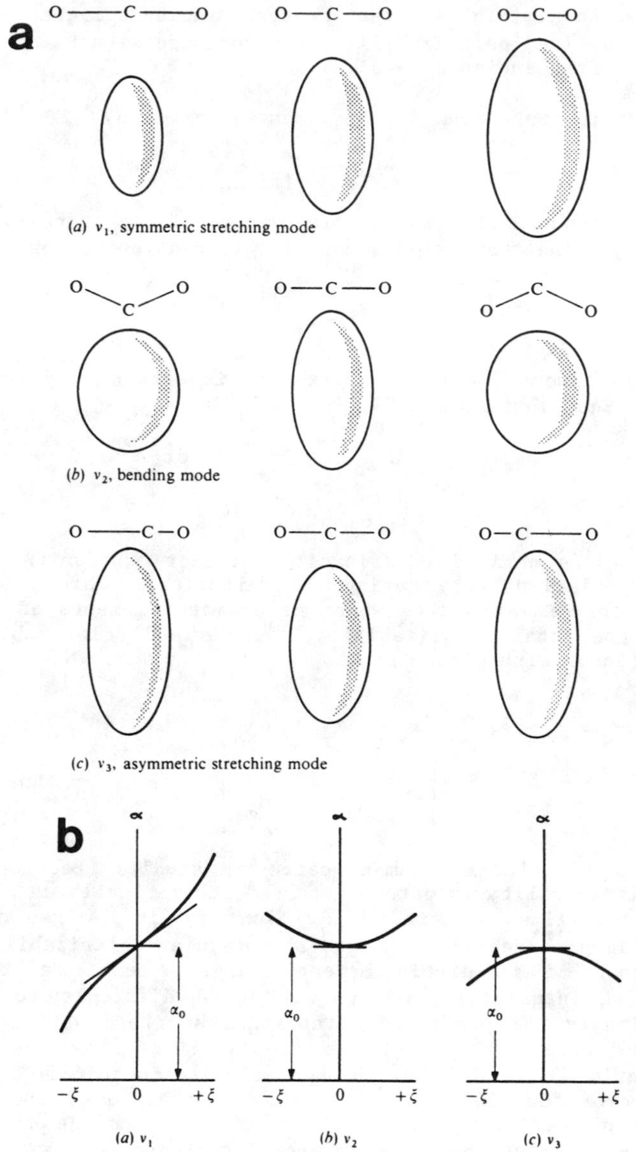

Figure 6. (a) Changes in the polarizability ellipsoid for the CO_2 molecule during its three modes of vibration (taken from Banwell, 1972, p. 135-136). (b) Variation of the polarizability α for CO_2 during the vibrations in (a). α_0 is the undeformed polarizability, and ξ is a generalized displacement coordinate. For small distortions, there is no change in α for the bending and asymmetric stretching modes ν_1 and ν_2, hence these vibrations are not Raman active. The symmetric stretch ν_1 is associated with a change in polarizability, hence this mode is Raman active.

give 3N solutions corresponding to the total degrees of freedom of its constituent atoms. Three of these solutions always correspond to translations of the entire system, and three (or two in the case of linear molecules) to its rotations. The remaining 3N-6 (or 3N-5) solutions are the vibrational normal modes for the system. The equations of motion are usually expressed and solved in the following way (Wilson et al., 1955).

The kinetic energy of the system (T) is given by

$$T = \frac{1}{2} \sum_{\alpha=1}^{N} m_\alpha \left[\left(\frac{d\Delta x_\alpha}{dt}\right)^2 + \left(\frac{d\Delta y_\alpha}{dt}\right)^2 + \left(\frac{d\Delta z_\alpha}{dt}\right)^2 \right] , \tag{33}$$

where m_α is the mass of the αth atom, and Δx_α, Δy_α, Δz_α are its small Cartesian displacements. The 3N Cartesian displacements are often replaced by generalized mass-weighted displacement coordinates

$$q_1 = \sqrt{m_1}\, \Delta x_1;\ q_2 = \sqrt{m_1}\, \Delta y_1;\ q_3 = \sqrt{m_1}\, \Delta z_1;\ q_4 = \sqrt{m_2}\, \Delta x_2;\ \ldots;q_{3N} = \sqrt{m_N}\, \Delta z_N \tag{34}$$

to give

$$T = \frac{1}{2} \sum_{i=1}^{3N} \dot{q}_i^2 \tag{35}$$

($\dot{q}_i$ is the time derivative of q_i). For small displacements q_i, the potential energy V may be expanded as a power series in q_i around $q_i = 0$ (r_0):

$$V = V_o + \sum_{i=1}^{3N} \left(\frac{\partial V}{\partial q_i}\right)_o q_i + \frac{1}{2} \sum_{i,j=1}^{3N} \left(\frac{\partial^2 V}{\partial q_i \partial q_j}\right)_o q_i q_j$$

$$+ \frac{1}{6} \sum_{i,j,k=1}^{3N} \left(\frac{\partial^3 V}{\partial q_i \partial q_j \partial q_k}\right)_o q_i q_j q_k + \ldots \tag{36}$$

The constant term V_o corresponds to the potential at the equilibrium configuration, and may be set to zero. At $q_i = 0$, there is no net force on the atoms, hence the second term in $(\partial V/\partial q_i)_o$ is also zero. Finally, in the harmonic approximation, only the quadratic term is retained, so

$$V = \frac{1}{2} \sum_{i,j=1}^{3N} \left(\frac{\partial^2 V}{\partial q_i \partial q_j}\right)_o q_i q_j = \frac{1}{2} \sum_{i,j=1}^{3N} f_{ij}\, q_i q_j \ . \tag{37}$$

This corresponds to a set of 3N independent harmonic oscillators i and j related by force constants f_{ij}. The Lagrangian equations of motion are written:

$$\frac{d}{dt}\frac{\partial T}{\partial \dot{q}_j} + \frac{\partial V}{\partial q_j} = 0 \qquad (j = 1, 2, \ldots, 3N) \quad . \tag{38}$$

Substitution for T and V from Equations 33 and 36 gives:

$$\ddot{q}_j + \sum_{i=1}^{3N} f_{ij}\, q_i = 0 \qquad (j = 1, 2, \ldots, 3N) \quad . \tag{39}$$

One solution for this set of 3N differential equations is of the form

$$q_i = A_i \sin (\lambda^{1/2} t + \phi) \tag{40}$$

(this is only a trial solution at this stage: the terms A_i, λ and ϕ are defined below), which on back-substitution into the second order differential Equation 39 gives

$$\sum_{i=1}^{3N} (f_{ij} - \delta_{ij}\lambda)\, A_i = 0 \qquad (j = 1, 2, \ldots, 3N) \tag{41}$$

with δ_{ij} the Kronecker delta function (δ_{ij} = 1 for i = j; δ_{ij} = 0 for i≠j). The values of λ which give non-vanishing solutions are those which satisfy the <u>secular determinant</u>

$$\begin{vmatrix} f_{11} - \lambda & f_{12} & f_{13} & f_{1,3N} \\ f_{21} & f_{22} - \lambda & \cdots & \cdots \\ f_{31} & \cdots & \cdots & \cdots \\ \cdots & \cdots & \cdots & \cdots \\ f_{3N,1} & \cdots & \cdots & f_{3N,3N} - \lambda \end{vmatrix} = 0 \tag{42}$$

This determinant has 3N roots λ_t which correspond to 3N values of the amplitude A_{it}, and phase angle ϕ_t in the solution (Eqn. 40). From the form of the solution and comparison with that for the simple harmonic oscillator (Equations 6 and 7), $\lambda^{1/2}$, A and ϕ may be respectively identified with the frequencies, Cartesian displacement amplitudes and phase of the vibrational normal modes. Six solutions (five for linear molecules) have λ_t = 0 and correspond to molecular translations and rotations. In each of the 3N-6 (or 3N-5) normal modes, all atoms are vibrating in phase about their equilibrium positions with frequency $\lambda_t^{1/2}$ and with Cartesian amplitudes A_{xt}, A_{yt}, A_{zt}. In general, this is the desired solution of a vibrational analysis for a molecular system: knowledge of the vibrational frequencies and atomic displacement vectors. This is generally achieved by calculation of or modelling the interatomic force field in terms of the force constants f_{ij} (see Section 6).

The first step in a vibrational analysis of a system is calculation of the number of vibrational normal modes, for example 3N-5 = 4 for CO_2 (a linear molecule), 3N-6 = 3 for H_2O, and 3N-6 = 9 for SiH_4. For crystalline systems, N is taken as the number of atoms in the primitive unit cell, and rotational degrees of freedom are ignored (however, these must be treated in vibrational calculations on crystals to avoid net resultant forces on the atoms) to give 3N-3 normal modes (at the Brillouin zone center: see Section 3.3). For example, crystalline silicon has N = 2 (face-centered cubic lattice), hence 3 normal modes, while forsterite (Mg_2SiO_4) has N = 28 and 81 normal modes.

3.2 Quantized polyatomic vibrations and the effects of anharmonicity.

The quantum mechanical description of a polyatomic system may be simply extrapolated from the treatment of the diatomic molecule (Section 2.2). In the polyatomic case, the 3N-6 (or 3N-5 for diatomic molecules) independent harmonic oscillators each have characteristic frequencies $\nu_t = \lambda_t^{1/2}$. Each oscillator is associated with a set of quantum numbers n_t defining its vibrational energy levels and corresponding wavefunctions. The energy levels for the entire system are then given by

$$E = (n_1 + \frac{1}{2}) h\nu_1 + (n_2 + \frac{1}{2}) h\nu_2 + (n_3 + \frac{1}{2}) h\nu_3 + \dots \quad (43)$$

The energy of the ground state (zero point energy) is obtained when all quantum numbers $n_t = 0$;

$$E_o = \frac{1}{2} h \sum_{t=1}^{3N} \nu_t . \quad (44)$$

The ground state wavefunction has an equal component of the atomic displacements for each vibrational normal mode t. A <u>fundamental</u> transition is one from the ground state to a level with $n_t = 1$, all other n_t' remaining zero. As for the diatomic molecule, <u>overtones</u> with $\Delta n_t = \pm 2, \pm 3$, etc. are weakly allowed for anharmonic models. However, in the polyatomic case, <u>combination</u> transitions where more than one quantum numbers n_t change at once are also possible for anharmonic systems.

The effects of anharmonicity on the energy levels of polyatomic systems are similar to those for the diatomic case: the zero point level drops in energy, energy levels close up, and spacings decrease for higher quantum numbers. However a further effect is apparent for polyatomic molecules. In the harmonic model, each normal mode forms an independent oscillator hence vibrational modes are discrete. In the anharmonic case, the oscillators are no longer independent, and energy transfer between modes is possible. Hence during a vibrational spectroscopic experiment where vibrational modes are excited by light, some of the incident light energy which is resonating with a particular vibration is dissipated through anharmonic effects, and appears as heat. Such energy transfer effects are extremely important in driving thermal phase transitions, for example that between α- and β-quartz (Section 7.3) However, although considerable work has been done on the anharmonicity of small molecules and simple crystals, there have been only a few studies for complex silicates (Gervais et al., 1973a,b; Iishi et al., 1979: see also Section 7.3).

3.3 Crystalline systems

The dynamics of crystals are treated in a similar way to polyatomic molecules, except that advantage is taken of the periodicity of the lattice. The equations of motion for crystalline systems are discussed in the chapters by Kieffer and Ghose (this volume). In general, a crystal will contain n primitive unit cells with N atoms per cell. Each unit cell treated as a molecule will have 3N normal modes, since molecular translations and rotations become vibrational modes (known as librations) when the molecule is embedded in a crystalline lattice. The vibrations of the entire crystal may be described in terms of the relative phase of the vibrational motion in adjacent unit cells. The atomic motions take the form of waves travelling through the crystal, known as lattice vibrations, with their wavelength determined by the phase relationships between adjacent unit cells (Fig. 8). In quantum mechanical terms, these lattice vibrations are termed phonons. These lattice waves, or phonons, may be completely described by their energy (often expressed in wavenumbers, cm^{-1}) and their wave vector $\tilde{k}$ in the direction of propagation (also unfortunately often given in units of reciprocal centimeters, cm^{-1}: see Section 2.4). The relation between vibrational mode energy and wave vector forms a dispersion relation. This reflects the effect of vibrational mode phase in adjacent unit cells on the frequency of that mode.

The Wigner-Seitz cell of a crystalline lattice is a type of primitive unit cell, and may be obtained by drawing lines from a lattice point to its nearest and next-nearest neighbours and drawing planes as perpendicular bisectors of these lines (Fig. 7). This type of cell is usually taken as the unit cell of the reciprocal lattice, with the origin at the center of the cell. In the reciprocal lattice, this cell is known as the first Brillouin zone. The dispersion relationships of lattice waves may be simply described within the first Brillouin zone. When all unit cells are in phase, the wavelength of the lattice vibration tends to infinity, and the wave vector $\tilde{k}$ approaches zero. Such modes are known as "zero-phonon" modes, and are present at the center of the Brillouin zone. When adjacent cells are exactly out of phase, the wave vector has a value $\pm \pi/a$ (where a is the direct lattice cell edge in the direction of $\tilde{k}$) at the edge of the first Brillouin zone (Fig. 8). The variation in phonon frequency as k-space is traversed is known as dispersion, and each set of vibrational modes related by dispersion is known as a branch. For each unit cell, three modes correspond to translation of all the atoms in the same direction. A lattice wave formed from displacements of this type is similar to the propagation of a sound wave through the lattice, hence these three branches are termed the acoustic branches (Fig. 8). At $\tilde{k} = \tilde{0}$ when all unit cells are in phase, these three branches become the three translations of the entire crystal, with zero frequency. The remaining 3N-3 branches involve relative displacements of atoms within each unit cell (Fig. 8). Since only vibrations of this type may interact with light, these are known as optical branches.

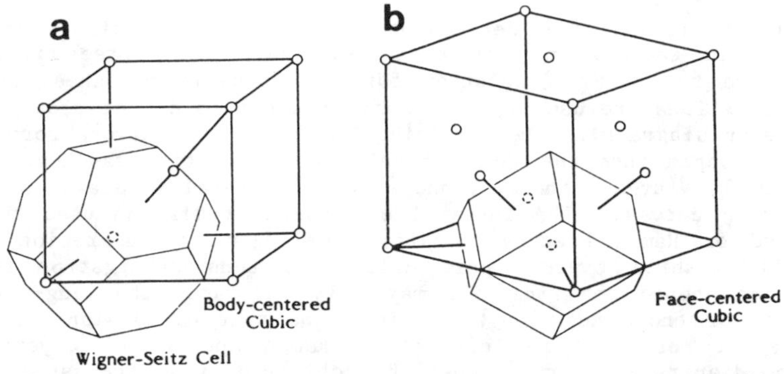

Figure 7. (a) The Wigner-Seitz cell for a body-centerd cubic lattice. (b) The reciprocal lattice for a body-centered cubic lattice is a face-centered cubic lattice, hence the Wigner-Seitz cell for the face-centered lattice shows the form of the first Brillouin zone for the body-centered cubic lattice.

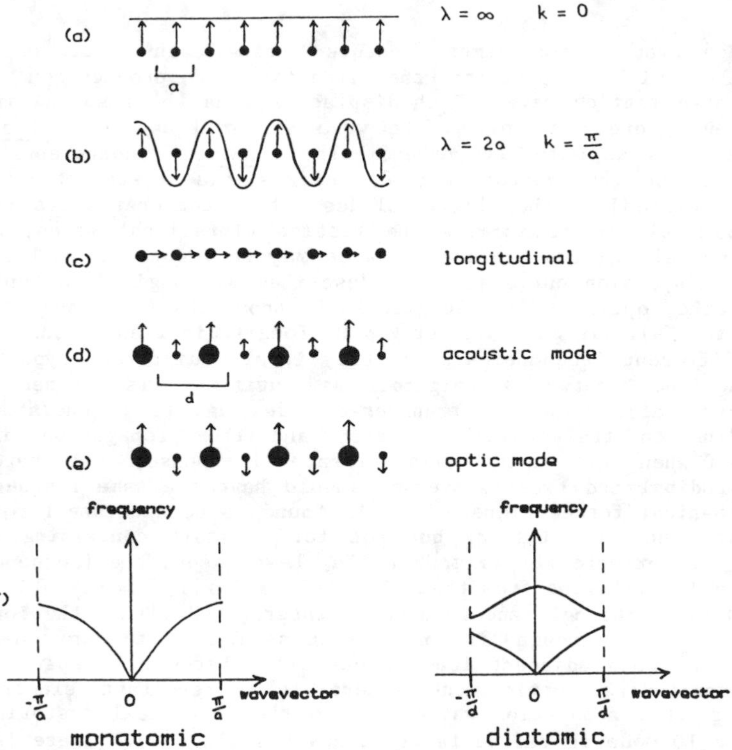

Figure 8. (a) and (b): Transverse modes for a monatomic chain of atoms with lattice spacing a. (a) shows a mode of infinite wavelength, hence zero wave vector. (b) shows the ode with wavelength $\lambda = 2a$, hence $k = \pi/a$ at the edge of the first Brillouin zone. (c) shows a longitudinal mode with $\lambda \sim 14a$ for the monatomic chain. (d) and (e) show transverse modes at $k = 0$ for a diatomic chain with lattice spacing d. The mode in (d) has both atoms within the unit cell moving in the same direction, hence is an acoustic mode. (e) shows an optic mode, with the atoms in the unit cell moving in opposite directions. The dispersion curves for the monatomic and diatomic chains are shown in (f) (see text and cf. chapters by Kieffer and Ghose for further discussion).

The total number of degrees of freedom for the entire crystal with N atoms in each of n unit cells is 3nN. The three translations correspond to the $\tilde{k} = \tilde{0}$ solution for the acoustic branches, while the three rotations are usually not considered (these do not plot easily on dispersion diagrams). The remaining 3nN-6 true vibrations form sets of points grouped into the 3N-3 optical and 3 acoustic branches. Since n is usually large, these branches are generally drawn as smooth dispersion curves. The infrared or visible radiation used to excite infrared or Raman spectra has its wavelength in the region 5000 to 100,000 Å, while typical unit cells are tens of Ångström units in dimension, hence the radiation may only interact with long-wavelength phonons, or those near $\tilde{k} = \tilde{0}$. This means that dispersion effects are not usually observed in infrared or Raman spectra of crystals, and phonon dispersion curves must be obtained via inelastic neutron scattering (see Ghose, this volume). The band broadening observed for disordered or amorphous materials may be related to loss of translational symmetry allowing dispersion effects to appear in infrared and Raman spectra (e.g., Shuker and Gammon, 1970; Piriou and Alain, 1979).

The atomic displacement vectors during each normal mode may be described with respect to the direction of propagation of their associated lattice wave. Each displacement may be resolved into three components, one parallel to the wave vector known as a *longitudinal* motion, and two mutually perpendicular *transverse* components at right angles to the wave vector (Fig. 8). For simple crystal structures, for example diamond or the alkali halides, these components are the actual atomic displacements forming the lattice vibrational waves, hence one longitudinal and two transverse waves may be identified. The branches of the dispersion curve are then described as longitudinal optic (LO), transverse optic (TO), longitudinal acoustic (LA) and transverse acoustic (TA). In general, for $\tilde{k} > \tilde{0}$, longitudinal and transverse modes have different frequencies since they involve different types of force interactions between neighbouring unit cells (shear versus compressional). The two transverse modes may be degenerate or not, depending on the crystal symmetry and the propagation direction. At $\tilde{k} = \tilde{0}$ when all unit cells vibrate in phase, all three modes (longitudinal and two transverse) should have the same frequency based on mechanical forces alone. This is found to be the case for homopolar crystals such as silicon, but not for crystals containing different atoms, for example KBr (Fig. 9). In these cases, the frequency of the LO mode is different from that of the TO modes at $\tilde{k} = \tilde{0}$. This is known as "TO-LO splitting", and is usually interpreted within the ionic model of crystals although it may be generalized to any heteropolar material. When adjacent ions of opposite charge are separated during the longitudinal optic mode, there is a resultant electric field opposing this separation which adds to the mechanical restoring force, and the LO mode is raised in frequency (see Fig. 9). There is no such resultant field for the corresponding transverse mode, which remains at the frequency defined by the mechanical force constant (see e.g., Born and Huang, 1954, p. 82-89; Rosenstock, 1961; Barron, 1961; Cochran, 1973, p. 82-94).

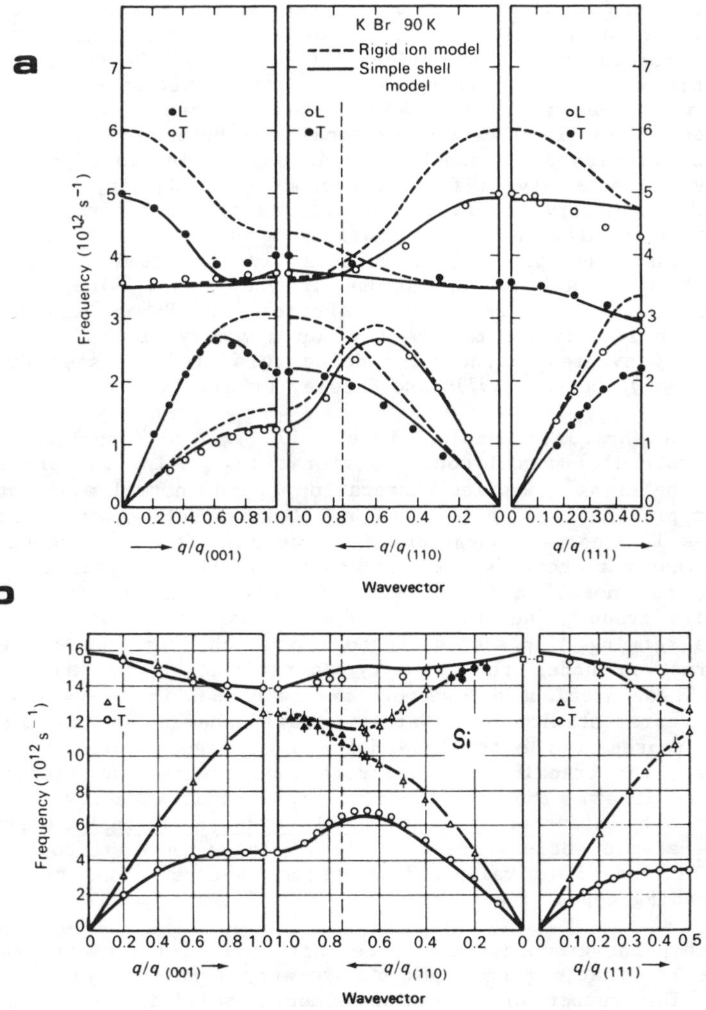

Figure 9. Dispersion curves in three directions in reciprocal space for (a) KBr and (b) Si, taken from Cochran (1973, p. 44 and 56). Circles and triangles are data points measured by inelastic neutron scattering. L and T refer to longitudinal and transverse branches. Solid curves are results of shell model calculations, while the dotted curves for KBr were calculated using a rigid ion model.

3.4 Prediction of infrared and Raman band activities

The application of symmetry and group theory to the vibrational modes of molecules and crystals is discussed in detail by Jaffé and Orchin (1965), Cotton (1971) and Fateley et al. (1972) (see also Herzberg, 1945, Chapter II; Wilson et al., 1955). In point group theory, five symmetry elements and their associated operations are considered: The identity E or I, the mirror plane σ, the inversion center i, proper rotation axes C_n of order n, and improper rotation axes S_n corresponding to a combination of $C_{n/2}$ and σ (the operation consists of a rotation $C_{n/2}$ followed by reflection in a mirror bisecting the axis (this is true for the Schoenflies system used by molecular spectroscopists; in the International or Hermann-Mauguin system the improper rotation is a combined rotation and inversion operation, and symmetry elements have different symbols (See Cotton, (1971), p. 343-345). The symmetry of all real molecules may be described by a set of these elements forming a mathematical group. The term point group implies that one point (not necessarily an atomic center) remains invariant under all the operations of the group. The first step in deciding the symmetries, hence infrared and Raman band activities, consists in identifying the point group symmetry of the molecule or the crystal. Convenient methods for doing this are discussed in Jaffé and Orchin (1965), Cotton (1971) and Fateley et al. (1972).

For an N-atom molecule, there will be 3N-6 (non-linear) or 3N-5 (linear molecule) normal modes of vibration. Using the primitive unit cell of the crystal implies enumeration of the normal modes at $\tilde{k} = \tilde{0}$, a useful approximation for infrared and Raman spectroscopy (Section 3.3), and gives 3N-3 normal modes for an N-atom unit cell. Since the symmetry of the nuclear centres defines that of the interatomic potential surface, the normal modes of vibration must also be described within this point group. Enumeration of vibrational mode symmetries generally involves setting displacement vectors on each atom and observing their transformation under the symmetry operations of the molecular point group. These displacement vectors may be chosen in many ways. The most general representation uses Cartesian displacement coordinates on each atom to represent the total 3N degrees of freedom of the system. In this case, the translational and rotational degrees of freedom must be subtracted later. The vibrational degrees of freedom may be considered explicitly by transformation to an internal coordinate system, where relative atomic motions are specified. These internal coordinates are usually expressed in valence bond terms, such as bond stretching and angle bending types of motion.

Each point group contains a certain number of symmetry species, or possible combinations of types of symmetry behaviour within the point group. The number of vibrational modes belonging to each symmetry species is fixed by the number of atoms and the molecular symmetry. Relatively simple methods for combining and reducing the atomic displacements into these symmetry species (known as the irreducible representations of the point group) are given in Jaffé and Orchin (1965), Cotton (1971) and Fateley et al. (1972). A worked example for forsterite, Mg_2SiO_4 is given at the end of this chapter (Section 8).

The set of true atomic displacements during each vibrational normal mode is described by the normal coordinate system. The above analysis assigns some combination of atomic displacements to a particular mode symmetry, to give a symmetry coordinate system. If the atomic masses and interatomic force constants are known (see Section 6), the normal coordinates, or absolute magnitude and direction of each atomic displacement, may be directly related to the symmetry coordinates. In most cases, this information is not available (Section 6), but symmetry coordinates may be used to constrain the possible types of vibrational motion. For example, when only one vibration belongs to a symmetry species (for example, the symmetric stretching of the SiO_4 tetrahedron), the normal mode must contain the atomic displacements from the symmetry coordinates (oxygen displacement away from the central silicon; no associated displacement of silicon); only the magnitude and in some cases the direction of the displacement remains undetermined. When several vibrations have the same symmetry, the normal mode displacements are linear combinations of the displacements involved in that symmetry species (see Herzberg, 1950, p. 145-148).

A useful tool in the analysis of crystal spectra is known as factor group analysis (Fateley et al., 1972). If a molecular unit, for example the SiO_4 group in olivine, may be identified in the unit cell, the vibrations of the free molecule can be simply related to modes appearing in the crystal spectrum (see Section 8). The mode symmetries of the free molecule are first enumerated. These are then related to the site symmetry of the molecule in the crystal which is a subgroup of the free molecular symmetry. Finally, these modes are transformed (and combined if there is more than one molecular group in the unit cell) into the full crystal symmetry via a further group correlation. This method can be very successful when (1) some of the molecular modes are only weakly perturbed by the crystal and (2) these fall in spectral regions relatively free from other lattice vibrations, and can allow reasonable identifications of bands in crystalline spectra.

The prediction of infrared and Raman band activities is relatively simple using a set of point group character tables (e.g., Salthouse and Ware, 1972: most books on vibrational spectroscopy also include a set of these tables, e.g., Turrell, 1972; Nakamoto, 1978. The reader should beware typographical errors in such tables!). Infrared radiation couples with a changing dipole moment during vibration: in Cartesian space, the components of this dipole moment have the same character as the three translations of the molecule. Any mode with symmetry corresponding to the x-, y-, or z-translations (labelled T_x, T_y, T_z in the first column to the right of most character tables) will be infrared active. Similarly, modes are Raman-active when they are associated with non-zero elements of the derived polarizability tensor (Expression 32; Section 2.4)

$$\alpha = \begin{bmatrix} \alpha_{xx} & \alpha_{yx} & \alpha_{zx} \\ \alpha_{xy} & \alpha_{yy} & \alpha_{zy} \\ \alpha_{xz} & \alpha_{yz} & \alpha_{zz} \end{bmatrix}$$

The symmetries associated with these elements are shown in the second column to the right of most character tables. Those modes with the same symmetry as one or more elements of the polarizability tensor will be Raman active. In most point groups, one or more modes are generally inactive. The mode types A and B refer to single modes, with A more symmetric than B; E is doubly degenerate, and T (or F) triply degenerate. For point groups with a center of inversion, the subscripts g (gerade = even) and u (ungerade = odd) refer respectively to modes symmetric and antisymmetric with respect to inversion. For such molecules, infrared bands are never Raman active and vice versa; this is known as the rule of mutual exclusion. Finally, the diagonal elements of the polarizability tensor are always associated with the Raman-active modes of highest symmetry: these usually give rise to the strongest Raman bands. Raman bands which are also infrared-active tend to contain off-diagonal elements of relatively low symmetry: these give rise to weaker Raman bands. In general, the more symmetric a vibration, the more intense it becomes in the Raman and weaker in the infrared, while strong infrared bands are generally associated with asymmetric modes of low Raman intensity.

4. INFRARED SPECTROSCOPY

4.1 Infrared reflection spectra

When an infrared beam is incident on a surface, proportions of the beam may be reflected, transmitted or absorbed by the material, while radiation may be emitted by the sample. This may be expressed through Kirchoff's law

$$T(\nu) = 1 - A(\nu) - R(\nu) - \kappa(\nu) , \tag{45}$$

where T, A, R and κ are the transmission, absorption, reflection and emission coefficients, all functions of the frequency of the radiation ν. The infrared reflection spectrum of a crystal with one vibrational absorption commonly has a form of the type shown in Figure 10 for ZnS. The position of the infrared absorption frequency is shown at 276 cm^{-1}. This is the frequency which would be measured in a thin film absorption experiment (Section 4.2). The form of this reflection curve and its relation to the atomic vibrations, may be obtained classically by considering the optical and dielectric characteristics of the material.

That part of the electromagnetic radiation which enters the material must be modulated within the sample to propagate as a wave. Due to the nature of the electromagnetic radiation, this modulation must involve a fluctuating electric (or magnetic: not considered here) dipole. The response of the sample to the radiation electric field may be considered in two parts: (1) an electronic contribution from movement of primarily the valence electrons relative to the nuclei, and (2) a contribution from the relative motions of the atomic cores (nuclei plus tightly bound inner electrons). The second contribution is generally treated within the ionic model for crystals. An ionic treatment is not really necessary, since any electronegativity difference in the consituent atoms will give rise to a change in electric dipole when the atoms are displaced relative to each other.

The refractive index (n) of the material is defined as

$$n = \frac{c}{v} , \tag{46}$$

where c is the speed of light in vacuum, and v is the velocity of propagation of the electromagnetic radiation within the material. A radiation beam propagating via electric dipole modulation within a material is often termed a polariton, to distinguish it from the photon in free space. The dielectric constant (ε) for the material is obtained through Maxwell's relations for electromagnetic waves (Fröhlich, 1958, p. 160-163) as

$$\varepsilon = n^2 \quad . \tag{47}$$

At this stage, absorption of the radiation by the material has not been considered, and ε and n are real quantities. Since the propagation velocity depends on the mechanism for modulation, v, n and ε are all dependent on the radiation frequency. At high frequency, the heavy nuclei remain stationary and modulation occurs via electronic motions. This defines the high frequency, or optical dielectric constant ε_∞.

At low frequency, in an "ionic" crystal, there is a dielectric contribution from the relative atomic displacements. Homonuclear crystals have no such contribution, and the dielectric constant is described by ε_∞ through the whole range of infrared to optical frequencies. The quantity ε_0 is defined as the low-frequency, or static dielectric constant, and may be obtained from capacitance measurements.

Consider the case of a crystal with one lattice vibration frequency ν_0. As the radiation frequency is scanned across the infrared region, the refractive index and dielectric constant will have the behavior shown in Figure 10. As the frequency approaches ν_0, the atomic displacements begin to resonate in a vibrational mode. However, lattice vibrational waves propagate much more slowly than electromagnetic radiation, and the polariton wave is slowed down by resonance with the lattice wave (this is known as retardation of the polariton). At the frequency ν_0, the polariton has become a phonon, but the light wave can no longer propagate through the lattice: its velocity has gone to zero, hence the refractive index n tends to infinity. The dielectric constant $\varepsilon(\nu)$ comes back from $-\infty$ on the high frequency side of the resonance, crossing the axis ε=0 at the frequency of the longitudinal optic mode, ν_L.

The reflectivity (R) of a crystal at normal incidence is given by the Fresnel formula,

$$R = \frac{(n-1)^2}{(n+1)^2} \quad , \tag{48}$$

to give the reflectivity curve shown in Figure 10. Since electromagnetic radiation is a transverse wave, the infrared beam may only interact with transverse optic vibrations (Section 3.3), and the frequency ν_0 corresponds to ν_{TO}, the frequency of the transverse optic mode at $\vec{k} = \vec{0}$. The longitudinal mode (ν_L) is defined by the frequency where the $\varepsilon(\nu)$ curve crosses the axis $\varepsilon = 0$ (Fig. 10). Between the frequencies ν_0 and ν_L, the crystal is totally reflecting (R = 1), i.e., the electromagnetic wave cannot propagate within the material.

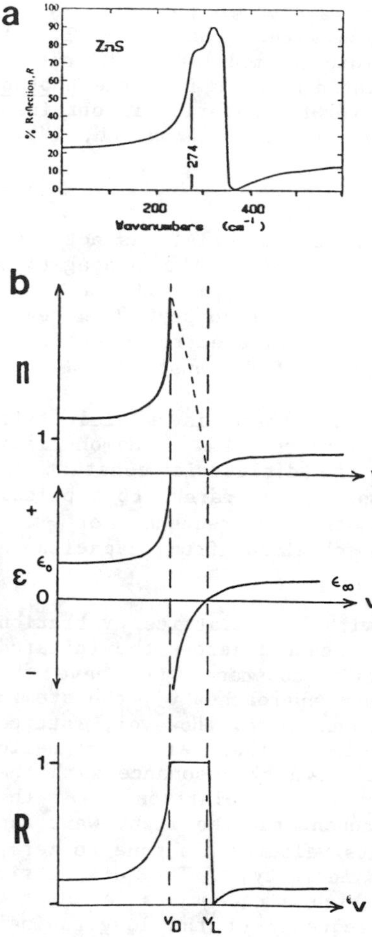

Figure 10.

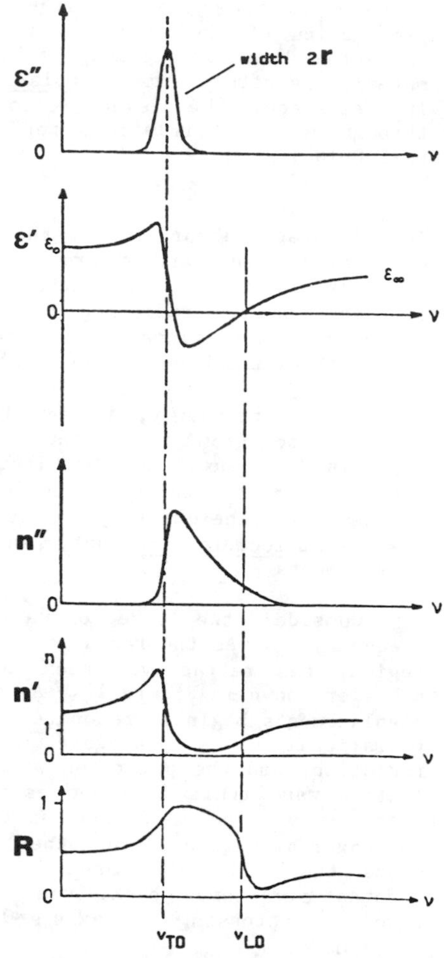

Figure 11.

Figure 10. (a) Typical infrared reflection spectrum for an ionic crystal. Reflectance of zincblende, ZnS, taken from Turrell (1972, p. 146). The infrared adsorption frequcy is shown at 274 cm^{-1}. (b) n(ν), ε(ν) and R(ν) curves for an ideal, non-absorbing crystal. The dielectric constant ε goes to positive infinity at the infrared resonant frequency (ν_o) which is also the frequency of the transverse lattice mode. The ε(ν) curve crosses the axis $\epsilon = 0$ at the frequency of the corresponding longitudinal mode, ν_L. The dielectric constant n is imaginary between ν_o and ν_L (dotted curve). The crystal is totally reflecting ($R = 1$) between ν_o and ν_L. This crystal would have no infrared absorption spectrum, since there is no dielectric loss.

Figure 11. Variation of optical constants with frequency for a typical real crystal. The dielectric loss function $\epsilon''(\nu)$ is the imaginary part of the complex dielectric function ($\epsilon'' = 2n'n''$). This function has a maximum at the infrared absorption frequency corresponding to the transverse optic mode (ν_{TO}), and has a width at half height 2Γ, where Γ is the damping coefficient due to anharmonicity. The real part of the dielectric function $\epsilon'(\nu)$ crosses the axis $\epsilon' = 0$ at the longitudinal optic frequency, ν_{LO}. n' and n" are the real and imaginary parts of the complex refractive index, and R is the reflection coefficient.

From the above, the width of the reflection band is determined by the dielectric character of the crystal; the more "ionic" the crystal, the wider its reflection band, and the greater its TO-LO separation (Section 3.3).

The infrared reflection spectra of real crystals never show complete reflection between frequencies ν_o and ν_L (ν_{TO} and ν_{LO}: (Fig. 11). This is due to some penetration of the radiation into the material, followed by a dissipation of energy among the vibrational degrees of freedom to appear as heat. The dissipation of radiative energy may be expressed by writing the refractive index as a complex quantity

$$n = n' + in'' \quad , \tag{49}$$

with a resulting complex dielectric constant

$$\varepsilon = \varepsilon' + i\varepsilon'' \ . \tag{50}$$

From Equation 47 and equating real and imaginary parts,

$$\varepsilon' = (n')^2 - (n'')^2 \quad \text{and} \quad \varepsilon'' = 2n'n'' \quad . \tag{51}$$

The function $\varepsilon''(\nu)$ is known as the dielectric loss function. This function has a maximum at the transverse frequency ν_{TO}, and the value of n'' determines the absorption coefficient. The magnitude of the dielectric loss is expressed via a damping coefficient Γ, and the variation of ε with frequency is written

$$\varepsilon(\nu) = \varepsilon_\infty + \frac{\nu_o^2(\varepsilon_o - \varepsilon_\infty)}{\nu_o^2 - \nu^2 - 2i\Gamma\nu} \quad , \tag{52}$$

where ν_o is the absorption frequency at ν_{TO}. The variation of the optical constants n' and n'', ε' and ε'' as a function of frequency (Fig. 11) may be obtained from the experimental reflectivity curve $R(\nu)$ using classical dispersion theory (Piriou and Cabannes, 1968), or through a Kramers-Kronig analysis (Turrell, 1972, p. 362-365). From this analysis, the frequencies ν_{TO} and ν_{LO} may be obtained with their oscillator strengths. The value 2Γ is the width at half-height of the absorption peak in $\varepsilon''(\nu)$, and determines the degree of anharmonicity of that mode (Fig. 11).

4.2 Pure absorption spectra

This method involves passing an infrared beam through a sample and observing the transmission minima corresponding to absorption by vibrational transitions within the sample. This is the general method for obtaining infrared spectra of gases and liquids. The gas or liquid sample is contained in a cell with infrared-transparent windows. These cells may be several metres long in the case of a dilute gas, but only several millimetres thick for dielectric liquids such as water. In the case of dielectric solids, pure infrared absorption becomes a very delicate experiment due to the extremely thin samples involved, and powdered samples are usually suspended in an inert matrix to obtain a powder absorption spectrum (Section 4.3).

Ignoring interference between incident and reflected beams, the transmission coefficient may be written as

$$T = \frac{(1-R)^2 e^{-\alpha \ell}}{(1-R^2 e^{-2\alpha \ell})} \quad , \tag{53}$$

where R is the reflectivity at the sample surface, ℓ is the sample thickness, and

$$\alpha = 4\pi\nu n'' \quad , \tag{54}$$

with ν the infrared frequency, and n" the imaginary part of the complex refractive index. When the film thickness is small relative to the wavelength of the exciting radiation, Equation 53 reduces to

$$T \approx 1 - 4\pi n'n''\nu\ell = 1 - 2\pi\varepsilon''\nu\ell \quad . \tag{55}$$

From Section 4.1, ε'' is maximized at the transverse mode frequency ν_{TO}, hence a transmission minimum occurs at $\nu = \nu_{TO}$ in a thin film infrared experiment. If a thicker slab of material is used, the approximate Equation 55 no longer holds, the band of low transmission broadens out asymmetrically about ν_{TO}, and subsidiary maxima due to interference effects may appear especially on the low frequency side of the band (Born and Huang, 1954, p. 123-128; Hadni, 1967, p. 585-596; Decius and Hexter, 1977, p. 190-193). The alkali halides have their absorption frequency at 100-300 cm^{-1}, or at 100-33μm. To obtain sharp absorption peaks, it was necessary to prepare films of 0.1-0.2 μm in thickness (e.g., Barnes and Czerny, 1931). As predicted above, the band became broader and asymmetric for thicker films, and subsidiary structure appeared, until at around 4-5 μm it was difficult to distinguish the absorption peak frequency. Silicate minerals commonly have strong infrared absorptions up to around 1000 cm^{-1}, or 10 μm. Comparing with the alkali halides, it is likely that sharp absorption peaks would only be observed for films 0.01-0.05 μm in thickness; quite a difficult experiment! The above arguments no longer hold for weak absorptions, and considerable work has been carried out using thick films for observation of weak bands in the far infrared (e.g., Wyncke et al., 1981), or the O-H stretching of hydrous species in low concentration in various minerals (Goldman et al., 1977; Rossman, 1979; Beran and Putnis, 1983; Aines and Rossman, 1984a,b).

4.3 Infrared powder spectra

Commonly known as infrared "transmission" or "absorption" spectra, these are obtained by grinding the sample and dispersing the powder in a supporting matrix, usually an inorganic liquid (e.g. nujol) or an alkali halide (commonly KBr). The resulting mull or pressed disc is placed in the beam, and an infrared transmission spectrum obtained. Since this method forms the most common technique for studies of minerals in the infrared, it is useful to note that there are a number of possible problems associated with this method.

First, the particle size is usually of the order of several microns as is the wavelength of the infrared radiation, and interference effects with the beam are possible. These are generally size and shape dependent, and may give rise to spurious structure in the spectrum (Tuddenham and Lyons, 1967; Luxon et al., 1969). Secondly, on contact with a given particle, the infrared beam is partially absorbed and partially reflected depending on the optical match between the powder and the support matrix, and the "transmission" curve may contain large components of both reflection and absorption spectra. If this is the case, the transmission minima (absorbance maxima) may not correspond to the vibrational frequencies ν_o but be shifted (up to several tens of wavenumbers in some cases) down-frequency due to the effect of the reflection band (see Section 4.1; also Piriou, 1974). Finally, the small particles have a relatively large surface to volume ratio. A number of types of surface mode are known with frequencies between the ν_{TO} and ν_{LO} bulk modes, and which may be strongly infrared active. These may give rise to extra absorptions in the spectrum, and may in some cases be dominant over the bulk modes (Sherwood, 1972, p. 116-134).

Despite these possible problems, infrared powder spectra have been applied with considerable success to many problems in mineralogy (e.g., Tarte, 1962, 1963a,b; Farmer, 1974; Rossmann, 1979). Weak bands are generally associated with small dipole changes, hence have a narrow reflectance band with little TO-LO splitting. These tend to show a single sharp absorption in powder spectra which corresponds well with the true absorption peak. This accounts for the success of the powder method in organic chemistry, with the essentially covalent hydrocarbons and related species. The technique has also been applied extensively to analysis of narrow bands in mineral spectra, for example the O-H stretching modes of hydrous minerals (Strens, 1974). Some care should be exercised with the strong major bands of most minerals. Some spectra do show more structure than expected from symmetry analysis (Section 3.4) which could be partly due to some of the effects noted above (Jeanloz, 1980; Akaogi et al., 1984). Whenever possible, it is likely to be useful to carry out control experiments, such as comparing spectra obtained with different particle sizes, or obtaining polarized thin film absorption or reflective spectra for one member of a series.

5. RAMAN SPECTROSCOPY

It was noted earlier that Raman scattering may be described by interaction of the incident beam with the molecular polarizability

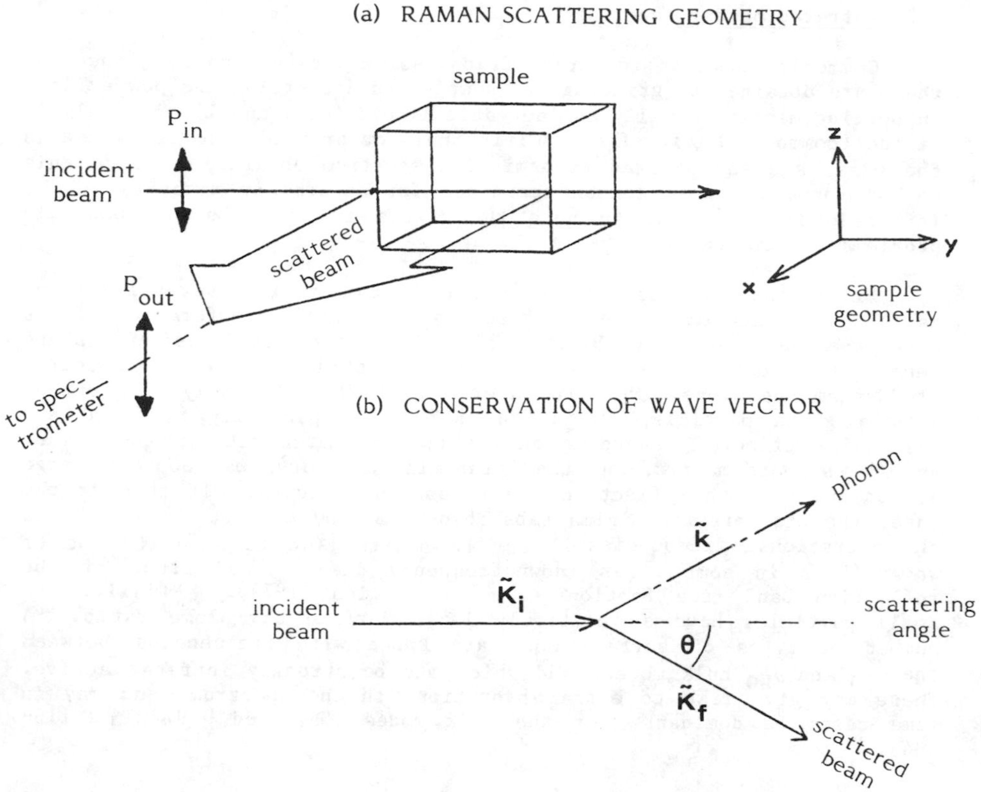

Figure 12. (a) The Raman scattering geometry for scattering at 90°. P_{in} and P_{out} refer to the polarizations of the incident and scattered beams respectively. This sample orientiation and polarization combination corresponds to y(zz)x in the Porto notation (see text) and will selct the component α_{zz} of the derived polarizability tensor. (b) Conservation of wave vector during the Raman scattering process. K_i and K_f refer to the wave vectors of the incident and scattered beams, and k to the phonon wave vector. These are related by $k = (K_i^2 + K_f^2 - 2K_iK_f \cos\theta)^{1/2}$. Since k is much smaller than K_i or K_f, this may be approximated by $k = 2K \sin\theta/2$. This diagram is commonly termed the "Raman scattering triangle".

tensor, and that a given mode symmetry is associated with particular elements of the derived polarizability tensor (Section 2.4). If polarized incident light (normally from a laser) is used and the scattered beam is also polarized, the laboratory geometry may be used to defined particular polarizability elements and the corresponding band symmetries obtained. For example, an orthogonal crystal may be cut following {001}, {010} and {100} with the laser beam incoming along y, the scattered radiation collected along x, and both the laser and the scattered beam polarized parallel to z (Fig. 12), the scattering geometry may be written as y(zz)x in the Porto notation (Damen et al., 1966). This will select the polarizability element α_{zz}. For an orthorhombic crystal D_{2h} this corresponds to modes of symmetry $A_{1g}(zz)$;

for cubic crystals T_d this gives modes A_1 (e.g., Salthouse and Ware, 1972). If the scattered polarization is changed to lie along y, the Porto symbol is written y(zy)x, the tensor element becomes α_{zy} and modes observed will have symmetry B_{3g} for the orthorhombic case and T_2 for the cubic. Materials which are completely isotropic such as liquids or glasses have only two such spectra, where the polarization of incoming and scattered beams may be parallel (VV, HH or parallel spectra) or crossed (VH, HV or perpendicular spectra). Relationships have been derived to relate the intensity in a given polarization to the mode symmetry, with the depolarization ratio defined as

$$\rho = \frac{I_1}{I_{11}} \quad , \qquad (56)$$

where I_1 and I_{11} are the intensities for perpendicular and parallel polarization of incident and scattered beams. For totally symmetric modes of cubic point groups, $\rho = 0$ and the band is completely polarized. Vibrations of symmetry E or T have $\rho = 0.75$, and these bands are fully depolarized. Bands of other symmetry have intermediate depolarization ratios, and are commonly termed partially polarized (see Herzberg, 1945, p. 246-249; Long, 1977, p. 62-64).

The frequency shift of the incident laser beam during Raman scattering reflects the <u>conservation of energy</u>. Total momentum must also be conserved during the process, described as <u>conservation of wave vector</u>. If the incident beam has wave vector $\tilde{K}$, the scattered beam has wave vector

$$\tilde{K}' = \tilde{K} - \tilde{k} \quad , \qquad (57)$$

where K' and K are much larger than k. From the <u>scattering triangle</u> (Fig. 12) the magnitudes of these vectors are given by

$$k \approx 2K \sin \frac{\Theta}{2} \quad , \qquad (58)$$

where Θ is the scattering angle. For the sake of standardization, most Raman spectra are taken at or near $\Theta = 90°$, although the advent of high-pressure diamond cell work and micro-Raman spectroscopy is increasing the number of <u>back-scattering</u> studies with $\Theta \approx 0°$. If Θ is scanned from 180° to 90° at constant $\tilde{K}$, the phonon wave vector $\tilde{k}$ is also scanned. Laser wavelengths are generally around 5000 Å while most crystal unit cell dimensions are 10-50 Å. In these cases, the region in $\tilde{k}$-space scanned is small, from $\tilde{k} = \tilde{0}$ to a maximum of 0.01 $\tilde{k}_{max}$, where $\tilde{k}_{max}$ is at the edge of the first Brillouin zone. This experiment is known as <u>polariton scattering</u>, and maps out the radiative part of the photon-phonon (or rather, polariton-phonon: see Section 4.1) dispersion curve as the photon frequency approaches that of the phonon and both become mixed (Born and Huang, 1954, p. 90; Sherwood, 1972, p. 98). Many mineralogical materials may have true unit cells much larger than 50 Å, for example due to disorder, and polariton scattering might be applied to obtain useful optical dispersion information over a larger region in k-space.

Brillouin scattering is closely related to Raman spectroscopy. It was noted earlier (Section 3.3) that the three acoustic branches rise from the translational degrees of freedom with zero frequency at $\tilde{k} = \tilde{0}$ and are related to propagation of sound waves in the lattice. Brillouin scattering takes advantage of the scattering triangle and that photon-phonon interactions take place around $\tilde{k} = \tilde{0}$, not exactly at the Brillouin zone centre. At some low frequency, the laser beam is scattered off the acoustic modes which appear as peaks in the Brillouin spectrum. The spectrometers used for this type of experiment have extremely high resolution (0.001 cm^{-1}) and generally work over a much more restricted range (generally 0 ± 5 cm^{-1}) than Raman spectrometers. Such Brillouin spectra give peaks corresponding to the propagation of acoustic lattice waves, hence information on the sound wave velocity within the material which may be usefully related to many material properties.

In normal Raman spectroscopy, care is taken that the incident beam wavelength is far from optical absorptions of the sample, since fluorescence and related electronic effects may interfere with or swamp the weak Raman spectrum. In resonance Raman spectroscopy the incident beam is tuned (usually via a dye laser) to match an optical absorption within the sample (see Burns, this volume), and the Raman intensity profile of selected bands monitored as the optical frequency is scanned to give an excitation profile. In general, selected vibrational modes are enhanced (often by one or more orders of magnitude) as optical resonance is approached. These tend to be symmetric modes involving the optically active centre, and long series of otherwise inactive overtones may appear in the spectrum, or weak combination bands involving those modes appear strong in the resonance spectrum (Long, 1977, p. 180-188). The theory of this process is still being developed (see e.g., Fujimara and Lin, 1979; Page and Tonks, 1981), and has as yet found little application in the mineral sciences. This technique could have a considerable impact if applied to analysis of trace amounts of molecular species containing absorbing centres, for example transition metal complexes in solution or trapped in crystal matrices. Finally, a number of light scattering techniques have evolved with the development of high field, pulsed lasers. These include stimulated Raman scattering (SRS), hyper-Raman spectroscopy, coherent anti-Stokes Raman scattering (CARS), and other related spectroscopies (see Long, 1977, p. 219-243). Again, these relatively sophisticated techniques have not yet been assimilated into the geosciences, but it is certain that in time they will have many useful mineralogical or geochemical applications.

6. INTERPRETATION OF VIBRATIONAL SPECTRA

6.1 Introduction

One of the major applications of vibrational spectroscopy has been in structural studies, where changes in the number, activities or positions of vibrational bands are analyzed in terms of structural changes. In order to make a full correlation between vibrational spectra and molecular structure, it would be necessary to know the atomic displacements associated with each vibrational mode. Since vibrational motions are governed by the interatomic forces within the molecule, a complete knowledge of the molecular structure and its potential energy as a function of nuclear coordinates would uniquely specify all of its vibrational properties. This level of understanding is now becoming available for some molecules through the development and increased feasibility of accurate ab initio calculations (e.g., Pople et al., 1981). Such calculations have been used to gain vibrational information on silicate molecular clusters which has been extrapolated to condensed silicates (O'Keeffe et al., 1980; Gibbs, 1982; O'Keeffe and Gibbs, 1984), but a direct application to the overall vibrational properties of complex silicate structures has not yet been carried out. At present, interpretations of vibrational spectra range from phenomenological studies involving the use of band symmetries, and observation of spectral changes on changing some structural parameter, to dynamical calculations based on model force fields.

6.2 Characteristic group frequencies and symmetry analysis

A considerable list of characteristic vibrational frequencies for specific structural groups such as C=O, C≡N, C-H, N-H and O-H, has been established in organic chemistry (e.g., Bellamy, 1958, 1968). Similar characteristic group frequencies have been identified for use by inorganic chemists, for example P=O, S-H, B-H or Si-H (e.g., Ebsworth, 1963; Nakamoto, 1978), or compounds with discrete anionic units such as carbonate, sulphate or nitrate (Ross, 1973). These identifications rely on the particular vibrational bands remaining relatively constant from one molecule to another, and occurring in a spectral region free from other interfering peaks. The first condition is most easily met when the vibrational mode is highly localized within a small group of for example two or three atoms. This is true for most hydrogenic stretching vibrations which involve mainly motion of the light hydrogen against its heavier neighbour, and for most of the other groups noted above. Once a vibrational mode becomes to be strongly coupled to motions of a significant part of the molecule, its identity is lost hence its use in identifying a particular structural group. In silicates, it seems likely that stretching motions of "non-bridging" oxygens against silicon are also highly localized and have vibrational frequencies characteristic of the particular silicate species. These observations have played a central role in the application of vibrational spectroscopy in structural studies of silicate melts and glasses (Brawer and White, 1975; Mysen et al., 1982; Matson et al., 1983; McMillan, 1984). A modified version of this approach has been used extensively in mineralogical studies, and involves association of characteristic spectral regions with particular cation coordination geometries, for example AlO_4 versus AlO_6; SiO_4 versus SiO_6; TiO_4 versus TiO_6 groups (Tarte, 1962, 1963a, 1967; Lazarev, 1972, pp. 167-176). Although this approach may be valid and useful in some cases, an increasing body of work suggests that its application may be limited and often fraught with danger (see for example, McMillan and Piriou, 1982; Ross and McMillan, 1984; Dickinson and Hess, 1985).

In cases where polarized infrared or Raman spectra are available, the band symmetries and polarization behavior can provide a wealth of structural information. For example, a number of studies have used polarized infrared absorption spectroscopy to characterize the type and absolute orientation of molecular water or hydroxyl species in minerals with a hydrous component (e.g., Goldman et al., 1977; Rossman, 1979; Beran and Putnis, 1983; Aines and Rossman, 1984a,b). Even when polarized spectra are not available, the number of bands observed for a particular species and their relative intensities can give useful structural information. For example, the disappearance of one band of molecular CO_2 for carbon dioxide dissolved in a silicate glass led Mysen and Virgo (1980a,b) to suggest that the CO_2 molecule was distorted in the glass structure.

6.3 Systematic spectral changes

A large number of vibrational studies have been carried out in which changes in spectra were noted as a function of some changing parameter such as sample composition, structure, or preparation conditions. Even though in most cases the detailed nature of the vibrational modes in such studies is not known, useful models may often be proposed to rationalize the vibrational behavior in terms of known structural changes. Even when there is no obvious interpretation of the systematic vibrational behavior, the data provide valuable input for calculation of systematic changes in thermodynamic properties (Kieffer, this volume). Some examples of such systematic vibrational studies are given below. Moore and White (1971) obtained the infrared and Raman spectra of a range of natural silicate garnets and analyzed the variations in bands derived from tetrahedral Si-O stretching motions in terms of the distortion of the tetrahedral groups. Similar correlations have been made for olivines and related orthosilicates (Tarte, 1963b; Burns and Huggins, 1972; White, 1975; Piriou and McMillan, 1983). There have been several systematic studies of the infrared spectra of pyroxene and pyroxenoid chain silicates, and the number of bands appearing in the 500-700 cm^{-1} region has been correlated with the number of Si-O-Si linkages in the repeat distance along the silicate chain (Lazarev, 1972, p. 95-107; Rutstein and White, 1971). The observed appearance and disappearance of particular Raman bands as a function of silica content in silicate glass series has allowed the association of these bands with structural groups in the glass (Brawer and White, 1975; Mysen et al., 1982; Matson et al., 1983; McMillan, 1984). Finally, a number of studies have used vibrational spectroscopy to follow changes in ordering or disordering behavior, for example in albite (Laves and Hafner, 1956; Martin, 1970), cordierite (Langer and Schreyer, 1969; Putnis and Bish, 1983; McMillan et al., 1984) or metamict zircons (Rossman, 1979).

6.4 Isotopic and chemical substitution experiments

If one atom in a molecule is isotopically substituted (for example, deuterium for hydrogen), the vibrational modes involving motion of that atom will shift in frequency depending on the isotopic mass difference and the type of motion involved (Müller, 1977). For many small molecules, this type of substitution may lead to an unambiguous assignment of vibrational modes or a unique determination of the molecular force field (Section 6.5). For more complex molecules, isotopic substitution can help determine the degree of participation of constituent atoms in particular vibrations, which places constraints on the nature of those vibrational modes. Common stable isotopes of use in mineralogy include $^{16}O/^{18}O$, $^{28}Si/^{20}Si$, $^{24}Mg/^{26}Mg$, $^{1}H/^{2}H$ and $^{12}C/^{13}C$, giving vibrational band shifts of several wavenumbers (or several hundred wavenumbers in the case of hydrogen). Only a few isotopic substitution studies have been carried out for silicates: these include Pâques-Ledent and Tarte (1963; Mg_2SiO_4), Galeener and Mikkelsen (1981), and Galeener and Geissberger (1983) (both vitreous SiO_2).

Chemical substitution experiments are similar in philosophy, in that one element is substituted into the site of another. However, in these studies the substituent may have different bonding characteristics to the original atom, hence affect the local force field. The substituted atom may also induce longer range structural changes in the surrounding matrix, and these may be difficult to analyze. This technique was pioneered by Tarte (1962, 1963a,b), who used it to identify modes associated with metal cation sites in olivines, garnets and other orthosilicates. Iishi et al. (1971) substituted Ge and Ga for Si and Al in vibrational studies of alkali feldspars, while similar substitutions have recently been used to help interpret the Raman spectra of aluminosilicate glasses (Fleet et al., 1984; Matson and Sharma, 1985).

6.5 Vibrational calculations

In the harmonic approximation, the force field of a molecular system is usually expressed via a matrix of force constants f_{ij} (Section 3.1). The diagonal force constants f_{ii} often refer to terms such as Si-O stretching or OSiO angle bending while the off-diagonal terms f_{ij} express interactions between these. For an N-atomic molecule, there are $n = 3N-6$ ($3N-5$ for linear molecules) vibrational modes, but $n(n + 1)/2$ independent force constants (from the square matrix form of the secular determinant 41). This implies that in general the vibrational problem is underdetermined, with more force constants to be evaluated than there are observed frequencies, except for the diatomic molecule with $n = 1$. This situation is often compounded by experimental difficulties: many normal modes are spectroscopically inactive due to their symmetry, while others are often weak or unresolvable. In reality, the situation is not nearly so bleak, at least for small molecules. Many of the force constants may be identical or simply related by the molecular symmetry, allowing these to be determined from the vibrational spectra. This is the case for CO_2, N_2O, and many other linear triatomic molecules (Mills, 1963; Jones, 1971). For more complex molecules, a unique harmonic force field may be established using data from isotopic substitution experiments (Section 6.4), or from measurements of vibrational-rotational coupling (Herzberg, 1945, pp. 370-500). This has been carried out for molecules including H_2O and SiH_4 (Mills, 1963; Jones, 1971). When there is still insufficient information to determine the complete force field, some terms may be evaluated for particular vibration symmetries (see Jones, 1971).

For most molecules, either the structure is too complex or too little information is available to determine a unique force field, and various model force fields have been developed. The two major types used are the valence force field, in which force constants are expressed in bond-stretching and angle-bending terms, and the central force field, where the internal coordinates are bonded and non-bonded interatomic distances. When the force field is uniquely specified by the available data, these models are known as the general valence and general central force fields, and are mathematically equivalent although they may be subject to different physical interpretations. Since the modelling is usually carried out within the harmonic approximation, these are also termed quadratic force fields. Anharmonicity corrections have been calculated from the vibrational data for a few molecules. These can be quite important, especially for hydrogenic vibrations (Jones, 1971; Zerbi, 1977), while there is evidence that modes involving the bridging oxygen of polymerized silicates may show considerable anharmonicity (Durig et al., 1977; see also Section 7.2). There are many variations on the central and valence force fields, for example the Urey-Bradley force field considers bond-stretching and angle-bending terms and also terms for non-bonded repulsions, while later modifications also include stretch-stretch and other interaction terms (Mills, 1963; Jones, 1971; Zerbi, 1977).

In general, the model force constant set is used to compute the vibrational frequency spectrum. This is compared with the available observed data, which are used to refine the force constants. In order to reduce the size of the problem, many force constants (especially interaction terms) are often initially set to zero, but may be included in further refinements to improve the fit. At this level of modelling, the force constants are often simply adjustable parameters, and it may be difficult to attach any physical significance to them. It is also difficult to compare absolute values of individual force constants between different studies, since the magnitudes of force constants depend on (a) how they are defined within a model force field, and (b) which other terms are include in the force constant set. For example, Jones (1971, p. 73) compares the force constants calculated for H_2O using three different general force fields. McMillan (1984) found that estimates of the Si-O stretching force constant ranged from 300-700 Nm^{-1} depending on the model used, while O'Keeffe et al. (1980) discussed the effect of inclusion or neglect of O...O interaction terms on the magnitude of k_{SiO}. In general, a given vibrational spectrum can always be reproduced by suitable refinement of any set of force constants if this set is large enough, hence a good match of calculated with observed frequencies is not usually sufficient to test the force field (Mills, 1963; Leigh et al., 1971; Szigeti, 1971; Müller and Mohan, 1977; Zerbi, 1977). If the resulting force constant set does not closely model the true interatomic potential, the calculated normal mode displacements may not resemble the true vibrational motions, resulting in significant problems for structural interpretation of the vibrational spectra. There have been a number of vibrational calculations for silicates (see McMillan, 1984). For the most part, the force fields used have been relatively unconstrained, and further work will be needed to test whether these provide reliable approximations to the silicate force field.

The case of crystalline materials is even more complex. Treated as a single large molecule, the number of vibrational modes ($n = 3N-6$, where N is the total number of atoms in the crystal) hence independent force constants to be evaluated is a very large number (see Leigh et al., 1971; Szigeti, 1971). The individual frequencies form points along the branches of the dispersion curves (Section 3.3). In general, infrared and Raman spectra only give information near $\tilde{k} = \tilde{0}$, and inelastic neutron scattering must be used to obtain vibrational data across the Brillouin zone (Ghose, this volume). Only a few inelastic neutron studies have been carried out for silicates, hence in general there is little information to constrain a full vibrational calculation including the effects of dispersion. Many theoretical studies have calculated the vibrational modes at $\tilde{k} = \tilde{0}$, but do not necessarily extrapolate to vibrations with $\tilde{k} > \tilde{0}$. A few calculations for silicates have included dispersion (e.g., Elcombe, 1967; Striefler and Barsch, 1975; Etchepare and Merian, 1978), and fit to infrared, Raman and inelastic neutron scattering data, and to the observed elastic strain constants.

Calculations for crystals have been carried out using both valence and central force fields. The valence force fields are similar to those used for molecules, but long range Coulomb forces must also be included to reproduce TO-LO splitting (Section 3.3). For ionic crystals, the potential energy of the crystal is often expressed as the sum of an attractive Coulomb term between ions of the opposite charge, and a short-range repulsive potential between like ions (see Burnham, this volume). This potential defines a central force field, and the vibrational modes may be calculated. The simplest model assumes that the ions are hard spheres, giving the *rigid ion* model. Refinements of this treatment include the *polarizable ion* model, where the ions are allowed to be deformed in the crystal field, and the *shell model*, where ions are considered as hard cores with a polarizable outer shell (Burnham, this volume; Cochran, 1979). Although these methods were developed for "ionic" crystals, they have been applied with some success to silicates (Elcombe, 1967; Striefler and Barsch, 1975; Iishi, 1978a,b; Iishi et al., 1979; Wolf and Jeanloz, 1984). Although the valence force fields and the central force models based on ionic interactions appear very different, they simply represent different ways of expressing the potential energy as a function of interatomic separation within the crystal, and in many cases, parameters within each model can be shown to be equivalent (e.g., Elcombe, 1967).

In general, when the force field is well-constrained, vibrational calculations can give a reliable description of the atomic displacements during each normal mode, which allows a unique structural interpretation of the vibrational spectra. If the force field is not well constrained, the calculated motions may or may not reflect true vibrational modes depending on the physical significance of the force model used, and the magnitude of any terms omitted. However, even such trial calculations are useful in that they can give insights into the nature of certain modes and help in constructing structural models. Finally, well-constrained vibrational calculations can lead to reliable values for elastic constants and thermodynamic properties (chapters by Burnham, and Kieffer, this volume), and give an atomistic interpretation of these parameters. When such calculations are possible for a wide range of minerals, they will form a true link between microscopic and macroscopic properties of the material.

7. APPLICATIONS OF VIBRATIONAL SPECTROSCOPY

7.1 Introduction

Three of the major applications of vibrational spectroscopy have been in qualitative and quantitative analysis, in studies of molecular and crystal structure, and for calculation of elastic and thermodynamic properties. All of these have been applied in mineralogy and geochemistry, and the following represent a few examples of each type of study.

There has been considerable discussion as to the mechanism for dissolution of water in silicate melts as a function of pressure and temperature (Burnham, 1967, 1979; Mysen, 1977; Holloway, 1981). Stolper (1982a,b) has carried out an elegant series of quantitative infrared absorption studies on thin films of hydrous glasses. These allowed both identification of hydroxyl groups and molecular water in the glasses, and measurement of their relative concentration as a function of total water content. These observations have led to a new model for the water dissolution in silicate glasses and melts, and to a thermodynamic description of the dissolution process based on the spectroscopic results (Stolper, 1982a,b).

The past decade has seen a great upsurge in the study of fluid inclusions in minerals, and the Raman microprobe has emerged as an instrument of choice for non-destructive in-situ analysis of inclusions down to several micrometers in diameter (Dhamelincourt et al., 1979; Rosasco and Roedder, 1979). The technique is sensitive to fluid species such as CO, CO_2, CH_4, H_2S, H_2O, H_2 and to daughter crystalline products. There is currently considerable interest in rendering the technique quantitative through extensive calibration (Dubessy et al., 1983; Cheilletz et al., 1984). This type of study has been used to interpret the fluid history of widely different rock types (e.g., Guilhamou et al., 1981; Bergman and Dubessy, 1984), and is contributing greatly to our understanding of fluids in the earth.

7.2 Structural studies

The major considerations in using vibrational spectroscopy as a structural tool have been discussed in Section 6. One of the major recent applications in geochemistry has been in structural studies of silicate melts and glasses (e.g., Brawer and White, 1975; Mysen et al., 1982; Matson et al., 1983; McMillan, 1984). The glass and melt spectra have been interpreted by systematic observations as a function of composition, pressure or temperature, by isotopic substitution (Galeener and Mikkelsen, 1981; Galeener and Geissberger, 1983), vibrational calculations (e.g., Furukawa et al., 1981), and many other methods. Much of our current understanding of silicate glass and melt structure at the molecular level is based on models derived from such vibrational studies. There is general agreement on many features of these models, such as changing polymerization of the silicate unit as a function of silica content, while a number of major questions still remain and are being actively discussed in the literature. It is in fact likely that vibrational spectroscopy alone is drawing near the limit of its capacity

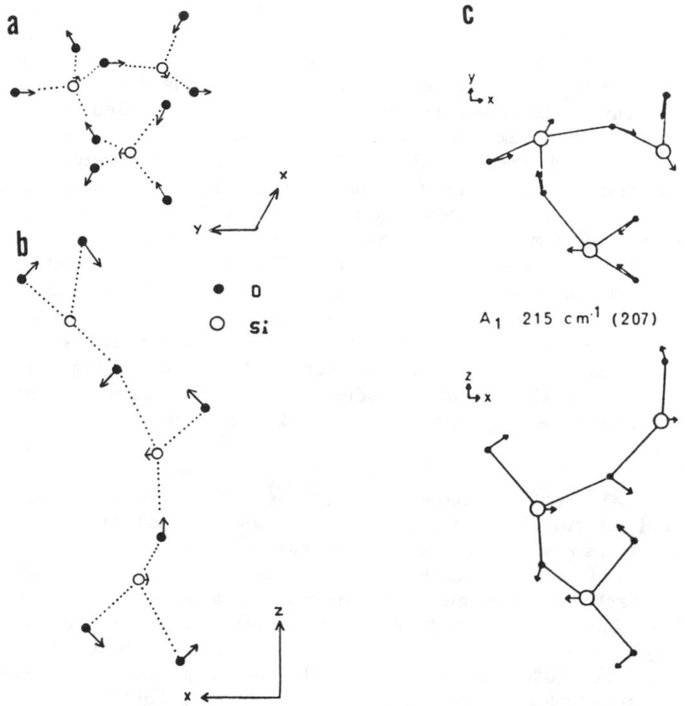

Figure 13. Displacement vectors relating atomic positions in α-quartz with those in β-quartz, projected on (a) the x-y and (b) the x-z plane. The atoms are drawn at their positions in α-quartz, and the tips of the arrows represent their positions in β-quartz (Wyckof, 1982, p. 312-313). (c) shows the calculated atomic displacements for the 207 cm^{-1} mode of α-quartz, taken from Etchepare et al. (1974) (see also Ghose, this volume).

to answer these questions, and further advances in the field will require input from other types of study.

Displacive phase transitions. A different type of structural study is found in the application of vibrational spectroscopy to phase transitions, illustrated by the classic studies of the α-β quartz transition (see also Ghose, this volume). On heating, α-quartz with trigonal symmetry transforms to the hexagonal β-phase via a non-first-order transition at 573°C. In 1940 Raman and Nedungadi showed by what is now known as Raman scattering that a mode near 220 cm^{-1} at room-temperature for α-quartz (Fig. 4) approached zero frequency increasingly rapidly as the transition temperature was approached. These authors suggested that the atomic displacements associated with the 220 cm^{-1} mode were those required for the α-β transformation. This type of behaviour is now known as "soft mode" behavior and is associated with displacive transitions (Cochran, 1960, 1961; Rao and Rao, 1978, p. 204-223). Since then, a considerable number of experimental and theoretical studies have been carried out of the structure and dynamics of α- and β-quartz as function of temperature (e.g., Simon and Shapiro et al., 1967; Axe and Shirane, 1970; McMahon, 1953; Scott, 1968; Höchli and Scott, 1971; Gervais and Piriou, 1975; Iishi, 1978a). The atomic displacements required to transform α-quartz into β-quartz are shown in Fig. 13. These may be compared with the displacement vectors calculated for the soft mode at 207 cm^{-1} (room-temperature, α-quartz) in a number of studies (e.g., Mirgorodskii et al., 1970; Etchepare et al., 1974: Figure 13). There is seen to be a close correspondence between the calculated mode displacements and the α-β transformation vectors. Then as α-quartz is heated and the 207 cm^{-1} mode is increasingly populated, the mode anharmonicity causes the average atomic positions to approach the β-quartz form with a corresponding decrease in restoring force hence mode frequency. Thermal population of this soft mode may then be said to drive the displacive transformation. There is an analogous mode for the β-quartz structure which increases to 74 cm^{-1} above the transition temperature. This mode is both Raman- and infrared-inactive (Axe and Shirane, 1970). In fact, when the transition is considered in detail, the above description is more complex due to coupling of the 207 cm^{-1} soft mode with other modes of the same symmetry in the α-quartz structure, (Scott, 1968, 1974; see also Ghose, this volume).

Iishi et al. (1979) have carried out a dynamical analysis for andalusite and measured its Raman spectrum as a function of temperature up to 600°C. They observed a considerable frequency decrease in two modes with increasing temperature (some degree of "mode softening"), which could perhaps presage the thermal transition to sillimanite. However, this behavior cannot reflect classical soft mode behavior, since although the andalusite and sillimanite structures are very similar, they are not simply related by a displacive transition and significant reconstruction must take place (Ribbe, 1980).

Order-disorder processes. A second type of transition of interest to mineralogy is that associated with diffusive processes, especially those related to Al-Si order. It has long been known that vibrational spectra are sensitive to the degree of order of a material (e.g., White, 1974), with broadened bands being observed for more disordered phases. A number of careful studies have followed the variation of vibrational spectra with degree of order, for example for alkali feldspars (Laves and Hafner, 1956; Hafner and Laves, 1957; Martin, 1970) and cordierite (Langer and Schreyer, 1969), allowing the calibration of vibrational spectroscopy as a simple tool for estimating the degree of order. However, the dynamics of disordered materials in general are not well understood, and detailed structural interpretations of such vibrational spectra in terms of atomic-scale order-disorder mechanisms are not yet possible. For example, recent work on Al-Si ordering in cordierite using a variety of techniques (including x-ray diffraction, electron microscopy, solution calorimetry, nuclear magnetic resonance, infrared and Raman spectroscopies: Putnis, 1980a,b; Putnis and Bish, 1983; Carpenter et al., 1983; Fyfe et al., 1983; McMillan et al., 1984) has shown that each individual technique highlights a different aspect of the ordering process, and that only consideration of the ensemble of studies begins to show a coherent picture. Even then, the vibrational data cannot be simply interpreted, although it is hoped that similar studies in the future will lead to a detailed understanding of the dynamics of such disordering systems.

7.3 Elastic properties and thermodynamic calculations

The stress-strain relationships for a crystal are usually expressed in the harmonic approximation by the 6x6 matrix of elastic constants C_{ij} (e.g., Born and Huang, 1954, p. 129-165). The number of independent constants is fixed by the crystal symmetry; for example for a cubic crystal, there are only three, c_{11}, c_{12} and c_{44}. The magnitudes of these elastic constants may be obtained from the initial slopes of the acoustic phonon branches near $\tilde{k} = \tilde{0}$ (Section 3.3), measured by inelastic neutron scattering or Brillouin spectroscopy, or may be calculated from first principles (Section 6.5).

The elastic constants are derived by considering small finite deformations of the crystal from its equilibrium structure. Since the elastic constants are simply related to the acoustic mode frequencies, if these frequencies were to become imaginary, the associated elastic constants would drive a corresponding deformation of the crystal (Born and Huang, 1954, p. 140-165). In this case, the lattice is termed dynamically unstable with respect to the deformed configuration. Such considerations can be useful in analyzing polymorphic phase transitions. Recently Wolf and Jeanloz (1984) have calculated dispersion curves for a cubic model of $MgSiO_3$ perovskite, a high-pressure phase considered as a good model for the mineralogy of the lower mantle. The calculation predicted that the cubic lattice would be dynamically unstable, in agreement with observation, and the form of the normal mode responsible for the instability corresponded well with the observed distortion of the cubic lattice. Such calculations should prove increasingly useful in studies of high pressure phases, where direct experiments may be difficult with current technology.

One of the major applications of vibrational spectroscopy has classically been in calculation of the vibrational heat capacity and related thermodynamic properties (Herzberg, 1945, p. 501-530). This goal provided much of the impetus for the development of lattice dynamical theory for crystals (Born and Huang, 1954; see Kieffer, this volume). When the vibrational density of states (the vibrational spectrum averaged over all points in the Brillouin zone) is known, for example from inelastic neutron scattering data, the specific heat as a function of temperature may be calculated directly (see Cochran, 1973, p. 69-72). Inelastic neutron scattering experiments have only been carried out for a few silicates (Ghose, this volume), and heat capacity calculations must rely on some type of modelling. Striefler and Barsch (1975) carried out a vibrational calculation for α-quartz including the effects of dispersion, and obtained a heat capacity curve in good agreement with experiment from their calculated vibrational density of states. Recently, Kieffer (1979a,b,c) has developed models to approximate the effects of dispersion on data at $\tilde{k} = \tilde{0}$, allowing the vibrational density of states to be modelled from infrared and Raman spectra and measurements of acoustic wave velocities. These models have been quite successful in their application to silicates and other minerals (e.g., Kieffer, 1979c, 1980, 1982; Akaogi et al., 1984). For example, Akaogi et al, (1984) used this approach to obtain heat capacities for the α-, β- and γ-phases of Mg_2SiO_4. In conjunction with calorimetric measurements of enthalpies, these values were used to predict phase relationships for Mg_2SiO_4 as a function of pressure and temperature, in good agreement with experiment. Further examples and discussion of this method are given in chapters by Kieffer and Navrotsky (this volume).

8. FORSTERITE - A WORKED EXAMPLE

8.1 Crystal structure and number of vibrations.

The structure of forsterite (Mg_2SiO_4) has been discussed in detail by Brown (1980). Forsterite has space group Pbnm ($\equiv D_{2h}^{16}$), with Z = 4 for a total of N = 28 atoms per unit cell. Since each atom has three degrees of translational freedom, the total number of vibrational modes at $\tilde{k} = \tilde{0}$ will be 3N-3 = 81, with three acoustic modes corresponding to translations of the entire crystal. The vibrational mode symmetries will be determined by considering the symmetry species of x, y and z excursions of equivalent atoms within each site group, and correlating these with the symmetry of the unit cell. The site occupancies for forsterite are:

Equivalent atoms	Site symmetry	Wyckoff notation
4 Mg (M1)	C_i ($\equiv S_2$)	a
4 Mg (M2)	C_s ($\equiv C_{1h}$)	c
4 O1	C_s	c
4 O2	C_s	c
4 Si	C_s	c
8 O3	C_1	d

8.2 Determination of mode symmetries

A number of definitions are in common use: γ and ζ refer respectively to site species and factor group (i.e., the unit cell point group) species, hence t^γ is the number of translation vectors belonging to a site species γ. For example, consider the four equivalent Si atoms on sites with point group C_s. The x and y excursions belong to symmetry species A', and z displacements to species A". If n is the number of atoms in an equivalent set, $f^\gamma = nt^\gamma$ forms the total degrees of freedom over that site species. Hence, for the Si atoms in forsterite:

t^γ	f^γ	Site species in C_s	
2(x,y)	8	A'	(59)
1 (z)	4	A"	

If the Cartesian axes for site and factor group symmetries are chosen to coincide, the site group species may be correlated with the factor group species (Fateley et al., 1972). In the present case, the correlation between sites C_s and the factor group D_{2h} of the olivine structure is:

<u>Site group species</u> C_s <u>Factor group species</u> D_{2h}

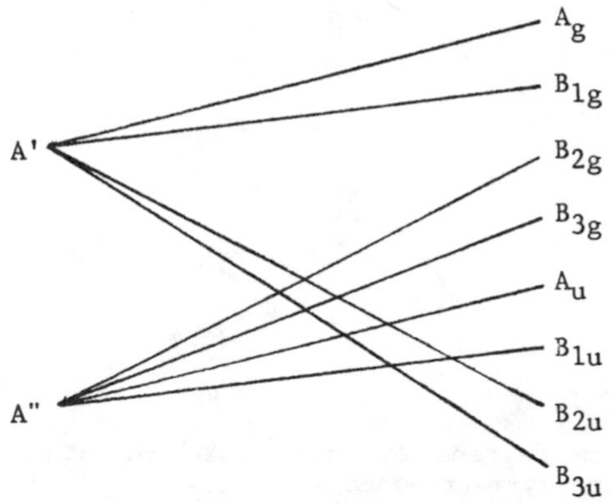

(60)

(Note: for the setting Pbnm, the σ_{xy} correlations must be chosen; see symmetry tables.) This correlation means that for C_s sites in olivine, A' species contribute to A_g, B_{1g}, B_{2u} and B_{3u} species of the factor group D_{2h}, and A" species to factor group species B_{2g}, B_{3g}, A_u and B_{1u}.

A quantity a_γ is then defined which specifies the number of degrees of freedom contributed by each site species γ to a factor group species ζ. This is given by

$$a_\gamma = \frac{f^\gamma}{\sum_\zeta {}^\gamma C_\zeta} \quad , \tag{61}$$

where ${}^\gamma C_\zeta$ is the degeneracy of each factor species ζ which receives a contribution from that site species γ. Degeneracies are defined as $C_\zeta = 1$ for species type A or B, 2 for E, and 3 for T ($\equiv$ F). For example, each site species A' of site group C_s correlates with species A_g, B_{1g}, B_{2u} and B_{3u} of D_{2h}, hence

$$a_{A'} = \frac{8}{1+1+1+1} = 2 \quad , \tag{62}$$

and similarly $a_{A''} = \frac{4}{4} = 1$. (63)

Finally, the quantity a_ζ is defined to determine the degrees of freedom for each factor group species ζ contributed from all site group species which correlate into that particular species ζ :

$$a_\zeta = \sum_\gamma a_\gamma \quad . \tag{64}$$

The sum is over all species γ which "feed into" ζ. For example, species A_g contains only contributions from A' site species, hence $a^{A_g} = a^{A'} = 2$. The quantitative correlation corresponding to (60) may then be written

$$\underline{f}^\gamma \qquad \underline{\gamma(C_s)} \qquad \underline{\zeta(D_{2h})} \qquad \underline{a}_\zeta$$

		A_g	2
		B_{1g}	2
8	A'	B_{2g}	1
		B_{3g}	1
		A_u	1
		B_{1u}	1
4	A"	B_{2u}	2
		B_{3u}	2

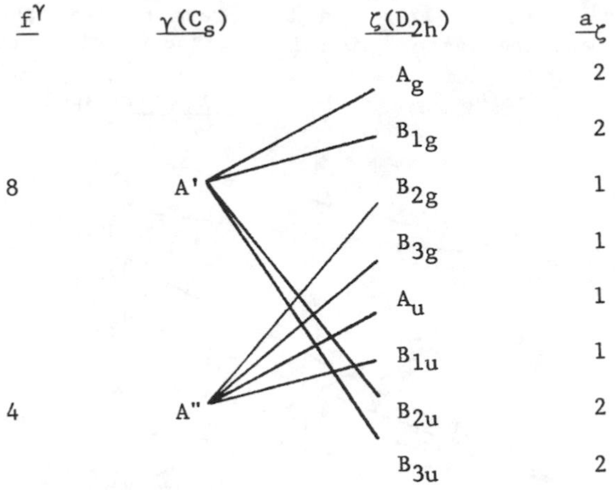

$$(65)$$

Hence, the Cartesian degrees of freedom of Si atoms on sites C_s in forsterite contribute symmetry species

$$\Gamma_{Si} = 2A_g + 2B_{1g} + B_{2g} + B_{3g} + A_u + B_{1u} + 2B_{2u} + 2B_{3u} \qquad (66)$$

to the vibrational degrees of freedom of the forsterite crystal spectrum.

The same operations may then be carried out over all remaining equivalent sets of atoms to give the total degrees of vibrational freedom

$$\Gamma^{tot} = 11A_g + 11B_{1g} + 7B_{2g} + 7B_{3g} + 10A_u + 10B_{1u} + 14B_{2u} + 14B_{3u} \qquad (67)$$

a total of 84 "modes". However, included in these are the three translations of the D_{2h} factor group,

$$\Gamma_{trans} = B_{1u} + B_{2u} + B_{3u} \quad , \qquad (68)$$

giving the 81 true vibrational modes as

$$\Gamma_{vib} = 11A_g + 11B_{1g} + 7B_{2g} + 7B_{3g} + 10A_u + 9B_{1u} + 13B_{2u} + 13B_{3u} \cdot \qquad (69)$$

8.3 Infrared and Raman activity

As discussed in Section 2.4, the infrared active species transform similarly to the Cartesian translation vectors, while Raman active modes transform as the quadratic combinations of x, y and z within the factor group D_{2h}. The infrared and Raman active modes are then

$$\Gamma_{IR} = 9B_{1u} + 13B_{2u} + 13B_{3u} \qquad (70)$$

$$\Gamma_{Raman} = 11A_g + 11B_{1g} + 7B_{2g} + 7B_{3g} \qquad (71)$$

i.e., 35 peaks are expected in the infrared, and 36 in the Raman spectrum, with the indicated symmetries. The ten A_u modes are spectroscopically inactive.

This method allows prediction of the band polarizations for polarized infrared reflection and absorption and Raman studies. The olivine crystal may be cut following {100}, {010} and {001}. For infrared work, the sample would be mounted with the incident beam normal to these faces. Then spectra taken with the incident beam polarized parallel to x, y or z would give the following spectra:

Orientation	Polarization	Symmetry species
{100}	y	B_{2u}
	z	B_{1u}
{010}	x	B_{3u}
	z	B_{1u}
{001}	x	B_{3u}
	y	B_{2u}

If the sample were mounted for 90° Raman scattering, with the incident beam following x, y or z, the following combinations of sample orientation and incident and scattered polarizations (Section 5) would give:

Orientation	Symmetry species
x(yy)z	A_g (yy)
x(yx)z	B_{1g}
x(zz)y	A_g (zz)
x(zx)y	B_{2g}
y(xx)z	A_g (xx)
y(xy)z	B_{3g}

These predictions may be compared with the results of experimental studies for forsterite (e.g., Servoin and Piriou, 1973; Iishi, 1978b: see Figs. 14 and 15).

8.4 Correlation of molecular group vibrations.

The above method allows enumeration of the number and symmetry type of the vibrational modes expected for a given crystal, and predicts their infrared and Raman activities and polarization characteristics, but does not permit any further interpretation of the spectra in terms of the form of the normal modes. Some band assignments may be reasonably carried out if the method is extended to consider molecular groups present within the crystal structure, and correlating vibrations of the free molecular group with those of the crystal. For example, the forsterite unit cell contains four identifiable SiO_4 groups. A hypothetical free SiO_4^{4-} ion with tetrahedral symmetry belongs to point group T_d and has the following vibrational modes (Herzberg, 1945, p. 100):

T_d species	Assignment
A_1	ν_1 symmetric stretching
T_2	ν_3 asymmetric stretching
E	ν_2 symmetric bending
T_2	ν_4 asymmetric bending

These may be correlated into the D_{2h} factor group via the site symmetry of the SiO_4 units in the crystal, which is C_s. Since these units are not free in the crystal, their "translations" and "rotations" (respectively T_2 and T_1 species of group T_d) become librational modes of the crystal. This correlation gives:

	Free ion SiO_4^{4-} (T_d) Rotation	Translation	Internal
A_1	--	--	$1(\nu_1)$
E	--	--	$1(\nu_2)$
T_1	1	--	--
T_2	--	1	$2(\nu_3,\nu_4)$

Site group (C_s)

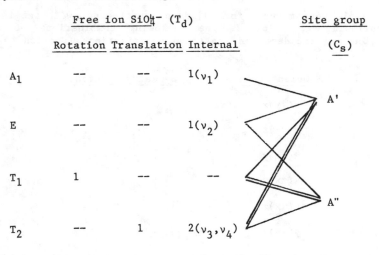

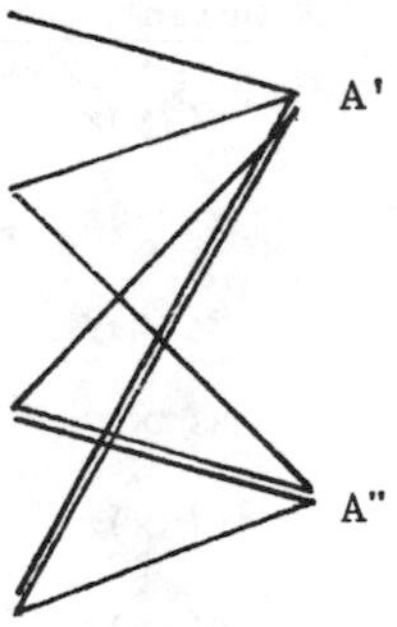

(71)

Note that in calculating an equivalent "f^γ " for free ion species, the degeneracies of these must be taken into account, i.e.,

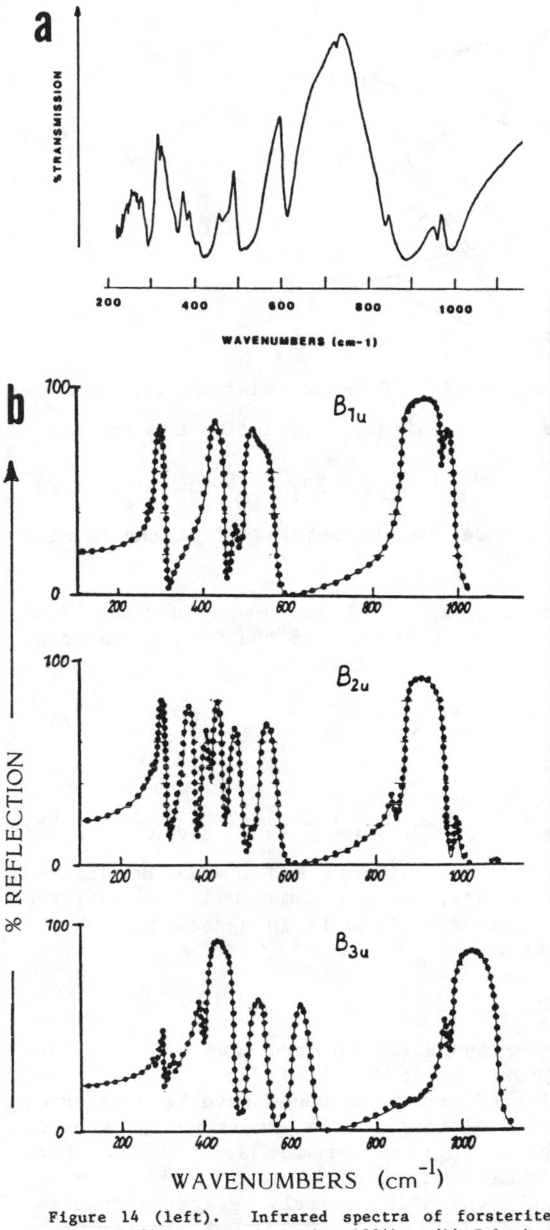

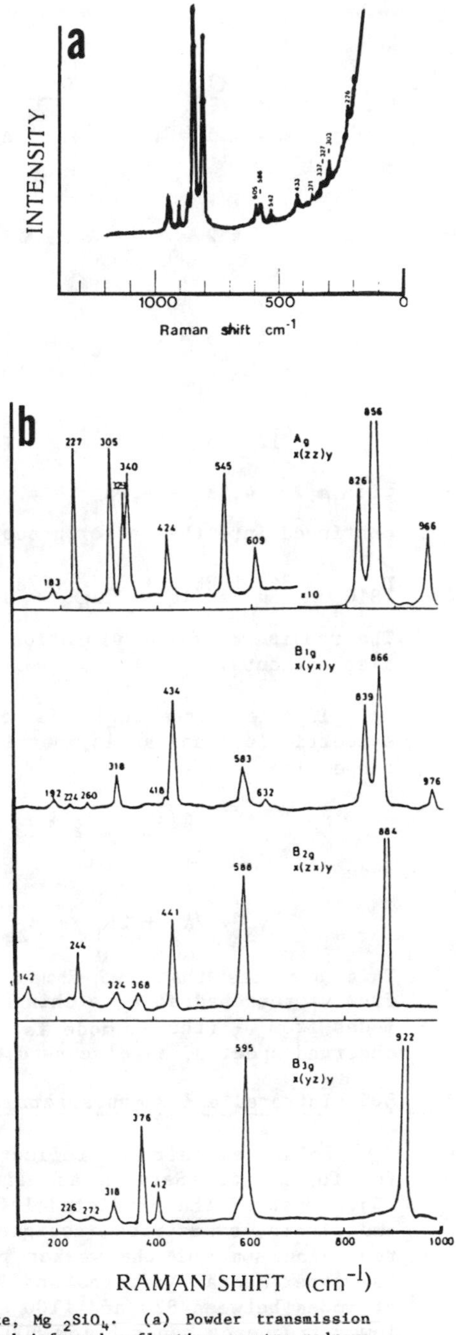

Figure 14 (left). Infrared spectra of forsterite, Mg_2SiO_4. (a) Powder transmission spectrum (from Akaogi et al., 1984). (b) Polarized infrared reflection spectra redrawn from Servoin and Piriou (1972).

Figure 15 (right). Raman spectra of forsterite. (a) Unpolarized spectrum of polycrystalline sample (Piriou and McMillan, 1983). (b) Polarized single crystal spectra from Iishi (1978) showing scattering geometries. Note that spectrum (a) runs right to left, and spectra in (b) run left to right.

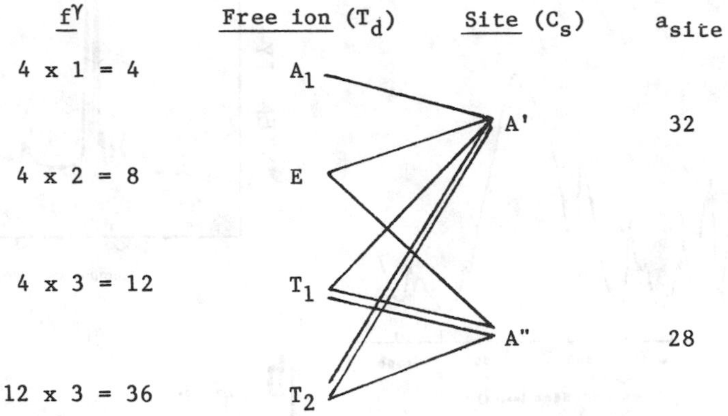

$$\underline{f^\gamma} \qquad \underline{\text{Free ion}} \ (T_d) \qquad \underline{\text{Site}} \ (C_s) \qquad a_{\text{site}}$$

$$4 \times 1 = 4 \qquad A_1$$

$$A' \qquad 32$$

$$4 \times 2 = 8 \qquad E \tag{72}$$

$$4 \times 3 = 12 \qquad T_1$$

$$A'' \qquad 28$$

$$12 \times 3 = 36 \qquad T_2$$

with $a_{A_1} = 4$, $a_E = 4$, $a_{T_1} = 6$, $a_{T_2} = 18$. This correlation may then be continued into the factor group D_{2h} as before (Correlation 65) to give

$$\Gamma_{SiO_4} = 8A_g + 8B_{1g} + 7B_{2g} + 7B_{3g} + 7A_u + 7B_{1u} + 8B_{1u} + 8B_{3u} \quad . \tag{73}$$

The remainder of the vibrational modes may be obtained from the Mg site displacements.

If the correlation is restricted to those modes derived from symmetric (ν_1) and antisymmetric (ν_3) stretching of the SiO_4 tetrahedra, these give

$$\nu_1 \rightarrow A_g + B_{1g} + B_{2u} + B_{3u} \tag{74}$$

and

$$\nu_3 \rightarrow 2A_g + 2B_{1g} + B_{2g} + B_{3g} + A_u B_{1u} + 2B_{2u} + 2B_{3u} \quad . \tag{75}$$

This predicts that two Raman and two infrared modes will be derived from ν_1 tetrahedral stretching motions, and six Raman and five infrared modes from ν_3 (the A_u mode is inactive). This is in agreement with the observed spectra, as discussed below.

8.5 Infrared and Raman spectra.

Polarized infrared reflection and Raman spectra have been obtained for forsterite (Servoin and Piriou, 1973; Iishi, 1978b: Figs. 14 and 15). Most of the predicted infrared and Raman bands have been observed and their symmetries assigned, although some controversy remains regarding some of the weaker peaks (Piriou and McMillan, 1983). Both the experimental infrared and Raman spectra of forsterite show a group of modes between 820 and 1100 cm^{-1} which are clearly separated from the lower frequency bands below 600 cm^{-1} (Figs. 14 and 15). A similar high-frequency group is observed for all simple silicates, and these bands are generally assigned to silicon-oxygen stretching motions. This is probably a reasonable assignment for the following reasons. The diatomic molecule SiO has its $\omega_e = 1253$ cm^{-1}, (Huber and Herzberg, 1979,

p. 606) while the molecule $H_6Si_2O_7$ has strong infrared and Raman bands in the 900-1250 cm^{-1} region corresponding to Si-O stretching (Emeleus et al., 1955; Curl and Pitzer, 1958; Durig et al., 1977). Further support for this assignment is provided by recent ab initio molecular orbital calculations of silicon-oxygen stretching force constants (e.g., O'Keeffe et al., 1980; Gibbs, 1982; O'Keeffe and Gibbs, 1984). The eight Raman and seven infrared modes observed in the high-frequency region for forsterite may then be confidently assigned to lattice vibrations derived from the ν_1- and ν_3- modes of the individual SiO_4 tetrahedral groups. It may be noted that in the Raman spectra, only ν_3-derived vibrations give rise to B_{2g} and B_{3g} modes (75), but both ν_1- and ν_3- type motions contribute to A_g and B_{1g} species. A considerable literature has been devoted to attempting to differentiate between these contributions (see Piriou and McMillan, 1983), but the $^{28}Si/^{30}Si$ isotopic substitution work of Pâques-Ledent and Tarte (1973) has shown that there is considerable vibrational coupling between ν_1- and ν_3- derived motions. This coupling is due to the distortion of the SiO_4 units from tetrahedral symmetry within the olivine structure, and has the result that the true atomic displacements in these high-frequency modes are likely to be combinations of ν_1- and ν_3- type motions (Piriou and McMillan, 1983).

8.6 Isotopic and chemical substitution and vibrational calculations.

As noted above, Pâques-Ledent and Tarte (1973) carried out $^{28}Si/^{30}Si$ isotopic substitution experiments for Mg_2SiO_4. These authors also made the substitution $^{24}Mg/^{26}Mg$. The combined results supported the assignment of the bands above 800 cm^{-1} to Si-O stretching motions with little Mg participation. In the infrared spectra, peaks at 466.5, 425.5, 415, 364, 320.5, 300.5 and 277.5 cm^{-1} showed a large shift on Mg substitution with little associated silicon shift, but most other peaks showed comparable silicon and magnesium shifts suggesting complex vibrations involving both silicon and magnesium displacements.

In an earlier study, Tarte (1963b) measured powder infrared spectra for the solid solution series Mg_2SiO_4-Ni_2SiO_4 in order to identify bands associated with the metal-oxygen polyhedra. He observed little change in the spectra above 475 cm^{-1}, but noted systematic changes in bands at 362, 298 and perhaps 422 cm^{-1}, and assigned the peak at 362 cm^{-1} to a vibration of the MgO_6 octahedral groups. The 362 cm^{-1} band may be identified with the 364 cm^{-1} peak measured by Pâques-Ledent and Tarte (1973) which showed a large Mg isotope effect, supporting Tarte's (1963b) assignment, while the 298 and 422 cm^{-1} peaks could correspond to the 300.5 and 425.5 cm^{-1} peaks of the isotopic study. Tarte (1963b) also suggested that the peak at 422 cm^{-1} might have two components: one from MgO_6 motions and one from the ν_2 deformation of the SiO_4 groups: this also seems to be supported by the isotopic substitution (Pâques-Ledent and Tarte, 1973, Table 6).

Both Devarajan and Funck (1975) and Iishi (1978b) have carried out vibrational calculations for forsterite. Devarajan and Funck (1975) found reasonable agreement between observed and calculated frequencies, but did not include the effects of the Coulomb field nor calculate atomic displacements. Iishi (1978b) did include Coulomb terms to calculate TO and LO mode frequencies, and presented the atomic displacements calculated for selected normal modes. There is some controversy remaining over the assignment of the experimental spectra (see Piriou and McMillan, 1983), and it is not yet certain what effect this might have on these calculations which are fit to particular interpretations of the spectra. The high-frequency modes calculated by Iishi (1978b) are in reasonable agreement with the expected motions derived from ν_1- and ν_3- stretching of the SiO_4 groups with little associated magnesium motion. The infrared modes at 400 and 280 cm^{-1} were calculated (Iishi, 1978b, Fig. 5) to involve essentially Mg motion, and the mode at 421 cm^{-1} to be associated with libration of the SiO_4 units, in good agreement with the studies of Tarte (1963b) and Pâques-Ledent and Tarte (1973). However, Iishi (1978b) also calculated a mode at 465 cm^{-1} to involve little Mg participation and one at 353 cm^{-1} to have large magnesium displacements, which does not agree with the isotopic results. It is evident that more work will be required before a reliable force field may be established for forsterite, and that our present understanding of its crystalline dynamics is far from complete.

8.7 Dispersion and heat capacity calculations

To the author's knowledge, no inelastic neutron scattering experiments have been carried out for forsterite, so the dispersion curves are not known, while the vibrational calculations (Section 8.6) did not include dispersion in their models. Kieffer (1979a, 1980) has applied her model to estimate the vibrational density of states from infrared, Raman and acoustic wave velocity data. This model was used to calculate the heat capacity curve for forsterite in good agreement with experiment (Kieffer, 1980; Akaogi et al., 1984). This approach was extended to higher pressure polymorphs of Mg_2SiO_4, and used to help establish a pressure-temperature phase diagram (Akaogi et al., 1984). Further details of these calculations and examples of their use are given in the chapters by Kieffer and Navrotsky (this volume).

In summary, the infrared and Raman spectra of forsterite are well known, and band symmetries have been assigned. The isotopic and chemical substitution experiments have given some insight into the nature of a number of the modes. The high-frequency modes derived from silicon-oxygen stretching are quite well understood, although details of the coupling between ν_1- and ν_3-derived motions and its relation to distortion of the SiO_4 group are not yet clear. The vibrational calculations have had some success, but some discrepancies remain with experimental results, and further refinement of the force field will be necessary. The major remaining step in defining the vibrational properties of forsterite is measurement of vibrational frequencies across the Brillouin zone via inelastic neutron spectroscopy. Until then, the effects of dispersion on the vibrational heat capacity cannot be reliably evaluated. However, when such measurements have been made and a well-constrained force field established, the vibrational contribution to the thermodynamic properties of forsterite will truly be understood in terms of its microscopic behavior.

REFERENCES

Aines, R.D. and Rossman, G.R. (1984a) Water in minerals? A peak in the infrared. J. Geophys. Res. 89, 4059-4071.

____ and ____ (1984b) The hydrous component in garnets. Am. Mineral. 69, 1116-1126.

Akaogi, M., Ross, N.L., McMillan, P. and Navrotsky, A. (1984) The Mg_2SiO_4 polymorphs (olivine, modified spinel and spinel) - thermodynamic properties from oxide melt solution calorimetry, phase relations, and models of lattice vibrations. Am. Mineral. 69, 499-512.

Ashcroft, N.W. and Mermin, N.D. (1976) "Solid State Physics", Holt-Saunders, Philadelphia.

Atkins, P.W. (1980) "Molecular Quantum Mechanics." Parts I and II. Clarendon Press, Oxford.

____ (1982) "Physical Chemistry," 2nd ed. W.H. Freeman & Co., San Francisco.

Axe, J.D. and Shirane, G. (1970) Study of the α-β quartz phase transformation by inelastic neutron scattering. Phys. Rev. B 1, 342-348.

Badger, R.M. (1934) A relation between internuclear distances and bond force constants. J. Chem. Phys. 2, 128-131.

______ (1935) The relation between the internuclear distances and force constants of molecules and its application to polyatomic molecules. J. Chem. Phys. 3, 710-714.

Banwell, C.N. (1972) "Fundamentals of Molecular Spectroscopy", 2nd ed. McGraw-Hill, N.Y.

Barnes, A.J. and Orville-Thomas, W.J. (1977) "Vibrational Spectroscopy - Modern Trends." Elsevier, N.Y.

Barnes, R.B. and Czerny, M. (1931) Messungen am NaCl und KCl im Spektralbereich ihrer ultraroten Eigen schwingungen. Z. Phys. 72, 447-461.

Barron, T.H.K. (1961) Long-wave optical vibrations in simple ionic crystals. Phys. Rev. 123, 1995-1998.

Bellamy, L.J. (1958) "The Infrared Spectra of Complex Molecules." 2nd ed. Methuen and Co., London.

____ (1968) "Advances in Infrared Group Frequencies." Methuen and Co., London.

Beran, A. and Putnis, A. (1983) A model of the OH positions in olivine, derived from infrared-spectroscopic investigations. Phys. Chem. Minerals 9, 57-60.

Bergman, S.C. and Dubessy, J. (1984) CO_2-CO fluid inclusions in a composite peridotite xenolith: implications for upper mantle oxygen fugacity. Contrib. Mineral. Petrol. 85, 1-13.

Berry, R.S., Rice, S.A. and Ross, J. (1980) "Physical Chemistry," John Wiley & Sons, N.Y.

Brawer, S.A. and White, W.B. (1975) Raman spectroscopic investigation of silicate glasses. I. The binary alkali silicates. J. Chem. Phys. 63, 2421-2432.

Brown, G.E. (1980) Olivines and silicate spinels. In "Orthosilicates:, P.H. Ribbe, ed. Reviews in Mineralogy, 5, 275-381.

Burnham, C.W. (1967) Hydrothermal fluids at the magmatic stage. In "Geochemistry of Hydrothermal Ore Deposits," H.L. Barnes, ed., pp. 34-76. Holt, Rinehart and Winston.

____ (1979) The importance of volatile constituents. In "The Evolution of the Igneous Rocks: Fiftieth Anniversary Perspectives" H.S. Yoder, Jr., ed., pp. 439-479. Princeton University Press, Princeton, N.J.

Burns, R.G. and Huggins, F.E. (1972) Cation determinative curves for Mg-Fe-Mn olivines from vibrational spectra. Am. Mineral., 57, 967-985.

Carpenter, M.A., Putnis, A., Navrotsky, A. and McConnell, J.D.C. (1983) Enthalpy effects associated with Al/Si ordering in anhydrous Mg-cordierite. Geochim. Cosmochim. Acta 47, 899-906.

Cheilletz, A., Dubessy, J., Kosztolanyi, C., Masson-Perez, N., Ramboz, C. and Zimmermann, J.L. (1984) Les fluides moléculaires d'un filon de quartz hydrothermal: comparaison des techniques analytiques ponctuelles et globales, contamination des fluides occlus par des composés carbonés. Bull. Minéral. 107, 169-180.

Cochran, W. (1960) Crystal stabilitv and the theory of ferroelectricity. Adv. Phys. 9, 387-423.

____ (1961) Crystal stability and the theory of ferroelectricity. Part II. Piezoelectric crystals. Adv. Phys. 10, 401-420.

____ (1973) "The Dynamics of Atoms in Crystals: Edward Arnold Ltd., London.

Cotton, F.A. (1971) "Chemical Applications of Group Theory," 2nd ed. John Wiley and Sons, N.Y.

Curl, R.F. and Pitzer, K.S. (1958) The spectrum and structure of disiloxane. J. Am. Chem. Soc. 80, 2371-2373.

Damen, T.C., Porto, S.P.S. and Tell, B. (1966) Raman effect in zinc oxide. Phys. Rev. 142, 570-574.

Decius, J.C. and Hexter, R.M. (1977) "Molecular Vibrations in Crystals." McGraw-Hill, N.Y.

Devarajan, V. and Funck, E. (1975) Normal coordinate analysis of the optically active vibrations (k=0) of crystalline magnesium orthosilicate Mg_2SiO_4 (forsterite). J. Chem. Phys. 62, 3406-3411.

Dhamelincourt, P., Beny, J.M., Dubessy, J. and Poty, B. (1979) Analyse d'inclusions fluids à la microsonde MOLE à effect Raman. Bull. Minéral. 102, 600-610.
Dickinson, J.E. and Hess, P.C. (1985) Rutile solubility and titaniium coordination in silicate melts. Geochim. Cosmochim. Acta (in press).
Dubessy, J., Geisler, D., Kosztolanyi, C. and Vernet, M. (1983) The determination of sulphate in fluid inclusions using the M.O.L.E. Raman microprobe. Application to a Keuper halite and geochemical consequences. Geochim. Cosmochim. Acta 47, 1-10.
Durig, J.R., Flanagan, M.J. and Kalasinsky, V.F. (1977) The determination of the potential function governing the low frequency bending mode of disioloxane. J. Chem. Phys. 66, 2775-2785.
Ebsworth, E.A.V. (1963) Inorganic applications of infra-red spectroscopy. In "Infrared Spectroscopy and Molecular Structure," M. Davies, ed., pp. 311-344. Elsevier, N.Y.
Elcombe, M.M. (1967) Some aspects of the lattice dynamics of quartz. Proc. Phys. Soc. 91, 947-958.
Emeleus, H.J., MacDiarmid, A.G. and Maddock, A.G. (1955) Sulphur and selenium derivatives of monosilane. J. Inorg. Nuclear Chem. 1, 194-201.
Etchepare, J. and Merian, M. (1978) Vibrational normal modes of SiO_2. III. α-berlinite ($AlPO_4$) and its relations to Γ-A phonon dispersion curves in α-quartz. J. Chem. Phys. 68, 5336-5341.
________, ________, and Smetankine, L. (1974) Vibrational normal modes of SiO_2. I. α and β quartz. J. Chem. Phys. 60, 1873-1876.
Farmer, V.C. (1974) "The Infrared Spectra of Minerals." Mineralogical Society, London.
Fateley, W.G., Dollish, F.R., McDevitt, N.T. and Bentley, F.F. (1972) "Infrared and Raman Selection Rules for Molecular and Lattice Vibrations: The Correlation Method." Wiley-Interscience, N.Y.
Fleet, M.E., Herzberg, C.T., Henderson, G.S., Crozier, E.D., Osborne, M.D. and Scarfe, C.M. (1984) Coordination of Fe, Ga and Ge in high pressure glasses by Mössbauer, Raman and x-ray absorption spectroscopy, and geological implications. Geochim. Cosmochim. Acta 48, 1455-1466.
Fluendy, M.A.D. and Lawley, K.P. (1973) "Chemical Applications of Molecular Beam Scattering." Chapman and Hall, London.
Freund, F. (1981) Mechanism of the water and carbon solubility in oxides and silicates and the role of O^-. Contrib. Mineral. Petrol. 76, 474-482.
_____ and Wengeler, H. (1981) The infrared spectrum of OH-compensated defect sites in C-doped MgO and CaO single crystals. J. Phys. Chem. Solids 43, 129-145.
_____, _____, Kathrein, H., Knobel, R. Oberheuser, G., Maiti, G.C., Reil, D., Kniffing, U. and Kötz, J. (1983) Hydrogen and carbon derived from dissolved H_2O and CO_2 in minerals and melts. Bull. Minéral. 106, 185-200.
Fröhlich, H. (1958) "Theory of Dielectrics. Dielectric Constant and Dielectric Loss." Oxford University Press, Oxford.
Fujimara, Y. and Lin S.H. (1979) A theoretical study of resonance Raman scattering from molecules. J. Chem. Phys. 70, 247-262.
Furukawa, T., Fox, K.E. and White, W.B. (1981) Raman specrocopic investigation of the structure of silicate glasses. III. Raman intensitites and structural units in sodium silicate glasses. J. Chem. Phys. 75, 3226-3237.
Fyfe, C.A., Gobbi, G.C., Klinowski, J., Putnis, A. and Thomas, J.M. (1983) Characterization of local atomic environments and quantitative determination of changes in site occupancies during the formation of ordered synthetic cordierite by ^{29}Si and ^{27}Al magic-angle spinning N.M.R. spectroscopy. J. Chem. Soc., Chem. Commun., 556-558.
Galeener, F.L. and Mikkelsen, J.C. (1981) Vibrational dynamics in ^{18}O-substituted vitreous SiO_2. Phys. Rev. B 23, 5527-5530.
____ and Geissberger, A.E. (1983) Vibrational dynamics in ^{30}Si-substituted vitreous SiO_2. Phys. Rev. B 27, 6199-6204.
Gervais, F. and Piriou, B. (1975) Temperature dependence of transverse and longitudinal optic modes in the α and β phases of quartz. Phys. Rev. B 11, 3844-3950.
_____, _____ and Cabannes, F. (1973a) Anharmonicity of infrared vibration modes in the nesosilicate Be_2SiO_4. Phys. Stat. Solidi (b) 55, 143-154.
_____, _____, and ____ (1973b) Anharmonicity in silicate crystals: temperature dependence of A_u type vibrational modes in $ZrSiO_4$ and $LiAlSi_2O_6$. J. Phys. Chem. Solids 34, 1785-1796.
Gibbs, G.V. (1982) Molecules as models for bonding in silicates. Am. Mineral. 67, 421-450.
Goldman, D.S., Rossman, G.R. and Dollase, W.A. (1977) Channel constituents in cordierite. Am. Mineral. 62, 1144-1157.
Guilhamou, N., Dhamelincourt, P., Touray, J-C. and Touret, J. (1981) Etude des inclusions fluides du système N_2-CO_2 de dolomites et de quartz de Tunisie septentrionale. Données de la microcryoscopie et de l'analyse à la microsonde à effet Raman. Geochim. Cosmochim. Acta 45, 657-674.
Hadni, A. (1967) "Essentials of Modern Physics Applied to the Study of the Infrared." Pergamon Press, Oxford.

Hafner, S. and Laves, F. (1957) Ordnung/Unordnung und Ultrarotabsorption. II. Variation der Lage und Intensität einiger Absorptionen von Feldspäten. Zur Struktur von Orthoklas und Adular. Z. Kristallogr. 109, 204-225.

Herzberg, G. (1950) "Molecular Spectra and Molecular Structure. I. Spectra of diatomic molecules," 2nd ed. Van Nostrand Reinhold, N.Y.

____ (1945) "Molecular Spectra and Molecular Struture. II. Infrared and Raman spectra of polyatomic molecules." Van Nostrand Reinhold, N.Y.

Höchli, U.T. and Scott, J.F. (1971) Displacement parameter, soft-mode frequency, and fluctuations in quartz below its α-β phase transition. Phys. Rev. Lett. 26, 1627-1629.

Holloway, J.R. (1981) Volatile interactions in magmas. In "Thermodynamics of Minerals and Melts," Advances in Physical Geochemistry, Vol. I, R.C. Newton, A. Navrotsky and B.J. Wood, eds., p. 273-293. Springer-Verlag, N.Y.

Huber, K.P. and Herzberg, G. (1979) "Molecular Spectra and Molecular Structure. IV. Constants of Diatomic Molecules." Van Nostrand Reinhold, N.Y.

Iishi, K. (1978a) Lattice dynamical study of the α-β quartz phase transition. Am. Mineral. 63, 1190-1197.

____ (1978b) Lattice dynamics of forsterite. Am. Mineral. 63, 1198-1208.

____ (1978c) Lattice dynamics of corundum. Phys. Chem. Minerals 3, 1-10.

____ Tomisaka, T., Kato, T., Yamaguchi and Umegaki, Y. (1971) Isomorphous substitution and infrared and for infrared spectra of the feldspar group. N. Jahrb. Mineral. Abh. 115, 98-119.

____ Salje, E. and Werneke, Ch. (1979) Phonon spectra and rigid-ion model calculations on andalusite. Phys. Chem. Minerals 4, 173-188.

Jaffé, H.H. and Orchin, M. (1965) "Symmetry in Chemistry". John Wiley & Sons, N.Y.

Jeanloz, R. (1980) Infrared spectra of olivine polymorphs: α,β phase and spinel. Phys. Chem. Minerals 5, 327-341.

Jones, L.H. (1971) "Inorganic Vibrational Spectroscopy," Vol. 1. Marcel Dekker, Inc.

Karr, C. (1975) "Infrared and Raman Spectroscopy of Lunar and Terrestrial Materials." Academic Press, N.Y.

Kieffer, S.W. (1979a) Thermodynamics and lattice vibrations of minerals: 1. Mineral heat capacities and their relationships to simple lattice vibrational models. Rev. Geophys. Space Phys. 17, 1-19.

____ (1979b) Thermodynamics and lattice vibrations of minerals: 2. Vibrational characteristics of silicates. Rev. Geophys. Space Phys. 17, 20-34.

____ (1979c) Thermodynamics and lattice vibrations of minerals: 3. Lattice dynamics and an approximation for minerals with application to simple substnaces and framework silicates. Rev. Geophys. Space Phys. 17, 35-58.

____ (1980) Thermodynamics and lattice vibrations of minerals: 4. Application to chain and sheet silicates and orthosilicates. Rev. Geophys. Space Phys. 18, 862-886.

____ (1982) Thermodynamics and lattice vibrations of minerals. 5. Applications to phase equilibria, isotope fractionation and high-pressure thermodynamic properties. Rev. Geophys. Space Phys. 20, 827-849.

Kittel, C. (1976) "Introduction to Solid State Physics", 5th edition, John Wiley & Sons, N.Y.

Langer, K. and Schreyer, W. (1969) Infrared and powder x-ray diffraction studies on the polymorphism of cordierite, $Mg_2(Al_4Si_5O_{18})$. Am. Mineral. 54, 1442-1459.

Laves, F. and Hafner, S. (1956) Ordnung/unordnung und Ultrarotabsorption. I. (Al,Si) - Verteilung in Feldspäten. Z. Kristallogr. 108, 52-63.

Lazarev, A.N. (1972) "Vibrational Spectra and Structure of Silicates." Consultants Bureau, N.Y.

Leigh, R.S., Szigeti, B. and Tewary, V.K. (1971) Force constants and lattice frequencies. Proc. Royal Soc. A320, 505-526.

Long, D.A. (1977) "Raman Spectroscopy." McGraw-Hill, N.Y.

Martens, R. and Freund, F. (1976) The potential energy curve of the proton and the dissociation energy of the OH^- ion in $Mg(OH)_2$. Phys. Stat. Solidi (a) 37, 97-104.

Martin, R.F. (1970) Cell parameters and infrared absorption of synthetic low to high albites. Contrib. Mineral. Petrol. 26, 62-74.

Matson, D.W. and Sharma, S.K. (1985) Structures of the sodium alumino- and gallosilicate glasses and their germanium analogs. Geochim. Cosmochim. Acta (in press).

____, ____, and Philpotts, J.A. (1983) The structure of high-silica alkali-silicate glasses - a Raman spectroscopic investigation. J. Non-Crystalline Solids 58, 323-352.

McMillan, P. (1984) Structural studies of silicate glasses and melts - applications and limitations of Raman spectroscopy. Am. Mineral. 69, 622-644.

____ and Piriou, B. (1982) The structures and vibrational spectra of crystals and glasses in the silica-alumina system. J. Non-Crystalline Solids 53, 279-298.

_____ Putnis, A. and Carpenter, M. A. (1984) A Raman spectroscopic study of Al-Si ordering in synthetic magnesium cordierite. Phys. Chem. Minerals 10, 256-260.
Mills, I.M. (1963) Force constant calculations for small molecules. In "Infrared Spectroscopy and Molecular Structure," (ed. M. Davies), pp. 166-198, Elsevier.
Mirgorodskii, A.P., Lazarev, A.N. and Makarenko, I.P. (1970) Calculation of the limiting lattice vibrations for alpha-quartz based on valence type force fields. Optics and Spectros. 29, 289-292.
Moore, R.K. and White, W.B. (1971) Vibrational spectra of the common silicates: I. The garnets. Am. Mineral. 56, 54-71.
Moore, W.J. (1972) "Physical Chemistry," 5th ed. Longman Group Ltd., London.
Müller, A. (1977) Isotopic substitution. In "Vibrational Spectroscopy - Modern Trends", A.J. Barnes and W.J. Orville-Thomas, eds., p. 139-166, Elsevier, N.Y.
_____ and Mohan, N. (1977) Some comments on the use of constraints and additional data besides frquencies in force constant calculations. In "Vibrational Spectroscopy - Modern Trend", A.J. Barnes and W.J. Orville-Thomas), eds., p. 243-259, Elsevier, N.Y.
Mysen, B.O. (1977) The solubility of H_2O and CO_2 under predicted magma genesis conditions and some petrologic and geophysical implications. Rev. Geophys. Space Phys. 15, 351-361.
_____ and Virgo, D. (1980a) Solubility mechanisms of carbon dioxide in silicate melts: a Raman spectroscopic study. Am. Mineral. 65, 885-899.
_____ and _____ (1980b0 The solubility behavior of CO_2 in melts along the join $NaAlSi_3O_8$-$CaAl_2Si_2O_8$-CO_2 at high temperatures and pressures: a Raman spectroscopic study. Am. Mineral. 65, 1166-1175.
_____, _____, and F. Seifert (1982) The structure of silicate melts: implications for chemical and physical properties of natural magma. Rev. Geophys. Space Phys. 20, 353-383.
Nakamoto, K. (1978) "Infrared and Raman spectra of Inorganic and Coordination Compounds." 3rd ed. Wiley-Interscience, N.Y.
O'Keeffe, M. and Gibbs, G.V. (1984) Defects in amorphous silica; ab initio MO calculations. J. Chem. Phys. 81, 876-879.
_____ Newton, M.D. and Gibbs, G.V. (1980) Ab initio calculation of interatomic force constants in $H_6Si_2O_7$ and the bulk modulus of α quartz and α cristobalite. Phys. Chem. Minerals 6, 305-312.
Page, J.B. and Tonks, D.L. (1981) On the separation of resonance Raman scattering into orders in the time correlator theory. J. Chem. Phys. 75, 5694-5709.
Pâques-Ledent, M.Th. and Tarte, P. (1973) Vibrational studies of olivine-type compounds - I. The i.r. and Raman spectra of the isotopic species of Mg_2SiO_4. Spectrochim. Acta 29A, 1007-1016.
Piriou, B. (1974) Etude des modes normaux par réflexion infrarouge. Ann. Chimie 9, 9-17.
_____ and Alain, P. (1979) Density of states and structural forms related to physical properties of amorphous solids. High Temperatures-High Pressures 11, 407-414.
_____ and Cabannes, F. (1968) Validité de la methode de Kramers-Kronig et application à la dispersion infrarouge de la magnésie. Optica Acta 15, 271-286.
_____ and McMillan, P. (1983) The high-frequency vibrational spectra of vitreous and crystalline orthosilicates. Am. Mineral. 68, 426-443.
Pople, J.A., Schlegel, H.B., Krishnan, R., DeFrees, J.S., Binkley, J.S., Frisch, M.J. and Whiteside, R.A. (1981) Molecular orbital studies of vibrational frequencies. Int'l. J. Quantum Chem.: Quantum Chem. Symp. 15, 269-278.
Putnis, A. (1980a) Order-modulated structure and the thermodynamics of cordierite reactions. Nature 287, 128-131.
_____ (1980b) The distortion index in anhydrous Mg-cordierite. Contrib. Mineral. Petrol. 74, 135-141.
_____ and Bish, D. L. (1983) The mechanism and kinetics of Al,Si ordering in Mg cordierite. Am. Mineral. 68, 60-65.
Raman, C.V. and Nedungadi, T.M.K. (1940) The α-β transformation of quartz. Nature 145, 147.
Rao, C.N.R. and Rao, K.J. (1978) "Phase Transitions in Solids," McGraw-Hill, N.Y.
Ribbe, P.H. (1980) Aluminum silicate polymorphs (and mullite). In "Orthosilicates", P.H. Ribbe, ed., Reviews in Mineralogy, 5, p. 189-214.
Rosasco, C.J. And Roedder, E. (1979) Application of a new Raman microprobe spectrometer to nondestructive analysis of sulphate and other ions in individual phases in fluid inclusions in minerals. Geochim. Cosmochim. Acta 43, 1907-1915.
Rosenstock, H.B. (1961) Nature of vibrational modes in ionic crystals. Phys. Rev. 121, 416-424.
Ross, N.L. and McMillan, P. (1984) The Raman spectrum of $MgSiO_3$ ilmenite. Am. Mineral. 69, 719-721.
Ross, S.D. (1972) "Inorganic Infrared and Raman Spectra." McGraw-Hill, London.

Rossman, G.R. (1979) Structural information from quantitative infrared spectra of minerals. In "Proceedings of the Symposium on Chemistry and Physics of Minerals." (ed. G. E. Brown). Trans. Am. Crystallogr. Association 15, 77-91.
Rutstein, M.S. and White, W.B. (1971) Vibrational spectra of high-calcium pyroxenes and pyroxenoids. Am. Mineral. 56, 877-887.
Salthouse, J.A. and Ware, M.J. (1972) "Point Group Character Tables and Related Data." Cambridge University Press.
Scott, J.F. (1968) Evidence of coupling between one- and two-phonon excitation in quartz. Phys. Rev. Letts. 21, 907-910.
_______ (1974) Soft-mode spectroscopy: Experimental studies of structural phase transitions. Rev. Mod. Phys. 46, 83-128.
Servoin, J.L. and Piriou, B. (1973) Infrared reflectivity and Raman scattering of Mg_2SiO_4 single crystal. Phys. Status Solidi (b) 55, 677-686.
Shapiro, S.M., O'Shea, D.C. and Cummins, H.Z. (1967) Raman scattering study of the alpha-beta transition in quartz. Phys. Rev. Lett. 19, 361-364.
Sherwood, P.M.A. (1972) "Vibrational Spectroscopy of Solids," Cambridge University Press, Cambridge.
Shuker, R. and Gammon, R.W. (1970) Raman scattering selection-rule breaking and the density of states in amorphous materials. Phys. Rev. Lett. 25, 222-225.
Simon, I. and McMahon, H.O. (1953) Study of the structure of quartz, cristobalite and vitreous silica by reflection in infrared. J. Chem. Phys. 21, 23-30.
Stolper, E. (1982a) Water in silicate glasses: an infrared spectroscopic study. Contrib. Mineral. Petrol. 81, 1-17.
_____ (1982b) On the speciation of water in silicate melts. Geochim. Cosmochim. Acta 46, 2609-2620.
Strens, R.G.J. (1974) The common chain, ribbon and ring silicates. In "The Infrared Spectra of Minerals," V.C. Farmer, ed., p. 305-330. Mineralogical Society, London.
Striefler, M.E. and Barsch, G.R. (1975) Lattice dynamics of α-quartz. Phys. Rev. B 4553-4566.
Szabo, A. and Ostlund, N.S. (1982) "Modern Quantum Chemisty: Introduction to Advanced Electroni Structure Theory", Macmillan Publishing Co., N.Y.
Szigeti, B. (1971) Force constants and experimental data. In "Phonons" M.A. Nusimovici, ed., p. 43-47. Flammarion Sciences, Paris.
Tarte, P. (1962) Etude infra-rouge des orthosilicates et des orthogermanates. Une nouvelle méthode d'interprétation des spectres. Spectrochim. Acta 18, 467-483.
__________ (1963a) Applications nouvelles de la spectrométrie infrarouge à la cristallochimie. Silicates Industriels 28, 345-354.
________ (1963b) Etude infra-rouge des orthosilicates et des orthogermanates - II. Structures du type olivine et monticellite. Spectrochim. Acta 19, 25-47.
_______ (1967) Infrared spectra of inorganic aluminates and characteristic vibrational frequencies of AlO_4 tetrahedra and AlO_6 octahedra. Spectrochim. Acta 23A, 2127-2143.
Tessman, J.R, Kahn, A.H. and Shockley, W. (1953) Electronic polarizabilities of ions in crystals. Phys. Rev. 92, 890-895.
Tuddenham, W.M. and Lyon, R.J.P. (1960) Infrared techniques in the identification of and measurement of minerals. Anal. Chem. 32, 1630-1634.
Turrell, G. (1972) "Infrared and Raman Spectra of Crystals." Academic Press, N.Y.
White, W.B. (1974) Order-disorder effects. In "The Infrared Spectra of Minerals," V.C. Farmer, ed., p. 87-110. Mineral. Soc. London.
_______ (1975) Structural interpretations of lunar and terrestrial minerals by Raman spectroscopy. In "Infrared and Raman Spectroscopy of Lunar and Terrestrial Materials", C. Karr, ed., p. 325-358. Academic Press, N.Y.
Wilson, E.B., Decius, J.C. and Cross, P.C. (1955) "Molecular Vibrations." McGraw-Hill, N.Y. Re-published in 1980 by Dover Publications, N.Y.
Wolf, G.H. and Jeanloz, R. (1984) Vibrational frequency spectrum of $MgSiO_3$ perovskite (abstr.). Trans. Am. Geophys. Union, EOS 65, 1105.
Wyncke, B., McMillan, P.F., Brown, W.L., Openshaw, R.E. and Bréhat, F. (1981) A room-temperature phase transition in maximum microcline. Absorption in the far infrared (10-200 cm^{-1}) in the temperature range 110-300 K. Phys. Chem. Minerals 7, 31-34.
Zerbi, G. (1977) Limitations of force constant calculations for large molecules. In "Vibrational Spectroscopy - Modern Trends," A.J. Barnes and W.J. Orville-Thomas, eds., p. 261-284, Elsevier, N.Y.

Chapter 3. Susan Werner Kieffer

HEAT CAPACITY and ENTROPY: SYSTEMATIC RELATIONS to LATTICE VIBRATIONS

1. INTRODUCTION

The goal of this volume is to interpret macroscopic thermodynamic properties of minerals in terms of their atomic structures and the motions of particles in these structures. To reach this goal, we must examine the structure and volume relations of crystals (chapters by Hazen, Jeanloz), the configurational and structural energies and entropies (Carpenter, McConnell, Navrotsky and Burnham), the enthalpies (Navrotsky), the electronic contributions to energies (Burns), and the atomic vibrations (McMillan, Kieffer, Ghose). We must also develop the theoretical links between microscopic and macroscopic properties, such as statistical mechanics, quantum mechanics, and lattice dynamical theory. In this paper I develop the theoretical concepts that connect atomic vibrations to thermodynamics; abstract simplifying relations from the vibrational spectra of silicates; and show how some obvious systematic properties of the spectra ease the rather formidable problem of enumerating and quantifying the modes of vibration of crystals.

The approach is pedagogic. As concepts are developed, a single mineral, quartz, is analyzed in detail; results are given for other minerals to emphasize various points throughout the text. A reader interested in understanding the thermodynamic behavior of a particular mineral might find it useful to simply reproduce the discussion and calculations of the text for that mineral. Tables are given so that a student need not do any computer programing to learn the basic concepts; however, for "playing" with variables that will be shown of importance, use of the tables is time-consuming, and use of a computer is recommended. The equations are simple integrals and are easily done on a wide variety of computers or hand-calculators; some FORTRAN computer programs by the author are available in a U.S.G.S. Open-file Report.

2. THERMODYNAMIC DATA: GRIST FOR MENTAL MILLS

The fundamental thermodynamic quantity that must be understood is the heat capacity; C_P is measured, C_V is calculated directly from lattice dynamics. To relate C_P and C_V at a given temperature, the thermal expansion, volume, and bulk modulus of the mineral must be known. Heat capacity is now measured as a function of temperature with great accuracy (an excellent source of data and references to original work is Robie et al., 1978). Heat capacities for several substances are shown in Figure 1. (For comparative purposes, heat capacities and entropies are normalized to atom-mole quantities throughout this paper.) Entropy is obtained from measured heat capacities from the equation:

$$S_T = \int_0^T \frac{C_P}{T'}\, dT'. \qquad (1)$$

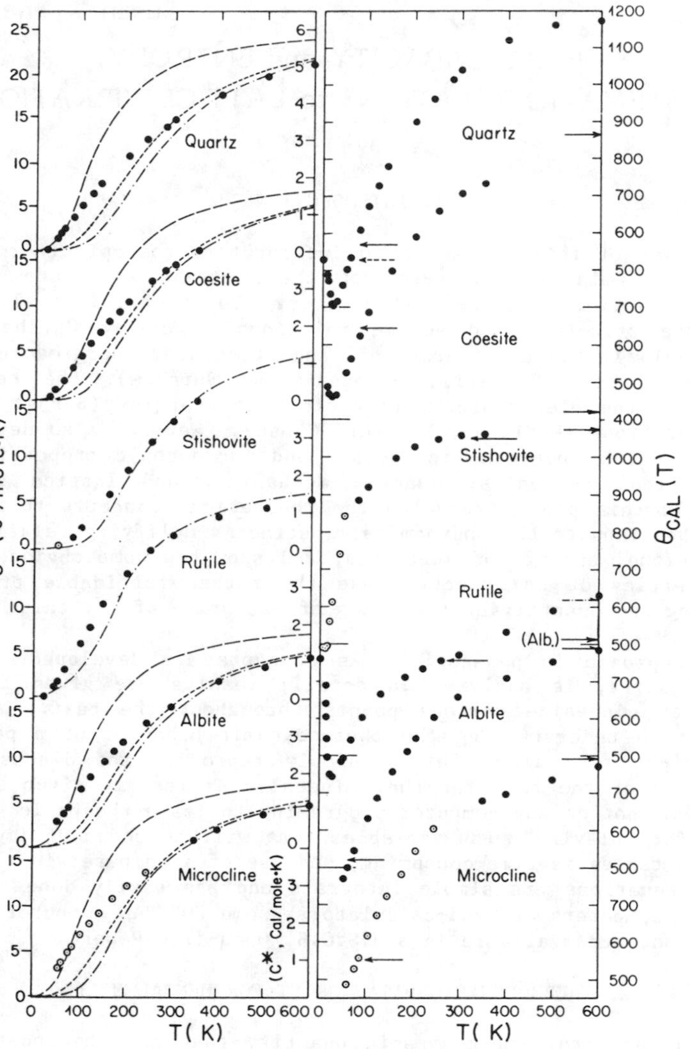

Figure 1. (left side) The specific heat, C_V, versus temperature for various minerals. (right side) $\theta_{cal}(T)$ versus T. C_V is normalized to that for one atom. C_P is graphed for all minerals except halite below 300 K and quartz, for which C_V are given. Dots are values from experimental data; references are in Kieffer (1979a,b,c; 1980; 1982). Long-dashed lines are values for a Debye model with the elastic Debye temperature. Short dashed lines are for a Debye model with the characteristic Debye temperature taken as the value that fits the data at 300 K. Dash-dot lines are for an Einstein model with the characteristic temperature taken as the mean frequency of the spectrum, approximated as described in Kieffer (1979a,b,c; 1980). (Continued on next page.)

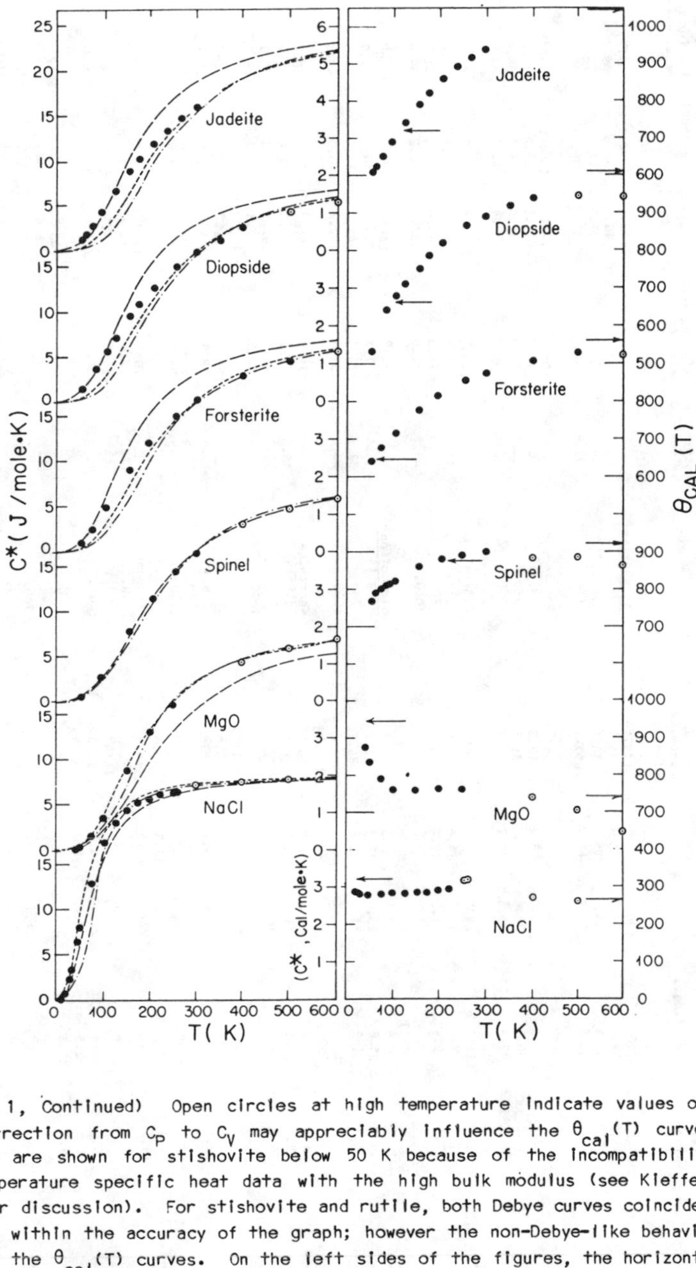

(Figure 1, Continued) Open circles at high temperature indicate values of C where the correction from C_P to C_V may appreciably influence the $\theta_{cal}(T)$ curves. Open circles are shown for stishovite below 50 K because of the incompatibility of the low-temperature specific heat data with the high bulk modulus (see Kieffer, 1979c, 1982 for discussion). For stishovite and rutile, both Debye curves coincide with the data to within the accuracy of the graph; however the non-Debye-like behavior can be seen in the $\theta_{cal}(T)$ curves. On the left sides of the figures, the horizontal arrows pointing to the left show the room-temperature elastic value of the Debye temperature, assumed to be close to the value at 0 K. Arrows pointing to the right indicate the high-temperature asymptotic values of $\theta_{cal}(T)$.

Table 1. Harmonic[1] entropy data and model predictions (S^*, J/mole-K).

Mineral (T, K)	Experimental values	Einstein (θ_E for arith. mean freq.)	Debye (θ_{el})	Kieffer (see 1980)	Einstein (θ_E for geom. mean freq.)
Halite		(210°)[5]	(306°)		(197°)
298.15	34.8	34.2	33.3	35.1	35.7
700.	55.0	55.1	54.0	55.9	56.6
1000.	63.1	63.9	62.8	64.7	65.5
Periclase		(567°)	(942°)		(547°)
298.15	13.2	12.4	9.9	13.2	13.0
700.	31.2	30.9	26.9	31.2	31.7
1000.	39.2	39.4	35.3	39.7	40.3
Brucite		(1184°)	(697°)		(661°)
298.15	12.6	2.4	15.2	14.4	9.6
400.	17.5	5.4	21.2	19.4	15.1
Corundum		(705°)	(1026°)	(Model 1)	(662°)
298.15	10.0	8.6	8.6	10.3	9.6
700.	26.8	25.8	25.0	27.3	27.2
1000.	34.7	34.2	33.3	35.6	35.7
Spinel		(680°)	(879°)	(Model 2)	(628°)
298.15	11.3	9.2	11.0	11.3	10.5
700.	28.5	26.6	28.5	28.5	28.5
1000.	36.9	35.0	36.9	36.9	37.7
Quartz		(815°)	(567°)		(617°)
298.15	13.7	6.4	19.3	13.9	10.8
700.	29.1	22.5	38.9	29.7	28.9
α-Cristobalite		(819°)	(519°)		(623°)
298.15	14.3	6.4	21.2	13.8	10.6
400.	19.1	11.0	27.8	18.5	16.2
Silica		(815°)	(491°)		(579°)
298.15	16.0	6.4	22.4	15.4	12.0
700.	31.8	22.5	42.4	31.0	30.4
1000.	39.8	30.8	51.1	39.0	38.9
Coesite		(779°)	(676°)	(Model 1)	(605°)
298.15	13.5	7.0	15.8	14.0	11.2
700.[2]	29.5	23.5	34.7	30.2	29.3
1000.[2]	37.4	31.8	43.3	38.2	37.9

Mineral (T, K)	Experimental values	Einstein (θ_E for arith. mean freq.)	Debye (θ_{el})	Kieffer (see 1980)	Einstein (θ_E for geom. mean freq.)
Stishovite		(757°)	(921°)	(Model 1)	(677°)
298.15	9.0	7.5	10.2	10.6	9.2
700.[2]	24.5	24.2	27.5	26.9	26.7
1000.[2]	32.0	32.5	35.2	35.1	35.1
Rutile		(580°)	(781°)	(Model 1)	(490°)
298.15	16.6	11.9	13.1	16.6	15.2
700.	34.0	30.3	31.3	34.5	34.4
1000.	42.1	38.8	39.8	42.9	43.0
Albite		(753°)	(472°)		(559°)
298.15	15.9	7.6	23.3	15.9	12.6
700.	32.3	24.3	43.4	32.1	31.2
1000.	40.2	32.6	52.1	40.2	39.8
Microcline		(746°)	(460°)		(550°)
298.15	16.5	7.7	23.9	16.2	12.9
700.	32.7	24.5	44.0	32.4	31.6
1000.	40.8	32.8	52.7	40.5	40.2
Anorthite		(757°)	(518°)		(582°)
298.15	15.3	7.5	21.3	14.8	11.9
700.	32.3	24.2	41.1	31.0	30.3
1000.	40.7	32.5	49.8	39.2	38.8
Clinoenstatite		(691°)	(719°)		(598°)
298.15	13.4	8.9	14.6	13.6	11.4
700.	29.9	26.2	33.2	30.6	29.6
(1000.)	37.9	34.6	41.8	38.9	38.1
Orthoenstatite		(705°)	(719°)		(598°)
(298.15	13.1	8.6	14.6	13.1	11.4
(700.)	NA[4]	25.8	33.2	30.1	29.6
1000.	NA	34.2	41.8	38.3	38.1
Diopside		(680°)	(654°)		(562°)
298.15	14.2	9.2	16.4	14.5	12.5
700.	31.2	26.6	35.5	31.5	31.1
1000.	39.5	35.0	44.1	39.8	39.7
Jadeite		(711°)	(724°)		(590°)
298.15	13.2	8.4	14.5	13.7	11.6
700.	29.7	25.6	33.1	30.5	29.9
1000.	37.5	34.0	41.6	38.7	38.4

Table 1 (Continued). Harmonic(1) entropy data and model predictions (S^*, J/mole-K).

Mineral (T, K)	Experimental values	Einstein (θ_E for arith. mean freq.)	Debye (θ_{el})	Kieffer (see 1980)	Einstein (θ_E for geom. mean freq.)
Tremolite		(763°)	(547°)		(590°)
298.15	13.3	7.4	20..1	14.1	11.6
700.	30.1	24.0	39.8	30.8	29.9
1000.	38.3	32.3	48.5	38.9	38.4
Muscovite		(860°)	(520°)		(620°)
298.15	14.6	5.7	21.2	13.8	10.7
700.(3)	31.0	21.3	41.0	30.0	28.7
1000.(3)	39.2	29.4	49.7	38.0	37.2
Talc		(882°)	(525°)		(611°)
298.15	12.4	5.4	21.0	12.5	11.0
700.	28.6	20.8	40.8	28.5	29.1
Calcite		(766°)	(468°)		(495°)
298.15	18.3	7.3	23.5	19.2	15.0
700.	35.3	23.9	43.6	35.3	34.1
1000.	43.3	32.2	52.3	43.3	42.7
Zircon		(694°)	(601°)		(569°)
298.15	14.0	8.82	18.1	14.4	12.3
700.	31.0	26.2	37.5	31.2	30.8
1000.	39.3	34.5	46.2	39.5	39.3
Forsterite		(670°)	(747°)		(569°)
298.15	13.4	9.4	13.9	14.0	12.3
700.	30.2	27.0	32.3	30.2	30.8
1000.	38.1	35.4	40.9	39.5	39.3
Pyrope		(697°)	(794°)		(589°)
298.15	13.3	8.8	12.8	13.6	11.7
700.	28.2	26.1	30.9	30.5	30.0
1000.	36.5	34.4	39.4	38.7	38.5
Grossular		(704°)	(821°)		(613°)
298.15	12.7	8.6	12.2	12.6	10.9
700.	29.7	25.8	30.1	29.5	29.0
1000.	37.9	34.2	38.6	37.7	37.5
Almandine		(693°)	(731°)		(579°)
298.15	NA	8.8	14.3	13.9	12.0
700.	NA	26.2	32.8	30.9	30.4
1000.	NA	34.6	41.4	39.2	38.9
Spessartine		(685°)	(742°)		(569°)
298.15	NA	9.0	14.0	14.2	12.3
700.	NA	26.4	32.5	31.2	30.8
1000.	NA	34.8	41.0	39.5	39.3
Andradite		(657°)	(738°)		(556°)
298.15	NA	9.7	14.1	14.5	12.7
700.	NA	27.4	32.6	31.8	31.3
1000.	NA	35.9	41.2	40.1	39.9
Kyanite		(757°)	(916°)		(674°)
298.15	10.4	7.5	10.3	10.8	9.3
700.	26.6	24.2	27.6	27.2	26.8
1000.	34.2	32.5	36.0	35.4	35.2
Andalusite		(754°)	(761°)		(644°)
298.15	11.5	7.5	13.5	12.0	10.1
700.	27.5	24.3	32.0	28.4	27.9
1000.	35.1	32.6	40.4	36.5	36.3
Sillimanite		(789°)	(782°)		(644°)
298.15	12.0	6.9	13.0	12.3	10.1
700.	28.3	23.2	31.3	28.5	27.9
1000.	36.1	31.5	39.8	36.6	36.3

TABLE NOTES

(1) Correction from total entropy to harmonic entropy given in Table 3 of Kieffer (1980).

(2) Metastable

(3) Experimental values possibly affected by dehydration.

(4) NA = not available.

(5) In data columns (2), (3), and (5) the numbers in parentheses are the Einstein (or Debye) temperatures used.

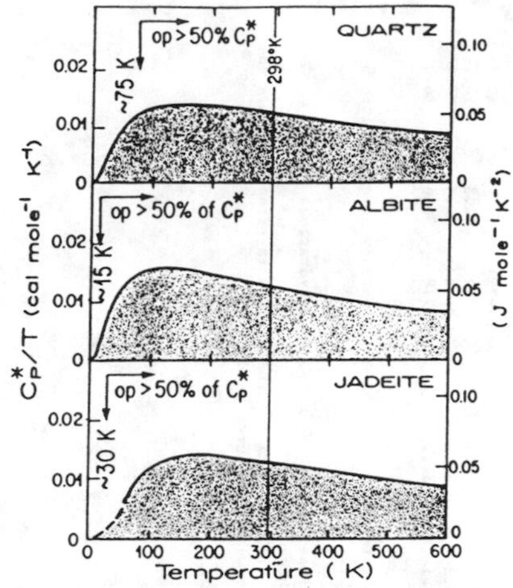

Figure 2. C_P^*/T versus T for quartz, albite, and jadeite. The third-law entropies are simply the areas under the curves. The temperatures at which optical modes contribute more than 50% of the heat capacity are shown with orthogonal arrows, and it can be seen that the entropy is dominated by the contributions of the optic modes. It can also be seen that only a small fraction of the total entropy at moderate to high temperatures arises below 50 K, and that the behavior of C_P^*/T in the range of 100-200 K determines a substantial part of the entropy at 298 K.

Because of the appearance of 1/T in this equation, reliable calculations of entropy depend on accurate measurement of C_P at low temperature. The sensitivity of entropy to low-temperature heat capacities is illustrated in Figure 2 and values of entropy for a number of minerals are given in Table 1. Explanation of the absolute values of the heat capacity and entropy, and of their trends, is the goal of this paper.

Notice in Figure 1 that the heat capacities all approach a limiting value at high temperature. When corrected from C_P to C_V, the limiting value for C_V was found to be 24.94 J mole^{-1} K^{-1}, i.e., 3R, where R is the gas constant. In the 19th century, the approach of C_V to a limiting value, the Dulong-Petit value, at high temperature had been noted, and explained in terms of the Equipartition Principle of classical physics.- Classical physics, however, predicted a constant specific heat and could not explain the decrease in C_V toward zero at zero Kelvin.

The discrepancy at low temperature between the values predicted by classical theory and measured values was one of the mysteries driving the development of quantum mechanics; another was the problem of radiation of an isothermal cavity. Although these two problems appear rather different, the same linear differential equation, that of the harmonic oscillator, appears in their analysis. This equation governs many apparently unrelated physical phenomena: the motion of a mass on a string; the vibrations of electrons in an atom, which generate light waves; the vibrations of atoms in a quadratic potential; and, as Feynman et al. (1963) point out, the growth of populations of foxes eating rabbits eating grass (Chapter 21 in Feynmann et al. has an excellent discussion of the harmonic oscillator). Therefore, the resolution of the heat capacity dilemma was intimately tied to the quantized harmonic oscillator behavior discovered in resolution of the black-body radiation problem by Planck.

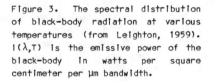

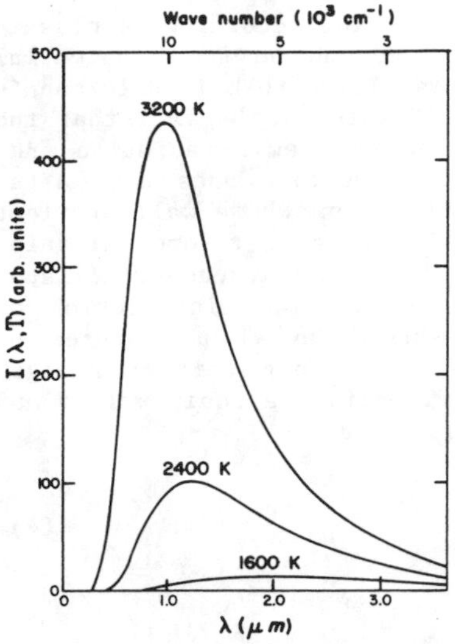

Figure 3. The spectral distribution of black-body radiation at various temperatures (from Leighton, 1959). $I(\lambda,T)$ is the emissive power of the black-body in watts per square centimeter per μm bandwidth.

2.1 Planck: The quantization of electromagnetic energy

Classical statistical mechanics could not explain the observed radiation spectrum of an isothermal cavity (Fig. 3) (the same radiation spectrum is emitted by a black body, and such radiation is therefore referred to as a black-body spectrum). Planck (1906) knew that empirically the spectrum could be represented by the following law:

$$I(\lambda,T) = \frac{c_1 \left(e^{c_2/\lambda T} - 1\right)^{-1}}{\lambda^5}. \tag{2}$$

λ is the wavelength of the radiation, I is its intensity (emissive power, e.g., in watts per sq cm per μm) and c_1 and c_2 are empirical constants. An especially important point from Figure 3 and this equation is the decay of spectral intensity at short wavelengths (high frequencies). Prediction of this decay was a major problem of classical statistical mechanics.

Consider the classical analysis of radiation being produced by electrons oscillating around an atom. To a first approximation, the electrons move in a quadratic potential, and their motion is therefore described by the harmonic oscillator equation:

$$m\,\frac{d^2x}{dt^2} = -bx, \tag{3}$$

where m is the electron mass and b is the force constant associated with the potential. In equilibrium, an electron should, according to classical physics, have a kinetic energy of 1/2 kT and an equal potential energy of 1/2 kT, so that the total energy is kT, where k is Boltzmann's constant, and T is absolute temperature in Kelvins. The fact that the electrons have an electric charge means that they emit radiation and, therefore, lose energy. However, the radiation loss does not matter because the electrons are in a perfectly black box whose walls reflect all emitted light back, and thermal equilibrium is therefore maintained. According to the classical mechanics of Boltzmann, each oscillator has the energy kT for each degree of freedom. The distribution of energy as a function of wavelength (λ) should be kT per degree of freedom times the number of degrees of freedom per unit wavelength interval. The number of normal modes of vibration per unit volume and per unit wavelength, $N(\lambda)$, can be worked out to be

$$N(\lambda) = 8\pi/\lambda^4, \tag{4}$$

so that

$$I(\lambda,T) = \frac{c}{4} N(\lambda)\ kT = \frac{2\pi ckT}{\lambda^4}. \tag{5}$$

This is known as <u>Rayleigh's Law</u>, or the <u>Rayleigh - Jean's Law</u>. In this equation, c is the speed of light and the factor of c/4 arises from the relation between power emitted per unit area of a black body and the energy density (Leighton, 1959, p. 62). According to this law, which is <u>rigorously</u> derived from classical statistical mechanics, the energy intensity of a radiating cavity should become infinitely large as wavelengths approach zero (i.e., at very high frequencies, such as x-ray frequencies). As Feynman et al. humorously point out -- we know that this is false because when we open a furnace door, we do not burn our eyes out with x-rays!

Planck found the solution to this dilemma by assuming that harmonic oscillators could only absorb and radiate energy from the radiation field in the cavity in exact integral multiples of a certain unit of energy. The permitted energy levels were assumed to be equally spaced at multiples of the unit energy $h\upsilon = \hbar\omega$ (Fig. 4), where $\omega = 2\pi c/\lambda$ and the constant $h = 6.6261 \times 10^{-34}$ J s is the <u>Planck quantum of action</u>, or <u>Planck's constant</u> ($\hbar = h/2\pi$). The probability that the energy will be absorbed or emitted is governed by statistical chance, and Planck demonstrated that the probability, P, of an energy level ε being occupied was

$$P(\varepsilon) = \alpha\, e^{-\varepsilon/kT}, \tag{6}$$

where α is the proportionality constant.

If there are a large number of oscillators, each with frequency ω, then some will be in the bottom energy state (quantum state), some will be in the first state with energy $\hbar\omega$, etc. Planck showed that the

Level	Energy	Number	Probability
4	$4\hbar\omega$	$N_4 = N_0 \exp(-4\hbar\omega/kT)$	$P_4 = A \exp(-4\hbar\omega/kT)$
3	$3\hbar\omega$	$N_3 = N_0 \exp(-3\hbar\omega/kT)$	$P_3 = A \exp(-3\hbar\omega/kT)$
2	$2\hbar\omega$	$N_2 = N_0 \exp(-2\hbar\omega/kT)$	$P_2 = A \exp(-2\hbar\omega/kT)$
1	$\hbar\omega$	$N_1 = N_0 \exp(-\hbar\omega/kT)$	$P_1 = A \exp(\hbar\omega/kT)$
0	0	N_0	$P_0 = A$

Figure 4. Schematic diagram of the energy levels of a quantized harmonic oscillator, showing the energy of each level, the number of atoms in the excited states, and the probability of occupancy.

average energy of the collection of oscillators was then

$$\langle E \rangle = \frac{\hbar\,\omega}{(e^{\hbar\omega/kT} - 1)}. \qquad (7)$$

This is to be compared with the energy of a classical oscillator, kT. When the average value of the quantum energy, Equation 7 is substituted into the Rayleigh formula, Equation 5, the spectral distribution for isothermal radiation is correctly predicted to be

$$I(\lambda,T) = \frac{2\pi c^2 h}{\lambda^5(e^{ch/\lambda kT} - 1)}. \qquad (8)$$

Because of the exponential term, the intensity does not approach infinity at short wavelengths, but correctly diminishes to zero.

Planck's expression for $\langle E \rangle$, Equation 7, above was the first quantum formula known, and it provided the foundation for modern physics -- for Einstein's work on the photoelectric effect (which proved Planck's theory), for Bohr's theory on the structure of the hydrogen atom, and, of particular interest to us, for Einstein's work on heat capacities.

2.2 Einstein: The quantization of particle energy

Einstein (1907) made the dramatic step of applying the theory of quantization of electromagnetic energy to quantization of particle energy -- particles vibrating in a solid. Each particle was envisioned as existing in an isotropic potential field due to all other particles, that is, each particle was a harmonic oscillator with three degrees of freedom, each having the same vibrational frequency, ω, or characteristic temperature, θ. The mean energy of each degree of freedom was then (Fig. 4):

$$\bar{\varepsilon} = \frac{\hbar\,\omega}{(e^{\hbar\omega/kT} - 1)}, \qquad (9)$$

Table 2. The Einstein heat capacity function.

x_E	$E(x_E)$	x_E	$E(x_E)$	x_E	$E(x_E)$	x_E	$E(x_E)$
0.1	0.999161	5.1	0.160528	10.1	0.004191	15.1	0.000063
0.2	0.996670	5.2	0.150827	10.2	0.003867	15.2	0.000058
0.3	0.992535	5.3	0.141624	10.3	0.003568	15.3	0.000053
0.4	0.986771	5.4	0.132901	10.4	0.003292	15.4	0.000049
0.5	0.979424	5.5	0.124641	10.5	0.003036	15.5	0.000045
0.6	0.970532	5.6	0.116827	10.6	0.002800	15.6	0.000041
0.7	0.960148	5.7	0.109442	10.7	0.002581	15.7	0.000037
0.8	0.948331	5.8	0.102466	10.8	0.002379	15.8	0.000034
0.9	0.935148	5.9	0.095885	10.9	0.002193	15.9	0.000031
1.0	0.920674	6.0	0.089679	11.0	0.002021	16.0	0.000029
1.1	0.904985	6.1	0.083833	11.1	0.001862	16.1	0.000026
1.2	0.888170	6.2	0.078329	11.2	0.001715	16.2	0.000024
1.3	0.870314	6.3	0.073151	11.3	0.001580	16.3	0.000022
1.4	0.851508	6.4	0.068284	11.4	0.001455	16.4	0.000020
1.5	0.831848	6.5	0.063712	11.5	0.001340	16.5	0.000019
1.6	0.811429	6.6	0.059419	11.6	0.001233	16.6	0.000017
1.7	0.790345	6.7	0.055392	11.7	0.001135	16.7	0.000016
1.8	0.768693	6.8	0.051616	11.8	0.001045	16.8	0.000014
1.9	0.746567	6.9	0.048078	11.9	0.000962	16.9	0.000013
2.0	0.724061	7.0	0.044764	12.0	0.000885	17.0	0.000012
2.1	0.701266	7.1	0.041662	12.1	0.000814	17.1	0.000011
2.2	0.678268	7.2	0.038761	12.2	0.000749	17.2	0.000010
2.3	0.655154	7.3	0.036048	12.3	0.000689	17.3	0.000009
2.4	0.632002	7.4	0.033513	12.4	0.000633	17.4	0.000008
2.5	0.608890	7.5	0.031145	12.5	0.000582	17.5	0.000008
2.6	0.585890	7.6	0.028935	12.6	0.000535	17.6	0.000007
2.7	0.563067	7.7	0.026872	12.7	0.000492	17.7	0.000006
2.8	0.540486	7.8	0.024949	12.8	0.000452	17.8	0.000006
2.9	0.518203	7.9	0.023155	12.9	0.000416	17.9	0.000005
3.0	0.496269	8.0	0.021484	13.0	0.000382	18.0	0.000005
3.1	0.474732	8.1	0.019927	13.1	0.000351	18.1	0.000005
3.2	0.453633	8.2	0.018478	13.2	0.000322	18.2	0.000004
3.3	0.433010	8.3	0.017129	13.3	0.000296	18.3	0.000004
3.4	0.412894	8.4	0.015874	13.4	0.000272	18.4	0.000003
3.5	0.393313	8.5	0.014707	13.5	0.000250	18.5	0.000003
3.6	0.374290	8.6	0.013621	13.6	0.000229	18.6	0.000003
3.7	0.355843	8.7	0.012613	13.7	0.000211	18.7	0.000003
3.8	0.337987	8.8	0.011676	13.8	0.000193	18.8	0.000002
3.9	0.320732	8.9	0.010806	13.9	0.000178	18.9	0.000002
4.0	0.304087	9.0	0.009999	14.0	0.000163	19.0	0.000002
4.1	0.288055	9.1	0.009249	14.1	0.000150	19.1	0.000002
4.2	0.272637	9.2	0.008554	14.2	0.000137	19.2	0.000002
4.3	0.257832	9.3	0.007909	14.3	0.000126	19.3	0.000002
4.4	0.243635	9.4	0.007311	14.4	0.000116	19.4	0.000001
4.5	0.230040	9.5	0.006756	14.5	0.000106	19.5	0.000001
4.6	0.217038	9.6	0.006243	14.6	0.000097	19.6	0.000001
4.7	0.204620	9.7	0.005767	14.7	0.000089	19.7	0.000001
4.8	0.192773	9.8	0.005326	14.8	0.000082	19.8	0.000001
4.9	0.181485	9.9	0.004918	14.9	0.000075	19.9	0.000001
5.0	0.170742	10.0	0.004540	15.0	0.000069	20.0	0.000001

so that the total energy, E, of a solid composed of N atoms is

$$E = 3N \frac{\hbar\omega}{(e^{\hbar\omega/kT} - 1)}. \tag{10}$$

The specific heat, C_V, is then obtained from

$$C_V = \left(\frac{\partial E}{\partial T}\right)_V = \frac{3\ Nk\ x^2\ e^x}{(e^x - 1)^2} = \frac{3\ Nk\ x^2}{4\ \sin^2\left(\frac{x}{2}\right)}, \tag{11a}$$

where $x = \hbar\omega/kT$. For molar heat capacity, N is taken as Avogadro's number, N_A, $6.022 \times 10^{23}\ mol^{-1}$. The function

$$\frac{x^2}{4\ \sin^2\left(\frac{x}{2}\right)} \tag{11b}$$

is called the Einstein function, and is denoted here as $\mathbf{E}(x)$ (Table 2). The product $N_A k$ equals R, the gas constant per atom. At high temperature

$$\mathbf{E}(x) \to \left[1 - \frac{1}{12}(x)^2 + \ldots\right]. \qquad (12)$$

The Einstein model therefore predicts that C_V approaches the Dulong-Petit value, 3R, at high temperatures, demonstrating that quantization is unimportant at high temperatures and that the specific heat has the same value as if the mean energy of each oscillator was kT. At low temperature

$$\mathbf{E}(x) \to 3\ R\ x^2\ e^{-x}. \qquad (13)$$

Therefore, at low temperatures the heat capacity approaches zero, a substantial improvement over the classical statistical mechanics formulation that gave constant specific heat. However, within a few years Einstein (1911) had found that the theory did not predict the correct form of C_V at low temperatures, and suggested that the assumption of a single characteristic vibrational frequency had to be abandoned. Madelung (1910) had reached a similar conclusion independently and had successfully explained some of the newly observed features of an infrared spectrum of NaCl.

Heat capacity curves for Einstein models for several minerals are shown in Figure 1. For each substance a characteristic frequency has been obtained as the arithmetic mean frequency of the lattice vibrational spectra proposed by Kieffer (1979a,b,c; 1980) to predict the thermodynamic functions; such spectra will be discussed later in this paper. Only for the substances of relatively simple structure, halite and periclase, is the heat capacity well-described by an Einstein model. The difference between measured heat capacities and those predicted by an Einstein model is substantial over a wide temperature range for most minerals considered.

Einstein's work was directed toward explaining variations of C_V from the Dulong-Petit law at moderately high temperatures. If the Einstein temperature is chosen arbitrarily to maximize the fit to high temperature data, the characteristic temperature, Θ, that best accomplishes this is high, and, as a consequence, the theory gives heat capacities that are consistently too low at low temperature. Nernst and Lindemann (1911) noted this and arrived at an empirical relation that fit the measured low-temperature heat capacities rather well:

$$C_V = \frac{3N}{2}\left[\ E\left(\frac{\Theta}{T}\right) + E\left(\frac{\Theta}{2T}\right)\right]. \qquad (14)$$

Much later (1935) Blackman provided a basis for this formula with lattice vibrational calculations on structures with simple geometry.

However, the failure of the Einstein model at low temperatures stimulated two important breakthroughs on the low-temperature, low-frequency form of the vibrational spectrum in 1912: the work of Debye (1912) from a continuum view, and the work, independently, of Born and von Karman (1912, 1913) on lattice vibrations.

2.3 Debye: The role of long wavelength vibrations

Debye was not aware of the theory of reciprocal lattices because the first crystal structures were not determined until 1912, and because Ewald did not invent the theory of reciprocal lattices until 1921; he developed his theory in the context of continuum mechanics. The derivation below is not the original derivation of Debye, but it gives recognition to the underlying crystal lattice and emphasizes the approximations made to actual lattice structure (Brillouin 1953).

Consider a primitive Bravais lattice with basis vectors $\mathbf{a}_1$, $\mathbf{a}_2$, $\mathbf{a}_3$, forming the edges of the cells. A corresponding reciprocal lattice is defined by the reciprocal basis vectors

$$\mathbf{b}_1 = \frac{2\pi\mathbf{a}_2 \times \mathbf{a}_3}{|\mathbf{a}_1 \cdot \mathbf{a}_2 \times \mathbf{a}_3|},\ \mathbf{b}_2 = \frac{2\pi\mathbf{a}_3 \times \mathbf{a}_1}{|\mathbf{a}_2 \cdot \mathbf{a}_3 \times \mathbf{a}_1|},\ \mathbf{b}_3 = \frac{2\pi\mathbf{a}_1 \times \mathbf{a}_2}{|\mathbf{a}_3 \cdot \mathbf{a}_1 \times \mathbf{a}_2|}. \tag{15}$$

The unit cell of the direct lattice, with volume V_L, may be represented in the reciprocal lattice by a unit cell whose volume V_R is inversely proportional to the volume of a unit cell of the direct lattice, $V_R = (2\pi)^3/V_L$. The reciprocal unit cell is called the Brillouin zone. Its boundaries are usually chosen symmetrically about the origin of the reciprocal lattice, $\mathbf{K} = \mathbf{0}$. The Brillouin zone is identical to the usual reciprocal cell adopted by crystallographers, except that the factor of 2π is included, that the origin is taken at the center of the cell rather than at a corner, and that the wave vector $\mathbf{K}$ is a coordinate rather than a reciprocal length.

The wave vectors for lattice vibrational waves are conveniently represented in reciprocal space in the form (e.g., Kittel, 1968, p. 53)

$$\mathbf{K}(\eta) = \eta_1\mathbf{b}_1 + \eta_2\mathbf{b}_2 + \eta_3\mathbf{b}_3, \tag{16}$$

where the η_i are integers and $\mathbf{K}$ is a reciprocal lattice vector.

Let us now consider the propagation of elastic waves through a lattice. It is well known (e.g., Brillouin, 1953) that all elastic vibrations of a lattice propagate as waves of the form

$$\boldsymbol{\mu}(\mathbf{x},t) = A\,\mathbf{n}\sin(\mathbf{K}\cdot\mathbf{x} - \omega t + \delta), \tag{17}$$

where $\boldsymbol{\mu}(\mathbf{x},t)$ is the displacement of the medium at the point $\mathbf{x}$ and time t; ω is the frequency; $\mathbf{n}$ is the direction in which the displacements

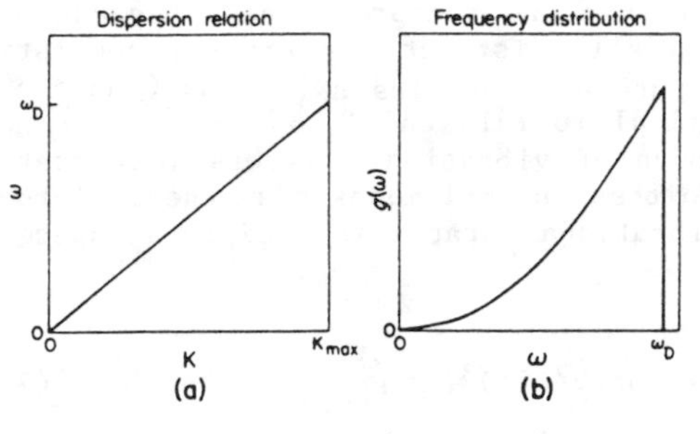

Figure 5. (a) Schematic dispersion relations for a Debye solid. (b) Schematic spectral density of states $g(\omega)$ for a Debye solid.

occur; **K** is the wave vector; A is an amplitude; and δ is a phase factor. In general, the frequency ω of lattice vibrations is a function of the direction and magnitude of the wave vector **K**. The wave velocity (the group velocity) **v** is given by

$$|\mathbf{v}| = \partial\omega/\partial|\mathbf{K}|. \tag{18}$$

The relation between ω and **K** (or $\nu = \omega/2\pi$ and $\mathbf{y} = \mathbf{K}/2\pi$) is called the <u>dispersion relation.</u> The Debye theory assumes that all modes of vibration are acoustic and all have the same wave velocity **v** (Fig. 5a). These waves are dispersionless; i.e., phase velocities and group velocities are the same.

The Debye theory also assumes that the vibrational states of the crystal correspond to wave vectors **K** whose tips are uniformly distributed in reciprocal space. Because the available volume in reciprocal space increases proportionally to $|\mathbf{K}^2|d\mathbf{K}$, the density of vibrational states, $f(\mathbf{K})$, has the form

$$f(\mathbf{K})d\mathbf{K} = 4\pi\mathbf{K}^2\ d\mathbf{K}\cdot d, \tag{19}$$

where d is the density (assumed uniform) of wave vectors in reciprocal space. The density of states $g(\omega)$, expressed in terms of the vibrational frequency $\omega = vK$, can therefore be written

$$g(\omega)\ d\omega = f\left[K(\omega)\right]\frac{dK}{d\omega}\ d\omega = a\omega^2 d\omega, \tag{20}$$

where a is a constant that depends on the elastic wave velocities. Here $g(\omega)\ d\omega$ is the number of vibrational states lying between ω and $\omega + d\omega$. The parabolic form of $g(\omega)$ is illustrated in Figure 5b. The simple parabolic form of the vibrational spectrum in Equation 20 is the central feature of the Debye model, and it is valid for any crystal in the limit as $\omega \to 0$ (long-wavelength phonons).

The value of a in Equation 20 is determined by the density of vibrational states in reciprocal space. All normal modes of the crystal are represented by wave vectors within one reciprocal cell because only those vibrational modes are physically distinct. A crystal consisting of one mole of the substance of interest contains nN_A atoms (where n is the number of atoms in the chemical formula and N_A is Avogadro's number), and it has $3nN_A$ normal modes of vibration. As was just stated, the wave vectors corresponding to these normal modes span one reciprocal cell. The molar density d of vibrational states in reciprocal space is thus

$$d = 3nN_A/V_R = 3nN_AV_L/(2\pi)^3 = 3nZV/(2\pi)^3, \tag{21}$$

where V is the molar volume of the crystal and Z is the number of formula units in the unit cell ($V_L = VZ/N_A$). When it is transformed to the frequency representation by Equation 20, this density corresponds to a value of a given by

$$a = 3nZV/2\pi^2 v_M{}^3. \tag{22}$$

As was previously stated, this assumes that all acoustic waves have the same speed v_M and hence that the crystal is elastically isotropic. In reality, the compressional (P) and shear (S) waves have separate speeds, so that for an isotropic crystal there are two separate spectral contributions of the type in Equation 22, one with $a_P = nZV/2\pi v_P{}^3$, and the other with $a_S = 2nZV/2\pi^2 v_S{}^3$. For simplicity, in Debye theory this more complicated spectrum is often replaced by a single spectrum with an averaged value of a:

$$a = \frac{3nZV}{2\pi^2 v_M{}^3} = \frac{nZV}{2\pi^2}\left(\frac{1}{v_P{}^3} + \frac{2}{v_S{}^3}\right). \tag{23}$$

Most crystals are elastically anisotropic, so that for each wave propagation direction **K** there are three, rather than two, distinct wave velocities, $v_1(\mathbf{K})$, $v_2(\mathbf{K})$, and $v_3(\mathbf{K})$. In a simple Debye theory, the separate spectral contributions from these different waves are averaged by using a mean velocity

$$\frac{3}{v_M{}^3} = \frac{1}{4\pi}\int\left(\frac{1}{v_1{}^3} + \frac{1}{v_2{}^3} + \frac{1}{v_3{}^3}\right)d\Omega, \tag{24}$$

where the integration is over all directions of **K**, $d\Omega$ being an increment of solid angle about the origin of the Brillouin zone. In lieu of the true mean velocities required by this equation (e.g., see Robie and Edwards, 1966 for an example) it has become common practice to use the Voigt-Reuss-Hill (VRH) average velocity for v_M in Equation 24 [see O.L. Anderson, 1963]. For most minerals the differences obtained by the two methods are not significant except at temperatures of a few Kelvins or few tens of Kelvins (Kieffer, 1979c).

Table 3. The Debye heat capacity function (Reissland, 1973).

x	0·0	0·1	0·2	0·3	0·4	0·5	0·6	0·7	0·8	0·9
0	1·000	0·9995	0·9980	0·9955	0·9920	0·9876	0·9822	0·9759	0·9687	0·9606
1	0·9517	0·9420	0·9315	0·9203	0·9085	0·8960	0·8828	0·8692	0·8550	0·8404
2	0·8254	0·8100	0·7943	0·7784	0·7622	0·7459	0·7294	0·7128	0·6961	0·6794
3	0·6628	0·6461	0·6296	0·6132	0·5968	0·5807	0·5647	0·5490	0·5334	0·5181
4	0·5031	0·4883	0·4738	0·4595	0·4456	0·4320	0·4187	0·4057	0·3930	0·3807
5	0·3686	0·3569	0·3455	0·3345	0·3237	0·3133	0·3031	0·2933	0·2838	0·2745
6	0·2656	0·2569	0·2486	0·2405	0·2326	0·2251	0·2177	0·2107	0·2038	0·1972
7	0·1909	0·1847	0·1788	0·1730	0·1675	0·1622	0·1570	0·1521	0·1473	0·1426
8	0·1382	0·1339	0·1297	0·1257	0·1219	0·1182	0·1146	0·1111	0·1078	0·1046
9	0·1015	0·0985	0·0956	0·0928	0·0901	0·0875	0·0850	0·0826	0·0803	0·0780
10	0·0758	0·0737	0·0717	0·0697	0·0678	0·0660	0·0642	0·0625	0·0609	0·0593

In the Debye model, the Brillouin zone is simplified by replacing the actual zone with a sphere of the same volume in reciprocal space, centered at the origin of the reciprocal lattice. Its radius $K_{max} = |\mathbf{K}_{max}|$ thus is given by

$$\frac{4}{3}\pi K^3_{max} = (2\pi)^3/V_L. \tag{25}$$

Corresponding to the maximum wave vector K_{max} is a maximum frequency $\omega_D = v_M K_{max}$. In the simplest form, using Equation 23, a sharp cutoff frequency ω_D is given by

$$\omega_D = v_M K_{max} = v_M\left(6\pi^2 \frac{N_A}{ZV}\right)^{1/3} = \left(\frac{9nN_A}{a}\right)^{1/3} v_M, \tag{26}$$

where N is the number of atoms in one mole of the crystal; $N = nN_A$. This cutoff frequency is the quantity normally used to characterize the Debye spectrum of a crystal, and it is usually given in terms of the Debye temperature θ_D, defined as follows:

$$\theta_D = \hbar\omega_D/k, \tag{27}$$

where $\hbar$ is Planck's constant h divided divided by 2π, and k is the Boltzmann constant. The Debye temperature calculated from Equations 26 and 27 on the basis of acoustic velocities is called the elastic Debye temperature to distinguish it from Debye temperatures estimated on the basis of specific heat data or by other means. From Equations 22, 26, and 27 it follows that the Debye temperature can be calculated from

$$\theta_D = \frac{\hbar}{k}\left(\frac{6\pi^2 N_A}{ZV}\right)^{1/3} v_M. \tag{28}$$

Table 4. Expressions for the thermodynamic functions according to a Debye model.

Function	Low-temperature limit	High-temperature limit
$E = -3nN_A\beta + 3nN_AkT\,\mathbf{D}(\theta_D/T)$	$E = -3nN_A\beta + (3\pi^4/5)\,nN_Ak\,(T^4/\theta_D{}^3)$	$E = -3nN_A\beta + 3nN_AkT$
$C_V = 3nN_Ak\,D(\theta_D/T)$	$C_V = (12\pi^4/5)\,nN_Ak\,(T/\theta_D)^3$	$C_V = 3nN_Ak$
$S = -3nN_Ak\,\ln(1 - e^{-\theta_D/T}) + 4nN_Ak\,\mathbf{D}(\theta_D/T)$	$S = (4\pi^4/5)\,nN_Ak\,(T/\theta_D)^3$	$S = -3nN_Ak\,\ln(\theta_D/T) + 4nN_Ak$
$F = -3nN_A\beta + 3nN_AkT\,\ln(1 - e^{-\theta_D/T})$ $-3nN_AkT\,D(\theta_D/T)$	$F = -3nN_A\beta - (3\pi^4/5)\,nN_Ak\,(T^4/\theta_D{}^3)$	$F = -3nN_A\beta +$ $3nN_AkT\,\ln(\theta_D/T) - nN_AkT$

For a Debye solid the partition function Z is given by

$$\ln Z = 3kTnN_A\beta - 3nN_A\,\ln(1 - e^{-\theta_D/T}) + 3nN_A(T^3/\theta_D{}^3)\int_0^{\theta_D/T}(x^3dx)/(e^x - 1) = 3kTnN_A\beta - 3nN_A\,\ln(1 - e^{-\theta_D/T}) + nN_A\mathbf{D}(\theta_D/T).$$

This equation contains the definition of $\mathbf{D}(\theta_D/T)$, which is the derivative of the Debye function $D(\theta_D/T)$ [Reif, 1965]:

$$\mathbf{D}(\theta_D/T) = (3T^3/\theta_D{}^3)\int_0^{\theta_D/T}(x^3dx)/(e^x - 1) \quad\text{and}\quad D(\theta_D/T) = 3(T^3/\theta_D{}^3)\int_0^{\theta_D/T}(x^4e^xdx)/(e^x - 1)^2.$$

The specific heat of a solid is obtained by considering the heat capacity contribution $C_V{}^E$ of the individual lattice vibrational oscillators (Einstein oscillators). As discussed in the previous section, the specific heat of one such oscillator of frequency ω is the Einstein function:

$$\mathbf{E}\left(\frac{\hbar\omega}{kT}\right) = \frac{(\hbar\omega/kT)^2\exp(\hbar\omega/kT)}{[\exp(\hbar\omega/kT) - 1]^2}. \tag{29}$$

The molar heat capacity is a summation of Einstein functions for all of the oscillators:

$$C_V = \int_0^{\omega_D} g(\omega)\,\mathbf{E}\left(\frac{\hbar\omega}{kT}\right)\,d\omega. \tag{30}$$

For a Debye solid, with $g(\omega)$ given by Equations 20 and 22, the specific heat in Equation 30 can be written with the use of Equation 29 as

$$C_V = 9nN_Ak\,\frac{T^3}{\theta_D{}^3}\int_0^{\theta_D/T}\frac{e^x}{(e^x - 1)^3}\,x^4dx = 3nN_AkD(\theta_D/T), \tag{31}$$

where $x = \hbar\omega/kT$. The function $D(\theta_D/T)$ defined by this equation, frequently called the Debye (heat capacity) function, is given here in Table 3. From the Debye function and a knowledge of the acoustic velocities (which give the Debye temperature θ_D), the heat capacity as a function of temperature can be predicted if the vibrational frequencies of the solid follow a parabolic frequency distribution.

All thermodynamic functions, not just the heat capacity, can be calculated for a Debye model. These functions, and their low- and high-temperature limits are given in Table 4.

Recalling now that Debye was trying to predict the form of the heat capacity function at low temperatures, we note that

$$C_V \to \frac{12\pi^4}{5} nN_A k \left(\frac{T}{\theta_D}\right)^3 = 1943.7n \left(\frac{T}{\theta_D}\right)^3 \text{ as } T \to 0. \qquad (32)$$

This equation is referred to as the Debye T^3 law, and it correctly describes the low-temperature behavior of the heat capacity. The Debye model is typically found to be valid for $T < \theta_D/50$ to $\theta_D/100$. At high temperatures, $D(\theta_D/T)$ approaches 1, and the molar heat capacity approaches a constant value $3nN_Ak = 3nR$, the Dulong-Petit limit.

The heat capacity of a number of minerals according to a Debye model is given in Figure 1 for two different assumptions about θ_D: for one curve, θ_{el} is used for the Debye temperature; for the other curve, a value of θ_D was chosen that fits the measured heat capacity at 300 K. Neither curve reproduces the measured heat capacity data for silicates and, when the Einstein curves are examined, we reach the conclusion that the data cannot be accounted for with a simple one-parameter model, either Debye or Einstein.

2.4 The need to know vibrational spectra

As mentioned above Einstein concluded that a single frequency was not an adequate representation of the vibrational spectrum, a conclusion supported by Debye's work and its success for simple substances. In this section, we explore what mineral heat capacity data tell us about mineral vibrational spectra compared to a Debye spectrum.

Graphs of measured versus predicted heat capacities, such as those in Figure 1, often cannot show the relative accuracy of a model clearly: for example, at low temperature where the heat capacities are small substantial differences between predicted and experimental values are not visible. A convention that has been adopted to express the relative values of heat capacities compared to those predicted by a Debye model is to express the heat capacities in terms of the Debye temperature that would be required to reproduce the heat capacity at various temperatures. This parameter is called the calorimetric Debye temperature, shown in the left columns of the two parts of Figure 1. It is convenient to normalize it to the elastic Debye temperature, as in Figure 6. In this figure, the silicates have been grouped by degree of

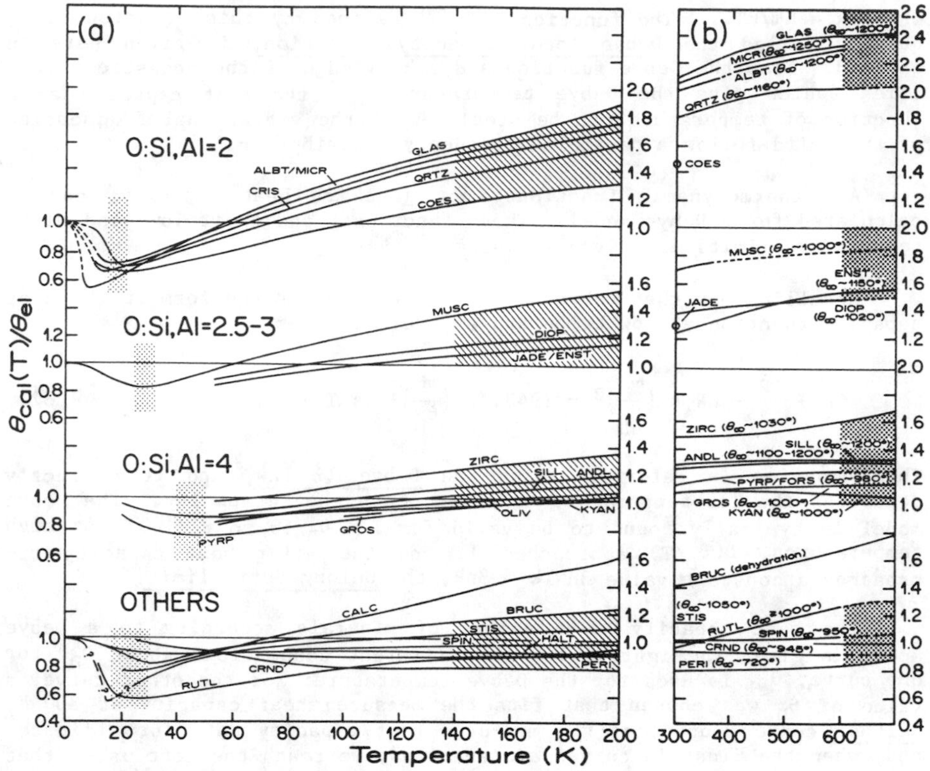

Figure 6. Curves of $\theta_{cal}(T)/\theta_{el}$ for (a) low-temperature data, and (b) high-temperature data. Note change in abscissa scale from (a) to (b). Extrapolated curves are dashed. Detailed explanation of these curves and the references for their data sources can be found in Kieffer (1979a, and 1982).

polymerization, expressed by the oxygen-silicon ratio. (Worked problem 1 in Section 7 explores some relations between C_V and $\theta_{cal}(T)$.)

If a substance were a perfect Debye solid, then the calorimetric Debye temperature would equal the elastic Debye temperature and the ratio of calorimetric to elastic Debye temperature would be unity at all temperatures. (For brevity, this ratio will simply be called the calorimetric ratio in this section.) If there is heat capacity in excess of that predicted by a Debye model, the calorimetric ratio is less than unity; if there is less heat capacity than predicted by a Debye model, the ratio is greater than unity. By implication, if there is excess heat capacity at a given temperature, there are more oscillators in some region of the spectrum than given by the parabolic Debye distribution; and, correspondingly, if the heat capacity is deficient at a given temperature, there are fewer oscillators in some region of the spectrum than given by the Debye distribution. These relations are the basis for looking for correlations between the thermodynamic behavior and the spectral behavior.

At low temperatures (see the dotted zones on the left of Figure 6) the calorimetric ratios drop below unity. The magnitude of the drop, and the temperature at which the minimum in the calorimetric ratio occurs, are both correlated with crystal structure. The minimum in the calorimetric ratio is the deepest for the framework silicates (Si:O = 1:2), and occurs at a lower temperature than for the other minerals. The minimum becomes progressively shallower for sheet (Si:O = 1:2.5), chain (Si:O = 1:3), and orthosilicates (Si:O = 1:4), and occurs at progressively higher temperatures in this sequence.

At high temperatures (see dotted zones on the right side of Figure 6), the calorimetric ratios exceed unity. The ratio is highest in framework silicates, lower in chain and sheet silicates, and lowest in orthosilicates.

At intermediate temperatures (see striped zones in the center of Figure 6) the calorimetric ratios rise smoothly from the low-temperature values to the high-temperature values. Because of the systematic behaviors at low and high temperatures mentioned above, the steepness of the rise decreases from framework through sheet and chain silicates to orthosilicates.

If only harmonic contributions to the heat capacity are considered (e.g., anharmonic, magnetic or electronic contributions are eliminated), then the implication of a low θ_{cal}/θ_{el} ratio ($\theta_{cal}/\theta_{el} < 1$) is that the real mineral vibrational spectrum has more degrees of freedom in the modes that are active and contribute to the heat capacity at the given temperature than does a Debye spectrum. Thus, from Figure 6 we conclude that for most silicates there are more oscillators than predicted by a Debye model at low frequencies. If the ratio θ_{cal}/θ_{el} is greater than unity, then the real vibrational spectrum has fewer active modes than a Debye spectrum. Thus, from Figure 6 we see that, compared to a Debye spectrum, most silicates have excess degrees of freedom at low temperatures, and are deficient in active modes at high temperatures. In the next section, we see that these inferences are confirmed by trends in the observed vibrational spectra.

3. SYSTEMATIC TRENDS IN VIBRATIONAL CHARACTERISTICS OF MINERALS

Thermodynamic and spectroscopic data on minerals indicate that lattice vibrations are controlled by structure and composition. <u>If</u> we could accurately describe the vibrational modes in crystals, <u>if</u> we could define their frequency ranges, and <u>if</u> we could specify the number of modes associated with different types of vibrations in different frequency ranges, then we would have a powerful tool for interpreting and predicting thermodynamic behavior because, as shown in Section 2, the thermodynamic functions are integrals over the vibrational spectrum. It is therefore imperative to try to understand the structural controls on the thermodynamic and vibrational characteristics.

As a rule of thumb, the behavior of the calorimetric ratios at a given temperature is related to the behavior of the vibrational modes at a frequency whose value, specified in cm^{-1}, is about numerically equal

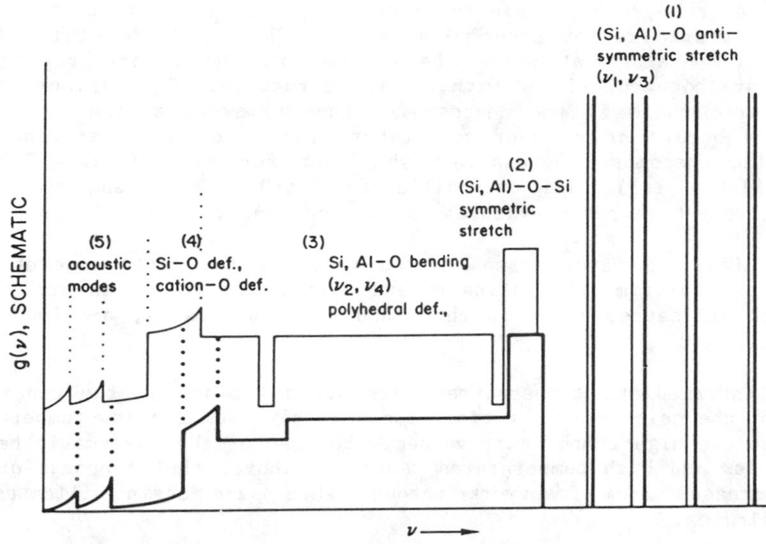

Figure 7. Schematic vibrational spectrum used for discussion of mineral vibrations. The five labeled groups are discussed in the text. The three acoustic branches appear as the upwardly curved sections at low frequency; in this schematic spectrum, the longitudinal acoustic mode overlaps the optical continuum. The light curve offset from the heavy spectrum shows schematically how a spectrum might be changed by the substitution of O^{18} for O^{16}, discussed in Kieffer (1982).

to the temperature, specified in Kelvins (readers can convince themselves of this by examining Table 2). Therefore, trends in the heat capacity implied in Figure 6 should correspond to trends in observed lattice vibrations at the following frequencies: (1) about 1000 cm^{-1} to relate to the high-temperature asymptotes attained at about 1000 K; (2) about 500 cm^{-1} to relate to the slopes of the curves in the middle temperature ranges; and (3) about 100 cm^{-1} to relate to the very-low temperature behavior.

The terminology used in the following discussion is introduced in Figure 7, a schemmatic diagram of an idealized vibrational spectrum for silicates. Different terminology is used by different investigators, and no attempt is made here to be rigorous in description of the detailed atomic motions (see McMillan, this volume). The terminology used here generally follows Lazarev (1972), who classifies modes, such as those associated with silicon-oxygen linkages, by reference to bonds of Si-O-Si bridges, and terminal Si-O bonds. Framework silicates, in which all, or nearly all, vibrations are of the type Si-O-Si, represent the one end member of this series. Orthosilicates represent the other end member in which all vibrations are of the type Si-O.

Using this terminology, five groups of modes can be distinguished in many silicates: (1) antisymmetric Si-O-Si stretching and Si-O stretching modes; (2) symmetric Si-O-Si stretching and/or Si-O-Si bridging deformations; (3) Si, Al-O bending deformations in isolated tetrahedra; (4) other polyhedral deformations and lattice distortions; and (5) acoustic modes. Their frequency ranges are generally as shown in Figure 7, but there can be significant changes in mode frequencies depending on crystal symmetry and atomic bonding (see the discussion of quartz below).

Enumeration of the modes follows from two assumptions: (a) for each Si-O-Si bond there is one symmetric stretch and one antisymmetric stretch; and (b) that for each Si-O bond there is one stretching and one bending mode. Consider the frequency ranges and numbers of these modes in approximate order of decreasing frequency.

The frequency of antisymmetric Si-O-Si stretching and Si-O stretching modes varies from 800 to 1200 cm^{-1} (Fig. 8). These modes correspond in a general way to the ν_1 and ν_3 modes of an isolated $[SiO_4]$ tetrahedron. Numerous spectroscopists have shown that the frequencies of these modes shift systematically upwards as the degree of polymerization increases from ortho to sheet, ring, chain, and framework silicates. Enumeration of the modes is accomplished by counting one mode per bridging Si-O-Si bond and one mode per terminal Si-O bond in the minerals. In Figure 8 (and successive figures of this type) the enumeration is expressed as a fraction of the total degrees of freedom associated with the primitive unit cell, e.g., one mode in quartz would be 1/(3x9) = 0.037 of the total degrees of freedom. Typically the antisymmetric Si-O-Si and Si-O stretching modes comprise 16-22% of the total in the crystal.

If modes in one mineral are at lower frequency than they are in another mineral, they contribute to the heat capacity at relatively lower temperatures. Thus, it follows from Figure 8 that the high-frequency modes in orthosilicates should contribute more to the heat capacity at a given temperature than the same modes in framework silicates. The high temperature calorimetric ratios should be lowest for the minerals with the lowest antisymmetric stretching mode frequencies (the orthosilicates), and highest for the minerals with the highest stretching frequencies (the framework silicates), exactly the trend shown in Figure 6. Thus there is a good, easily-interpreted correspondence between the observed vibrational spectra and thermodynamic behavior at high temperature.

The symmetric stretching modes and Si-O bending modes are more difficult to identify and enumerate than the antisymmetric Si-O-Si and Si-O stretching modes, because they lie at intermediate frequencies that begin to overlap the frequency range of other polyhedral deformations. Where they have been identified and counted, they comprise from 5 to 25% of the modes and lie between about 400 and 800 cm^{-1} (Fig. 9). The range of these frequencies is systematically lower as the polymerization decreases, with the one exception being muscovite. This trend is the same as for the antisymmetric stretching frequencies and contributes to the systematic variation seen in the middle temperature range in the slopes of the calorimetric ratios. For example, the bending modes for orthosilicates, tend to be squeezed to lower frequencies than the sym-

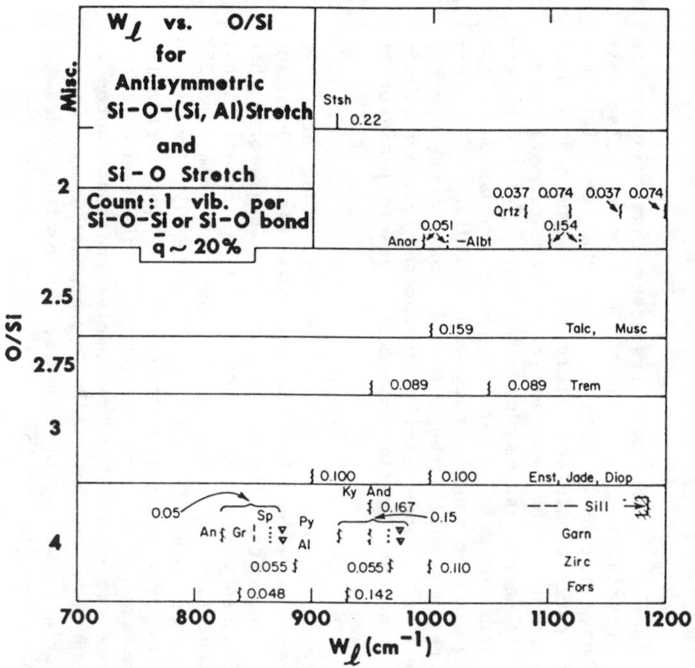

Figure 8. Summary of the frequencies (expressed as wave numbers) at which antisymmetric Si-O-Si,Al stretching bands occur in silicates of different polymerization, as expressed by the O/Si ratio. The decimal fractions by each band represent the fraction of degrees of freedom of the unit cell for that mode. In this and following figures, the mineral abbreviations are (from top to bottom) : Stsh, stishovite; Anor, anorthite; Albt, albite; Qrtz, quartz; Talc; talc; Musc, muscovite; Trem, tremolite; Enst, enstatite; Jade, jadeite; Diop, diopside; Ky, kyanite; And, andalusite; Sill, sillimanite; Sp, spessartine; Py, pyrope, An, andradite, Gr, grossular; Al, almandine; (Garn; garnets); Zirc, zircon; Fors, forsterite. The various symbols with the lines are simply to distinguish lines appropriate to different minerals.

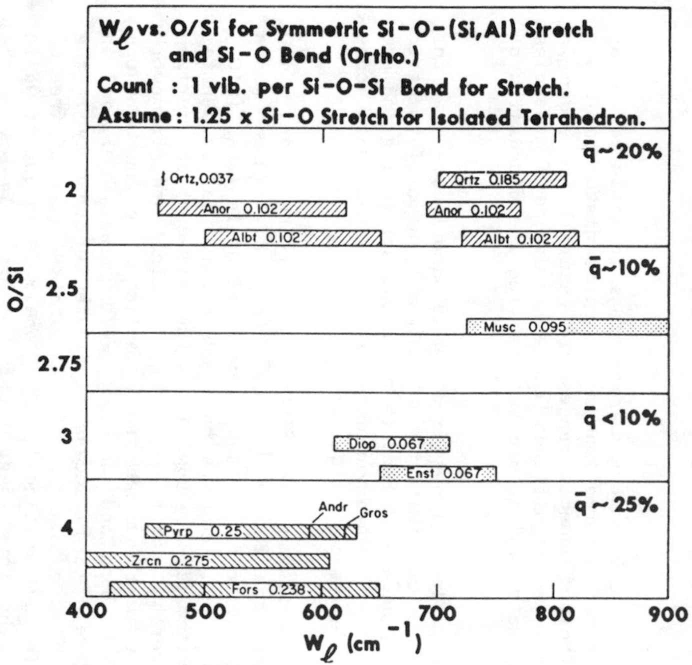

Figure 9. Summary of the frequencies at which symmetric Si-O-Si,Al stretching modes and Si-O bending modes occur in silicates of different polymerization, as expressed by the O/Si ratio. These frequencies span ranges indicated by the boxes; geometric patterns distinguish different O/Si ratios. The decimal fractions in the boxes represent the fraction of degrees of freedom of the unit cell for those modes.

metric stretching modes of the feldspars. These modes give the orthosilicates higher heat capacities in the middle temperature range. The orthosilicates therefore have systematically lower calorimetric ratios and lower slopes than the feldspars.

Before addressing the behavior of the lowest frequency optic modes, a difficult problem, consider the lowest frequency, long wavelength modes -- the acoustic modes. These modes comprise 3 out of the 3s total degrees of freedom for the unit cell, typically only a few percent of the total modes for silicates. There are very few data showing the frequency of acoustic modes across the Brillouin zone. At long wavelengths, their behavior is given by their elastic wave velocities, but in the region of interest, at $\mathbf{K} > 0$ where the frequency increases toward far-infrared optical frequencies, data are sparse. Therefore, to relate the acoustic properties to lattice vibrational spectra for comparison with the optical modes discussed above, acoustic wave velocities must be converted to characteristic frequencies and extrapolated across the Brillouin zone by some sort of model. Anisotropy and dispersion both influence the behavior of acoustic vibrations and are accounted for as described later in this paper. Acoustic modes thus modelled typically range up to an upper cutoff frequency of about 100 cm^{-1}, although a few minerals show values substantially higher than this (Fig. 10). Acoustic modes often overlap the lowest optic modes in frequency, and therefore it is only at the very lowest temperatures (a few to a few tens of K; $T < \theta_D/100$ in most cases) that correlations between acoustic modes and heat capacities can be sought. The correlation is an inverse one: high acoustic velocities give high characteristic frequencies and low heat capacities, whereas low acoustic velocities give high heat capacities. (In Figure 6, this correlation is masked by the normalization of the calorimetric ratio to the elastic Debye temperature, but where this ratio is unity near 0 K, the crystal behaves as a Debye solid.)

The modes discussed thus far account for roughly 50% of the total. The remainder can be loosely termed "lattice distortions" and include stretching and bending in the weaker polyhedra and librations of discrete structural units. Infrared and Raman spectra show that these modes may extend up to about 600 cm^{-1} (Fig. 10). The lowest frequency of these modes is much more difficult to specify because such modes are weak in infrared and Raman spectra, and because impurities in samples can give confusing data. The lowest mode observed in spectra, and its estimated dispersion across the Brillouin zone, are shown in Figure 10. There is a suggestion in the data that framework silicates have the lowest optical mode frequencies and that the lowest mode frequency increases as the degree of polymerization decreases. It is clear that in the absence of any low-frequency spectral data, the optic modes could be assumed to extend into the range of the acoustic modes, particularly when dispersion is estimated. Thus, to a very rough approximation, the frequency of the lowest optic mode could could be estimated solely from acoustic data and knowledge of the crystal structure and volume; such an estimate can be useful in the absence of other data, e.g., for very high pressure minerals that cannot be synthesized, but whose elastic constants can be estimated.

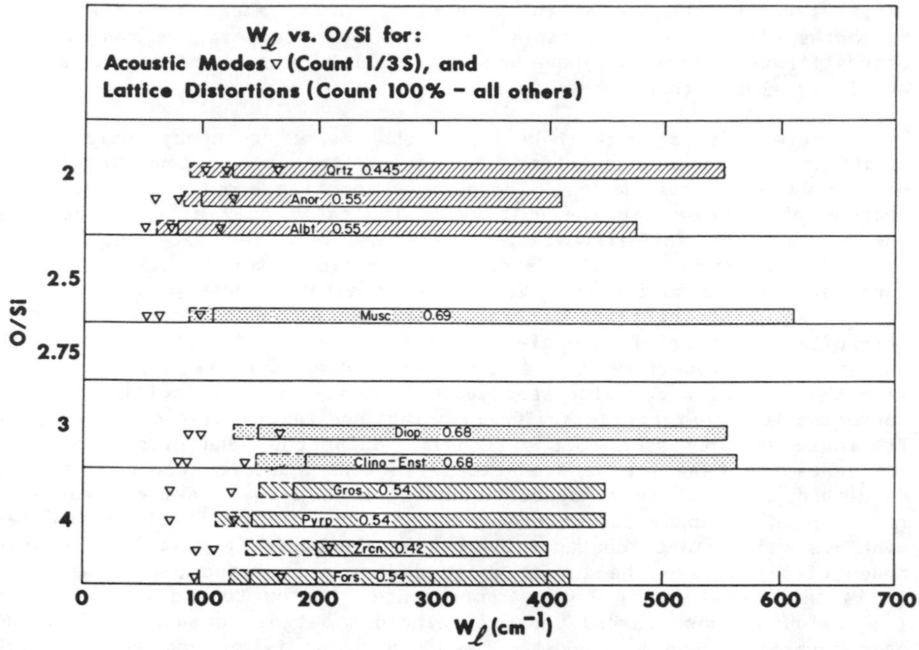

Figure 10. Summary of frequencies of acoustic modes and lattice distortions for the different mineral structures as represented by the O/Si ratio. The inverted triangles represent the zone-boundary (**K** = 0) acoustic mode frequencies of the two shear and longitudinal modes, calculated with an assumed dispersion relation as described in Kieffer (1979c). The bands represent the range of the distortion modes. The dashed extension on the low-frequency end of each band is an estimate of the lowering of the observed K = 0 frequency due to dispersion (estimated as described in Kieffer, 1979c). The decimal fraction gives the fraction of degrees of freedom associated with these modes, obtained by subtracting all other enumerable modes from 1.00.

However, an estimate made in this way would not be sufficiently accurate to predict the thermodynamic properties required for many geological problems because it is exactly in this range of frequencies that the vibrational modes critically influence the entropy. This point is emphasized in Figure 11, in which the entropy of minerals at 298 K (normalized to one atom) is plotted against the frequency of the lowest optical mode as observed at **K** = 0. The stippled area encloses many of the minerals with silicon in 4-fold coordination, a good sub-group to focus on for discussion. Note that as the frequency of the lowest optical mode decreases from approximately 200 cm^{-1} to about 70-80 cm^{-1}, the entropy at 298 K increases by about 30%. It is not, of course, just the position of one low mode that determines the entropy, but rather the distribution of all acoustic and low-frequency optic modes. Of particular importance is the fact that the frequency of the lowest optical mode may decrease substantially as **K** increases away from the center of the Brillouin zone, so the average mode frequency may be substantially lower than seen at **K** = 0. (This effect will be illustrated analytically for the linear chain in the next section.) This substantially influences

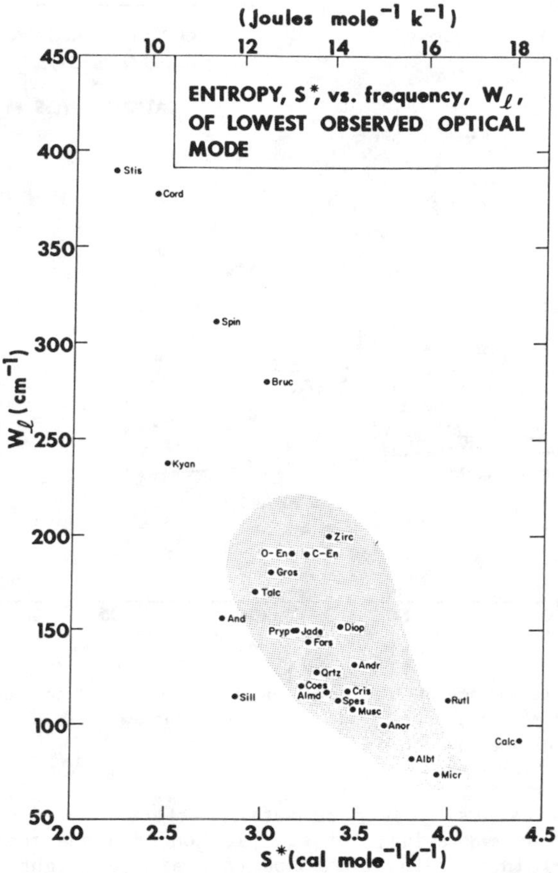

Figure 11. The dependence of harmonic entropy on the frequency of the lowest observed optical mode. Entropies are experimentally measured except almandine, spessartine, and andradite for which calculated values of Kieffer (1980) were used. Stippled area encloses silicates with Si in four-fold coordination, discussed in the text.

the thermodynamic properties. However, because the acoustic and low-frequency optic modes both reflect the same structural properties -- namely the motions of the heavier structural units bound by the weaker bonds -- they have the same systematic trends and combine to cause the low-temperature calorimetric trends.

If the lowest-frequency modes reflect the vibrations of the heaviest, most weakly bound cations in the crystal, then a graph of lowest optical mode frequency versus cation mass might be helpful in predicting vibrational trends (Fig. 12). Such a graph shows that, within given structural groups, there are rather simple trends that do depend on cation mass, but that different structural groups have different dependences. The vibrational frequencies are lowest in the monovalent feldspars, higher in divalent feldspars; next in the sheet minerals, then progressively higher in the orthosilicates, chain and ring silicates,

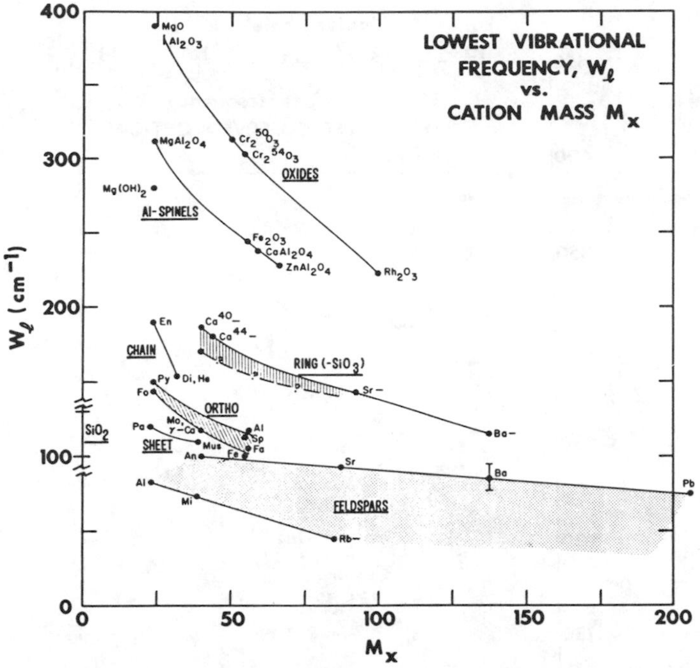

Figure 12. The dependence of the lowest known optical mode frequency on cation mass (expressed in atomic mass units) and on mineral structure. Fe_2O_3 falls on the Al-spinel line by coincidence.

and highest in oxides. Many competing effects can be seen on this graph: focusing on magnesium, for example, one can see that in periclase magnesium in 6-fold coordination bound to a simple, light-weight oxygen, has a cutoff frequency of nearly 400 cm^{-1}. The frequency decreases as the coordination changes and as the effective anion mass increases down through spinels, brucite and the chain and orthosilicates. This trend suggests that a characteristic reduced mass may be important (see next section). It is gratifying that the systematic trends seen in the vibrational modes on this graph are indeed those reflected in the low-temperature heat capacity. Notice that the feldspars have the lowest vibrational frequencies, followed by the heavy orthosilicates and quartz, then the light orthosilicates and heavy chain silicates, and finally the light chain silicates, spinels, and oxides. Then, referring back to Figure 6, note that within the limited data, the thermodynamic anomalies follow roughly the same order. The temperatures at which the calorimetric ratio minima occur (the heat capacity excesses) increase systematically in the same order as the position of the lowest optic modes. The fact that no set of parameters -- such as bond length, cation mass, anion mass, or coordination number -- seems to appear that allows a reliable quantitative prediction of mode frequency probably indicates that no single type of vibration, e.g., cation-oxygen bending, accounts for the lowest frequency vibration, but that different deformations may be responsible within different structural types.

In summary, there is a direct and detailed dependence of the thermodynamic behavior of minerals on the number and position of various vibrational modes. In particular, the high temperature behavior directly reflects the position of about 20% of the modes which can be called antisymmetric Si-O-Si or Si-O stretching modes; the middle temperature range behavior depends on more modes than explicitly considered here, but is strongly influenced by the frequencies of vibration of the 10 to 20% of the modes that can be called Si-O-Si symmetric stretching and Si-O bending modes; and the lowest temperature behavior reflects primarily the structural control, and secondarily, the cation compositional control, on the low-frequency modes.

4. THE INFLUENCE OF WAVE VECTOR: THE BORN-VON KARMAN THEORY

Neither the Debye nor Einstein models have fundamental structural properties in them, e.g., interatomic forces or atomic masses. This is both their strength and weakness. It is a strength because, at the time of the development of these theories, little was known about the atomic structures of even the simplest cubic compounds. These theories are useful even today for minerals for which there is a similar lack of data, for certain minerals whose structure gives them a relatively simple thermodynamic behavior, and for analyses of high temperature behavior where certain universal limits are approached (see Jeanloz, this volume). However, the simplicity of the Debye and Einstein models is also a weakness because the dependence of the thermodynamic properties of minerals on detailed atomic properties cannot be quantified. The original work of Born and von Karman (1912, 1913) that examined the influences of atomic structure went nearly unnoticed because of the initial success of the simple, one-parameter Debye model in contrast to the complexity of the Born-von Karman model. It was not really until the later work of Blackman (1935), Born (1942), and Leighton (1948) that the calculation of the characteristic vibrational frequencies of a crystal from structural properties became a common endeavor.

The basic assumption of the Born-von Karman theory, and the many theories derived from it, is that the displacements of atoms from their equilibrium positions are small so that the potential energy of a crystal can be written as an expansion in terms of the displacements. The full three-dimensional treatment of lattice vibrations by the Born-von Karman theory is complex, and even a brief summary requires introduction of a substantial amount of notation. A summary would also need to include a discussion of the ways in which different ions are treated at the atomic level in efforts to include realistic descriptions of atomic environments. For example, in a rigid-ion model, the ionic charges are approximated by point charges centered at the nuclei; in a shell model the polarizablity of ions is accounted for by regarding each ion as being composed of a rigid core and a charged shell which is bound to the core by a spring; and in the bond-charge and valence-charge models covalency is accounted for. Because the use of the Born-von Karman model and these variations requires a substantial commitment of research effort to a specific mineral and is usually better directed toward understanding of fundamental bonding characteristics than the bulk thermodynamic properties, I am not going to review this theory, but refer the reader to Born and Huang (1954) and Brüesch (1982) for general

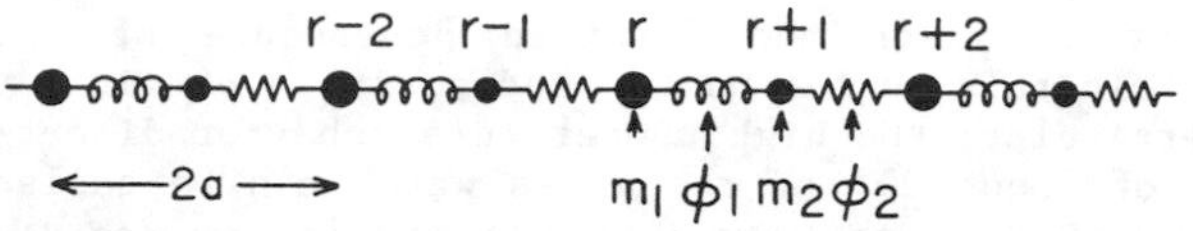

Figure 13. A linear diatomic chain with two different force constants.

concepts, to Elcombe's (1966, 1967) work on quartz, and the review of this work by Barron et al. (1976). Elcombe, like many other workers who have presented detailed vibrational models, stresses the difference in results that can be obtained by different treatments, and the great amount of data needed to constrain the models. Brüesch (1982) points out that it is conceptually possible to apply the Born-von Karman model to complex crystals, but that the practical difficulties are numerous: the computations are tedious because of the large number of degrees of freedom, the number of unknown parameters is large, and the physical picture of the ionic groups as entities executing rotations and translations is lost because only the motions of individual atoms (not groups or clusters) are given by such calculations. These are serious difficulties in the use of full lattice dynamical treatments for minerals.

Many (but not all) aspects of the full three-dimensional Born von-Karman treatment, can be illustrated by consideration of simple linear chains: monatomic and diatomic chains are discussed in most introductory solid state texts (e.g., Kittel, 1968; Brillouin, 1953). The essential results are illustrated here with two examples -- the diatomic linear chain with two force constants, and the chain with a basis. Another example is given as Worked Problem 2 in Section 7.

4.1 The diatomic chain with two different force constants

Consider the diatomic chain with a basis as shown in Figure 13; some mathematics of this derivation are omitted and can be found in Kieffer (1979a). Although it is somewhat difficult to imagine a real crystal in which such a chain arises, one might consider a chain of -(Si-O)=(Mg=O)-(Si-O)=(Mg=O)- as an example, where the Si-O units comprise m_1 and the Mg-O units comprise m_2. Because of the different valences of Si and Mg, the force constants between alternating units are different.

Denote the two masses $m_r = m_1$ and $m_{r+1} = m_2$, and the two force constants ϕ_1 and ϕ_2. Let the subscript $_1$ denote the larger mass and the larger force constant. Let the interatomic distance be a, so that 2a is the lattice repeat distance. The range of the first Brillouin zone is therefore $-\pi/2a \leq K \leq \pi/2a$; the vector notation **K** is dropped here because the problem is one-dimensional. Assume that only nearest-neighbor interactions are significant. Denote the position of the rth atom by μ. Then the equation of motion, for particle r is

$$F_r = \phi_1 (\mu_{r+1} - \mu_r) + \phi_2 (\mu_{r-1} - \mu_r) = m_1 (d^2\mu_r / d^2t), \qquad (33)$$

and for particle r+1 is

$$F_{r+1} = \phi_1 (\mu_r - \mu_{r+1}) + \phi_2 (\mu_{r+2} - \mu_{r+1}) = m_2 (d^2\mu_{r+1} / dt^2). \quad (34)$$

Solutions to these equations have the form of traveling waves with different amplitudes (designated ξ_1 and ξ_2) on the even $(r, r+2, \ldots)$ and odd $(r+1, r+3, \ldots)$ atoms:

$$\mu_r = \xi_2 \exp\{irKa - i\omega t\}, \quad (35)$$

$$\mu_{r+1} = \xi_1 \exp\{i(r+1)Ka - i\omega t\}. \quad (36)$$

These equations have a solution only if the determinant of the coefficients of ξ_1 and ξ_2 vanishes:

$$\begin{vmatrix} m_1\omega^2 - \phi_1 - \phi_2 & \phi_1\exp(iKa) + \phi_2\exp(-iKa) \\ \phi_1\exp(-iKa) + \phi_2\exp(iKa) & m_2\omega^2 - \phi_1 - \phi_2 \end{vmatrix} = 0. \quad (37)$$

This determinant is called the <u>secular determinant</u>, and the matrix of its coefficients is the <u>dynamical matrix</u>.

The general solution to the secular determinant is:

$$\omega^2 = \frac{(\phi_1 + \phi_2)}{2}\left(\frac{1}{m_1} + \frac{1}{m_2}\right) \mp \left\{\left(\frac{\phi_1 + \phi_2}{2}\right)^2 \left(\frac{1}{m_1} + \frac{1}{m_2}\right)^2 - 4\frac{\phi_1\phi_2}{m_1 m_2}\sin^2 Ka\right\}^{1/2}. \quad (38)$$

This is sometimes written in the form

$$\omega^2 = \frac{\phi_1 + \phi_2}{2\mu}\left[1 \mp \left(1 - \frac{16\phi_1\phi_2\,\mu^2}{(\phi_1 + \phi_2)^2\, m_1 m_2}\sin^2\frac{Ka}{2}\right)^{1/2}\right], \quad (39)$$

where $$\mu = \frac{m_1\, m_2}{(m_1 + m_2)} \quad (40)$$

is called the <u>reduced mass</u>. These solutions are shown in Figure 14. For small K, $\sin^2 Ka \cong K^2a^2$, and the quadratic Equation 39 has the two roots:

$$\omega_o^2 \sim (\phi_1 + \phi_2)(m_1 + m_2)^{-1} K^2 a^2 \quad (41)$$

and $$\omega_1^2 \sim (\phi_1 + \phi_2)\left(\frac{1}{m_1} + \frac{1}{m_2}\right). \quad (42)$$

$(\phi_2/\phi_1, m_2/m_1) = (m_2/m_1, \phi_2/\phi_1)$

(1,1)
(0.5,1)
(0.25,1)
(0.5,0.5) (0.10,1)
(0.5,0.25)
(0.5,0.1)
(0.1,0.1)

(1,1)
(0.5,0.5)
(0.5,0.5)
(0.5,0.25) (0.25,1)
(0.10,1)
(0.5,0.1)
(0.1,0.1)

$\omega\sqrt{m_2/\phi_1}$

K/a

$\pi/4$ $\pi/2$

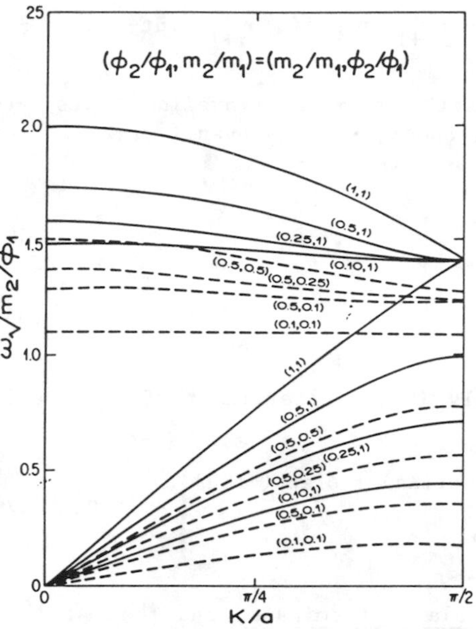

Figure 14. Generalized dispersion relations for the linear diatomic chain with two different force constants. Numbers shown in parentheses on the curves refer to the ratios ϕ_2/ϕ_1 and m_2/m_1.

The first root, ω_0, is called the <u>acoustic mode</u>, and the second, ω_1 is called the <u>optic</u> mode. At the Brillouin zone boundaries, $K = \mp\pi/2a$, the limiting values of frequency are determined by both the mass ratio m_1/m_2, and the force constant ratio ϕ_1/ϕ_2. If $\phi_1 \geqslant \phi_2$, the two branches approach the limits:

$$\omega_2^2 \sim \frac{4\,\phi_2}{(1 + \phi_2/\phi_1)\,(m_1 + m_2)} \qquad \text{(acoustic branch)} \quad (43)$$

and

$$\omega_3^2 \sim \phi_1\,(1 + \phi_2/\phi_1)\,(\frac{1}{m_1} + \frac{1}{m_2}). \qquad \text{(optic branch)} \quad (44)$$

If $m_1 = m_2$, then the ratio of frequencies at the Brillouin zone boundary for the optic and acoustic branches is given by $(\phi_1/\phi_2)^{1/2}$ in this case where ϕ_1 is larger than ϕ_2. If the force constants are equal, $\phi_1 = \phi_2 = \phi$ (Kittel, 1968, p. 149.), then the ω^2 has the following limiting values:

$$\omega_0^2 \simeq \frac{2\,\phi\,K^2\,a^2}{(m_1 + m_2)} \qquad \text{(acoustic branch at small K)} \quad (45a)$$

$$\omega_1^2 \simeq 2\phi(\frac{1}{m_1} + \frac{1}{m_2}) \qquad \text{(optic branch at small K)} \quad (45b)$$

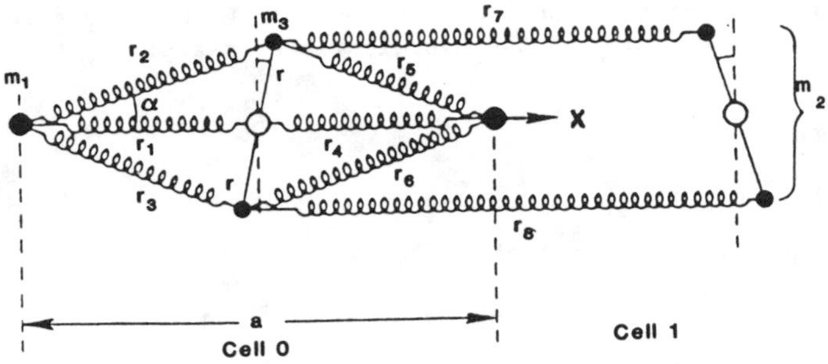

Figure 15. Linear chain of alternating atoms (small circles) and linear rigid XY_2 molecules. (Brüesch, 1982). Notation described in text.

$$\omega_2^2 \simeq (2\phi/m_1) \quad \text{(acoustic branch, Brillouin zone boundary)} \tag{46}$$

$$\omega_3^2 = (2\phi/m_2) \quad \text{(optic branch, Brillouin zone boundary)} \tag{46b}$$

The reader should explore the quantitative differences in the branch shapes as the force constant and mass ratios vary -- Figure 14 shows that the effect is quite significant, and shows why extrapolation of observational data taken at K = 0 to the zone boundary $K = K_{max}$ may be difficult.

4.2 The chain with a basis

When well-bound atomic clusters can be identified and treated as rigid bodies at the outset, their translational and rotational oscillations, referred to collectively as the external vibrations, can be examined by approximate methods. The intra-cluster vibrations, referred to as the internal vibrations, must be treated by separate methods -- often by comparison with a gas of a composition that contains the clusters. Let us then examine Brüesch's (1983) model for the linear chain consisting of alternating atoms and linear, rigid XY_2 molecules (the bases). The chain is assumed to have N unit cells of length a, identical to the one shown in Figure 15. The atoms have mass m_1; the rigid XY_2 ions have total mass m_2 and moment of inertia I; the mass of each Y atom is m_3. The bond lengths within the ion are r. The coordinate system that is most suitable for description of the atomic displacements is the internal coordinate system of the distances between particles r_1, r_2,... (this is known as the "internal coordinate formalism"). Motion is considered only in the chain direction, x. There are 8 stretching coordinates, and 3 force constants (f,g,h). The potential energy per unit cell is

$$\frac{2\Phi}{N} = f(r_1^2 + r_4^2) + g(r_2^2 + r_3^2 + r_5^2 + r_6^2) + h(r_7^2 + r_8^2) \tag{47}$$

In this expression Φ is the potential energy of the chain, and N is the total number of unit cells. To describe the motions of the chain, the analysis follows the same form as for the simple diatomic chain above; the mathematics are complex and are omitted here (details are in Brüesch, 1983). The elements, M_{ii} of the dynamical matrix are:

$$M_{11} = 2(f+2c^2g)$$

$$M_{12} = -f(1 + \psi)$$

$$M_{13} = M_{14} = -c^2g(1 + \psi)$$

$$M_{22} = 2f$$

$$M_{23} = M_{24} = 0$$

$$M_{33} = 2\{c^2g + h\ [1 - \cos(qa)]\} = M_{44}$$

$$M_{34} = 0 \tag{48}$$

and, for the remaining elements, $M_{k'k} = M_{kk'}$.

In these elements, $c = \cos\alpha$, where α is the angle from the atom to the ends of the XY_2 cluster, and $\psi = \exp(iqa)$. The eigenfrequencies are

$$\omega_{1,2}^{\ 2} = \frac{f + 2c^2g}{\mu}\ \{1 \mp [1 - 4\frac{\mu^2}{M_1M_2}\ \sin^2\ (\frac{aq}{2})]^{1/2}\}, \tag{49}$$

$$\text{where } \mu = \frac{m_1m_2}{(m_1 + m_2)} \quad \text{and } \omega_3^{\ 2} = \frac{4r^2}{I}[c^2g + 2h\ \sin^2(\frac{aq}{2})]. \tag{50}$$

The eigenfrequencies $\omega_{1,2}$ in Equation 49 are identical to those of the diatomic linear chain Equation 39 (with $\phi_1 = \phi_2$), except that the force constant f has been replaced by the effective force constant $f + 2c^2g$. (In comparing Equations 39 and 49, note that the unit cell dimension for the diatomic chain was taken as 2a whereas it was taken as a in this example.) The force constants f and g describe interactions between the atom and the linear molecule (Fig. 15). Note that the frequencies of the acoustic and longitudinal optic modes (ω_1 and ω_2) are independent of the force constant h describing the interactions between adjacent bases. ω_3 describes the librational motion of the XY_2 molecules. The frequency of this mode depends on the interaction of the XY_2 molecule with the atom (through g) and with the adjacent XY_2 molecules (through h), as well as on the moment of intertia of the basis. The lack of coupling between the translational and rotational oscillations arises from the simple symmetry of the problem.

To obtain a qualitative idea of the atomic properties that influence the relative frequencies of the stretching and librational modes, consider the expressions for ω_2 and ω_3 at $K = 0$:

Figure 16. Schematic dispersion relations for a complex unit cell with 3s degrees of freedom.

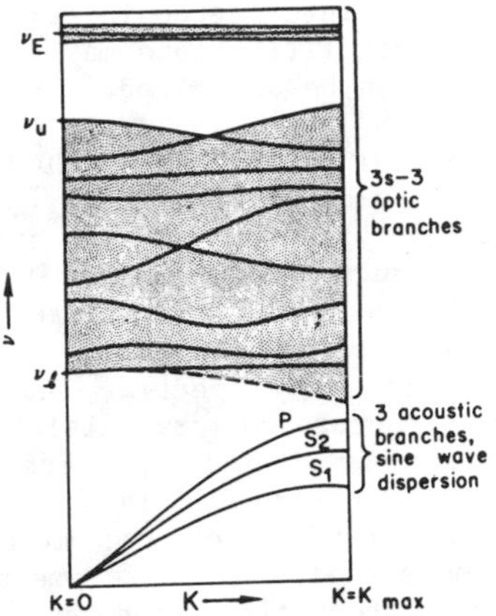

$$\omega_2{}^2 = \frac{2(f + 2c^2g)}{\mu} \tag{51}$$

and
$$\omega_3{}^2 = \frac{4r^2c^2g}{I}, \tag{52}$$

so
$$\frac{\omega_2{}^2}{\omega_3{}^2} = \frac{2(f + 2cg)^2 I}{4\mu r^2 c^2 g} \tag{53}$$

If we assume that the linear XY_2 molecule has $I = 2\ m_3\ r^2$, where m_3 is the mass of Y, this simplifies to

$$\frac{\omega_2{}^2}{\omega_3{}^2} = \frac{(f + 2c^2g)m_3}{\mu c^2 g} = 2(\frac{f}{2c^2g} + 1)\ \frac{m_3}{\mu}. \tag{54}$$

The quantity, $f/2c^2g$ is always positive; therefore the ratio ω_2/ω_3 takes on a value greater or less than 1 depending on the value m_3/μ. The behavior of the dispersion curves is shown in Worked Problem 3 in Section 7.

The concepts of a linear chain introduced here can be extended to a three-dimensional unit cell. If such a cell contains s atoms, then it has 3s degrees of freedom (Fig. 16). Each mode of vibration appears as a separate branch on a dispersion plot. The dispersion relations can be different in each direction of the crystal because of anisotropy. Of the 3s degrees of freedom, three are acoustic modes which approach zero frequency as the wave vector K approaches zero.

These simple analyses give a qualitative idea of the atomic properties that should be considered in trying to interpret the behavior of the low-frequency external modes in crystals; i.e., (1) the effective force constants, c,f, and g; (2) the reduced mass, μ; (3) the moment of inertia, I; and (4) the ion size, through r and c.

Although we rarely know the absolute values of these quantities, the discussion in Section 3 suggests that trends in vibrational characteristics can be interpreted in terms of different structural parameters and that some simplifications may be made so that approximate vibrational properties can be described.

5. MICROSCOPIC KNOWLEDGE NEEDED VERSUS MICROSCOPIC FACTS AVAILABLE

The papers in this volume stress in many ways that details of microscopic structure significantly affect macroscopic properties. It is thus truly inspiring to realize how many conceptual developments of the theory of lattice vibrations were developed before before Moseley's (1913) paper on x-ray spectra, and the beginning of explorations into the atomic structure of crystals! We now know that for prediction of many of the thermodynamic properties, we can ignore the microscopic structure of a substance only if it is essentially monatomic and isotropic. Understanding of the microscopic structure of a mineral is therefore essential for interpretating the lattice vibrations and thermodynamic properties, and for making suitable approximations to allow thermodynamic properties to be calculated. In this section, relevant properties of quartz are given as an example of data needed.

5.1 Bulk and acoustic properties

Knowledge of the crystal acoustic properties is required for analysis of low-frequency, low-temperature behavior, and knowledge of certain bulk properties (density, volume, bulk modulus, and thermal expansion) is required to make anharmonic corrections to calculated harmonic values for comparison with experimental data.

As emphasized in Section 2c, there are three characteristic velocities in each direction of the crystal (that is, for each value of **K**), and these velocities are characteristic of the slopes of the frequency-wave vector curves (Figs. 5a and 16). The discussion of the Debye model suggests that we can ignore the dependence of these wave velocities on **K** and recognize three average velocities in a crystal: two shear velocities, and one longitudinal velocity. As detailed calculations below show, this approximation is reasonable, because the acoustic branches themselves typically only comprise a few percent of the total vibrational modes, and uncertainties in the elastic wave velocities are small compared to uncertainties in the dispersion relations for both the acoustic and optic modes. Details of one method of elastic wave averaging are given in Kieffer (1979c). For the example of quartz, the elastic wave velocities are: u_1 = 3.76 km/s; u_2 = 4.46 km/s, $u_{VRH,S}$ = 4.05 km/s, u_3 = 6.05 km/s, and $\bar{u}$ = 4.42 km/s, where u_1 and u_2 are the maximum and minimum measured shear velocities, $u_{VRH,S}$ is the Vogt-Reuss-Hill averaged shear velocity, u_3 is the compressional velocity, and $\bar{u}$ is the average velocity.

Other bulk properties needed for thermodynamics are: density, ρ = 2.65 g/cm^3; compressibility: $K_o = 2.662 \times 10^{-12}$ Mbar^{-1}, or bulk modulus, B_o = 0.377 Mbar; temperature derivative of the bulk modulus, $\partial B/\partial T = -0.100 \times 10^{-3}$ Mbar/K; and the pressure derivative of the bulk modulus, $\partial B/\partial P = 6.2$. The pressure derivatives are needed only for calculating thermodynamic properties at high pressures; see Kieffer (1982).

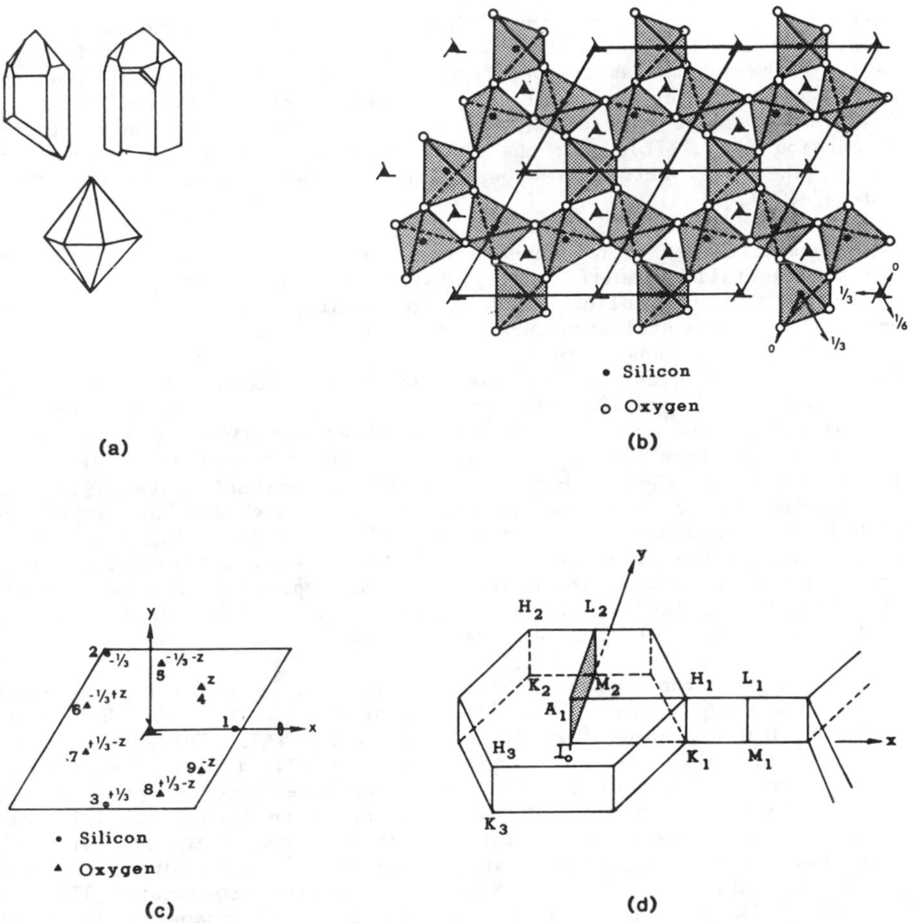

Figure 17. (a) Common habits of quartz crystals that reflect the trigonal trapezoidal symmetry. (b) Structure of α-quartz projected on (0001), illustrating the positions of atoms and SiO^{-4}-polyhedra (Deer et al., 1963, v. 4, p. 184). (c) Structure of left-handed α-quartz, referred to coordinates used by Elcombe (1967) for analysis of inelastic neutron scattering data and Born-von Karman models. (d) Part of reciprocal space for quartz, including the upper half of the first Brillouin zone (Barron et al., 1976). Notation shows directions for the dispersion relations shown in Figure 18.

5.2 Structural data

Knowledge of the structure of a crystal is necessary to account for the effect of the discrete structure of the lattice on continuum properties, e.g., dispersion of the acoustic and optic branches toward the Brillouin zone boundary, and for interpretation of vibrational spectra.

Crystals of quartz (Fig. 17a) consist of atoms of silicon (atomic weight = 28.09) and oxygen (atomic weight = 16). The atoms sit, on the average, in an orderly arrangement on a lattice (Fig. 17b) in which the pattern of atoms repeats after a characteristic distance in each direction. The smallest unit that is repeated is called the primitive unit cell. Alpha-quartz has trigonal symmetry, nine atoms per primitive unit cell, and characteristic unit cell dimensions of a = 4.913 Å, and c = 5.405 Å. I adopt the conventions of Elcombe (1967) for the following discussion (Fig. 17c). In the three crystallographic directions x,y, and z, the wave vectors are respectively $(2\pi/a)(\zeta,0,0)$; $(2\pi/a)(0,\zeta,0)$; and $2\pi/c(0,0,\zeta)$.

Identification of fundamental structural units within the unit cell can substantially simplify interpretation of mineral properties (e.g., as pointed out by Pauling, 1929; Bragg et al., 1965; Hazen, this volume). The fundamental structural units have been variously called ionic groups, complex ions, clusters, and cation coordination polyhedra. Ionic groups or clusters are groups of atoms more tightly bound than other units in a crystal; coordination polyhedra are the spaces enclosed by passing planes through each set of three coordinating anions of a central cation (see Hazen, this volume). Clusters sometimes correspond in concept to polyhedra, but not always. In orthosilicates $[SiO_4]$ is the fundamental cluster and polyhedron. The polyhedra in rutile are $[TiO_6]$ octahedra, but anisotropy of elastic constants has given rise to suggestions in the literature that the structure of rutile belongs to a transition type between ionic and molecular types; references to modified TiO_2 "molecules" as the basic cluster can be found in the literature, e.g., Seitz (1940, p. 50); Matossi (1951).

In quartz, the fundamental structural unit is the $[SiO_4]$ tetrahedron (Fig. 17b). The average Si-O distance in the tetrahedron is 1.6092 Å; O-O distances vary from 2.612 to 2.645 Å. Intra-tetrahedral angles vary from 108.81° to 110.52° (Levien et al., 1980). The tetrahedral volume at 1 atm is 2.138 Å^3; this decreases only slightly as the pressure increases to over 60 kbar, indicating that the tetrahedra are relatively incompressible. In fact, tetrahedra have their own compressibilities and thermal expansions: for the $[SiO_4]$ tetrahedron, the compressibility is greater than 5 Mbar (in comparison to about 0.377 Mbar for the bulk compressibility) and the thermal expansion is about 12×10^{-6} K^{-1} (see Hazen and Finger, 1982; also Levien et al., 1980). Unfortunately, as yet, the bulk mineral properties can rarely be related to the polyhedral properties because bond-bending properties when polyhedra are linked are not known.

The diffraction of x-rays by quartz crystals indicates that the atoms do not simply "sit" on fixed sites, but rather, vibrate about an average position because of thermal motions. In principle, therefore, there is information about lattice vibrations in x-ray data. The intensity of scattered x-radiation depends on the type of atom, its average position, and the amount of displacement from that position due to thermal motion. It appears that the thermal motions of atoms in complex crystals can be worked out from the anisotropic scattering factors of x-rays (see Finger, 1975, for a calculation of the librations and translations of the carbonate ion in calcite by this approach.) If this approach could be generally extended to silicates, a valuable link between x-ray photon and phonon behavior will have been established.

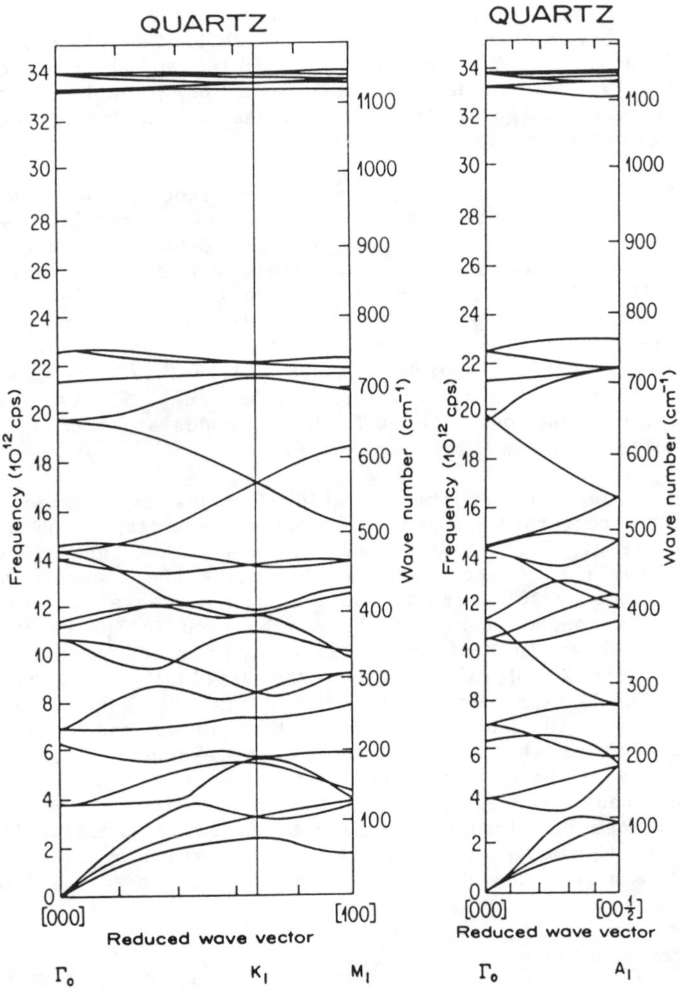

Figure 18. The dispersion curves of α-quartz (from Elcombe, 1967) calculated from a Born-von Karman model. The seven lowest curves with wave vectors in the z-direction were also measured by inelastic neutron scattering at room temperature. The calculated curves are in good qualitative agreement with the measured curves, but agree quantitatively for only a few branches. The six modes isolated at high frequencies correspond to internal modes of the SiO_4^{-4}-tetrahedron.

5.3 Spectroscopic data

Inelastic neutron scattering (INS) data show the behavior of the lattice vibrational modes through the reciprocal cell, from **K** = 0 to **K** = $\mathbf{K}_{max}$ (the reciprocal cell for quartz is shown in Fig. 17d, and INS spectra are shown in Fig. 18). The Raman and infrared frequencies of the modes are those at **K** = 0. There is one INS mode for each of the modes of vibration of the primitive cell. For quartz, the 9 atoms in a primitive cell have 27 degrees of freedom. Three of these are translations of the whole cell, at zero frequency and infinite wavelength, the acoustic modes discussed above. The remaining 24 modes have been assigned to various Si-O bending, stretching and deformations as shown in Figure 19. The frequency of each mode, as given by Elcombe is shown in the figure; slightly different values appear when frequencies are cited from different sources.

The vibrations fall into several distinct groups: those with the highest frequencies are (Si-O) antisymmetric stretching modes, often denoted as ν_1 and ν_3. In Figure 19 these modes show as motions of the silicons relative to the oxygens. There are six such modes in quartz (#1,6,11,12,20,21). At lower frequencies there are Si-O-Si modes whose deformations involve primarily the motion of Si; these are usually called Si-O-Si or Si-Si stretching modes. In Figure 18 five of these modes are at relatively high frequency (7,13,14,22,23); the sixth mode occurs at an anomalously low frequency, 466 cm^{-1} (#2) (see Moenke, 1974, p. 366 for explanation). Lower frequency modes are bending and deformational modes.

The frequencies of the vibrational modes can change dramatically within the reciprocal space, the effect of dispersion discussed in Section 2 (Fig. 17). Acoustic modes increase in frequency as K increases away from the zone center; in many cases the acoustic modes show a maximum frequency somewhere in the zone and decrease in value before the zone edge is reached (this effect can be seen in INS spectra of halite and rutile shown in Kieffer, 1979b). Optic modes do not, in general, maintain their $\mathbf{K} = 0$ values across the Brillouin zone, and may either increase or decrease in frequency through the zone. One important exception to this generalization is that optic modes associated with vibrations where the force constants are exceptionally strong can be relatively flat across the Brillouin zone (e.g., the Si-O stretching modes at about 1100-1200 cm^{-1}). The dispersion of modes across the Brillouin zone has important implications for thermodynamic properties because the thermodynamic properties are obtained by averaging the mode frequencies across the whole reciprocal space, not solely from the $\mathbf{K} = 0$ values.

5.4 Force constants

In principle, if the harmonic force constants in a crystal were known, a full lattice dynamics calculation could be done and the thermodynamic properties obtained (see the papers by Burnham and McMillan in this volume for discussions of parts of this problem.). Numerous investigators since Saksena (1940, 1942) have attempted to derive the harmonic force constants of quartz (Elcombe, 1967; Weidner and Simmons, 1972; Striefler and Barsch, 1975; Barron et al., 1976; Iishi, 1976, 1978). The derived force constants are quite model dependent, as can be seen from Table 4 in Barron et al. (1976). None of these authors calculated a full lattice-vibrational spectrum, $g(\omega)$. A spectrum was calculated for rutile by Traylor et al. (1971), using a shell model to fit the inelastic neutron scattering data. The spectrum thus derived did not give a particularly good fit to measured heat capacities, probably because the model cannot account for changes in the frequency of the low-frequency optic modes with temperature.

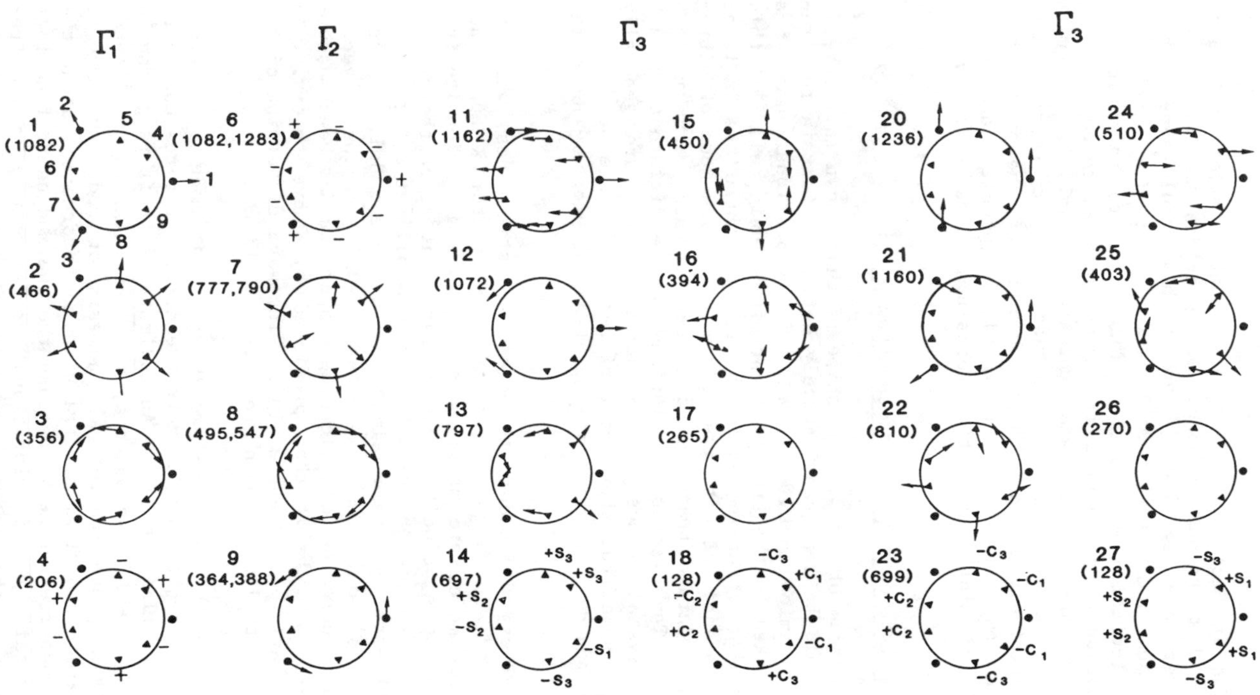

Figure 19. The symmetry modes of α-quartz (from Elcombe, 1966). The diagrams show the nine atoms of the unit cell, projected onto the basal plane as in Figure 17. Motion of the atoms in the x-y plane is indicated by arrows, and perpendicular to this plane by ±. The modes not shown (numbers 5,10, and 19) are acoustic modes. The numbers in parentheses are the wave numbers of the mode; for Γ_2, the transverse and longitudinal wave numbers are shown.

6. PREDICTION OF THE THERMODYNAMIC BEHAVIOR OF MINERALS

The discussion in Section 5 illustrates that a major problem is that no experimental technique unambiguously provides the vibrational spectrum $g(\omega)$. There are few minerals for which we could hope to have the abundant data and analyses that we have for quartz (or rutile, for which the full calculation was done). Therefore, to calculate the macroscopic thermodynamic properties from available microscopic data, the reader should be able to use, and evaluate the validity of, many different levels of approximation. Fortunately, the thermodynamic properties are relatively insensitive to details of the vibrational spectrum, except at low temperatures, i.e., low frequencies, and we can take advantage of this in approximating certain regions of the vibrational spectrum.

6.1 An approximate lattice vibrational spectrum

Consider a spectrum of the general shape of that shown in Figure 7, and how its form could be semiquantitatively defined. Certain requirements of lattice dynamics must be obeyed. The vibrational unit of a crystal should be taken as the Bravais or primitive unit cell. If this cell contains a number of atoms denoted by s, then there are 3s total degrees of freedom associated with the unit cell. Each mode of vibration appears as a separate branch on a dispersion plot such as that shown in Figure 16. The dispersion relations can be different in each direction of the crystal because of anisotropy. They are generally unknown; one must assume that they are confined to a range of frequencies that can be bounded by available spectra and guesses about the dispersion behavior.

Of the 3s degrees of freedom, three are acoustic modes, which by definition, approach zero frequency as the wave vector $\mathbf{K}$ approaches zero. Two of these modes are shear modes and one is a longitudinal mode. Anisotropy of wave velocities can be large but since these modes may only comprise a few percent of the total in silicates, it may be adequate to use VRH-averaged velocities. In a Debye model, the frequency of the modes would increase linearly as the wave vector increases. In cases where the acoustic mode dispersion has been measured, dispersion reduces ω below that predicted by Debye theory. To account for this behavior I suggest a simple sine wave dispersion of the modes in accordance with linear chain theory.

The remaining (3s-3) vibrational modes are optic modes. They can span a broad range of frequencies if there are different force constants and mass units in the primitive cell. An estimate of their range can be obtained from infrared and Raman spectra, but their distribution cannot be inferred unless a group theoretical analysis of mode activity is performed. Furthermore, optical spectra give no information about the dispersion of the mode frequencies. Because the thermodynamic functions are sensitive to details of the vibrational spectrum only at low frequencies, it is only at the lowest frequencies that a correct estimate of the dispersion relations appears to be essential for thermodynamic calculations. The simplest assumption would be that there is no dispersion, that is, that all modes are Einstein oscillators maintaining the value of the frequency at $\mathbf{K} = 0$ across the Brillouin zone (for internal consistency, dispersion should probably be eliminated in the acoustic modes if this assumption is used). The assumption that I prefer, because it recognizes the possibly strong influence of dispersion of the low-frequency modes on the thermodynamic properties, is that all of the branches behave as Einstein oscillators except the lowest optic mode. In the models published in Kieffer (1979c; 1980; 1982) I have assumed that the frequency of this mode decreases across the zone inversely as the square root of a characteristic mass ratio (as in the chain models discussed in Section 4). With such an assumption, the researcher must choose some characteristic masses believed to be involved in the lowest frequency vibrations, e.g., cations, oxygens, or polyhedra.

The optic modes are then assumed to follow the simplest possible frequency distribution, $g(\omega)$: they are assumed to be distributed uniformly between a lower cutoff frequency, ω_l, and an upper cutoff frequency, ω_u, both of which can be specified from spectroscopic data (infrared, Raman, or INS), with dispersion of the lower cutoff frequency across the Brillouin zone accounted for. The density of states (height of the continuum in Figure 7) is determined from normalization of the total number of optical modes to (3s-3) for the primitive unit cell.

If sufficient information is available about the frequency and distribution of some modes to allow them to be defined separately from the optic continuum, this information should be used to gain more accuracy in the calculated functions. For example, silicon-oxygen modes, hydroxyl stretching modes, or carbonate bending or stretching modes may be separably enumerable (see Kieffer 1979b, 1980). Such modes can then be placed in one or more Einstein oscillators at frequencies ω_{E1}, ω_{E2},.... The fractions of the total modes associated with each oscillator are q_1, q_2,....In cases where spectral data are very detailed, it may be possible to enumerate the $\mathbf{K} = 0$ frequencies for all (3s-3) optic modes, viz. the treatment of Lord and Morrow (1957) for quartz, used by Kieffer (1982) to calculate the pressure dependence of the thermodynamic properties of quartz.

It is important that the reader realize that in this treatment there are no free parameters for "fitting" data: (1) acoustic velocities determine the slopes of the acoustic branches; and (2) spectral data constrain the frequencies and enumeration of optic modes. However, because spectral data are rarely definitive, there is a certain arbitrariness in the form of the spectrum assumed, and in the resolution that can be obtained, but as data improve, the models will become more constrained.

The vibrational spectrum in Figure 7 has the following form:

$$g(\omega) = \sum_{1}^{3} 3N_A\left(\frac{2}{\pi}\right)^3 \frac{1}{Z} \frac{[\sin^{-1}(\omega/\omega_i)]^2}{(\omega_i^2 - \omega^2)^{1/2}} + g_0(\omega)^* \text{ for } 0 < \omega < \omega_1, \tag{55a}$$

$$g(\omega) = \sum_{2}^{3} 3N_A\left(\frac{2}{\pi}\right)^3 \frac{1}{Z} \frac{[\sin^{-1}(\omega/\omega_i)]^2}{(\omega_i^2 - \omega^2)^{1/2}} + g_0 \text{ for } \omega_1 < \omega < \omega_2, \tag{55b}$$

$$g(\omega) = 3N_A\left(\frac{2}{\pi}\right)^3 \frac{1}{Z} \frac{[\sin^{-1}(\omega/\omega_3)]^2}{(\omega_3^2 - \omega^2)^{1/2}} + g_0 \text{ for } \omega_2 < \omega < \omega_3, \tag{55c}$$

$$g(\omega) = g_0 \text{ for } \omega > \omega_3, \tag{55d}$$

$$g(\omega) = g_0 + g_E \text{ for } \omega = \omega_E. \tag{55e}$$

(The asterisk in the first line of this equation means that it is implicit henceforth that g_0 is zero outside of the bounds $\omega_l \leq \omega \leq \omega_u$.) ω_1, ω_2, and ω_3 can be found from equations (8 - 16) in Kieffer (1979c).

The total molar specific heat normalized to the monatomic equivalent, C_V^*, then becomes

$$C_v^* = \frac{3N_A k}{nZ}\left(\frac{2}{\pi}\right)^3 \sum_1^3 \int_0^{\omega} \frac{[\sin^{-1}(\omega/\omega_i)]^2(\hbar\omega/kT)^2 \exp(\hbar\omega/kT)d\omega}{(\omega^2 - \omega^2)^{1/2} \; [\exp(\hbar\omega/kT) - 1]^2}$$

$$+ 3N_A k\left(1 - \frac{1}{s} - q\right) \int_{\omega_i}^{\omega_u} \frac{(\hbar\omega/kT)^2 \exp(\hbar\omega/kT)d\omega}{(\omega_u - \omega_i)\,[\exp(\hbar\omega/kT) - 1]^2}$$

$$+ \frac{3N_A kq(\hbar\omega_E/kT)\exp(\hbar\omega_E/kT)}{[\exp(\hbar\omega_E/kT) - 1]^2}, \tag{56}$$

where $N_A k = R$, the gas constant, and $nZ = s$, the number of atoms in the unit cell.

This expression contains three functions which are similar to the Debye function:

"Dispersed acoustic function"

$$\mathbf{S}(x_i) = \left(\frac{2}{\pi}\right)^3 \int_0^{x_i} \frac{[\sin^{-1}(x/x_i)]^2 x^2 e^2 dx}{(x_i^2 - x^2)^{1/2} (e^x - 1)^2} \tag{57}$$

"Optic continuum function"

$$\mathbf{K}(x_u, x_u) = \int_{x_1}^{x_u} \frac{x^2 e^x dx}{(x_u - x_i)^x (e - 1)^2} \tag{58}$$

Einstein Function

$$\mathbf{E}(x_E) = x_E^2 e^{x_E}/(e^{x_E} - 1)^2 \tag{59}$$

In these expressions, $x = \hbar\omega/kT$. $\mathbf{S}(x_i)$ is the heat capacity function for a monatomic solid which has a sine dispersion function;

Table 5. The heat capacity function with a sine-dispersion relation.

[Important Addendum to Kieffer, 1979c: In order to have the tabulated values range from 0 to 1. (like a table of the Debye function), the tabulated quantity is $\Sigma_{i=1}^{3}$ $\mathbf{S}(x_i)$, with $x_1 = x_2 = x_3$. To use it for just one acoustic branch, divide the tabulated values by 3.]

x_i	$S(x_i)$	x_i	$S(x_i)$	x_i	$S(x_i)$	x_i	$S(x_i)$
0.1	0.999323	5.1	0.245435	10.1	0.027960	15.1	0.006660
0.2	0.997321	5.2	0.234640	10.2	0.026954	15.2	0.006514
0.3	0.993989	5.3	0.224270	10.3	0.025993	15.3	0.006373
0.4	0.989350	5.4	0.214316	10.4	0.025075	15.4	0.006236
0.5	0.983426	5.5	0.204767	10.5	0.024199	15.5	0.006103
0.6	0.976248	5.6	0.195613	10.6	0.023362	15.6	0.005974
0.7	0.967853	5.7	0.186844	10.7	0.022563	15.7	0.005849
0.8	0.958282	5.8	0.178448	10.8	0.021798	15.8	0.005727
0.9	0.947583	5.9	0.170413	10.9	0.021068	15.9	0.005609
1.0	0.935806	6.0	0.162728	11.0	0.020369	16.0	0.005494
1.1	0.923010	6.1	0.155382	11.1	0.019700	16.1	0.005383
1.2	0.909254	6.2	0.148363	11.2	0.019060	16.2	0.005274
1.3	0.894601	6.3	0.141659	11.3	0.018448	16.3	0.005169
1.4	0.879119	6.4	0.135259	11.4	0.017861	16.4	0.005066
1.5	0.862873	6.5	0.129151	11.5	0.017299	16.5	0.004966
1.6	0.845933	6.6	0.123324	11.6	0.016761	16.6	0.004869
1.7	0.828371	6.7	0.117767	11.7	0.016245	16.7	0.004775
1.8	0.810258	6.8	0.112470	11.8	0.015750	16.8	0.004683
1.9	0.791662	6.9	0.107420	11.9	0.015275	16.9	0.004593
2.0	0.772656	7.0	0.102609	12.0	0.014819	17.0	0.004506
2.1	0.753306	7.1	0.098025	12.1	0.014382	17.1	0.004421
2.2	0.733681	7.2	0.093659	12.2	0.013962	17.2	0.004338
2.3	0.713846	7.3	0.089502	12.3	0.013559	17.3	0.004257
2.4	0.693864	7.4	0.085543	12.4	0.013171	17.4	0.004179
2.5	0.673795	7.5	0.081775	12.5	0.012799	17.5	0.004102
2.6	0.653698	7.6	0.078188	12.6	0.012440	17.6	0.004028
2.7	0.633628	7.7	0.074775	12.7	0.012096	17.7	0.003955
2.8	0.613635	7.8	0.071526	12.8	0.011764	17.8	0.003884
2.9	0.593769	7.9	0.068434	12.9	0.011445	17.9	0.003814
3.0	0.574074	8.0	0.065492	13.0	0.011137	18.0	0.003747
3.1	0.554592	8.1	0.062693	13.1	0.010841	18.1	0.003681
3.2	0.535361	8.2	0.060029	13.2	0.010556	18.2	0.003616
3.3	0.516415	8.3	0.057495	13.3	0.010281	18.3	0.003553
3.4	0.497785	8.4	0.055083	13.4	0.010015	18.4	0.003492
3.5	0.479501	8.5	0.052788	13.5	0.009759	18.5	0.003432
3.6	0.461585	8.6	0.050604	13.6	0.009513	18.6	0.003373
3.7	0.444060	8.7	0.048525	13.7	0.009274	18.7	0.003316
3.8	0.426943	8.8	0.046547	13.8	0.009044	18.8	0.003260
3.9	0.410252	8.9	0.044664	13.9	0.008822	18.9	0.003206
4.0	0.393997	9.0	0.042871	14.0	0.008607	19.0	0.003153
4.1	0.378190	9.1	0.041164	14.1	0.008400	19.1	0.003100
4.2	0.362838	9.2	0.039539	14.2	0.008199	19.2	0.003049
4.3	0.347947	9.3	0.037991	14.3	0.008005	19.3	0.003000
4.4	0.333520	9.4	0.036517	14.4	0.007817	19.4	0.002951
4.5	0.319558	9.5	0.035112	14.5	0.007636	19.5	0.002903
4.6	0.306061	9.6	0.033773	14.6	0.007460	19.6	0.002857
4.7	0.293028	9.7	0.032497	14.7	0.007289	19.7	0.002811
4.8	0.280454	9.8	0.031281	14.8	0.007125	19.8	0.002767
4.9	0.268334	9.9	0.030121	14.9	0.006965	19.9	0.002723
5.0	0.256664	10.0	0.029015	15.0	0.006810	20.0	0.002680

$\mathbf{K}(x_u, x_l)$ is the heat capacity function of a continuum spanning the range (of nondimensionalized frequencies) from x_l to x_u; $\mathbf{E}(x_E)$ is the heat capacity function of an Einstein oscillator. These functions are given in Tables 5, 6, and 2 respectively. At low temperatures the dispersed acoustic function has the same dependence on the third power of the temperature as the Debye function (see Worked Problem 4 in Section 7).

Table 6a. The continuum heat capacity function (part 1 of 3).

x_u \ x_ℓ	0.1	0.2	0.3	0.4	0.5	0.6	0.7	0.8	0.9	1.0
0.5	0.991452	0.989251	0.986507	0.983228						
1.0	0.970066	0.966567	0.962543	0.958002	0.952957	0.947420	0.941407	0.934932	0.928014	
1.5	0.937303	0.932629	0.927454	0.921787	0.915644	0.909037	0.901984	0.894502	0.886611	0.878330
2.0	0.895655	0.889966	0.883803	0.877179	0.870109	0.862610	0.854697	0.846392	0.837715	0.828685
2.5	0.847944	0.841417	0.834447	0.827051	0.819242	0.811038	0.802459	0.793524	0.784254	0.774670
3.0	0.796922	0.789739	0.782147	0.774160	0.765798	0.757077	0.748017	0.738637	0.728961	0.719008
3.5	0.744989	0.737320	0.729275	0.720872	0.712127	0.703059	0.693687	0.684032	0.674117	0.663961
4.0	0.694023	0.686023	0.677679	0.669009	0.660031	0.650765	0.641228	0.631443	0.621430	0.611211
4.5	0.645361	0.637159	0.628645	0.619837	0.610752	0.601410	0.591829	0.582031	0.572036	0.561865
5.0	0.599836	0.591539	0.582961	0.574117	0.565025	0.555705	0.546176	0.536458	0.526570	0.516534
5.5	0.557876	0.549571	0.541010	0.532211	0.523190	0.513968	0.504562	0.494991	0.485277	0.475438
6.0	0.519609	0.511360	0.502880	0.494185	0.485293	0.476223	0.466992	0.457619	0.448124	0.438527
6.5	0.484955	0.476810	0.468456	0.459910	0.451188	0.442308	0.433288	0.424147	0.414902	0.405573
7.0	0.453706	0.445701	0.437506	0.429139	0.420614	0.411950	0.403164	0.394274	0.385298	0.376252
7.5	0.425589	0.417747	0.409733	0.401564	0.393254	0.384822	0.376283	0.367655	0.358955	0.350200
8.0	0.400301	0.392637	0.384818	0.376857	0.368773	0.360579	0.352293	0.343931	0.335510	0.327045
8.5	0.377539	0.370062	0.362444	0.354699	0.346843	0.338890	0.330857	0.322760	0.314614	0.306435
9.0	0.357014	0.349730	0.342316	0.334787	0.327158	0.319445	0.311661	0.303824	0.295947	0.288046
9.5	0.338464	0.331372	0.324161	0.316847	0.309443	0.301963	0.294423	0.286838	0.279222	0.271589
10.0	0.321652	0.314750	0.307740	0.300635	0.293450	0.286198	0.278894	0.271552	0.264186	0.256810
10.5	0.306368	0.299652	0.292838	0.285937	0.278964	0.271932	0.264856	0.257747	0.250621	0.243491
11.0	0.292428	0.285895	0.279270	0.272567	0.265799	0.258979	0.252120	0.245235	0.238338	0.231441
11.5	0.279675	0.273318	0.266877	0.260364	0.253792	0.247175	0.240524	0.233853	0.227174	0.220499
12.0	0.267970	0.261783	0.255518	0.249189	0.242805	0.236382	0.229930	0.223462	0.216990	0.210526
12.5	0.257194	0.251171	0.245076	0.238921	0.232719	0.226481	0.220218	0.213943	0.207668	0.201404
13.0	0.247244	0.241378	0.235446	0.229459	0.223429	0.217367	0.211285	0.205194	0.199106	0.193032
13.5	0.238030	0.232315	0.226539	0.220713	0.214848	0.208954	0.203044	0.197128	0.191218	0.185323
14.0	0.229475	0.223906	0.218279	0.212606	0.206898	0.201165	0.195418	0.189669	0.183927	0.178203
14.5	0.221512	0.216081	0.210598	0.205072	0.199513	0.193934	0.188343	0.182751	0.177170	0.171608
15.0	0.214081	0.208784	0.203438	0.198051	0.192636	0.187203	0.181760	0.176320	0.170890	0.165482
15.5	0.207132	0.201963	0.196747	0.191495	0.186217	0.180923	0.175622	0.170324	0.165040	0.159778
16.0	0.200620	0.195572	0.190483	0.185359	0.180211	0.175050	0.169883	0.164723	0.159576	0.154453
16.5	0.194503	0.189573	0.184604	0.179603	0.174580	0.169546	0.164508	0.159478	0.154462	0.149471
17.0	0.188749	0.183931	0.179077	0.174193	0.169290	0.164377	0.159462	0.154556	0.149665	0.144801
17.5	0.183325	0.178616	0.173871	0.169100	0.164311	0.159514	0.154716	0.149929	0.145158	0.140413
18.0	0.178204	0.173598	0.168960	0.164296	0.159616	0.154930	0.150245	0.145570	0.140914	0.136284
18.5	0.173361	0.168855	0.164318	0.159757	0.155183	0.150602	0.146025	0.141458	0.136910	0.132390
19.0	0.168774	0.164364	0.159924	0.155463	0.150989	0.146510	0.142035	0.137572	0.133129	0.128713
19.5	0.164425	0.160106	0.155759	0.151393	0.147015	0.142634	0.138258	0.133894	0.129550	0.125234
20.0	0.160293	0.156062	0.151806	0.147531	0.143246	0.138958	0.134676	0.130407	0.126159	0.121939
20.5	0.156364	0.152219	0.148049	0.143861	0.139665	0.135467	0.131275	0.127098	0.122941	0.118812
21.0	0.152624	0.148560	0.144473	0.140370	0.136259	0.132147	0.128042	0.123952	0.119883	0.115843
21.5	0.149058	0.145073	0.141066	0.137044	0.133015	0.128987	0.124965	0.120959	0.116974	0.113017
22.0	0.145656	0.141747	0.137817	0.133873	0.129923	0.125974	0.122032	0.118107	0.114202	0.110327
22.5	0.142406	0.138570	0.134714	0.130845	0.126971	0.123099	0.119234	0.115386	0.111560	0.107762
23.0	0.139298	0.135532	0.131748	0.127952	0.124151	0.120352	0.116562	0.112788	0.109036	0.105314
23.5	0.136324	0.132626	0.128910	0.125184	0.121453	0.117725	0.114007	0.110305	0.105625	0.102974
24.0	0.133474	0.129842	0.126193	0.122533	0.118871	0.115211	0.111562	0.107929	0.104318	0.100736
24.5	0.130742	0.127173	0.123588	0.119993	0.116396	0.112803	0.109219	0.105653	0.102109	0.098594
25.0	0.128119	0.124611	0.121089	0.117557	0.114023	0.110493	0.106974	0.103471	0.099992	0.096541
25.5	0.125601	0.122152	0.118689	0.115217	0.111744	0.108276	0.104818	0.101378	0.097960	0.094572
26.0	0.123180	0.119788	0.116383	0.112970	0.109556	0.106147	0.102749	0.099368	0.096010	0.092681
26.5	0.120851	0.117514	0.114165	0.110808	0.107451	0.104100	0.100759	0.097436	0.094136	0.090865
27.0	0.118609	0.115326	0.112031	0.108728	0.105426	0.102130	0.098845	0.095578	0.092334	0.089118
27.5	0.116450	0.113218	0.109975	0.106725	0.103476	0.100234	0.097003	0.093790	0.090599	0.087437
28.0	0.114367	0.111185	0.107993	0.104795	0.101597	0.098407	0.095228	0.092067	0.088929	0.085819
28.5	0.112359	0.109225	0.106082	0.102933	0.099785	0.096645	0.093516	0.090406	0.087318	0.084259
29.0	0.110420	0.107333	0.104237	0.101136	0.098037	0.094945	0.091865	0.088804	0.085765	0.082755
29.5	0.108547	0.105505	0.102455	0.099401	0.096349	0.093304	0.090271	0.087258	0.084266	0.081303
30.0	0.106736	0.103739	0.100733	0.097724	0.094717	0.091718	0.088732	0.085764	0.082818	0.079901

In terms of these functions the molar heat capacity, normalized to the monatomic equivalent, is

$$C_V{}^* = \frac{3N_A k}{nZ} \sum_{1}^{3} \mathbf{S}(x_i) + 3N_A k\left(1 - \frac{3}{3s} - q\right) \mathbf{K}\binom{x_u}{x_i} + 3N_A k\, q\mathbf{E}(x_E). \qquad (60)$$

It is important to emphasize that the heat capacity can be calculated from tables of these functions just as it can be calculated from tables of the Debye or Einstein functions. An example of the use of the tables is given in Worked Problem 5 in Section 7.

Table 6b. The continuum heat capacity function (part 2 of 3).

x_u \ x_ℓ	1.1	1.2	1.3	1.4	1.5	1.6	1.7	1.8	1.9	2.0
0.5										
1.0										
1.5	0.869681	0.850685	0.851365	0.841745						
2.0	0.819325	0.809657	0.799705	0.789491	0.779041	0.768377	0.757523	0.746504	0.735341	
2.5	0.764794	0.754651	0.744261	0.733650	0.722840	0.711855	0.700721	0.689458	0.678090	0.666639
3.0	0.708802	0.698364	0.687720	0.676891	0.665901	0.654773	0.643529	0.632193	0.620786	0.609331
3.5	0.653587	0.643018	0.632277	0.621386	0.610368	0.599246	0.588041	0.576775	0.565469	0.554145
4.0	0.600807	0.590240	0.579534	0.568708	0.557787	0.546791	0.535741	0.524658	0.513562	0.502473
4.5	0.551540	0.541082	0.530512	0.519851	0.509121	0.498343	0.487536	0.476720	0.465915	0.455137
5.0	0.506370	0.496099	0.485741	0.475318	0.464849	0.454353	0.443851	0.433360	0.422898	0.412484
5.5	0.465495	0.455468	0.445376	0.435240	0.425077	0.414907	0.404749	0.394618	0.384534	0.374511
6.0	0.428845	0.419099	0.409307	0.399488	0.389660	0.379841	0.370048	0.360297	0.350605	0.340987
6.5	0.396177	0.386734	0.377261	0.367776	0.358297	0.348840	0.339421	0.330056	0.320761	0.311548
7.0	0.367156	0.358027	0.348881	0.339736	0.330609	0.321515	0.312469	0.303486	0.294581	0.285766
7.5	0.341407	0.332594	0.323776	0.314969	0.306189	0.297452	0.288771	0.280161	0.271634	0.263203
8.0	0.318554	0.310053	0.301556	0.293080	0.284639	0.275247	0.267919	0.259667	0.251503	0.243439
8.5	0.298239	0.290042	0.281857	0.273700	0.265585	0.257526	0.249534	0.241624	0.233805	0.226089
9.0	0.280136	0.272232	0.264347	0.256497	0.248594	0.240950	0.233279	0.225692	0.218199	0.210812
9.5	0.263954	0.256331	0.248733	0.241175	0.233668	0.226224	0.218856	0.211574	0.204389	0.197309
10.0	0.249438	0.242083	0.234759	0.227477	0.220250	0.213090	0.206008	0.199013	0.192116	0.185326
10.5	0.236369	0.229269	0.222203	0.215184	0.208222	0.201329	0.194515	0.187791	0.181164	0.174644
11.0	0.224557	0.217699	0.210878	0.204106	0.197394	0.190753	0.184192	0.177720	0.171347	0.165081
11.5	0.213841	0.207212	0.200622	0.194084	0.187607	0.181203	0.174879	0.168545	0.162509	0.156479
12.0	0.204082	0.197669	0.191299	0.184981	0.178726	0.172543	0.166443	0.160432	0.154518	0.148710
12.5	0.195163	0.188955	0.182791	0.175681	0.170635	0.164662	0.158770	0.152969	0.147264	0.141663
13.0	0.186983	0.180968	0.175000	0.169086	0.163237	0.157461	0.151766	0.146161	0.140652	0.135246
13.5	0.179456	0.173625	0.167840	0.162112	0.156448	0.150858	0.145349	0.139928	0.134603	0.129379
14.0	0.172508	0.166850	0.161240	0.155687	0.150198	0.144783	0.139448	0.134201	0.129049	0.123997
14.5	0.166076	0.160583	0.155138	0.149750	0.144426	0.139177	0.134006	0.128924	0.123933	0.119042
15.0	0.160105	0.154768	0.149479	0.144247	0.139081	0.133937	0.128972	0.124043	0.119206	0.114467
15.5	0.154548	0.149358	0.144218	0.139134	0.134115	0.129169	0.124301	0.119518	0.114826	0.110229
16.0	0.149363	0.144313	0.139314	0.134371	0.129492	0.124685	0.119956	0.115311	0.110755	0.106294
16.5	0.144514	0.139598	0.134732	0.129922	0.125176	0.120502	0.115904	0.111390	0.106963	0.102629
17.0	0.139970	0.135181	0.130441	0.125758	0.121139	0.116590	0.112117	0.107726	0.103422	0.099209
17.5	0.135703	0.131034	0.126416	0.121853	0.117354	0.112924	0.108569	0.104296	0.100107	0.096009
18.0	0.131688	0.127135	0.122631	0.118183	0.113798	0.109481	0.105239	0.101077	0.096999	0.093009
18.5	0.127904	0.123460	0.119056	0.114727	0.110451	0.106242	0.102107	0.098051	0.094077	0.090191
19.0	0.124332	0.119993	0.115703	0.111468	0.107295	0.103190	0.099156	0.095201	0.091326	0.087538
19.5	0.120953	0.116714	0.112525	0.108389	0.104315	0.100307	0.096371	0.092511	0.088732	0.085037
20.0	0.117754	0.113610	0.109516	0.105476	0.101496	0.097582	0.093738	0.089970	0.086281	0.082675
20.5	0.114719	0.110667	0.106664	0.102715	0.098825	0.095000	0.091245	0.087564	0.083962	0.080441
21.0	0.111837	0.107873	0.103957	0.100095	0.096291	0.092552	0.088881	0.085284	0.081764	0.078324
21.5	0.109096	0.105216	0.101385	0.097605	0.093884	0.090227	0.086637	0.083120	0.079678	0.076316
22.0	0.106487	0.102688	0.098936	0.095236	0.091595	0.088016	0.084503	0.081063	0.077696	0.074408
22.5	0.103999	0.100278	0.096603	0.092980	0.089414	0.085910	0.082472	0.079105	0.075810	0.072593
23.0	0.101626	0.097978	0.094378	0.090828	0.087335	0.083903	0.080536	0.077239	0.074014	0.070865
23.5	0.099358	0.095782	0.092252	0.088773	0.085350	0.081988	0.078690	0.075460	0.072301	0.069217
24.0	0.097189	0.093682	0.090221	0.086810	0.083454	0.080158	0.076925	0.073760	0.070665	0.067644
24.5	0.095113	0.091672	0.088277	0.084931	0.081640	0.078408	0.075239	0.072136	0.069102	0.066141
25.0	0.093124	0.089747	0.086415	0.083132	0.079904	0.076733	0.073624	0.070581	0.067606	0.064703
25.5	0.091217	0.087901	0.084630	0.081408	0.078239	0.075128	0.072078	0.069092	0.066174	0.063326
26.0	0.089386	0.086129	0.082917	0.079754	0.076643	0.073589	0.070595	0.067665	0.064801	0.062007
26.5	0.087627	0.084428	0.081273	0.078165	0.075110	0.072111	0.069171	0.066295	0.063484	0.060742
27.0	0.085936	0.082792	0.079692	0.076639	0.073637	0.070692	0.067804	0.064980	0.062219	0.059527
27.5	0.084309	0.081218	0.078171	0.075171	0.072221	0.069327	0.066490	0.063715	0.061004	0.058359
28.0	0.082742	0.079703	0.076707	0.073758	0.070859	0.068014	0.065226	0.062499	0.059835	0.057237
28.5	0.081233	0.078244	0.075297	0.072397	0.069546	0.066749	0.064009	0.061329	0.058710	0.056156
29.0	0.079777	0.076836	0.073939	0.071085	0.068281	0.065531	0.062836	0.060201	0.057626	0.055116
29.5	0.078372	0.075479	0.072627	0.069820	0.067062	0.064356	0.061705	0.059114	0.056582	0.054113
30.0	0.077016	0.074168	0.071361	0.068598	0.065884	0.063222	0.060615	0.058065	0.055574	0.053146

6.2 Heat capacities and entropies calculated from the spectra

Values of specific heat are accurately predicted by this model, as shown in the six parts of Figure 20 (column a in each of these parts shows the heat capacity; column b shows the calorimetric Debye temperature). Predicted entropies are given in Table 1. A comparison of data and theory on 30 minerals showed that the entropy at 298 K was given to within 4% by the models; in many cases the uncertainty in prediction arises from the lack of data on thermal expansion needed to calculate C_P from C_V given by the model. A discussion of the implications of the model and causes for deviations from measured data for each mineral is given in Kieffer (1979c, 1980) and is not repeated here.

The model provides insight into the observed calorimetric trends shown in Figure 6, because it demonstrates quantitatively how systematic variations in various groups of vibrational modes influence the low- and

Table 6c. The continuum heat capacity function (part 3 of 3).

x_u \ x_ℓ	2.1	2.2	2.3	2.4	2.5	2.6	2.7	2.8	2.9	3.0
0.5										
1.0										
1.5										
2.0										
2.5	0.655128	0.643578	0.632008	0.620439						
3.0	0.597847	0.586356	0.574876	0.563425	0.552023	0.540684	0.529425	0.518261	0.507204	
3.5	0.542820	0.531516	0.520249	0.509037	0.497897	0.486844	0.475892	0.465054	0.454343	0.443771
4.0	0.491410	0.480390	0.469429	0.458545	0.447752	0.437064	0.426495	0.416057	0.405761	0.395616
4.5	0.444407	0.433738	0.423148	0.412652	0.402262	0.391993	0.381856	0.371862	0.362021	0.352342
5.0	0.402132	0.391859	0.381679	0.371606	0.361653	0.351831	0.342151	0.332624	0.323257	0.314060
5.5	0.364565	0.354710	0.344959	0.335326	0.325823	0.316459	0.307244	0.298188	0.289299	0.280583
6.0	0.331456	0.322027	0.312711	0.303520	0.294465	0.285556	0.276802	0.268209	0.259787	0.251539
6.5	0.302431	0.293423	0.284535	0.275778	0.267161	0.258694	0.250385	0.242239	0.234265	0.226467
7.0	0.277053	0.268455	0.259981	0.251642	0.243446	0.235403	0.227518	0.219798	0.212248	0.204874
7.5	0.254879	0.246673	0.238595	0.230654	0.222859	0.215216	0.207732	0.200412	0.193262	0.186285
8.0	0.235485	0.227653	0.219950	0.212385	0.204966	0.197699	0.190590	0.183645	0.176867	0.170260
8.5	0.218486	0.211005	0.203655	0.196443	0.189376	0.182461	0.175702	0.169105	0.162673	0.156409
9.0	0.203538	0.196388	0.189368	0.182486	0.175748	0.169160	0.162727	0.156452	0.150339	0.144392
9.5	0.190345	0.183503	0.176792	0.170217	0.163786	0.157502	0.151370	0.145394	0.139577	0.133921
10.0	0.178651	0.172098	0.165674	0.159386	0.153238	0.147237	0.141384	0.135685	0.130141	0.124754
10.5	0.168239	0.161956	0.155800	0.149778	0.143895	0.138154	0.132561	0.127117	0.121825	0.116686
11.0	0.158928	0.152895	0.146989	0.141215	0.135577	0.130080	0.124726	0.119518	0.114459	0.109549
11.5	0.150562	0.144764	0.139091	0.133547	0.128137	0.122865	0.117733	0.112744	0.107900	0.103203
12.0	0.143013	0.137434	0.131978	0.126649	0.121451	0.116388	0.111462	0.106677	0.102032	0.097530
12.5	0.136173	0.130798	0.125544	0.120415	0.115414	0.110546	0.105812	0.101215	0.096756	0.092435
13.0	0.129948	0.124754	0.119699	0.114757	0.109941	0.105254	0.100699	0.096277	0.091989	0.087837
13.5	0.124262	0.119258	0.114369	0.109602	0.104958	0.100441	0.096051	0.091793	0.087665	0.083669
14.0	0.119050	0.114213	0.109491	0.104887	0.100403	0.096044	0.091810	0.087704	0.083725	0.079875
14.5	0.114255	0.109576	0.105009	0.100558	0.096226	0.092014	0.087926	0.083961	0.080122	0.076408
15.0	0.109830	0.105299	0.100878	0.096571	0.092380	0.088307	0.084355	0.080524	0.076815	0.073228
15.5	0.105734	0.101342	0.097059	0.092887	0.088829	0.084887	0.081062	0.077356	0.073769	0.070301
16.0	0.101932	0.097672	0.093518	0.089473	0.085540	0.081721	0.078016	0.074427	0.070955	0.067599
16.5	0.098393	0.094257	0.090226	0.086302	0.082486	0.078782	0.075190	0.071712	0.068347	0.065096
17.0	0.095092	0.091073	0.087158	0.083347	0.079642	0.076047	0.072562	0.069187	0.065924	0.062772
17.5	0.092005	0.088098	0.084291	0.080587	0.076988	0.073496	0.070110	0.066834	0.063666	0.060608
18.0	0.089112	0.085310	0.081607	0.078004	0.074505	0.071110	0.067820	0.064636	0.061559	0.058587
18.5	0.086395	0.082693	0.079088	0.075582	0.072177	0.068873	0.065673	0.062578	0.059586	0.056698
19.0	0.083839	0.080232	0.076720	0.073305	0.069989	0.066774	0.063659	0.060646	0.057735	0.054926
19.5	0.081430	0.077913	0.074490	0.071162	0.067931	0.064798	0.061764	0.058831	0.055996	0.053262
20.0	0.079155	0.075725	0.072386	0.069140	0.065990	0.062936	0.059979	0.057120	0.054359	0.051695
20.5	0.077004	0.073656	0.070397	0.067230	0.064157	0.061178	0.058295	0.055507	0.052815	0.050218
21.0	0.074967	0.071697	0.068515	0.065423	0.062423	0.059516	0.056702	0.053982	0.051356	0.048823
21.5	0.073035	0.069840	0.066731	0.063711	0.060781	0.057941	0.055194	0.052539	0.049975	0.047504
22.0	0.071200	0.068076	0.065037	0.062086	0.059222	0.056448	0.053764	0.051171	0.048667	0.046254
22.5	0.069455	0.066399	0.063428	0.060541	0.057742	0.055030	0.052406	0.049872	0.047425	0.045068
23.0	0.067794	0.064803	0.061896	0.059072	0.056333	0.053681	0.051116	0.048637	0.046246	0.043941
23.5	0.066210	0.063282	0.060436	0.057672	0.054992	0.052397	0.049887	0.047462	0.045123	0.042869
24.0	0.064698	0.061831	0.059043	0.056337	0.053713	0.051173	0.048716	0.046343	0.044054	0.041848
24.5	0.063254	0.060444	0.057714	0.055062	0.052492	0.050004	0.047598	0.045275	0.043034	0.040875
25.0	0.061873	0.059119	0.056442	0.053844	0.051326	0.048888	0.046531	0.044255	0.042060	0.039946
25.5	0.060551	0.057850	0.055226	0.052679	0.050210	0.047821	0.045511	0.043281	0.041130	0.039058
26.0	0.059284	0.056635	0.054061	0.051563	0.049142	0.046799	0.044534	0.042348	0.040240	0.038209
26.5	0.058069	0.055469	0.052944	0.050493	0.048118	0.045820	0.043598	0.041454	0.039387	0.037396
27.0	0.056903	0.054351	0.051872	0.049466	0.047136	0.044880	0.042701	0.040598	0.038570	0.036617
27.5	0.055783	0.053277	0.050842	0.048481	0.046193	0.043979	0.041840	0.039776	0.037786	0.035870
28.0	0.054706	0.052244	0.049853	0.047533	0.045287	0.043113	0.041013	0.038986	0.037033	0.035152
28.5	0.053669	0.051250	0.048901	0.046623	0.044416	0.042281	0.040218	0.038228	0.036309	0.034463
29.0	0.052671	0.050294	0.047985	0.045746	0.043577	0.041480	0.039453	0.037498	0.035613	0.033800
29.5	0.051710	0.049372	0.047103	0.044901	0.042770	0.040708	0.038717	0.036795	0.034944	0.033162
30.0	0.050783	0.048484	0.046252	0.044088	0.041992	0.039965	0.038007	0.036119	0.034299	0.032547

high-temperature thermodynamic functions (column c in each figure shows the per cent contribution of the three major groups of vibrational modes -- acoustic, optic continuum, and Einstein oscillators -- to the heat capacity; column d in each figure shows the assumed vibrational spectra and observed Raman and infrared bands).

This model demonstrates the following ideas quantitatively:

(1) Trends at high temperatures are determined by the position and relative numbers of the highest-frequency modes (antisymmetric (Si,Al)-O stretching modes for most silicates and Al-O, O-H stretching or Si-O modes for some other minerals studied). The position of these modes varies systematically with degree of polymerization of tetrahedra, and therefore high-temperature calorimetric behavior is relatively systematic as a function of crystal structure and mineral composition.

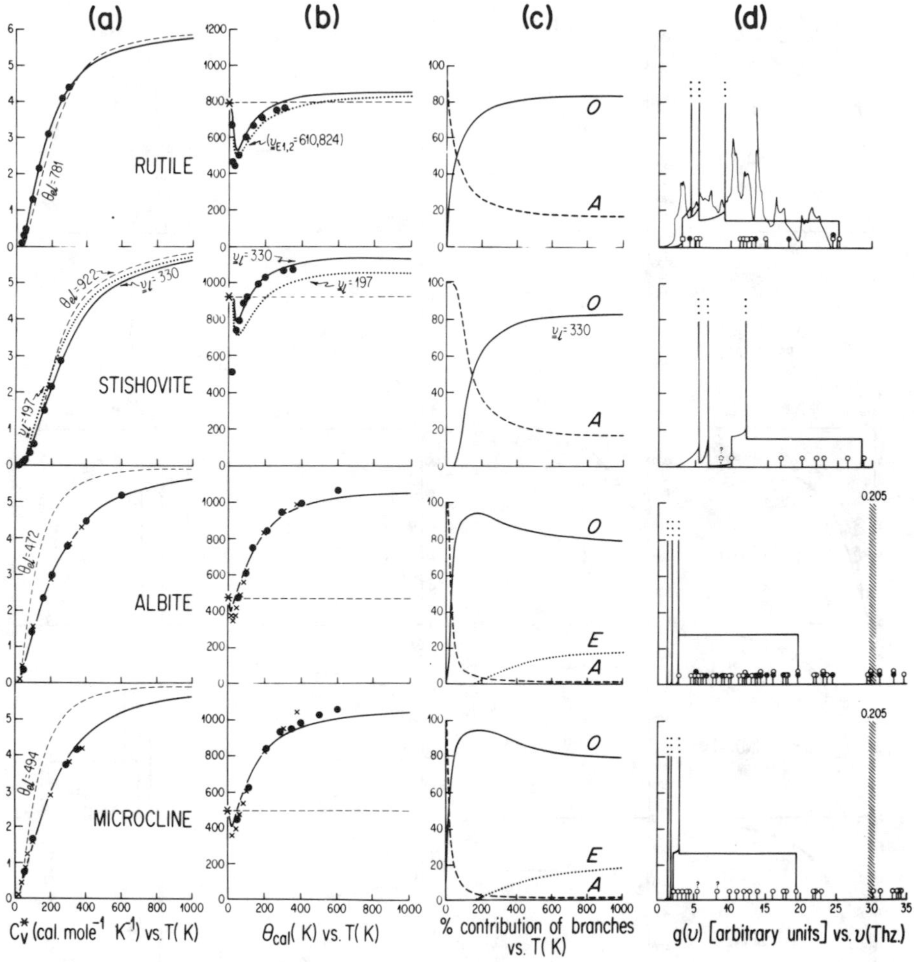

Figure 20. Data (points) and model (lines) values for halite, brucite, corundum, and spinel. (a) Atom-mole heat capacities as a function of temperature. (b) Calorimetric Debye temperatures as a function of temperature. (c) Percent contribution to heat capacity of acoustic modes (A), optic continuum (O), and Einstein oscillator(s) (E) as a function of temperature. (d) Model spectra compared to lattice dynamics calculations (for NaCl) or Raman (solid circles), infrared (open circles), or INS (squares) data. The vertical scale shown in (d) for Figures 20-25 is arbitrary. Frequencies are given as ν (tetrahertz) and as w (cm^{-1}). The singularities in the acoustic branches at ν_1, ν_2, and ν_3 are indicated by three dots on the acoustic branches. Parameters used in the models are given in Kieffer (1979c, 1980). References to heat capacity data are given in Figure 4 of Kieffer (1979a). References to spectral data are given in Kieffer (1979b).

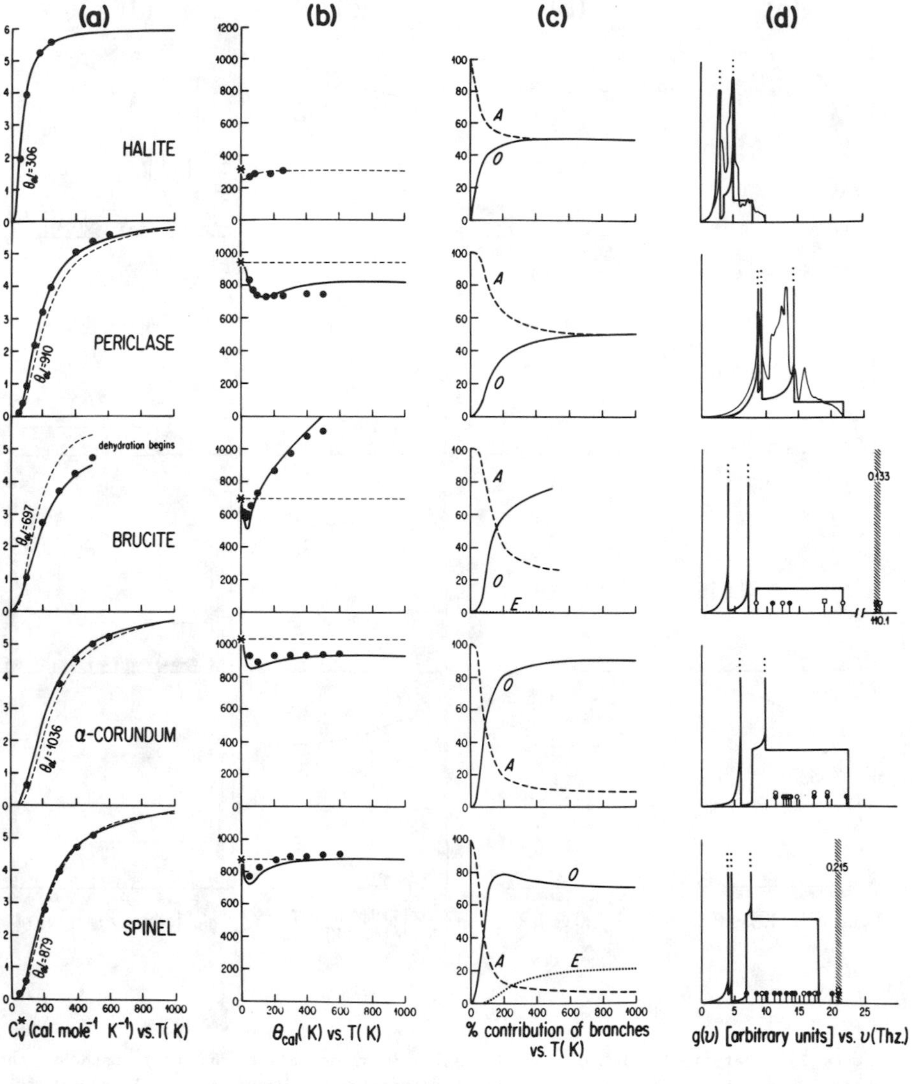

Figure 20, continued.

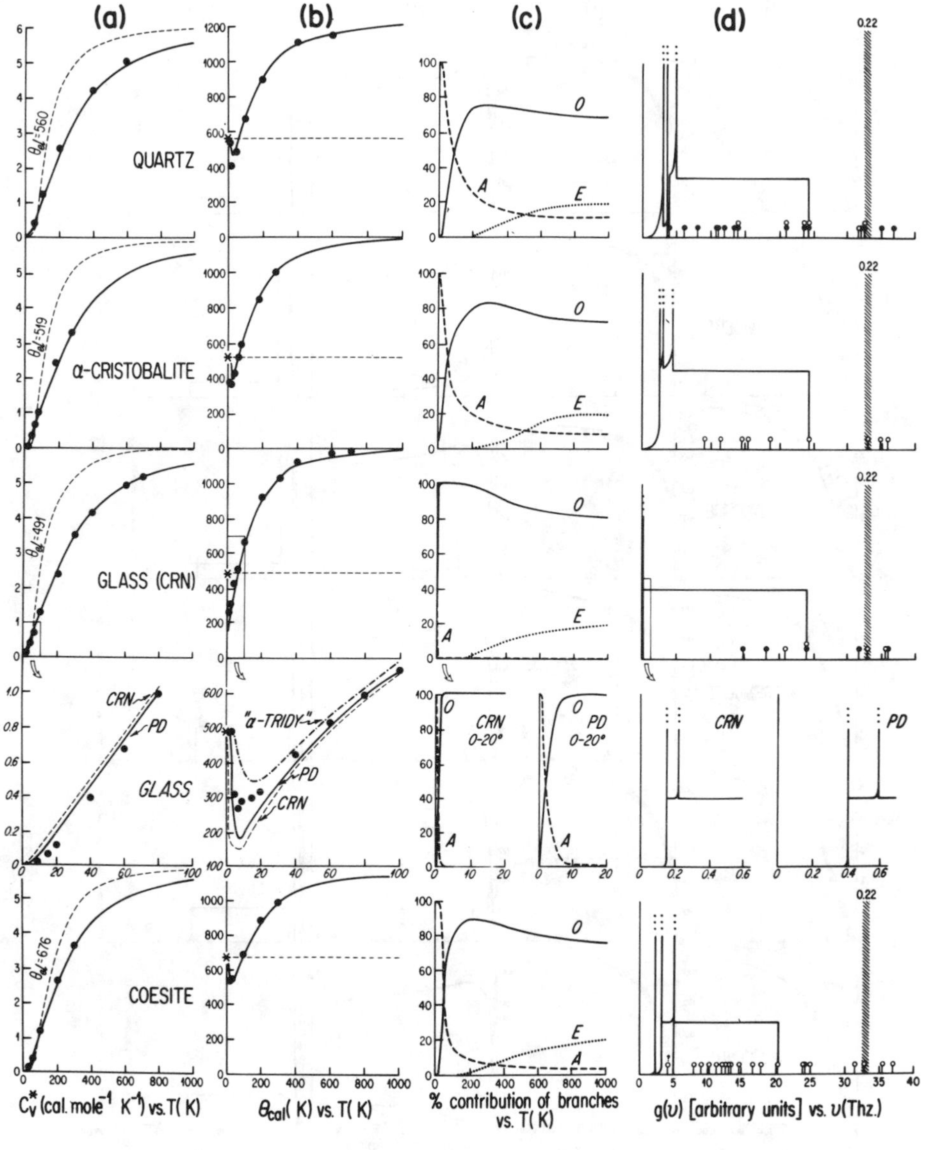

Figure 20, continued.

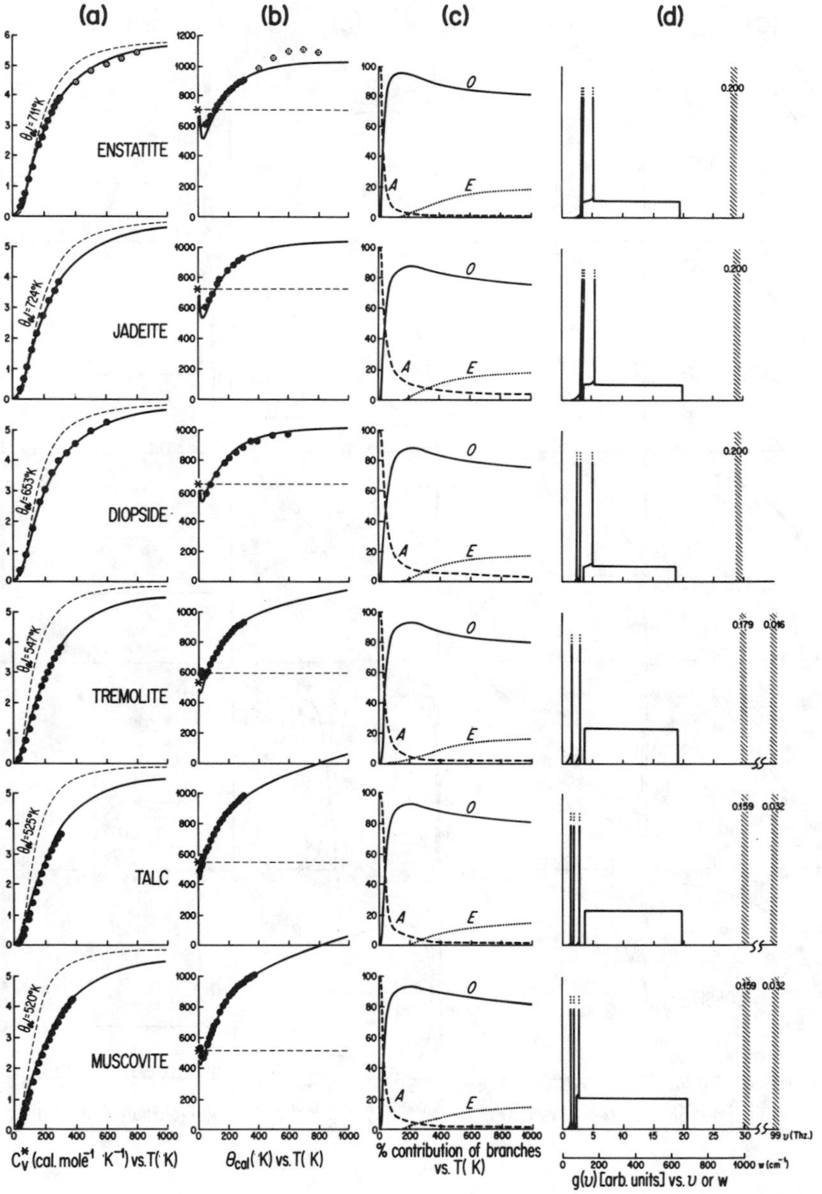

Figure 20, continued.

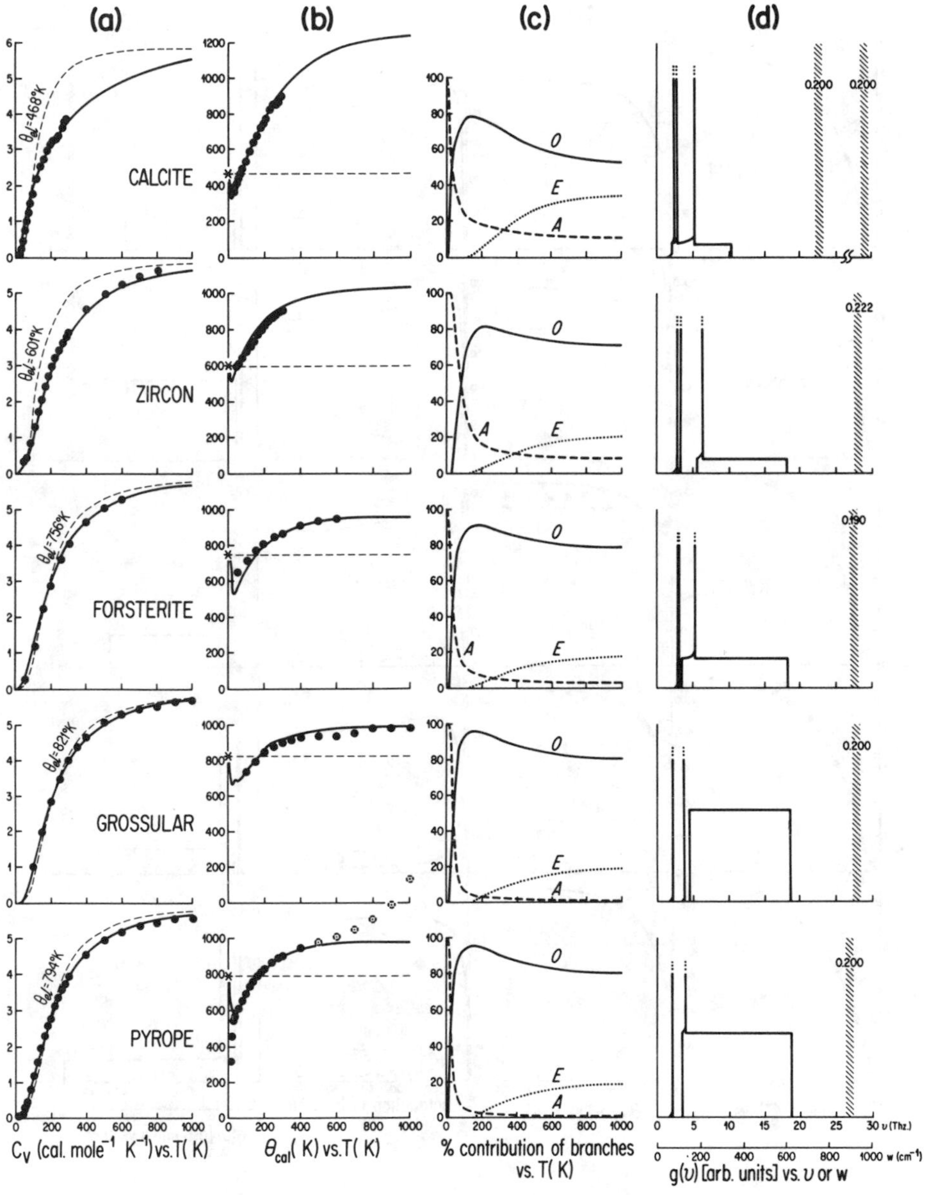

Figure 20, continued.

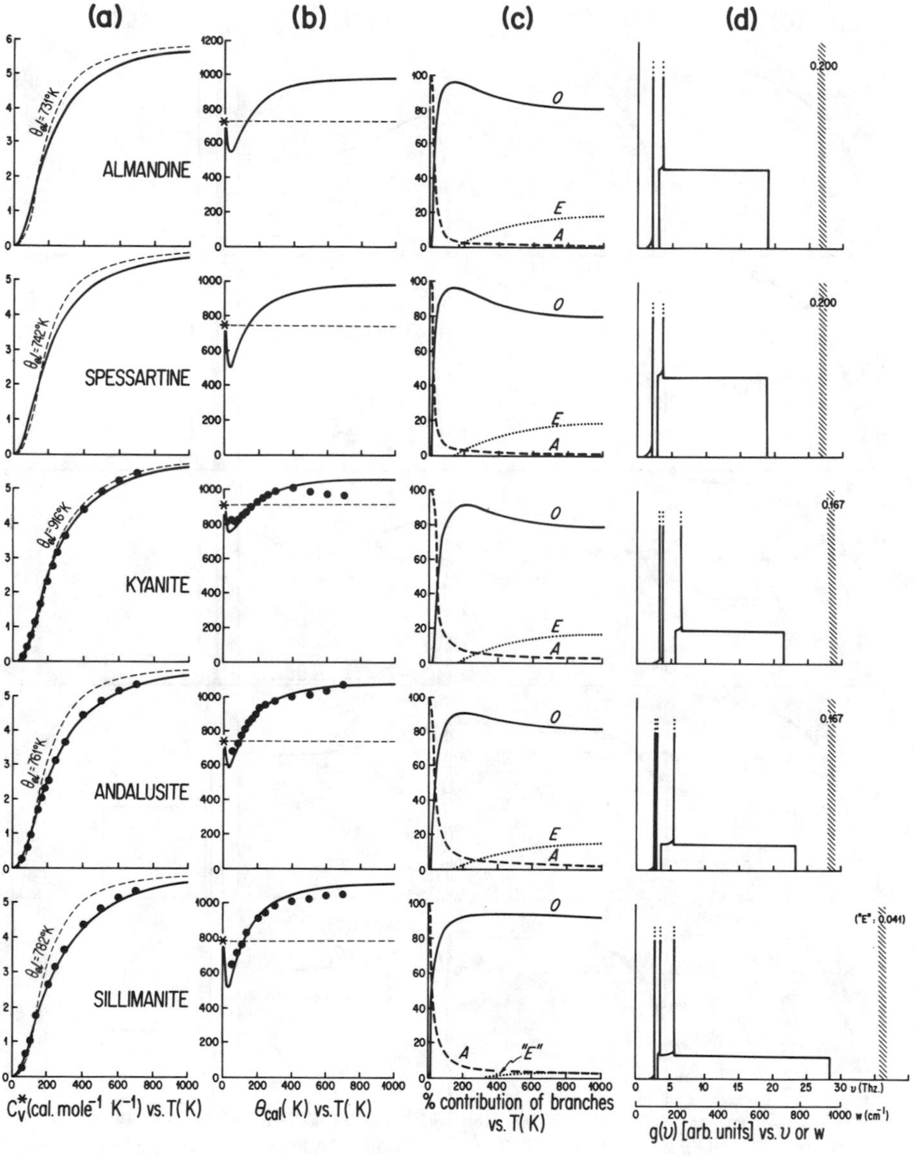

Figure 20, concluded.

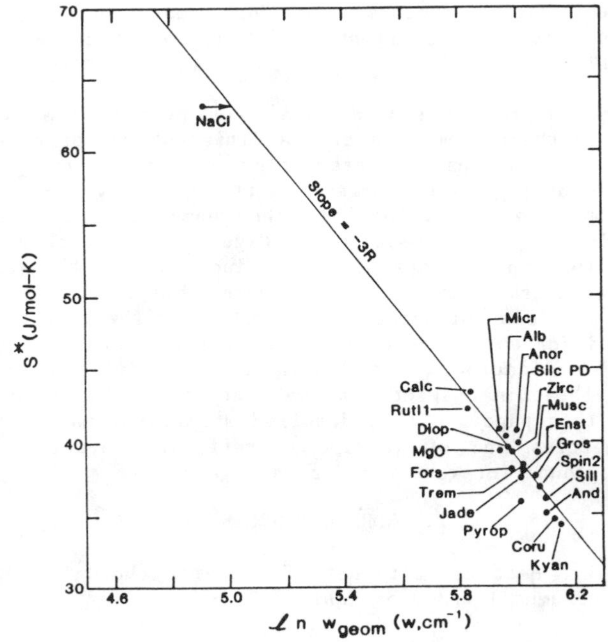

Figure 21. Harmonic entropy at 1000 K as a function of geometric mean spectral frequency. Minerals were selected for which the experimental values of entropy are well-known and for which the vibrational spectra, approximated as in Figure 7, are felt by the author to be well-constrained. The entropy is the measured value minus the estimated anharmonic contribution as listed in Kieffer (1980, Table 3). The geometric mean frequency was calculated from such spectra using the values given in Table 1 of Kieffer (1980). The left value for NaCl is from this method; the right value is that determined by numerous lattice dynamics calculations (e.g., as referenced in Wallace, 1972, p. 429). Values of S^* at 298, 700 and 1000 K are given in Table 1. The light line has the slope-3R predicted by the high-temperature limiting equation for entropy of an Einstein oscillator. Much of the scatter in the data may be due to the uncertainty in subtracting the anharmonic contribution out from the experimentally determined total entropy.

(2) Trends at low temperatures are determined primarily by the position of the lowest-frequency optic modes and secondarily by the magnitude and relative proportion of acoustic modes and their relationship to the optic mode frequencies. Because the position of the lowest-frequency optic modes is sensitive to the size and coordination of cations and various polyhedra in the minerals, low-temperature calorimetric trends are not nearly as systematic as high-temperature trends, and must be more cautiously interpreted and predicted in terms of individual mineral properties.

(3) The model is sufficiently accurate that phase equilibra involving rather small energy changes can be predicted. It is also useful for examining possible inconsistencies among various data sets. If additional approximations are made, the model can be used for calculating

isotopic fractionation factors, and for prediction of thermodynamic properties as a function of pressure. These applications are detailed in Kieffer (1982).

Although the reader should beware of placing too much physical significance on the assumed spectra, a different way of emphasizing the relation between the spectral properties and the thermodynamics is to calculate the average frequencies of the spectra. High-temperature entropies should be proportional to the geometric mean frequency (see Wallace, 1972; Jeanloz, this volume). Figure 21 shows this indeed to be the case, in that the entropy at 1000 K increases as the mean geometric frequency decreases. However, the figure also emphasizes one of the fundamental problems of mineralogy and petrology: phase relations amongst the silicates are determined by very small differences in energies and, by implication, by rather small differences in properties of the lattice vibrational spectra as well as by non-harmonic effects not discussed in this paper. Thus, detailed studies of macroscopic thermodynamic properties and of lattice vibrational characteristics should provide rich ground for research for many years.

7. WORKED PROBLEM SET

Problem 1. Heat capacity data for quartz (Lord and Morrow, 1957) are given in columns 1 and 2 of Table 7.

Table 7. Heat Capacity of α-Quartz.

T (K)	C_P (cal/mole-K)	C_V (cal/mole-K)	$C_V^*/3R$	x	$\theta_{cal} = xT$ (K)
10	0.010	0.010	0.0006	73	730
20	0.162	0.162	0.0091	29.5	590
30	0.507	0.507	0.0283	20.2	606
40	0.938	0.938	0.0523	16.4	657
50	1.392	1.392	0.078	9.9	495
60	1.857	1.856	0.104	8.9	534
70	2.328	2.326	0.130	8.2	574
80	2.802	2.799	0.157	7.6	608
100	3.744	3.639	0.204	6.8	680
120	4.660	4.652	0.260	6.1	732
140	5.534	5.522	0.309	5.5	770
160	6.347	6.330	0.354	5.2	832
180	7.101	7.078	0.396	4.8	864
200	7.803	7.774	0.435	4.5	900
220	8.453	8.416	0.471	4.2	924
240	9.061	9.017	0.504	4.0	960
260	9.637	9.585	0.536	3.8	988
280	10.189	10.137	0.567	3.6	1008
300	10.703	10.633	0.595	3.4	1020
400	12.726	12.591	0.704	2.8	1120
500	14.208	13.997	0.783	2.3	1150
600	15.452	15.097	0.844	1.9	1140
700	16.348	15.668	0.876	1.6	1120

(a) Use Table 3 in the text to obtain the calorimetric Debye temperature, $\theta_{cal}(T)$, and plot this versus T.

(b) Calculate θ_{el} and graph C_V(Debye) to compare with data. (Acoustic velocities for quartz are given in Section 5-1 of the text).

(c) Then take $\theta_{cal}(300)$ as an "average" Debye temperature, and plot C_V according to the Debye model with this value.

(d) Assume that dispersion of this single acoustic branch is given by a sine-wave instead of being linear, and use Table 5 to calculate the heat capacity. Does inclusion of this physically realistic effect improve the agreement with data?

(e) If you are ambitious, repeat part (d) including shear and longitudinal velocities (i.e., three branches instead of one); however, you should be able to simply sketch the effects on C_V and $\theta_{cal}(T)$.

Solution 1.

(a) Note that the data are in cal/mole-K; many older data are in these units. C_P must be converted to C_V (column 3 in Table 7), a correction typically important at a few hundred degrees, and, in this case, large near the α-β transition. Normalize C_V by $3nN_Ak$ (3nR) (column 4) to use Table 3 to get $x = \theta_{cal}(T)/T$ (column 5) from which θ(T) can be calculated (column 6). Table 3 does not cover low temperatures (θ/T > 11) so Equation 32 must be used. A handy form for it is

$$\left(\frac{\theta_D}{T}\right)^3 = \frac{233.781}{(C_V\ /\ 3R)}.$$

(b) Use Equation 28, which can be recast into the convenient form

$$\theta_{el} = 251.45\ \left(\frac{\text{density * atoms/formula weight}}{\text{molecular weight}}\right)^{1/3} v_m(\text{km/sec}).$$

Using a density of 2.65 g/cm^3, 3 atoms/formula weight, a molecular weight of 60, and a mean velocity of 4.425 km/sec, θ_{el} = 567 K. The heat capacity for θ_{el} is shown as the long dashed curve in Figure 1.

(c) From Table 7, $\theta_{cal}(300)$ = 1020 K. The short dashed curve on Figure 1 shows the heat capacity with θ_{cal} taken as this value.

(d) Sine wave dispersion will reduce the maximum frequency from ω_D to $\omega_{max} = \omega_D/\pi/2$ (see Kieffer 1979a, pp. 8-10). ω_D = 567/1.44 = 394 cm^{-1}; ω_{max} = 251 cm^{-1} (corresponding to ω_{max} = 361 K). For T = 50, 100, 200, 400, 800 K, x = 7.22, 3.61, 1.80, 0.90., 0.45, so from Table 5 C_V* = 0.56, 2.74, 4.82, 5.65, 5.88 cal/mol. The agreement with measured data becomes worse.

(e) This effect is discussed in Kieffer (1979a, p. 9, Fig. 6).

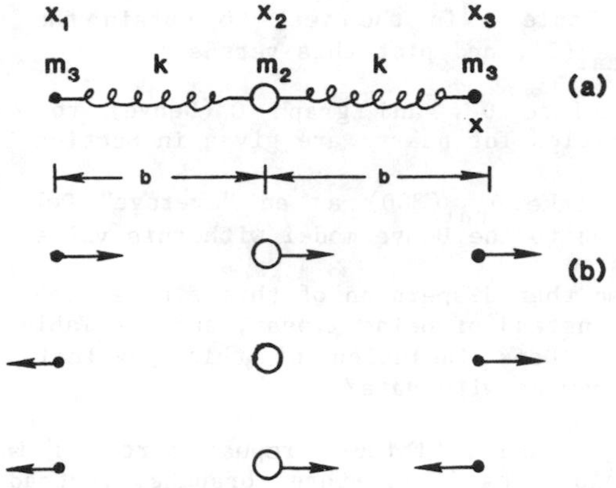

Figure 22. (a) Schematic representation of a CO_2 molecule. (b) Particle motions for the three longitudinal modes of vibration.

Problem 2. The CO_2 molecule can be likened to a system made up of a central mass m_2 connected by springs of spring constant k to two identical masses m_1 and m_3 (with $m_3 = m_1$) (Fig. 22).

(a) Set up and solve the equations for the normal modes in which the masses oscillate along the line joining their centers (the "longitudinal" modes).

(b) Putting m_3 = 16 units, m_2 = 12 units, what is the calculated ratio of the frequencies of the two "significant" modes? What are the values for an isolated CO_2 molecule?

Solution 2. Problems like this will give readers an intuitive feeling for particle vibrations, and numerous examples are given in to classical mechanics texts (from Goldstein (1950), p. 333).

(a) The equilibrium distance between the particles is b. Denote the instantaneous positions of the particles as x_1, x_2, and x_3. The potential energy of the system is

$$V = \frac{k}{2}(x_2 - x_1 - b)^2 + \frac{k}{2}(x_3 - x_2 - b)^2.$$

Introduce coordinates relative to the equilibrium positions,

$\eta_i = x_i - x_{0i}$, where $(x_{02} - x_{01} = x_{03} - x_{02} = b)$. The potential energy is then

$$V = \frac{k}{2}(\eta_2 - \eta_1)^2 + \frac{k}{2}(\eta_3 - \eta_2)^2$$

or $V = \frac{k}{2} (\eta_1{}^2 + 2\eta_2 + \eta_3 - 2\eta_1\eta_2 - 2\eta_2\eta_3)$.

The potential energy matrix has the form $\mathbf{V} = \begin{pmatrix} k & -k & 0 \\ -k & 2k & -k \\ 0 & -k & k \end{pmatrix}$.

The kinetic energy is $\mathbf{T} = \frac{m_3}{2} (\dot{\eta}_1{}^2 + \dot{\eta}_3{}^2) + \frac{m_2}{2} \eta_2{}^2$,

so the kinetic energy matrix is $\mathbf{T} = \begin{pmatrix} m_3 & 0 & 0 \\ 0 & m_2 & 0 \\ 0 & 0 & m_3 \end{pmatrix}$.

The secular equation is then easily constructed:

$$|\mathbf{V} - \omega^2 \mathbf{T}| = \begin{vmatrix} k - \omega^2 m_3 & -k & 0 \\ -k & 2k - \omega^2 m_2 & -k \\ 0 & -k & k - \omega^2 m \end{vmatrix}.$$

The determinant gives the cubic equation

$$\omega^2 (k - \omega^2 m_3) \left(k (m_2 + 2 m_3) - \omega^2 m_2 m_3\right) = 0$$

with three solutions:

$$\omega_1 = 0,\ \omega_2 = (k/m_3)^{1/2},\ \text{and}\ \omega = \left(\frac{k}{m_3} \left(1 + \frac{2m_3}{m_2}\right)\right)^{1/2}.$$

Goldstein shows how to obtain the relative particle motions; the results are shown in Figure 22b. ω_1 corresponds to rigid translation of the whole molecule. ω_2 is the classic equation for oscillation of a mass m suspended by a spring of force constant k, and each end atom moves with this frequency while the center atom remains stationary. In the third mode, the center atom participates in the motion. ω_2 and ω_3 are the modes referred to as "significant" in the problem.

(b) $\omega_3/\omega_2 = 1.91$. In the CO_2 spectrum, these bands are observed at 1340 and 2350 cm^{-1}, a ratio of 1.75.

Problem 3. Graph the dispersion curves for the chain with a basis shown in Figure 15 and discussed in the text for the following nominal values: f = 0.4, g = 0.5, h = 0.15 (all units mdyn/Å), $\alpha = 30^0$, r = 1.14 Å. Use m_2 = 107.88, and m_3 = 5, 14, and 20. Show how the choice of different values for the mass m_3 affects the behavior of the ω_2 and ω_3 modes.

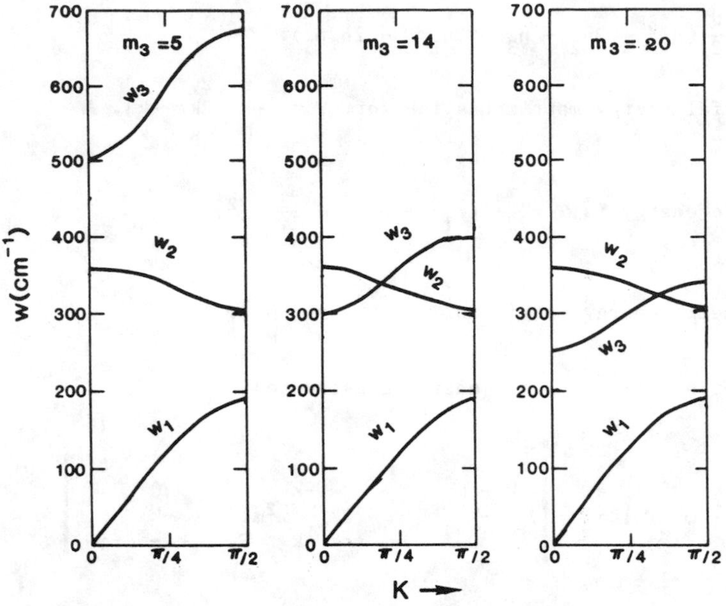

Figure 23. The behavior of the acoustic (1), longitudinal optic (2), and librational (3) modes as M_3 is varied in the chain-with-a-basis problem.

Solution 3. The results are shown in Figure 23. The reason for the discussion in the text and in this problem is to emphasize the difficulty in finding a universal mechanical behavior for the low-frequency modes. In this simple example, the lowest observed mode could be either a longitudinal optic mode or a librational mode depending only on the value of m_3.

Problem 4. Demonstrate that a spectrum obeying a sine-wave dispersion approaches Debye-like behavior in the low-temperature limit.

Solution 4. Let $x = h\nu/kT$. Then

$$\frac{C_V^*}{3R} = \left(\frac{2}{\pi}\right)^3 \cdot \frac{1}{nZ} \sum_{i=1}^{3} \int_0^{x_i} \frac{(\sin^{-1}\frac{x}{x_i})^2\, x^2\exp(x)}{(x_i^2 - x^2)^{1/2}\,[\exp(x) - 1]^2}\,dx,$$

where $x_i = h\nu_i/kT$, ν_i being the upper cutoff of the distribution for each branch. For $x \gg 1$ and $x \ll x_i$ (criteria for the low-temperature limit):

$$[\sin^{-1}(\frac{x}{x_i})]^2 \approx (\frac{x}{x_i})^2 + \ldots$$

and $[x_i^2 - x^2]^{-1/2} \approx \frac{1}{x_i}[1 + \frac{1}{2}(\frac{x}{x_i})^2]+....$

With some algebra

$$\frac{C_V^*}{3R} \approx (\frac{2}{\pi})^3 \frac{1}{nZ} \sum_{i=1}^{3} \frac{1}{x_i^3} \int_0^\infty \frac{x^4 \exp(x)}{[\exp(x) - 1]^2} =$$

$$(\frac{2}{\pi})^3 \frac{1}{nZ} \sum_{i=1}^{3} \frac{1}{x_i^3} \frac{4\pi^4}{15} =$$

$$(\frac{2}{\pi})^3 \frac{1}{nZ} \frac{4\pi^4}{15} (\frac{T}{\Theta_{sine}})^3$$

For the last equality, the isotropy of a Debye model has been invoked so that $x_1 = x_2 = x_3 = h\nu_i/kT$, and $h\nu_i/k$ has been defined as θ_{sine}. In Kieffer (1979a) it is shown that $\theta_{sine} = (2/\pi) \times \theta_D$, where θ_D is the Debye temperature associated with a dispersionless spectrum. Therefore,

$$\frac{C_V^*}{3R} \rightarrow \frac{1}{nZ} \frac{4\pi^4}{15} (\frac{T}{\Theta_D})^3$$

as in Debye theory.

Problem 5.

(a) To see how Tables 2, 5, and 6 can be used to calculate the heat capacity for a mineral with a spectrum of the form in Figure 7, calculate the heat capacity of quartz at 300 K and several other temperatures, using the following spectral parameters: w_l = 128 cm^{-1}, w_u = 809 cm^{-1}, and w_E = 1100 cm^{-1}. Assume that there are 0.22 of the modes at w_E. Keep track of the separate contributions of the acoustic, optic continuum, and Einstein oscillator modes to the heat capacity. Graph and compare with data.

(b) Graph $\theta_{cal}(T)$.

(c) Graph the contribution of the acoustic, continuum, and Einstein (intramolecular) modes to C_V.

(d) If you feel ambitious, program the equations for internal energy, entropy, and energy (these can be found in Kieffer, 1979c) and calculate the thermodynamic functions published by Robie et al. (1978). (The authors' Fortran computer program can be requested as a U.S.G.S. Open-file Report.)

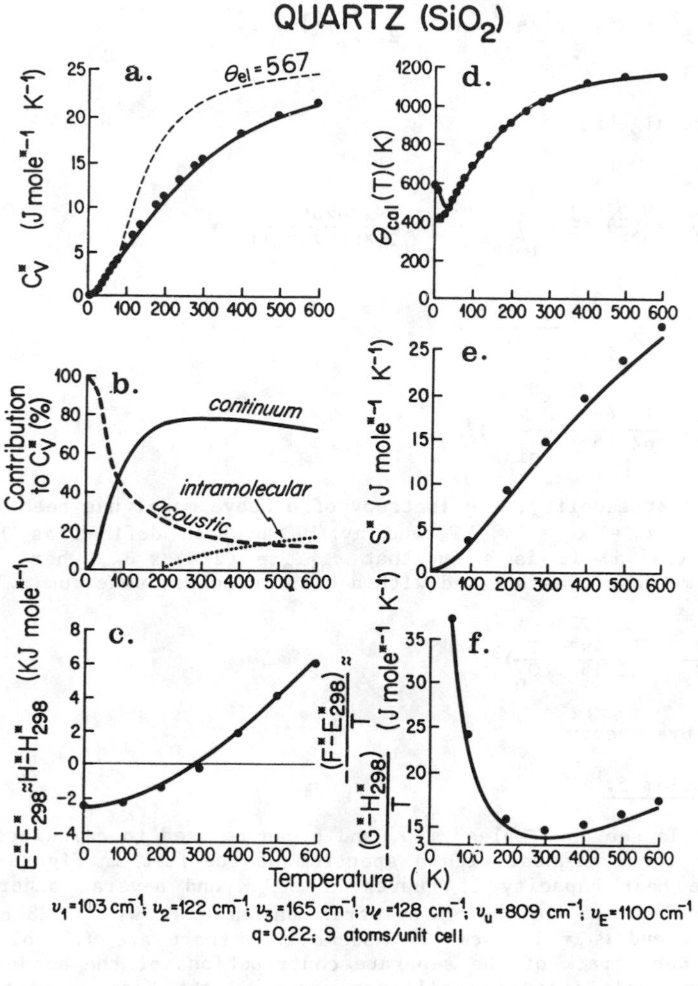

Figure 24. (a) The heavy line is C_V^* calculated from the parameters of Worked Problem (4) (listed at the bottom of the Figure). Dots are experimental data. The dashed curve is C_V^* predicted from a Debye model with θ_{el} = 567 (See Worked Problem 1). (b) θ_{cal} (T) versus temperature. (c) Percent contribution of the various parts of the assumed spectrum to the heat capacity. (d) Entropy versus temperature. (e) Internal energy at temperature T relative to 298.15 K (approximately equal to enthalpy at 1 bar pressure) versus temperature. (f) The Gibbs energy function versus temperature. $(G_T - H_{298})/T = (H_T - H_{298})/T - S_T$ (see Roble et al., 1978 for more detailed definitions).

Solution 5.

(a) The dimensionless frequencies are $x_l = 0.61$, $x_u = 3.88$, and $x_E = 5.28$. The maximum frequencies of the acoustic branches are $w_{1,max} = 103\ cm^{-1}$, $w_{2,max} = 122\ cm^{-1}$, and $w_{3,max} = 165\ cm^{-1}$, giving $x_{i,max}$ for the acoustic branches of 0.49, 0.59, and 0.79. To find the contribution of each acoustic branch, use Table 5: e.g., for $x_1 = 0.49$, the tabulated value is 0.984, and, since we are considering each acoustic branch individually, this value must be divided by 3, giving a contribution of 0.328. The other two values are, respectively, 0.977/3, and 0.948/3. The sum of these three tabulated values is 0.97, which must be multiplied by $3N_Ak/nZ$ to give an acoustic contribution to C_V^* of 0.64 cal $mole^{-1}K^{-1}$ (2.68 J $mole^{-1}K^{-1}$). Table 5 shows that the acoustic branches are contributing nearly their maximum heat capacity, normalized to 3/3s degrees of freedom, so that it is not necessary to go through the painful calculation of $w_{i,max}$ if the temperature is high relative to the acoustic mode frequencies (expressed in cm^{-1}); their contributin to the heat capacity can be assumed to be the saturated value, properly normalized. From Table 6, the continuum contributes (1-3/27-0.22) x 3.90 = 2.63 cal mol^{-1}. From Table 2, the Einstein oscillator contributes 0.84 x 0.22 = 0.185 cal $mol^{-1}\ K^{-1}$. C_V^* at 300 K is therefore 3.47 cal $mol^{-1}\ K^{-1}$ = 14.52 Joules $mol^{-1}\ K^{-1}$. The procedure is the same for other temperatures. The results are shown in Figure 24a. Note the apparent excellent agreement of theory (solid line) and data (dots) at all temperatures.

(b) $\theta_{cal}(T)$ is shown in Figure 24b. The purpose of this part of the problem is to emphasize how a C_V versus T plot "hides" differences between theory and experiment at low temperatures. These differences can be seen on a θ_{cal} (T) graph.

(c) The results are shown in Figure 24c. Note how different parts of the spectrum vary in importance at different temperatures.

(d) These results are shown in Figure 24d,e,f.

ACKNOWLEDGMENTS

The efforts and talent of Margie Dennis in preparation of the camera-ready copy of this manuscript and in organizing the author's computer files of thermodynamic data were invaluable. Publication authorized by U.S. Geological Survey March 8, 1985.

REFERENCES

Anderson, O.L. (1963) A simplified method for calculating the Debye temperature from elastic constants. J. Phys. Chem. Solids 24, 909-917.

Barron, T.H.K., Huang, C.C. and Pasternak, A. (1976) Interatomic forces and lattice dynamics of α-quartz. J. Phys. C, Solid. State Phys. 19, 3925-3940.

Blackman, M. (1955) The specific heat of solids. In: Handbuch der Physik 7(1), 325-382.

Born, M. (1942) Proc. Phys. Soc. London 54, 362.

_____ and Huang, K. (1954) Dynamical Theory of Crystal Lattices. Oxford, London.

_____ and von Karman, T. (1912) Zeit. Physik 13, 297.

_____, _____ (1913) Zeit. Physik 14, 65.

Brillouin, L. (1953) Wave Propagation in Periodic Structures. Dover, New York.

Brüesch, P. (1982) Phonons: Theory and Experiments I. Springer-Verlag, Berlin.
Bragg, W.L., Claringbull, G.F. and Taylor, W.A. (1965) Crystal Structure of Minerals. Cornell Univ. Press, Ithaca, New York.
Deer, W.A., Howie, R.A. and Zussman, J. (1963) Rock-Forming Minerals, vol. 4 (Framework Silicates). Longmans Press, London.
Debye, P. (1912) Zür Theorie der spezifischen Wärmen. Ann. Physik 39(4), 789-839.
Dulong, P.L. (1829) Annales Chimie et de Physique 41, 113-158.
Einstein, A. (1907) Ann. Physik 22, 180-190.
_____ (1911) Ann. Physik 35, 679-694.
Elcombe, M.M. (1966) The lattice dynamics of quartz and ionic semiconductors. Ph.D. Thesis, Cambridge Univ., Cambridge, England.
_____ (1967) Some aspects of the lattice dynamics of quartz. Proc. Phys. Soc. 91, 947.
Feynman, R.P., Leighton, R.B. and Sands, M. (1963) The Feynman Lectures on Physics, v. 1, 52 chapters. Addison-Wesley, Reading, Massachusetts.
Finger, L.W. (1975) Least-squares refinement of the rigid-body motion parameters of CO_3 in calcite and magnesite and correlation with lattice vibrations. Carnegie Inst. Wash. Yearbook 74, p. 572-579.
Goldstein, H. (1950) Classical Mechanics. Addison-Wesley, Reading, Massachusetts.
Hazen, R.M. and Finger, L.W. (1982) Comparative Crystal Chemistry. John Wiley, New York.
Iishi, K. (1976) The analysis of the phonon spectrum of α quartz based on a polarizable ion model. Z. Kristallogr. 144, 907-912.
_____ (1978) Lattice dynamical study of the α-β quartz phase transition. Am. Mineral. 63, 1190-1197.
Kieffer, S.W. (1979a) Thermodynamics and lattice vibrations of minerals: 1. Mineral heat capacities and their relationships to simple lattice vibrational models. Rev. Geophys. Space Phys. 17, 1-19.
_____ (1979b) Thermodynamics and lattice vibrations of minerals: 2. Vibrational characteristics of silicates. Rev. Geophys. Space Phys. 17, 20-34.
_____ (1979c) Thermodynamics and lattice vibrations of minerals: 3. Lattice dynamics and an approximation for minerals with application to simple substances and framework silicates. Rev. Geophys. Space Phys. 17, 35-59.
_____ (1980) Thermodynamics and lattice vibrations of minerals: 4. Application to chain and sheet silicates and orthosilicates. Rev. Geophys. Space Phys. 18, 862-886.
_____ (1982) Thermodynamics and lattice vibrations of minerals: 5. Applications to phase equilibria, isotopic fractionation, and high-pressure thermodynamic properties. Rev. Geophys. Space Phys. 20, 827-849.
Kittel, C. (1968) Introduction to Solid State Physics, 3rd ed. John Wiley, New York.
Lazarev, A.N. (1972) Vibrational Spectra and Structure of Silicates (translated from Russian). Consultants Bureau, New York.
Leighton, R.B. (1948) Rev. Mod. Phys. 20, 165, 1948.
_____ (1959) Principles of Modern Physics. McGraw-Hill, New York.
Levien, L., Prewitt, C.T. and Weidner, D.J. (1980) Structure and elastic properties of quartz at pressure. Am. Mineral. 65, 920-930.
Madelung, E. (1910) Zeit. Physik 11, 898.
Matossi, F. (1951) The vibration spectrum of rutile. J. Chem. Phys. 19, 1543-1546.
Moenke, H.H.W. (1974) Silica, the three-dimensional silicates, botosilicates, and beryllium silicates. In: The Infrared Spectra of Minerals, Mineral. Soc. Monogr., 4, V.C. Farmer, ed. Mineral. Soc., London.
Moseley, H.G.J. (1913) Phil. Mag.[6] 26, 1024.
Nernst, W. and Lindemann, F.A. (1911) Berl. Ber., 494.
Planck, Max (1906) Wärmestrahlung, Leipzig.
Pauling, L. (1929) Principles determining the structure of complex ionic crystals. J. Am. Chem. Soc. 51, 1010-1026.
Reif, F. (1965) Fundamentals of Statistical and Thermal Physics. McGraw-Hill, New York.
Reissland, J.A. (1973) The Physics of Phonons. John Wiley, London.
Robie, R.A. and Edwards, J.L. (1966) Some Debye temperatures from single-crystal elastic constant data. J. Appl. Phys. 37, 2659-2663.
_____, Hemingway, B. S. and Fisher, J. R. (1978) Thermodynamic prooperties of minerals and related substances at 298.15 K and 1 bar (10^5 pascals) presure and at higher temperatures. U.S. Geol. Surv. Bull. 1452, 456 pp.
Saksena, B.D. (1940) Analysis of the Raman and infra-red spectra of α-quartz. Proc. Indian Acad. Sci. 12A, 93.
_____ (1942) Force constants and normal modes of the totally symmetric vibratins in α-quartz at room temperature. Proc. Indian Acad. Sci. 16A, 270.
Seitz, F. (1940) The Modern Theory of Solids. McGraw-Hill, New York.
Striefler, M.E. and Barsch G.R. (1975) Elastic and optical properties of stishovite. Phys. Rev. 12B, 4553-4566.
Traylor, J.H., Smith, H.G., Nicklow, R.M. and Wilkenson, M.K. (1971) Lattice dynamics of rutile. Phys. Rev. B 3(10), 3457-3472.
Wallace, D.C. (1972) Thermodynamics of Crystals. John Wiley, New York.
Weidner, D.J. and Simmons, G. (1972) Elastic properties of α-quartz and the alkali halides based on an interatomic force model. J. Geophys. Res. 77, 826-847.

Chapter 4. Subrata Ghose

LATTICE DYNAMICS, PHASE TRANSITIONS and SOFT MODES

INTRODUCTION

The subject of lattice dynamics has its origin in the papers published by Max Born and von Kármán (1912), and P. Debye (1912) just prior to the discovery of x-ray diffraction by crystals by Max von Laue and his associates. Hence, the theory of lattice dynamics was developed before the existence of the crystal lattice was experimentally verified! The development of lattice dynamics is considered to have laid the foundation of solid state physics. In spite of its early development, only recently has it received tremendous attention by both theoretical and experimental physicists due to the possibility of its direct verification through the experimental determination of the phonon dispersion relations by inelastic neutron scattering. This technique was pioneered by B.N. Brockhouse in the early 1960's at Chalk River, Canada. The first international conference on lattice dynamics was held in Copenhagen in 1963, the proceedings of which contain fascinating reminiscences of the early history by Born and Debye (Wallis, 1965). Today, along with the study of phase transitions, it is one of the most active fields of research in modern solid state physics.

The first section of this chapter contains an outline of the basic ideas of the theory of lattice dynamics and the experimental determination of the phonon dispersion relations and their interpretation. The second section is a discussion of the dynamics of displacive phase transitions and critical phenomena, with examples. The theory of lattice dynamics and phase transitions is highly mathematical and fairly complicated. Further reading is suggested, with a caution that the mathematical notation varies from one book to another. In spite of this, a rich dividend to be gleaned is a more fundamental understanding of the dynamical phenomena in the crystalline world.

1. LATTICE DYNAMICS AND PHONON DISPERSION RELATIONS

1.1 Theory

We begin our discussion of lattice dynamics in terms of the traditional one-dimensional linear chain. (For examples of a diatomic chain and a diatomic chain with a basis, see Kieffer, this volume.) Consider a set of atoms of mass M along a line separated by a distance 'a', so that the one-dimensional lattice vectors are $\underset{\sim}{R} = na$, for integral values of n. Let u(na) be the displacement along the line from its equilibrium position of the atom that oscillates about na, i.e. longitudinal motion (Fig. 1). We assume for simplicity only interaction between neighboring atoms. The harmonic potential energy, U^{harm} is given by

$$U^{harm} = \frac{1}{2} f \sum_n [u(na) - u([n+1]a)]^2 \qquad (1.1)$$

where f is the force constant between two atoms.

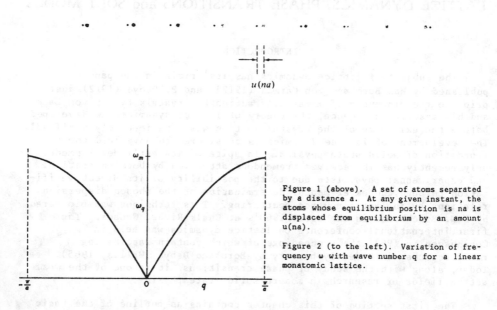

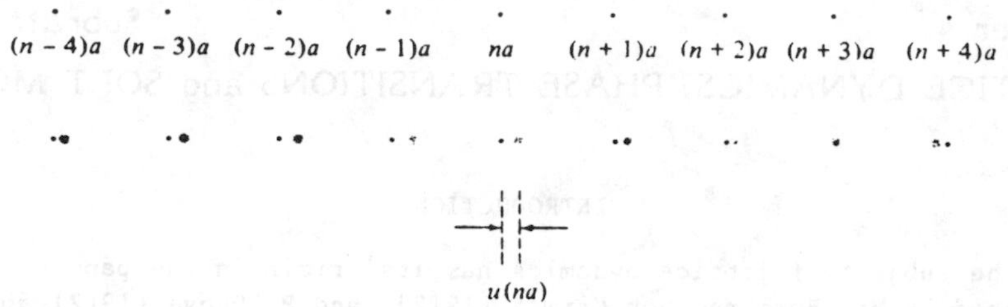

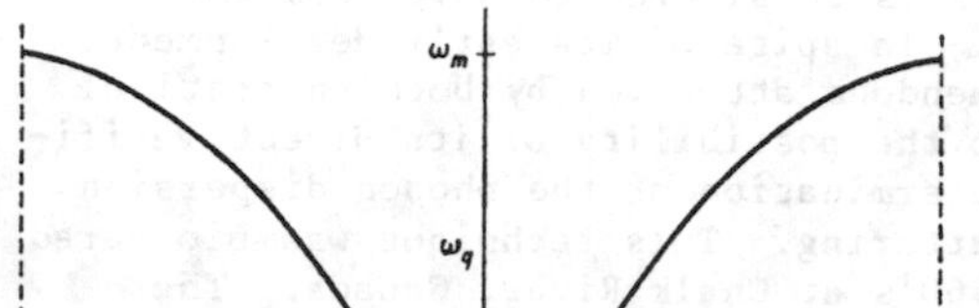

Figure 1 (above). A set of atoms separated by a distance a. At any given instant, the atoms whose equilibrium position is na is displaced from equilibrium by an amount u(na).

Figure 2 (to the left). Variation of frequency ω with wave number q for a linear monatomic lattice.

The equations of motion for this linear chain are:

$$M \frac{\partial^2 u}{\partial t^2} = - \frac{\partial U^{harm}}{\partial u(na)} = - f\,[2u(na) - u([n - 1]a) - u([n + 1)a)]. \qquad (1.2)$$

Assume the atoms to be connected to its nearest neighbors by massless springs with spring constant f. We impose now the periodic boundary condition for the linear chain known as Born–von Kármán boundary condition: we simply join the two ends together with an additional spring. This is necessary so that all the atoms in the chain have the same near neighbor situations and to simulate an infinite chain of atoms without surface effects.

Solutions to the equations of motion 1.2 can be found in terms of travelling waves with frequency ω:

$$u(na,t) \propto e^{i(qna-\omega t)} \qquad (1.3)$$

The periodic boundary condition requires that

$$e^{iqNa} = 1, \qquad (1.4)$$

where N = total number of atoms in the chain. We then have

$$q = \frac{2\pi}{a}\,\frac{n}{N} \text{ , where n is an integer} \qquad (1.5)$$

Note that when q is changed by $2\pi/a$, the displacement u(na) given by Equation 1.3 is unchanged. As a result, there are only N values of q consistent with Equation 1.5 that yield direct solutions, which lie between $-\pi/a$ and $+\pi/a$.

We can now find the dispersion relations of the frequency ω by substituting Equation 1.3 into 1.2:

$$\begin{aligned} - M\omega^2 e^{i(qna-\omega t)} &= - f[2-e^{-iqa}-e^{iqa}]e^{i(qna-\omega t)} \\ &= - 2 f (1- \cos qa)e^{i(qna-\omega t)} \end{aligned} \tag{1.6}$$

Hence, we have the dispersion relation, where $\omega = \omega(q)$,

$$M \omega^2(q) = 2f(1 - \cos qa) \tag{1.7}$$

giving the frequency ω of a wave of wave number q. The frequency ω is plotted as a function of the wave number q in Figure 2. When q is small compared to π/a (i.e., when the wavelength is large compared to the interatomic distance), ω is linear in q, that is there is no dispersion just as for a sound wave in a continuous medium. For larger q the frequency rises to a maximum at $q = \pi/a$ and then decreases to zero again at $q = 2\pi/a$. The reason for this is that $q = 2\pi/a$ is a wave of wavelength a, all atoms are displaced by the same amount and there is no restoring force. In fact ω(q) is periodic in a distance $2\pi/a$ which is the spacing of the one-dimensional reciprocal lattice. ω(q) then can be represented within a one dimensional <u>Brillouin zone</u> with zone boundaries at $+ \pi/a$ and $- \pi/a$.

We now suppose that the range of the interatomic force is increased and is represented by the force constant f_n for the interaction between atoms separated by a distance na. The relation between frequency and wave number is found to be:

$$M \omega^2(q) = 2 \sum_n f_n(1- \cos n q a) \tag{1.8}$$

This function is still periodic in reciprocal space, but the greater the range of the interatomic force, the more Fourier coefficients contribute to $\omega^2(q)$. If somehow $\omega^2(q)$ can be determined as a function of q, by Fourier analysis of Equation 1.8, the force constants f_n can be determined. This then is the basis of a method of investigating interatomic forces. In practice, the harmonic approximation involved in Equation 1.2, i.e., the force on an atom is directly proportional to the relative displacements of other atoms, is not strictly valid. However, for small displacements, we may take

$$f_n = \left[\frac{\partial^2\phi(r)}{\partial r^2}\right]_{r = na}, \tag{1.9}$$

assuming the existence of an interatomic potential of $\phi(r)$ between two atoms separated by distance r. This is the harmonic approximation where higher order terms in $\phi(r)$ are neglected.

We now have to generalize these results in various ways to understand the situation in a real crystal. First, we consider a unit cell containing several atoms, still in one dimension and only longitudinal displacements permitted. We need now two indices to identify each atom, ℓ for the unit cell and k for the type of atom. It can be shown that if the unit cell contains N atoms, there are N different modes of vibration having the same wave number q, but distinguished from one another by having different frequencies $\omega_j(q)$ (where j can have values 1, 2, ...n) and different patterns of movement of the atoms. In a one-dimensional crystal one of these "branches" of the dispersion relation, e.g., j - 1, has always the property that $\omega(q) \propto q$ for q small; such a mode of vibration is known as the <u>acoustic mode</u> because its dispersion relations is of the form $\omega = cq$ (where the phase velocity $c = \omega/q$) for small q, which is characteristic of sound waves. The remaining N-1 branches have finite frequencies at q = 0 and are known as <u>optic modes</u>, because the long wavelength optic modes in ionic crystals can interact with electromagnetic radiation, and are responsible for much of the characteristic optical behavior of such crystals (infrared and Raman spectroscopy). So far we have restricted each atom to vibrate longitudinally in the direction of the chain. If transverse displacements are also allowed, for which the restoring forces will usually be different, the number of modes of vibration for a given wave number q is increased to 3N where three are acoustic modes, and the rest optic modes.

These results can now be extended to three dimensions. In the Born-von Kármán theory, the force constants between any pair of atoms form a 3 x 3 symmetric tensor with a maximum of six independent components. The solution of the equations of motion gives again travelling waves of frequency ω_j, j = 1, 2, ...3N, with the displacement of an atom identified by ℓ and k given by

$$u(\ell k) = \sum_{\underset{\sim}{q}j} \underset{\sim}{e}_j(k\underset{\sim}{q})\, U_j(\underset{\sim}{q}) \exp i[\underset{\sim}{q}.\underset{\sim}{r}(\ell k) - \omega_j(\underset{\sim}{q})t] \tag{1.10}$$

where, $\underset{\sim}{e}_j(k\underset{\sim}{q})$ = lattice vibration eigenvector.

$\underset{\sim}{q}$ = lattice vibration wavevector,

U_j = amplitude, and $\underset{\sim}{r}(\ell k)$ = position vector.

As in one dimension, unique values of $\underset{\sim}{q}$ are confined to the unit cell in reciprocal space surrounding the origin (the first <u>Brillouin zone</u>) and $\omega(\underset{\sim}{q})$ is periodic in the reciprocal lattice. As before, the dispersion relation $\omega_j(\underset{\sim}{q})$ is determined by the force constants, as are the eigenvectors or polarization vectors $\underset{\sim}{e}_j(k\underset{\sim}{q})$ which determine the pattern of motion of the different atoms in the course of a mode of vibration identified by q and j. The mathematical formalism for the three-dimensional case is given in Appendix 1. (For details see Born and Huang, 1954; Cochran and Cowley, 1967; Donovan and Angress, 1971.)

We have so far considered lattice vibrations in terms of classical concepts. Each independent normal mode of vibration behaves like a simple harmonic oscillator and is quantized the same way, the quantum of energy $\hbar\omega_j(q)$ being known as a <u>phonon</u>. Although the language of phonons is more convenient than that of normal modes, the two nomenclatures are completely equivalent. The phonon dispersion relations are best

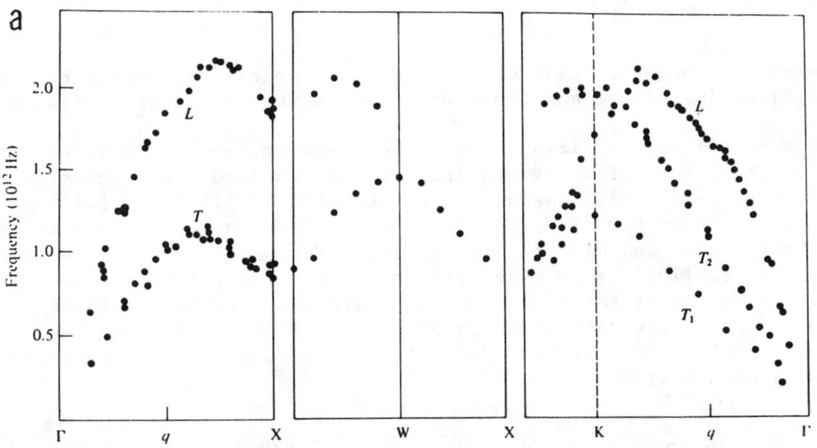

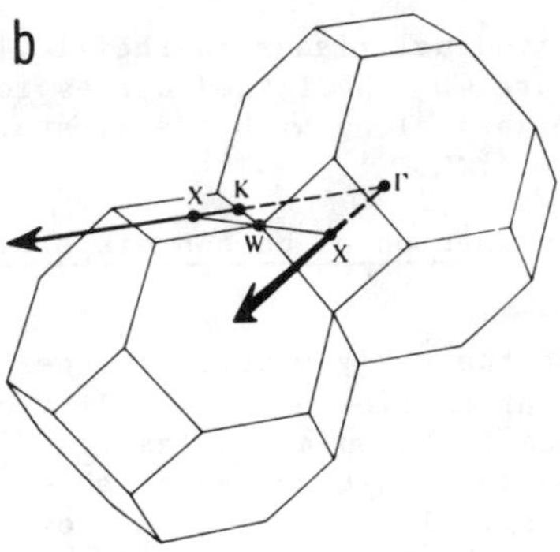

Figure 3 (above and left). (a) Typical dispersion curves for normal-mode frequencies in a monatomic Bravais lattice. The curves are for lead (face-centered cubic) at 100 K and are plotted along the reciprocal lattice directions for the Brillouin zone shown in (b). After Brockhouse et al. (1962).

Figure 4 (below). The interplanar force constants for longitudinal waves in the [100] direction in lead. The horizontal axis gives the distance of the plane from the reference plane. After Brockhouse et al. (1962).

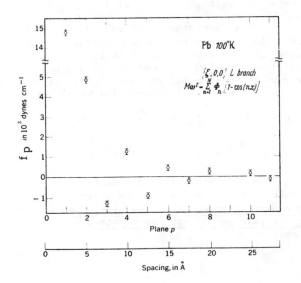

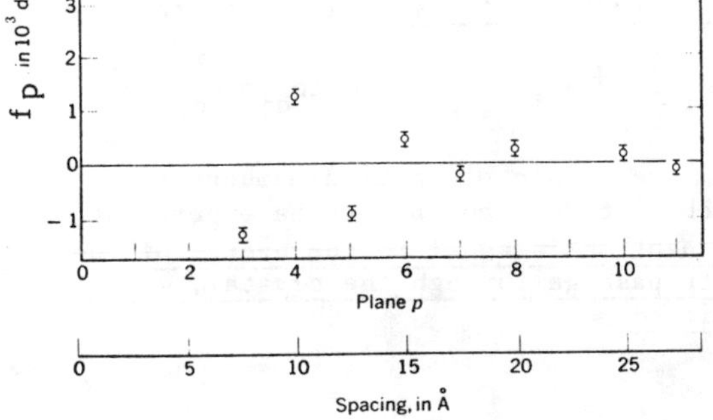

determined by inelastic neutron scattering. Typical dispersion curves for a three-dimensional monatomic Bravais lattice are shown in Figure 3.

For metals crystallizing in the cubic system, where the metal atoms occur at the points of a Bravais lattice such as lead (face centered cubic), monatomic linear chains exist along [100], [110] and [111] directions. In such a case, the results obtained on lattice vibrations of a linear chain can be directly applied. When the phonon wavevector $\tilde{q}$ is parallel to [100], successive planes of atoms play the role of the individual masses of the linear chain. Here the interplanar force constant f_p from the p th plane in any of the above directions can be determined from a Fourier transform of ω^2 as a function of $\tilde{q}$ using the following equation:

$$f_p = -\frac{Ma}{2\pi}\int_{-\pi/a}^{\pi/a} dq\ \omega(\tilde{q})^2 \cos pqa \qquad (1.11)$$

The interplanar force constant f_p for longitudinal planes in the [100] direction in lead at 100 K is shown in Figure 4. The fitted curves for values of $M\omega^2$ for five planes and twelve planes along with the first five Fourier components are shown in Figure 5.

1.2 Inelastic neutron scattering and determination of phonon dispersion relations

Slow neutron spectroscopy developed in the early sixties is the most powerful modern method for obtaining information about the lattice dynamics of crystalline solids. This method is known as inelastic neutron scattering to draw attention to the fact that an appreciable energy change of neutrons occur on scattering. This is because slow neutrons have both a wavelength comparable with interatomic distances and an energy comparable with phonon energies.

Consider a neutron of momentum $\tilde{p}$ and energy $E = p^2/2M_n$ (where M_n = neutron mass) that is incident upon a crystal. Since the neutron interacts strongly with the atomic nuclei in the crystal, it will pass through the crystal emerging with a momentum p' and energy $E' = (p')^2/2M_n$. (Here we are neglecting the interaction of neutrons with the magnetic moment of electrons in transition metal bearing compounds.)

The law of conservation of energy requires that

$$E' - E = -\sum_{\tilde{q}j} \hbar\omega_{\tilde{q}j}\Delta n_{\tilde{q}j}, \quad \text{where } \Delta n_{\tilde{q}j} = n'_{\tilde{q}j} - n_{\tilde{q}j} \qquad (1.12)$$

where n_{qj} and n'_{qj} are phonon occupation numbers (with phonon wavevector q and phonon branch j) before and after the experiment. It means that the energy of the neutron is equal to the energy of the phonons it has absorbed during its passage through the crystal, minus the energy of the phonons it has emitted.

The law of conservation of crystal momentum is given by:

$$\underset{\sim}{p}' - \underset{\sim}{p} = - \sum_{\underset{\sim}{q}j} \hbar \, \underset{\sim}{q} \Delta n_{\underset{\sim}{q}j} + \hbar \underset{\sim}{K} \tag{1.13}$$

where K is a reciprocal lattice vector. This law can be stated as: the change in neutron momentum is just the negative of the change in total phonon crystal momentum to within an additive reciprocal lattice vector. Crystal momentum of a phonon is simply a name for $\hbar$ times the phonon wavevector, $\underset{\sim}{q}$.

Neutrons that absorb or emit precisely one phonon give us the most important information. Relevant equations for the cases of absorption and emission of a single phonon are:

$$\frac{(p')^2}{2M_n} = \frac{p^2}{2M_n} + \hbar \, \omega_j \left(\frac{\underset{\sim}{p}' - \underset{\sim}{p}}{\hbar} \right) \quad \text{phonon absorbed} \tag{1.14}$$

$$\frac{(p')^2}{2M_n} = \frac{p^2}{2M_n} - \hbar\omega_j \left(\frac{\underset{\sim}{p}' - \underset{\sim}{p}}{\hbar} \right) \quad \text{phonon emitted} \tag{1.15}$$

These relations can be conveniently shown in a vector diagram for the case of one phonon scattering (Fig. 6). Let us denote the incident and scattered neutron wavevectors k and k' respectively, the phonon wavevector $\underset{\sim}{q}$ and the reciprocal lattice vector $\underset{\sim}{K}$. In terms of the vector diagram (Fig. 6), the momentum conservation relation can be expressed as:

$$\underset{\sim}{Q} = \underset{\sim}{k}' - \underset{\sim}{k} = \underset{\sim}{q} \pm \underset{\sim}{K} \tag{1.16}$$

and, the energy conservation relation,

$$\hbar\omega = \frac{\hbar^2}{2M_n} (k'^2 - k^2) = \pm \hbar\omega j(\underset{\sim}{q}) \tag{1.17}$$

where, k and k' are incident and scattered neutron wave numbers.

By using Equations 1.16 and 1.17 for the conservation of crystal momentum and energy, a scattered neutron beam can be analyzed to give the frequencies $\omega_j(q)$ of modes for any wavevector q. The different branches are distinguished by their different frequencies and by the fact that the cross-section for the scattering process depends on the polarization properties of the mode with which the beam is interacting (see Lovesey and Springer, 1977).

The triple-axis neutron spectrometer, first developed by B.N. Brockhouse (1962), consists essentially of a monochromating crystal on one axis, the specimen crystal on the second and an analyzing crystal on the third axis (Fig. 7). Note that it is a spectrometer and not a diffractometer, since neutron energies are measured. The maximum count rate for a low energy phonon is low (about 300 counts/min) and for a high frequency phonon as low as 10 counts/min using a High Flux Beam Reactor. To gain intensity the sample must be large (1 to 20 cm^3). Hence, a lattice dynamics experiment is very time consuming and

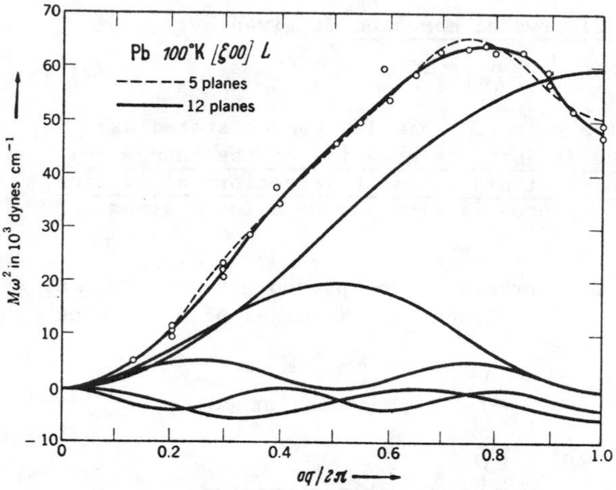

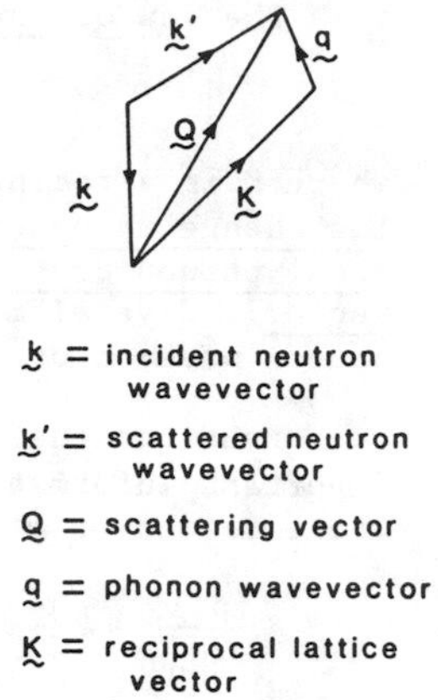

Figure 5 (left). Values of $M\omega^2$ for longitudinal branch in the [100] direction in lead plotted against reduced wavevector. The solid curve with 12 planes (good fit to data points) and the dashed curve with five planes are shown. The first five Fourier components are also plotted. After Brockhouse et al. (1962).

Figure 6 (right). Momentum conservation relation during an inelastic neutron scattering experiment.

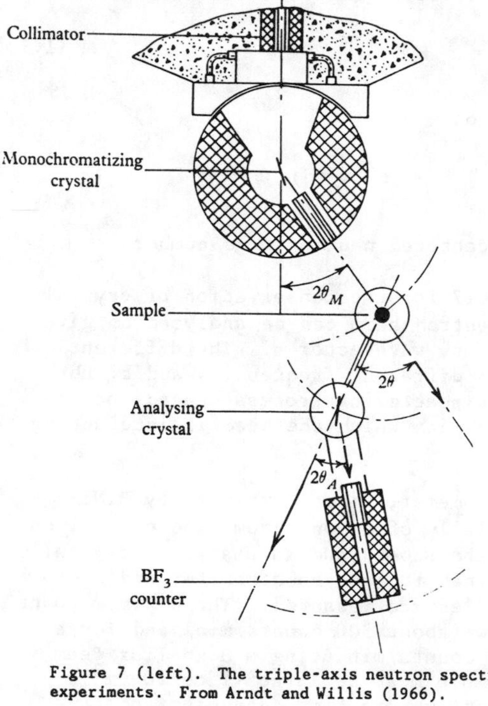

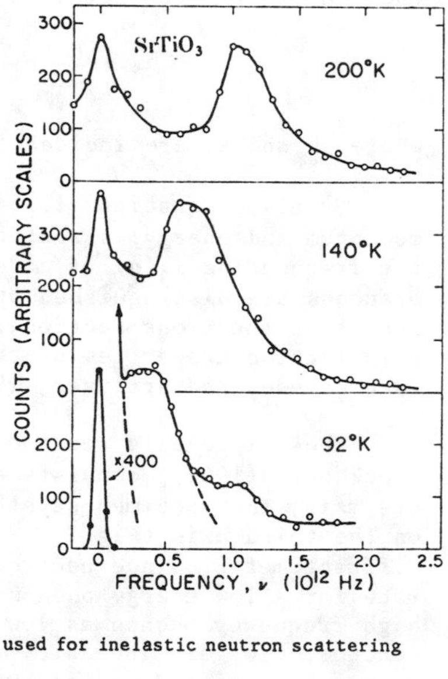

Figure 7 (left). The triple-axis neutron spectrometer used for inelastic neutron scattering experiments. From Arndt and Willis (1966).

Figure 8 (right). Typical inelastic neutron scattering spectra: soft mode in strontium titanate at three temperatures above and below the phase transition at 110 K. After Cowley et al. (1969).

expensive. Typical inelastic neutron scattering spectra are shown in Figure 8.

An atlas of phonon dispersion relations in insulators determined by inelastic neutron scattering is available (Bilz and Kress, 1979), which includes data on a number of minerals including MgO, MnO, FeO, PbS, NaCl, KCl, KBr, AgCl, CaF_2, D_2O (ice), C (graphite, diamond), ZnS, ZnO, TiO_2, SnO_2, MgF_2, $CaCO_3$ (calcite), $NaNO_3$, KNO_3, MoS_2, NiS, Cu_2O, SiO_2 (α quartz) and $MgAl_2O_4$ (spinel) along with their phonon density of states in some cases. The phonon dispersion relations in α quartz have been experimentally determined by Elcombe (1967) and Dorner, Grimm and Rzany (1980) (unpublished: see Boysen et al., 1980), which have been fitted by various model calculations (Elcombe, 1967; Barron, Huang and Pasternak, 1976).

1.3 Interpretation of phonon dispersion curves

We have already seen how the squared frequencies of a one-dimensional crystal can be analyzed by Fourier techniques to give the interatomic force constants. For a three-dimensional crystal with several atoms in the unit cell, in principle this is still possible if values of both phonon frequencies $\omega_j^2(\tilde{q})$ and eigenvectors $\tilde{e}_j(k,\tilde{q})$, are known for a sufficient range of values of $\tilde{q}$. Although some information about $\tilde{e}_j(k,\tilde{q})$, i.e., the polarization of the mode and the pattern of atomic displacement in this mode, in principle, can be obtained from the intensities of the neutron scattering, they are usually unknown in practice. This situation is reminiscent of the phase problem in crystal structure analysis. Given a model for the interatomic forces, we can test its correctness against the measured $\omega_j(\tilde{q})$. But only with the latter, we cannot deduce uniquely an interatomic potential or even the force constants. Hence, more than one model can be fit to the data, and the model giving a good fit may not necessarily be meaningful. However, if a model with no or few adjustable parameters fits the data, our confidence in it is increased.

As an example, we will discuss the phonon dispersion relations in a simple ionic crystal, namely KBr, which is cubic with the sodium chloride structure. The cohesive energy of an alkali halide crystal is very adequately described in terms of the predominantly attractive electrostatic (Coulombic) potential, and a short range repulsive potential between near neighbor ions. The potential between any ion pair in this model is given by:

$$\phi(r) = A\exp(-r/\rho) \pm \frac{e^2}{r} \tag{1.18}$$

where A and ρ are empirical constants, r the interionic distance and e is the electronic charge. The stability condition can be used to give ρ, a distance such that the only important repulsive forces are between nearest neighbors, and a value for A can be chosen which gives agreement with the measured cohesive energy and elastic constants. The electrostatic potential for the whole crystal is: $-N\alpha_M/r_o$, where r_o is the equilibrium distance between nearest neighbor ions, and α_M is the Madelung constant, a number which only depends on the geometry of the crystal structure (see Kittel, 1971). The contribution to the radial

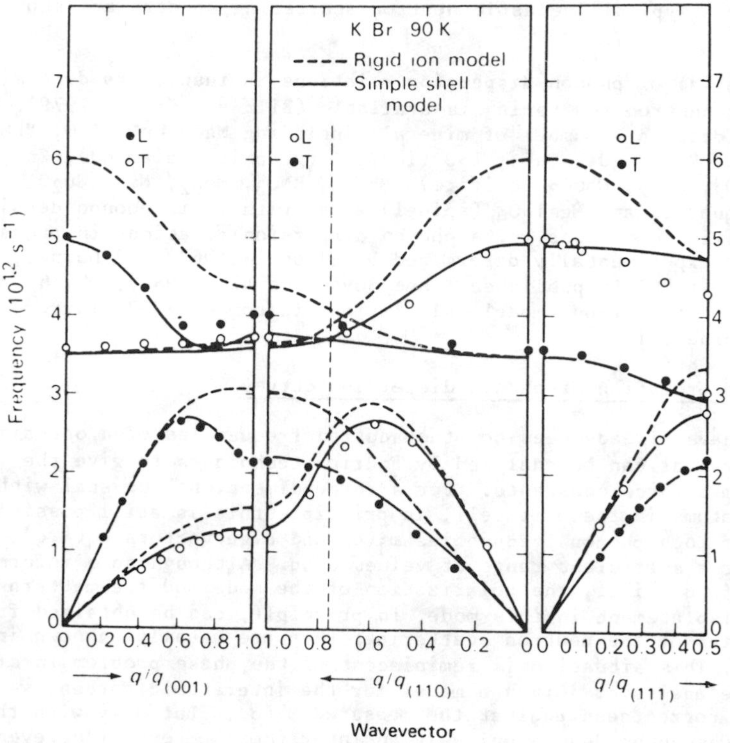

Figure 9. Phonon frequencies in KBr. Measured values are compared with calculations based on the rigid ion model and the simple shell model. After Woods et al. (1960).

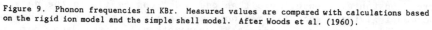

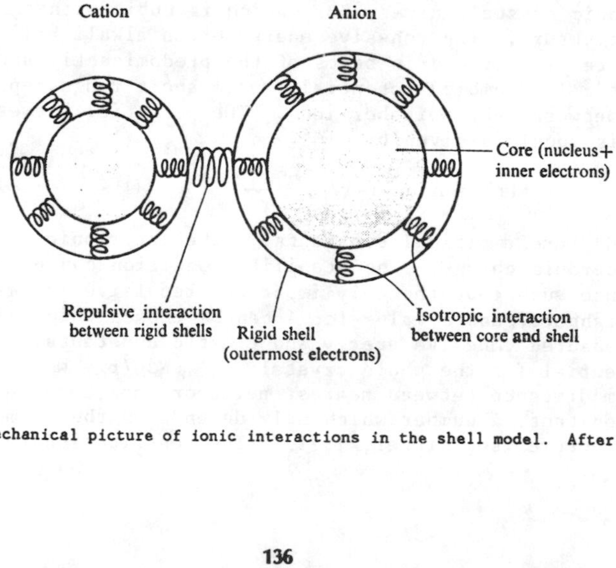

Figure 10. Mechanical picture of ionic interactions in the shell model. After Willis and Prior (1975).

and tangential force constants between any ion pair, which originates in the Coulomb potential, is completely determined. In calculating the phonon dispersion curves for KBr it was assumed that the repulsive potential extended only to nearest neighbors. Thus, its contribution to the force constants can involve only two parameters whether it is specified by the Born-Mayer exponential term in Equation 1.17 or in some other way. The contribution to the tangenitial force constant between nearest neighbors can be related to the Madelung constant, using the condition that the crystal is in equilibrium under two opposing interactions. In calculating the phonon dispersion curves for KBr the remaining parameter was chosen to give the observed value for the initial gradient of the longitudinal acoustic mode for which q is in the [100] direction. The calculated curves are the dotted lines in Figure 9. Considering that only one adjustable parameter was used, the agreement with experimental data is not bad and supports the assumption that KBr is ionic. However, there are obvious discrepancies, particularly for the longitudinal optic modes. The model tested is the rigid ion model, where the electronic polarization of the ions by the fields set up on the course of the lattice vibrations is not taken into account. The rigid ion model treats the ions as point charges. Kellerman (1940) has used this model to calculate the phonon dispersion curves for NaCl. The most serious defect of this model is that it is not consistent with the dielectric properties of crystals. At very high frequencies the electrons alone respond to the oscillating electric field and the dielectric constant is equal to the square of the optical refractive index, whereas the rigid ion model predicts the dielectric constant to be unity. This defect can be rectified by assigning electronic polarizability to the ions arising from polarization of the electrons around the ions in an electric field at high frequencies. This is the polarizable ion model, which is still not very satisfactory.

The currently popular model is the shell model (see Cochran, 1971). This model can be appreciated in terms of a mechanical model shown in Figure 10, where rigid spherical 'shells' representing the outermost electrons are connected to the ionic 'cores' by isotropic springs. A simple form of this model involving parameters which can be chosen to fit the dielectric properties gives good general agreement with measured values of $\omega_j(q)$ in KBr (Fig. 9) (Woods et al., 1960). Agreement within experimental error is obtained by using a more complicated model in which both positive and negative ions are polarizable, and the "short range" interaction extends to the second nearest neighbors. However, the trouble with this model is the large number of free parameters and the unreal values some of them sometimes assume when they are determined by least-squares analysis of experimental dispersion curves.

1.4 X-ray measurements of phonon spectra

For a lattice of ions, which are vibrating, it is possible for x-ray photons to be inelastically scattered with the emission or absorption of one or more phonons. A typical x-ray energy is several keV(10^3 eV), whereas a typical phonon energy is several meV(10^{-3} eV). As a result, the change in the energy of an inelastically scattered photon is extremely difficult to measure. The resolution of such minute photon frequency shifts is so difficult that one can usually measure the total scattered radiation of all frequencies, as a function of scattering angle, in the diffuse background of radiation found at angles away from those satisfying the Bragg condition (Thermal Diffuse Scattering, TDS). Because of this difficulty in energy resolution, the characteristic structure of the one-phonon processes is lost and their contribution to the total radiation scattered at any angle, cannot be simply distinguished from the contribution of the multiphonon processes. Provided that one can find some way of subtracting from the total intensity the contribution from the multiphonon processes, usually through a theoretical calculation, the phonon dispersion relations can be extracted from a measurement of the intensity of the scattered x-rays as a function of angle and the frequency of the incident x-rays. In addition to the difficulty mentioned above, one also has to correct for the contribution to the intensity due to inelastically scattered electrons (Compton background). As a result of these difficulties, x-ray scattering is a far less powerful probe of the phonon spectrum than neutron scattering.

1.5 Optical measurements of phonon spectra

If photons of visible light from a high intensity laser beam are scattered with the emission or absorption of phonons, the frequency shifts are still very small, but they can be accurately measured by interferometric techniques. One can then isolate the one-phonon contribution to the light scattering and extract values of $\omega(\underset{\sim}{q})$ for the phonons participating in the process. However, because the photon wavevectors (of the order 10^5 cm^{-1}) are small compared with Brillouin zone dimensions (of the order 10^8 cm^{-1}), information is provided only about phonons in the immediate neighborhood of $q = 0$. The process is known as Brillouin scattering, when the phonon emitted or absorbed is acoustic, and Raman scattering, when the phonon is optical.

When we consider the conservation laws for these processes, we have to take into account the fact that the photon wavevectors inside the crystal will differ from their free space values by a factor of the index of refraction of the crystal $\underline{n}$. Therefore, if the free space wavevectors of the incident and scattered photons are k and k', and the corresponding angular frequencies are ω and ω', conservation of energy and crystal momentum in a one phonon process requires

$$\hbar\,\omega' = \hbar\omega \pm \hbar\omega_j(\underset{\sim}{q}) \qquad (1.19)$$

and

$$\hbar n\underset{\sim}{k}' = \hbar n\underset{\sim}{k} \pm \hbar\underset{\sim}{q} + \hbar\underset{\sim}{K} \qquad (1.20)$$

Here, + sign refers to processes in which a phonon is absorbed (anti-Stokes component of the scattered radiation) and the - sign refers to processes in which the phonon is emitted (the Stokes component). Since the photon wavevectors k and k' are small in magnitude compared to the dimensions of the Brillouin zone, for phonon wavevectors q in the first zone, the crystal momentum conservation law can be obeyed only if the reciprocal lattice vector K is zero. The two types of processes are shown in Figure 11. Brillouin and Raman spectra of quartz are shown in Figures 12 and 13.

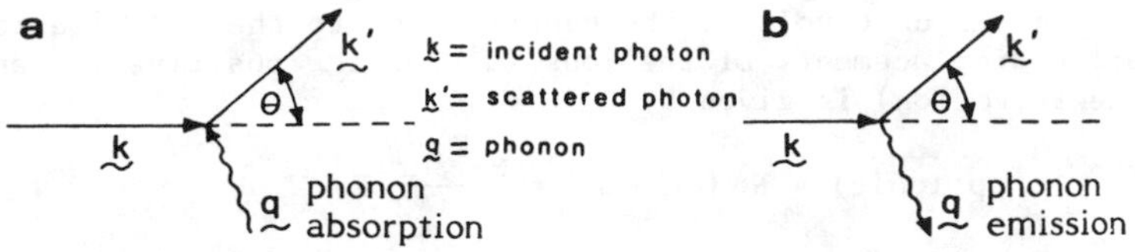

Figure 11. Conservation of energy and crystal momentum in a one phonon process.

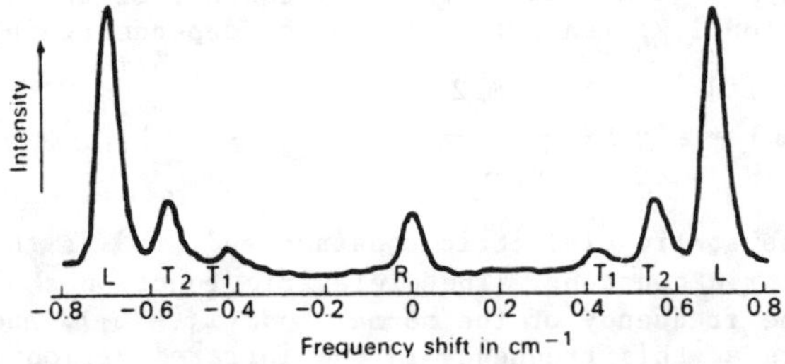

Figure 12. Brillouin scattering spectrum of quartz. After Shapiro et al. (1966).

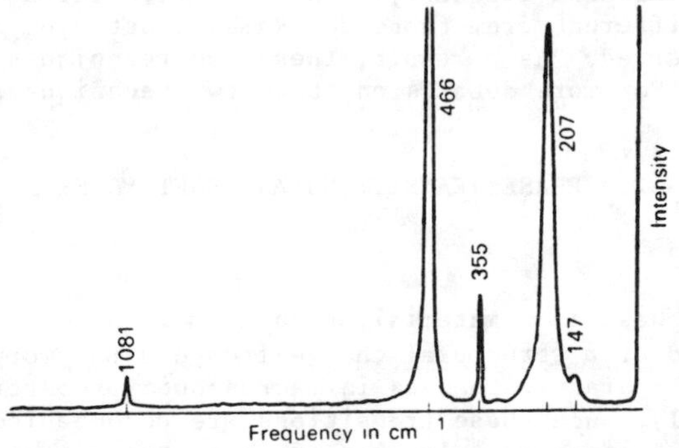

Figure 13. Raman spectrum of quartz. After Shapiro et al. (1967). The peak at 147 cm^{-1} is due to anharmonic processes.

Brillouin scattering allows a determination of the acoustic modes, from which the elastic constants can be determined. Elastic constants of diamond were determined this way by Krishnan (1947), using a strong light beam. The current technique involves a laser beam and a Fabry-Perot interferometer to analyze the scattered light. Elastic constants of very small crystals (~1 mm) can be determined this way. D.J. Weidner and his associates at Stony Brook have determined elastic properties of a large number of rock forming silicates using this technique (see Weidner et al., 1982).

The normal modes $\omega_j(\tilde{q})$ at $\tilde{q} = 0$ can also be determined by the interaction of infrared radiation with an ionic crystal. The application of the electric field, $\tilde{E}$ causes the positive and negative ions to be displaced in opposite directions, thereby polarizing the crystal. The dielectric polarization P is defined as the dipole moment per unit volume. For a diatomic crystal, where there are N positive and N negative ions per unit volume, the contribution to the polarization from the relative displacements of the ions (u^+ for the positive ion and u^- for the negative ion) is given by

$$p(\text{ionic}) = Ne\,(u^+ - u^-) = \frac{Ne^2/\mu}{\omega_T^2 - \omega^2}\,E \qquad (1.21)$$

where μ is the reduced mass of the ion-pairs with masses M^+ and M^- given by $1/\mu = 1/M^+ + 1/M^-$ and ω_T is the frequency of an optical mode at $q = 0$. Equation 1.21 leads to a frequency dependent dielectric function,

$$\varepsilon(\omega) = \varepsilon(\infty) + \frac{\omega_T^2}{\omega_T^2 - \omega^2} \cdot [\varepsilon(o) - \varepsilon(\infty)], \qquad (1.22)$$

where $\varepsilon(o)$ is the static dielectric constant and $\varepsilon(\infty)$ is the high frequency dielectric constant. The dielectric function shows a resonant behavior near the frequency of the normal mode $\omega_j (= \omega_T)$, and strong absorption occurs at this frequency in the infrared region of the electromagnetic spectrum. The most detailed information is obtained by measuring both the reflectivity and transmission of specimens as a function of the infrared frequency. The selection rules for infrared absorption are different from those for Raman scattering and different modes may be observed. As a result, these two techniques are usually complimentary. (For more details on these two techniques, see McMillan, this volume.)

2. PHASE TRANSITIONS AND SOFT MODES

2.1 Introduction

Changes of phase in a material, such as melting of a solid, evaporation of a liquid or a structural change from one polymorph to another, are accompanied by drastic changes in macroscopic properties of the material. Usually, such phase transitions are accompanied by changes in the arrangement of atoms or molecules of which the material is composed. There is another class of phase transitions, where the physical properties of the materials change continuously as a function of some scalar parameter, whereas the symmetry undergoes a sudden change to a lower symmetry at some critical value of that parameter. In the case of a solid, no dramatic reconstruction of the lattice takes place at the critical point. Many ferroelectric-paraelectric and paramagnetic-ferro- or antiferromagnetic phase transitions belong to this category. As structural phase transitions of this type in minerals, we can cite the cases of α-β transition in quartz, the $P2_1/c$ to $C2/c$ transition in $(Mg,Fe)_2Si_2O_6$ clinopyroxenes and $P\bar{1}$ to $I\bar{1}$ transition in anothite. Such second order (displacive) transitions are characterized by certain symmetry restrictions, first formalized by Landau (1937), and have been investigated extensively by theoretical and experimental techniques in the recent years (see Rao and Rao, 1978; Bruce and Cowley, 1981; Müller and Thomas, 1981).

In the preceding section, we have treated the crystal as a simple harmonic oscillator, a consequence of which is that the frequencies of the normal modes are independent of temperature. However, phenomena such as lattice thermal expansion and conductivity, ionic diffusion and phase transition indicate that anharmonic effects are present. A simplifying assumption is usually made ("quasi-harmonic" crystal), where the modes are still harmonic (regardless of compression or expansion) but force constants are a function of volume (pressure, temperature). This assumption will be implicit in the following discussion of displacive phase transition. A consequence of the presence of anharmonic effects is that frequencies of some phonon modes may change drastically with temperature and attain zero value at the transition temperature. These modes are known as soft modes.

2.2 Landau's theory of second order phase transition

During a second order phase transition, the entropy and volume of the system remain continuous, while the heat capacity and thermal expansion undergo a discontinuous change. Landau (1937) proposed a theory that can account for this behavior (see Landau and Lifshitz, 1969). In a second order phase transition, the properties of a crystal change gradually and only one phase exists at a given temperature. For this to happen, certain symmetry conditions must be satisfied.

Let us denote the positions of atoms k in the unit cell ℓ by a vector $R_0(\ell,k)$ in the high symmetry phase with symmetry G_0. If the crystal changes continuously, the new structure (low symmetry phase with symmetry G) can be described by a vector

$$\underset{\sim}{R}(\ell,k) = \underset{\sim}{R}_0(\ell,k) + \underset{\sim}{u}(\ell,k) \tag{2.1}$$

The continuous change implies that the symmetry group G must be a subgroup of G_0 (first condition).

The displacement vector u, which produces the new structure of lower symmetry G, belongs to only one irreducible representation of G_0 (second condition). [This irreducible representation has characters +1 for symmetry operations which exists in both phases R_0 and R. The characters for symmetry operations not present in R are -1. This explains why the soft mode in the less symmetric phase is always in the totally symmetric representation (A_1).]

The transition from high temperature to low temperature corresponds to the onset of an ordering process. In the ordered, less symmetrical (low temperature) phase, it is possible to identify a long range order parameter, which decreases continuously with temperature and becomes zero at the transition temperature. The free energy function can be expanded in a power series in the order parameter, η. For small values of η (near the transition temperature) the free energy, $F(P,T,\eta)$ may be written as:

$$F(P,T,\eta) = F_0(P,T) + a\eta + b\eta^2 + c\eta^3 + d\eta^4 + \cdots \tag{2.2}$$

where $F(P,T)$, a, b, c, d are constants. If the value of $F(P,T,\eta)$ is to remain unaltered by the change of sign of η during a continuous phase transition, coefficients of odd powers of η should be equal to zero (third condition). (For group theoretical arguments leading to this result, see Landau and Lifshitz, 1969.) Therefore,

$$F(P,T,\eta) = F_o(P,T) + b\eta^2 + d\eta^4 + \cdots \tag{2.3}$$

The equilibrium value of the long range order parameter is obtained by the minimization of the free energy and the following conditions are obtained:

$$\left(\frac{\partial F}{\partial \eta}\right)_{P,T} = \eta\,(b + 2d\eta^2) = 0 \tag{2.4}$$

$$\left(\frac{\partial^2 F}{\partial \eta^2}\right)_{P,T} = (b + 6d\eta^2) > 0 \tag{2.5}$$

From Equation 2.4, we obtain the solutions $\eta = 0$ and $\eta^2 = -b/2d$. Since $\eta = 0$ corresponds to the disordered state, it follows from the second equation that $b > 0$ on one side of the transition temperature. Similarly, using the value $\eta^2 = -b/2d$ in Equation 2.5, we find that $b < 0$ for the ordered phase. Thus, b should change sign through a second order transition. Since b is negative for the ordered phase in the vicinity of the transition, d should be positive ($-b/2d = \eta^2 > 0$). Assuming b to vary linearly with temperature, we find that near the transition point

$$b(P,T) = B(T-T_c) \tag{2.6}$$

where T_c is the "critical" or transition temperature; η^2 now becomes

$$\eta^2 = -b/2d = -B(T-T_c)\,/2d. \tag{2.7}$$

or

$$\eta = \eta_o(T-T_c)^{\frac{1}{2}} \tag{2.8}$$

The exponent 1/2 is characteristic of mean-field theories. Landau's theory is a mean-field theory, in which the existence of long range forces play a dominant role. In order to see how Landau's model is connected to a soft mode, let us assume the simple one-dimensional harmonic oscillator model. In such a case F would be given by

$$F = 1/2\ \omega^2 Q_m^2 + \text{other noncritical terms} \tag{2.9}$$

where ω is the vibration frequency and Q_m the mean value of the normal coordinate. Comparing Equations 2.2 and 2.9, we get

$$\eta = Q_m \text{ and } b = \omega^2.$$

Using Equation 2.6, we have

$$\omega^2 = B(T-T_c). \tag{2.10}$$

Thus, in a second order phase transition, where the assumption of the existence of long range force is reasonable, we would expect the existence of a vibrational mode frequency whose variation with temperature is approximately represented by Equation 2.10, which represents a soft mode. The idea behind a soft mode is that with decreasing frequency, the atomic displacements become larger and larger and at T_c with $\omega = 0$, the dynamic atomic displacements become frozen in as static displacements resulting in the low symmetry phase.

In second order phase transitions some symmetry elements of the high temperature phase are lost on cooling below T_c. This is a case of broken symmetry, where the symmetry that is broken involves a discrete symmetry group and the free energy in the low symmetry phase has a finite number of minima connected with the symmetry operation broken at T_c. The frequency of the symmetry mode will be non-zero at all temperatures below T_c. As $T \to T_c$, the free energy minima merge into a minimum corresponding to the high temperature phase (with zero order parameter); at $T = T_c$, the frequency of the soft mode will be zero. Thus, the existence of a soft mode $\omega \neq 0$ at $T < T_c$ which goes to zero at $T \to T_c$ is a consequence of breaking a discrete symmetry. The symmetry-breaking mode of a high temperature phase becomes the symmetry-restoring mode of the low temperature phase. We will illustrate this idea using the α-β quartz transition as an example in Section 2.4.

2.3 Soft modes and their investigation by inelastic neutron scattering

The idea of a soft mode was mentioned by Raman and Nedungadi (1940) with respect to the α-β transition in quartz. They stated "The 220 cm^{-1} line (in the Raman spectrum of quartz) behaves in an exceptional way, spreading out greatly towards the exciting line and becoming a weak diffuse band as the transition temperature is approached.... It appears therefore reasonable to infer that the increasing excitation of this particular mode of vibration with increasing temperature and the deformation of the atomic arrangement resulting therefrom are in a special measure responsible for .... inducing the transition from the α to the β-form." Saksena (1940) also associated this mode with the α-β transition in quartz. Cochran (1960) and Anderson (1960) have since developed the soft mode idea, which is becoming increasingly important in the study of second order phase transitions. Although the α-β phase transition in quartz is not strictly a second order transition, the soft mode concept has been very useful in understanding the dynamical aspects of this phase transition, which has been extensively studied by neutron and light scattering. We will discuss some details of this phase transition in Section 2.4.

Inelastic neutron scattering is a very powerful tool for phase transition studies for two reasons: first, it can cover many Brillouin zones and second, phonons are always coupled to neutron waves, whereas many soft phonons, particularly above T_c, are Raman and infrared inactive (see Shirane, 1974; Dorner, 1981).

For soft mode investigations the typical neutron has an energy of 14 meV (1meV = 8.07 cm^{-1}), wavelength $\lambda = 2.42$ Å and wavevector

$k = 2.60$ Å^{-1}, which permits the investigation of a large range of phonon wavevector q at different Brillouin zones. The equation used is:

$$Q = k_i - k_f = K(hkl) + q \quad (2.11)$$

where k_i and k_f are initial and final neutron wavevectors, q the wavevector of the phonon, and K (hkℓ) is the reciprocal lattice vector. The phonon energy is given by the change in the neutron energy:

$$\hbar\omega = \Delta E$$

$$\text{and hence, } \hbar\omega_j(q) = \frac{\hbar^2}{2M_n}(k_i^2 - k_f^2) \quad (2.12)$$

When the energy transfer $E' - E = \Delta E$ of the neutrons is zero, we have the elastic Bragg scattering, used for the "static" structure determination. In soft mode studies, one determines the relation between ω_1, and q as a function of temperature near the soft mode. The phonon energy of soft mode at a special symmetry point in the Brillouin zone decreases with lowering of temperature towards the transition point, and obeys the relation:

$$(\hbar\omega)^2 = A\,(T - T_c) \quad (2.13)$$

which may be called a condensing phonon mode. Such a material will undergo a second order (displacive) phase transition at T_c.

Once the location and temperature dependence of the soft mode is established, we would like to know what are the atomic motions involved in this soft mode and how they are related to the atomic positions of the low symmetry (low temperature) phase. The intensity of inelastic scattering for a one-phonon process is proportional to the square of the inelastic structure factor. Before we consider the inelastic structure factor, let us look at the elastic (Bragg) structure factor for the neutron case, which is given by

$$F_{el}(Q) = \sum_{k}^{\text{unit cell}} \overline{b}_k \exp[-W_k(Q) + iQ \cdot r_k] \quad (2.14)$$

where, $\overline{b}_k$ = neutron scattering length for the atom k, W_k = Debye-Waller (temperature) factor for atom k, Q is the scattering vector (k_i-k_f) and r_k the position vector to the equilibrium position of atom k in the unit cell.

The corresponding inelastic neutron structure factor is given by

$$G_{inel}(q,Q) = \sum_{k}^{\text{unit cell}} \overline{b}_k \exp[-W_k + iQ\cdot r_k][Q \cdot e_j(kq)]\, M_k^{-\frac{1}{2}} \quad (2.15)$$

This expression contains the eigenvector $e_j(kq)$, which is normalized to unity; it describes the pattern of displacements in the normal mode in one unit cell. For a known crystal structure where the temperature factors are known, this is the only unknown quantity that has to be

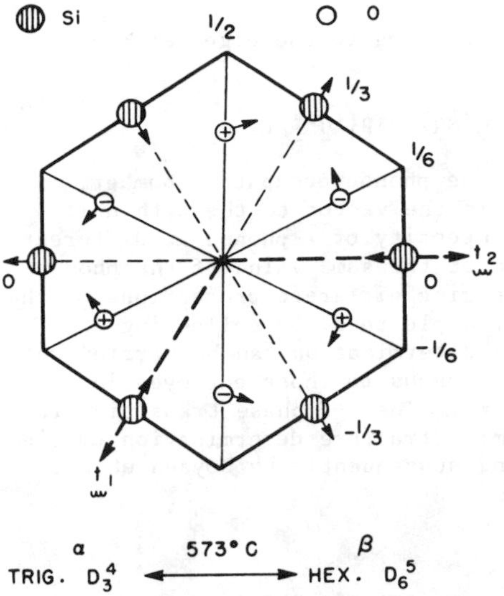

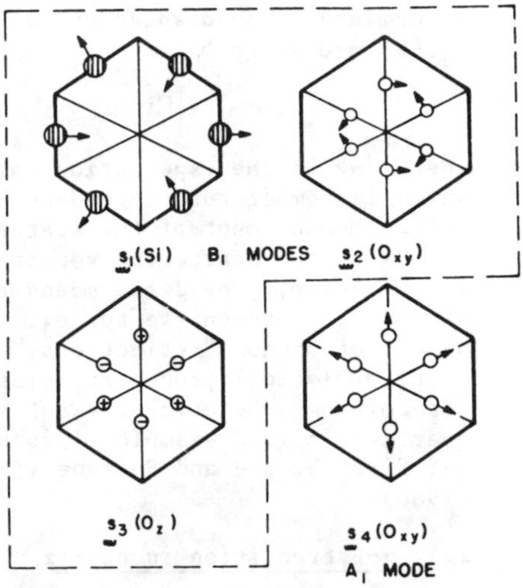

Figure 14 (left). Shifts in the atomic positions involved during the α-β transition in quartz.

Figure 15 (right). Symmetry mode vectors which describe the displacements associated with the α-β quartz transition.

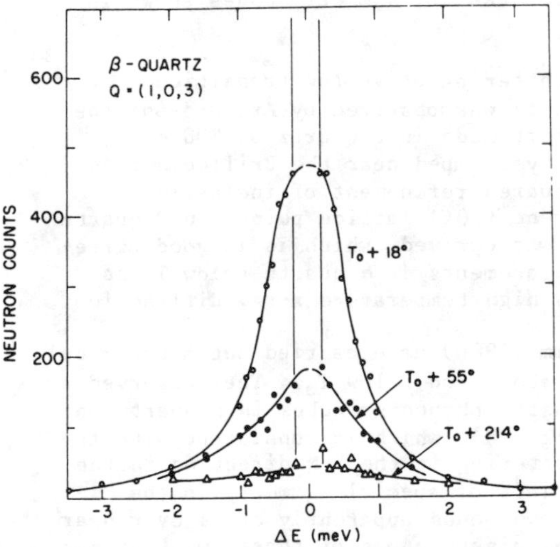

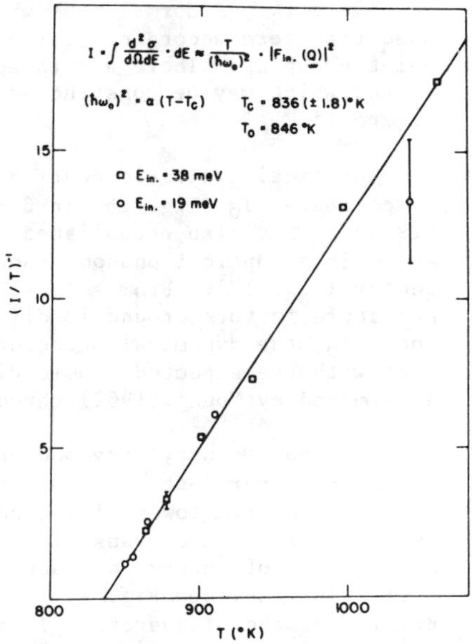

Figure 16 (left). Critical inelastic scattering about the (1,0,3) reflection in β-quartz at several temperatures above T_0. The central peak superimposed on the inelastic scattering is the Bragg peak (1,0,3).

Figure 17 (right). Temperature dependence of the integrated inelastic scattering intensity for the (1,0,3) reflection. Note that the temperature dependence obeys Curie-Weiss law.

Figures 14-17 from Axe and Shirane (1970).

determined. The displacements $\underset{\sim}{u}$ corresponding to the eigenvectors $\underset{\sim}{e}_j(k\underset{\sim}{q})$ are given by

$$\underset{\sim}{u}_k = [\hbar(\langle n \rangle + \tfrac{1}{2}) / M_k \omega]^{\frac{1}{2}} \; \underset{\sim}{e}(k\underset{\sim}{q}) \; \exp(i\underset{\sim}{q} \cdot \underset{\sim}{\ell}), \qquad (2.16)$$

where $\langle n \rangle$ is the expectation value of the phonon occupation number, which is temperature dependent, and ℓ is the vector to the ℓ th unit cell. Measurement of the scattering intensity of a phonon at different values of the scattering vector $\underset{\sim}{Q}$, but at the same value of the phonon wavevector $\underset{\sim}{q}$, provides a means of measuring different projections of the phonon displacement vector $\underset{\sim}{e}_j$. In principle then, by collecting a number of phonon "reflections," a mode determination can be carried out using analytical procedures closely analogous to those employed in determining a "static" crystal structure. The α-β phase transition in quartz is a good example of such dynamic structure determination carried out first by Axe and Shirane (1970) and subsequently by Boysen et al. (1980).

2.4 α-β transition in quartz

The α-β phase transition in quartz at 573.3°C involves changes from the high temperature hexagonal β form with space group symmetry $D_6^4(P6_222)$ [or its enantiomorph $D_6^5(P6_222)$] to the trigonal α form with symmetry $D_3^4(P3_12)$ [or $D_3^6(P3_22)$, resulting in a loss of 2-fold rotation symmetry parallel to $\underline{c}$. The associated atomic displacements are shown in Figure 14. A normal mode which breaks the symmetry of the β phase must transform according to the B_1 irreducible representation of the point group D_6. There are three different vibrational modes in quartz which may be constructed from the $3B_1$ symmetry modes shown in Figure 15.

Critical inelastic neutron scattering above the transition temperature, T_0 (Fig. 16) in β quartz was observed by Axe and Shirane (1970). They also established a soft mode in β quartz at 208 cm^{-1}, which is an optical phonon branch overdamped near the Brillouin zone center (Fig. 17). From a least squares refinement of inelastic structure factors around 13 different (h0ℓ) lattice points in β quartz above T_0, the dynamical structure was derived, which is in good agreement with the expected atomic displacements in α quartz below T_0 as determined by Young (1962) through high temperature x-ray diffraction.

Boysen, Dorner, Frey and Grimm (1980) have carried out a dynamical structure determination of quartz above and below T_0. They observed a softening of two lowest lying acoustic phonon branches in α quartz on approaching the α-β transition (Fig. 18), which is consistent with the observation of intense diffuse scattering in the Γ-M direction in the Brillouin zone (see Fig. 19 for points of special symmetry in the Brillouin zone of quartz). These two modes apparently cross over near the transition point, such that the eigenvectors of these two branches are interchanged above T_0.

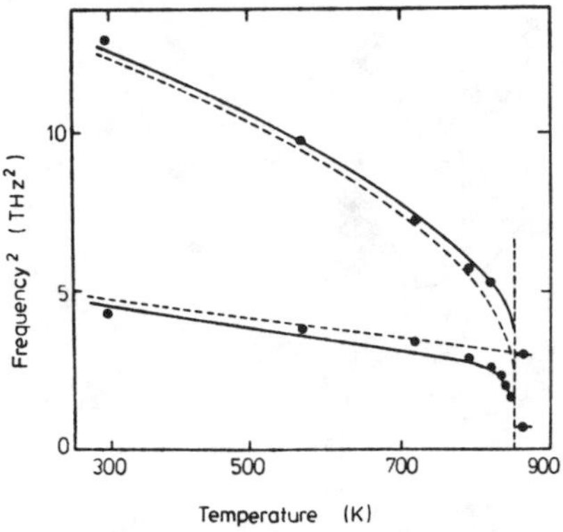

Figure 18 (left). The temperature dependence of two modes at the zone boundary (M point) in α-quartz. The broken curves are uncoupled frequencies. After Boysen et al. (1980).

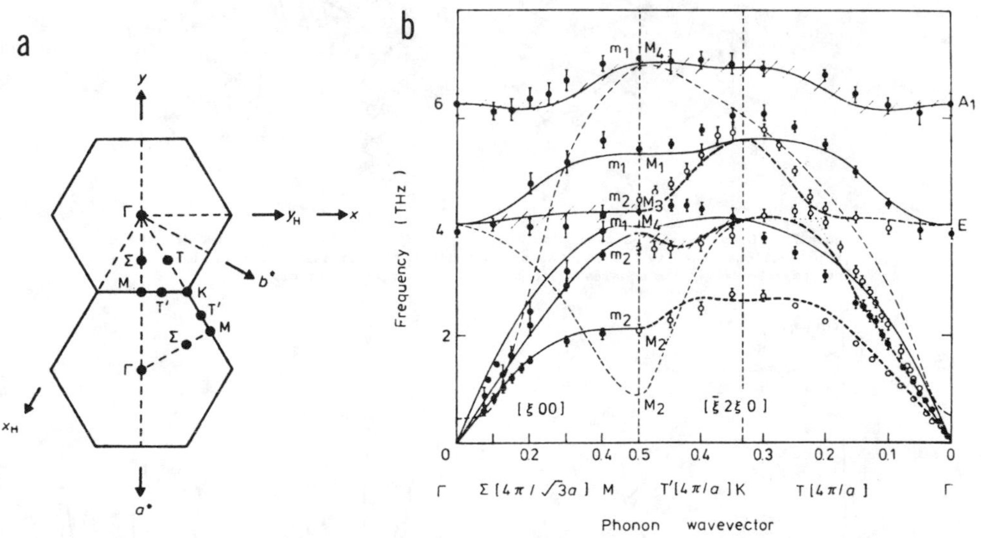

Figure 19 (below). (a) Part of the Brillouin zone (hk0 plane) of quartz showing points of special symmetry and the orthogonal coordinate system with respect to the hexagonal axes (Boysen et al., 1980). (b) Dispersion branches of six lowest modes in α-quartz at 295 K as measured by Dorner, Grimm and Rzany (1980, unpublished). The dashed curves indicate the softening of the strongly temperature dependent branches. After Boysen et al. (1980).

The results of the dynamic structure determination in α quartz at 573 K are shown in terms of displacements $\tilde{u}$ calculated from the eigenvectors $\tilde{e}_j$ in Figure 20. The lower mode can be described by an antiphase translation of the silicate chains, which is transmitted by a rotation of the coupling tetrahedra (Fig. 20a). In this motion, all [SiO_4] tetrahedra stay rigid, whereas the Si-Si distance is changed. The upper mode in β phase shows a similar motion. The upper mode in α quartz (correspondingly the lower mode in β phase) can be described by a libration of the silicate tetrahedra within the chain around axes parallel to y, connected to an up and down motion of Si1 parallel to z (Fig. 20b). This confirms the statement made earlier that the two modes change eigenvectors above T_0.

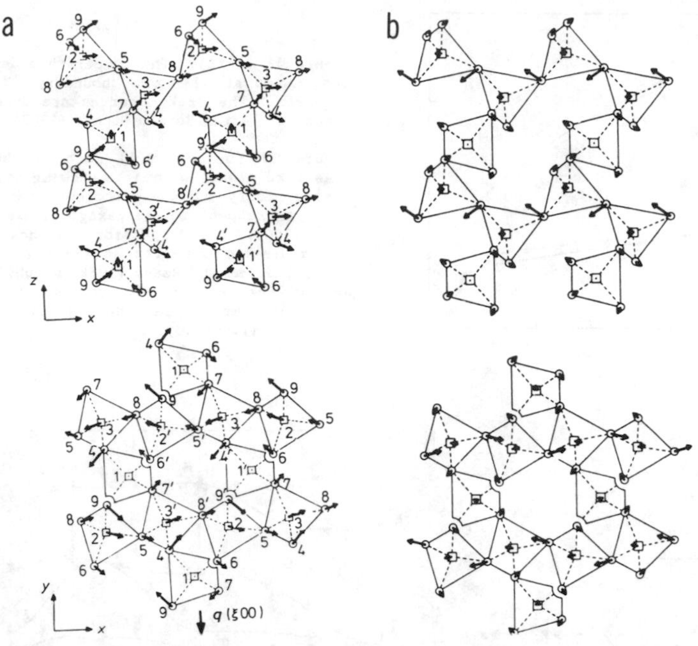

Figure 20. Displacement patterns for lower (a) and upper (b) mode at 573 K in x,y and x,z projections (see Fig. 19 for the orthogonal coordinate system used). Small circles indicate oxygen and squares silicon atoms. After Boysen et al. (1980).

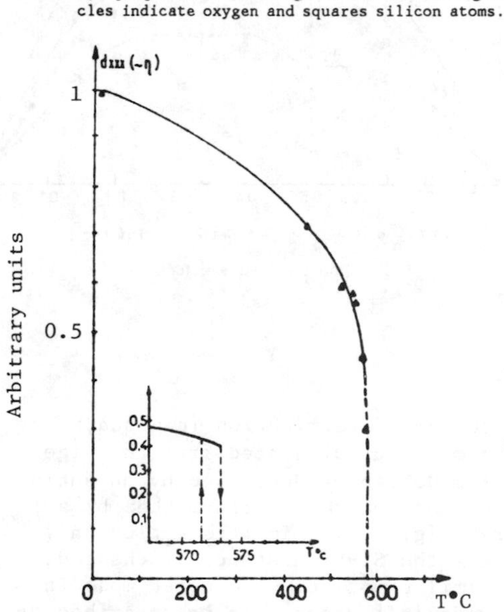

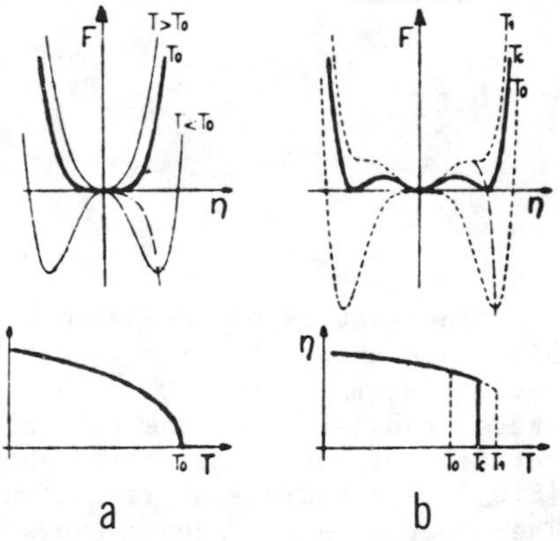

Figure 21 (left). Temperature dependence of the second harmonic piezoelectric coefficient d_{xxx} angle (normalized to 1 at ambient temperature). Filled triangles indicate rotation of silicate tetrahedra based on data from Young (1962). After Bachheimer and Dolino (1975).

Figure 22 (right). Variation of the free energy as a function of the order parameter according to the Landau theory; (a) second order transition (d > 0); (b) first order transition (d < 0). After Dolino et al. (1983).

We have treated the α-β phase transition in quartz so far in terms of a continuous (second order) transition driven by a soft mode mechanism. This is not strictly true. Variation of a number of physical properties with temperature which can be correlated with the order parameter η show small but definite discontinuities at the transition temperature T_o. For example, the piezoelectric coefficient d_{xxx} from second harmonic generation associated with the order parameter η (Fig. 21) or the thermal expansion coefficient and certain elastic constants (C_{44} and C_{66}) which vary with η^2 show definite discontinuities near T_o (see Dolino et al., 1983). These phenomena can be explained by an order-disorder mechanism.

Let us examine the Landau expression for the free energy in terms of the order parameter η (Eqn. 2.3)

$$F(P,T,\eta) = F_o(P,T) + b\eta^2 + d\eta^4 + \cdots \qquad (2.3)$$

In this expression, if the coefficient d is positive, $\eta \rightarrow 0$ at T_o continuously and the transition is second order. On the other hand, if d is negative, η shows a discontinuity at the phase transition and the transition is then first order, which is the case for quartz. The enthalpy of transition, ΔH has been estimated to be 360 J/mole by Dolino et al. (1983). The variations of free energy as a function of the order parameter for (a) second order transition ($d > 0$) and (b) first order transition ($d < 0$) are shown in Figure 22. In the case of the first order transition, T_c is the theoretical transition temperature; for quartz the transition temperature T_o is about 10°C above T_c and the free energy variation at T_c has three minima (Fig. 22). To understand this behavior, we have to examine the details of the quartz structure.

The α quartz structure is a framework of rigid $[SiO_4]$ tetrahedra consisting of double spirals of non-intersecting chains around the trigonal axis. The tetrahedra are tilted away from the symmetrical quartz configuration about the diad axes parallel to [100] by an angle ϕ (Megaw, 1973). This angle, which varies continuously with temperature with a value of 16° at room temperature to 7° near T_o, has been equated with the order parameter η and the frequency of the soft mode at 207 cm^{-1} (Höchli and Scott, 1971). Two possible enantiomorphs α_1 and α_2 (left- and right-handed) are transformed into each other by a diad axis coinciding with the trigonal axis. These two configurations correspond to the Dauphiné twin domains, which differ by the tilt angles ϕ of opposite directions. The Dauphiné twin component in α quartz increases with temperature (Young, 1962). Near T_o the triangular Dauphiné twin domains with strongly temperature dependent fluctuations of the domain walls have been observed by transmission electron microscopy (Van Tendeloo et al., 1976). β quartz then exists in two equilibrium configurations α_1 and α_2, which can be described as a particle in a double minimum potential well. Near T_o the α_1 configuration can switch over to the α_2 configuration through an intermediate β configuration resulting in a triple minimum potential well (cf. Fig. 22). Above T_o, the (averaged) hexagonal β quartz configuration contains equal amounts of α_1 and α_2. This interpretation of β quartz is favored on the basis of a neutron refinement at 590°C (Wright and Lehmann, 1981). From the above discussion, it is clear that the α-β quartz transition is a result of successive dynamic and order-disorder mechanisms.

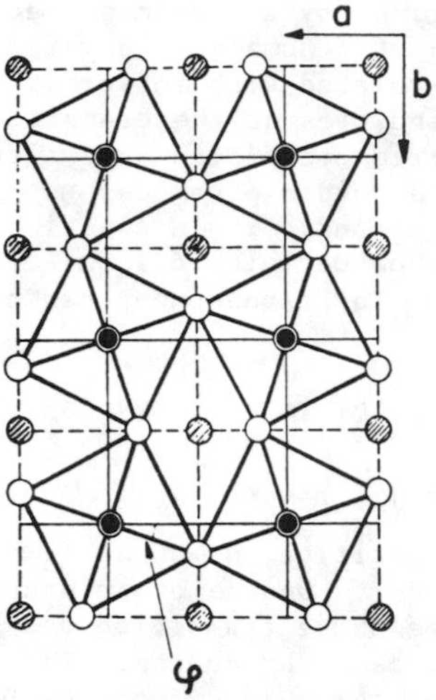

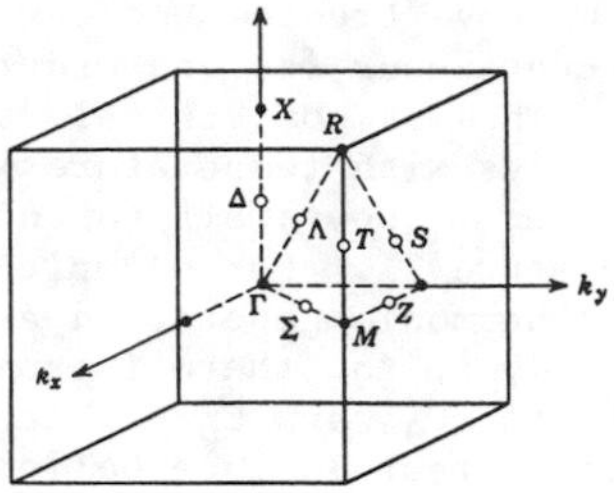

Figure 23 (left, above). The low temperature tetragonal structure of strontium titanate (left) as suggested by Unoki and Sakudo (1967). Note the clockwise and counterclockwise rotation of alternate [TiO_6] octahedra compared to the symmetric cubic phase. Reciprocal space (cubic) showing symmetry points R and M at the zone boundary and Γ at the origin. After Shirane (1974).

Figure 24 (below). (a) The zone boundary soft mode eigenvectors for cubic strontium titanate. (b) Variation of the soft mode frequency with temperature. After Shirane and Yamada (1969).

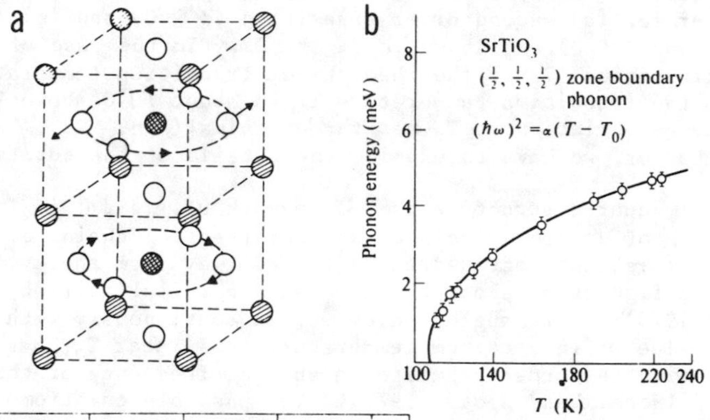

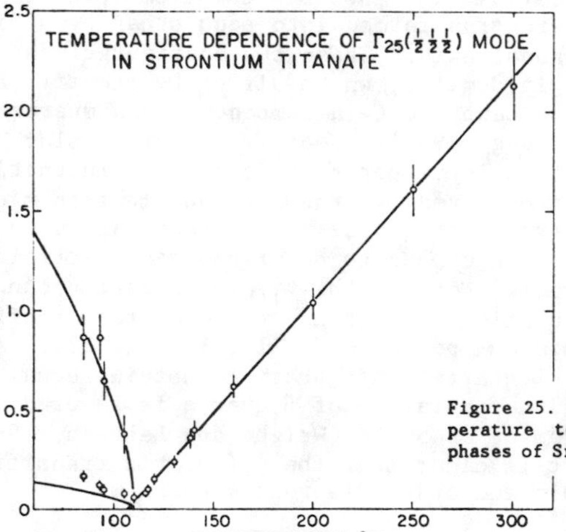

Figure 25. Soft mode frequencies in the low temperature tetragonal and high temperature cubic phases of $SrTiO_3$. After Cowley et al. (1969).

Dynamically, the transition is driven principally by a strongly coupled libration of the rigid silicate tetrahedra along chains parallel to axes [100], [010] and [110] (Boysen et al., 1980; Liebau and Böhm, 1982). When the libration amplitude in α quartz reaches a critical limit near T_0, the α_1 configuration switches over to the α_2 configuration resulting in Dauphiné twinning and the transition is driven by the order-disorder mechanism.

2.5 Cubic to tetragonal phase transition in perovskite type strontium titanate

Not all phase transitions are as complex as quartz. The transition in strontium titanate at 110 K from cubic to tetragonal symmetry is a simple and classic example of a structural phase transition, which is strictly second order. It has been extensively studied by a number of techniques including elastic and inelastic neutron scattering, electron paramagnetic resonance, elasticity and dielectric measurements, etc. At 110 K, the elastic constants show significant change, whereas no anomaly was observed for the dielectric constant. The latter shows a very strong temperature dependence, which follows the Curie-Weiss law. The cubic lattice constant shows a small splitting (<0.1%) below T_0 due to the tetragonal distortion. On the basis of EPR measurements, Unoki and Sakudo (1967) proposed a model, where the high temperature cubic perovskite phase is distorted by an anti-phase rotation of neighboring $[TiO_6]$ octahedra (Fig. 23). That the tetragonal distortion is the consequence of a soft mode instability was proven directly by the detection of a soft phonon at R point on the zone boundary in the cubic phase by inelastic neutron scattering (Shirane and Yamada, 1969) (Fig. 24). At the transition the zone boundary becomes a superlattice point, thereby doubling the unit cell. The soft mode energy $\hbar\omega_0$ follows the equation

$$(\hbar\omega_0)^2 = A(T-T_0)^\gamma,$$

where γ has a value greater than one. The soft mode in the cubic phase corresponds to a very simple rotation of the octahedra as shown in Figure 23, which is also the static structure of the tetragonal phase. The octahedra on the layer above and below rotate in the opposite way. The octahedral rotation angle ϕ then is the order parameter η of the phase transition. In the cubic phase a single normal mode of vibration condenses to give rise to the low temperature tetragonal phase. The Raman and inelastic neutron scattering spectra of the tetragonal phase reveal the presence of two soft phonon modes. The zone boundary soft phonon in the cubic phase is apparently the progenitor of these two soft modes (Fig. 25) (Fleury, Scott and Worlock, 1968; Cowley, Buyers and Dolling, 1969).

We have so far mentioned zone center and zone boundary phonons. In fact entire phonon branches can be soft, which may lead to incommensurate phase transitions as in K_2SeO_4 (Iizumi et al., 1977). We have also considered the phase transitions in terms of one order parameter, where in fact, there can be more than one in some cases. For example, in the phase transition of low albite, $NaAlSi_3O_8$ (triclinic, $C\bar{1}$) to monalbite (monoclinic, C2/m) at 980°C requires two equally important

Table 1. Some typical soft-mode transitions* (from Rao and Rao, 1978)

	Method of study	Soft mode	Structural change	Transition temperature
$PbTiO_3$	NS & RS	Zone center T.O phonon	Cubic to tetragonal	763 K
$BaTiO_3$	NS	Zone center T.O phonon	Cubic to tetragonal	403 K
$KTaO_3$	NS & RS	Zone center T.O phonon	Cubic to tetragonal?	~0 K
$KNbO_3$	NS	Zone center T.O phonon	Cubic to tetragonal	693 K
$LaAlO_3$	NS	Zone center T.O phonon	Cubic to rhombohedral	535 K
$SrTiO_3$	NS	Zone boundary T.O phonon	Cubic to tetragonal	110 K
$KMnF_3$	NS	Zone center T.O phonon	Cubic to tetragonal	186 K
$KMnF_3$	NS	Zone center T.O phonon	Tetragonal to orthorhombic	91 K
$KCoF_3$	RS	Magnon	Cubic to tetragonal	114 K
$RbCoF_3$	RS	Magnon	Cubic to tetragonal	101 K
$Tb(MoO_4)_3$	NS	Zone center T.O phonon	Tetragonal to orthorhombic	433 K
NbO_2	NS	Zone center T.O phonon	Rutile to body-centered tetragonal	1073 K
SiO_2	NS & RS	Zone center T.O phonon	Hexagonal to trigonal	846 K
$AlPO_4$	RS	Zone center T.O phonon	Hexagonal to trigonal	853 K
Na_2WO_3	NQR	Zone center T.O phonon	Cubic to tetragonal	400 K
GeTe	RS	Zone center T.O phonon	Cubic to rhombohedral	670 K
ND_4Br	NS	Zone boundary (coupled to flipping mode)	Cubic (disordered) to cubic (ordered)	215 K
KD_2PO_4	NS	Zone center T.O phonon	Tetragonal (disordered) to tetragonal (ordered)	220 K
KH_2PO_4	RS	Zone center T.O phonon	Tetragonal (disordered) to tetragonal (ordered)	122 K
$TbVO_4$	RS	Acoustic phonon electron coupled mode	Tetragonal to orthorhombic	34 K
Nb_3Sn	NS	Acoustic phonon electron coupled mode	Cubic to tetragonal	46 K

* Detailed references to the original literature may be obtained from two recent reviews of Scott, 1974. Shirane, 1974. NS = neutron spectroscopy; RS = Raman spectroscopy; NQR = nuclear quadrupole resonance spectroscopy.

order parameters: one to describe the degree of Si-Al order-disorder and another to describe the atomic displacements involved in the displacive transition from the Al-Si disordered triclinic phase to the monoclinic phase. The theory for this phase transition including the coupling of the order parameters with elastic strain parameters has been given by Salje (1985).

Possible ferroelectric phase transitions in paraelectric (cubic) $BaTiO_3$ according to the theory of Devonshire (1949) is given in Appendix 2. Some typical soft mode transitions are listed in Table 1.

2.6 Critical fluctuations and critical-point exponents

Some of the basic facts of critical phenomena, i.e., the behavior of systems near the critical point of a phase transition, have been known since the nineteenth century, particularly from the study of the behavior of CO_2 near its critical point (Andrews, 1869). A gas consists of a collection of molecules that move independently at high temperatures, but as the temperature is lowered towards the critical temperature, the molecules begin to come together to form small droplets which are continually forming and breaking up again. When the temperature becomes very close to the critical point, the dimensions of these droplets becomes of the order of the wavelength of light, and the scattering power of the gas is very much increased; this phenomena is called critical opalescence. In the liquid phase at a temperature just below the critical temperature there will be small pockets of vapor. In the same way that the formation of droplets in the gas just above T_c and of pockets of vapor in the liquid just below T_c lead to the phenomenon of critical opalescence in the scattering of light by a fluid near the critical point for the liquid-gas transition, it is reasonable to expect the fluctuations of atomic arrangements near a continuous (second order) structural phase transition or the orientation of magnetic moments near a magnetic phase transition to lead to similar increase near T_c in the scattering properties of the crystal undergoing phase transition. In particular, these critical fluctuations lead to anomalies in the scattering of neutrons and of ultrasonic waves. The cross section for the scattering of a beam of neutrons by a crystal shows a marked increase in the vicinity of T_c. The neutron intensity as a function of temperature for the (1/2 1/2 3/2) superlattice reflection in $SrTiO_3$ is shown in Figure 26 and the critical scattering in Figure 27. For $T > T_c$ the intensity above the incoherent background is due to scattering by fluctuations. Similarly, the attenuation of ultrasonic waves propagating through a crystal undergoing phase transition also shows a marked increase in the vicinity of the critical point. In magnetic crystals, the survival of short range order at temperatures not too much in excess of T_c can be demonstrated by observing a vestigial spin-wave behavior of the magnetic moments above T_c. Such spin-wave like excitations in the paramagnetic phase at temperatures not too much higher than T_c are described as paramagnons, and their existence has been detected experimentally both in the inelastic scattering of neutrons as well as in the Raman scattering of laser light (e.g., Fleury, 1969, on NiF_2). There are a number of other techniques that can also be used to study the various physical properties of a magnetic crystal at temperatures near T_c. These include measurements of magnetization and specific heat and the use of nuclear magnetic resonance and of the Mössbauer effect. For structural phase transitions specific heat measurements, nuclear magnetic resonance and electron paramagnetic resonance experiments are valuable (see Rao and Rao, 1978; Ausloos and Elliot, 1983; Bruce and Cowley, 1981; Müller and Thomas, 1981).

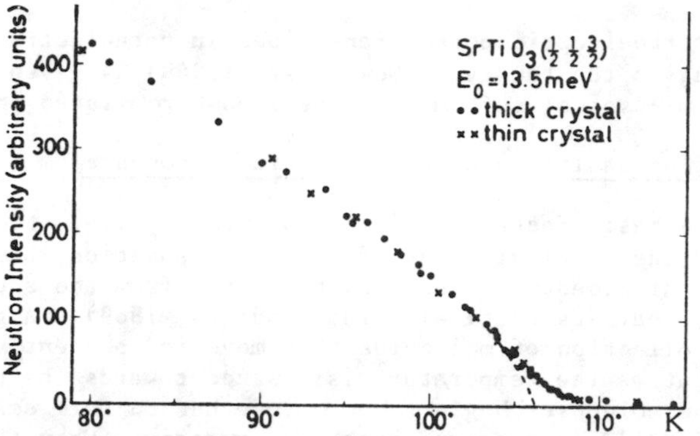

Figure 26. Temperature dependence of neutron intensity for the (1/2,1/2,3/2) superlattice reflection in $SrTiO_3$. After Riste et al. (1971).

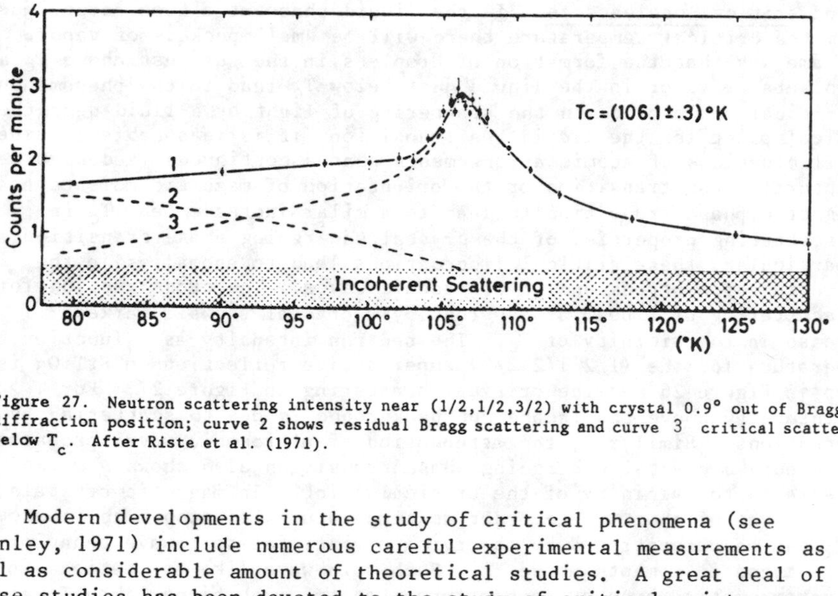

Figure 27. Neutron scattering intensity near (1/2,1/2,3/2) with crystal 0.9° out of Bragg diffraction position; curve 2 shows residual Bragg scattering and curve 3 critical scattering below T_c. After Riste et al. (1971).

Modern developments in the study of critical phenomena (see Stanley, 1971) include numerous careful experimental measurements as well as considerable amount of theoretical studies. A great deal of these studies has been devoted to the study of critical point exponents or critical exponents. As indicated earlier, any phase transition may be charcterized by an order parameter η. The critical exponents arise when one considers the behavior of the order parameter as a function of temperature near the critical point and assumes that the temperature dependence of any given order parameter η can be expressed as a power law in (T_c - T). It is conventional to write the order parameter as a power of ($-\varepsilon$), where

$$\varepsilon = (T - T_c)/T_c \tag{2.17}$$

η would then be expressed as B $(-\varepsilon)^\beta$, where B is a constant and β is the critical exponent for η. Guggenheim (1945) has shown for several simple examples of the liquid-gas system that the density in the region of T_c, $(\rho-\rho_c)$ obeyed a law of the form

$$(\rho-\rho_c) \propto (-\varepsilon)^\beta$$

with a value of β which is remarkably close to 1/3. This value of the critical exponent has also been found in diverse systems like structural and magnetic phase transitions, which indicate that the critical behavior is independent of the detailed nature of the interparticle interactions. The ESR studies on $SrTiO_3$ and $LaAlO_3$ indicates that in the region of 0.1T_c to T_c, the critical exponent changed form 1/2 to 1/3, characteristic of critical behavior (Müller and Berlinger, 1971). This universality of the critical exponent for three-dimensional systems is understandable, since in the critical region, fluctuations in the order parameter are enormous and the system senses only changes occurring over dimensions which are much larger compared to interatomic distances.

In α quartz, Höchli and Scott (1971) correlated the tilt angle ϕ of the silicate tetrahedra with the soft mode and certain elastic constants and expressed the relationships by the empirical formula

$$\eta(T) = \eta_o + k(T - T_o)^\beta$$

where the critical exponent $\beta \approx 1/3$. Just at the α-β transition temperature (within 0.1°C) an intense light scattering was observed by Yakolev et al. (1956), which was interpreted as critical opalescence. Later studies have shown that the scattered light does not change in frequency which definitely eliminates the existence of critical fluctuations. The scattering is produced by the Dauphiné twin domains which appear in the vicinity of the transition (Shapiro and Cummins, 1968). Furthermore, an incommensurate phase between α and β near T_o has been recently observed (Dolino, Bachheimer and Zeyen, 1983).

3. APPENDIX 1

THEORY OF LATTICE VIBRATIONS IN A PERFECT CRYSTAL IN HARMONIC APPROXIMATION

A comprehensive account of the formal theory of lattice dynamics is given by Born and Huang (1954). We present here a summary of the mathematical formalism for a perfect crystal in harmonic approximation. The potential energy of a crystal, ϕ, can be expanded in a power series of the displacements of the ions, u (ℓ,k), from their equilibrium positions $r(\ell k)$, where ℓ denotes the ℓ th unit cell (ℓ = 1, ...r), and

k is kth type of ion (k = 1, ...,N) within the unit cell. The potential is given by

$$\phi = \phi_0 + \tfrac{1}{2}\sum_{\ell\ell'}\sum_{kk'}\sum_{\alpha\beta} \phi_{\alpha\beta}\binom{\ell\ell'}{kk'} u_\alpha\binom{\ell}{k} u_\beta\binom{\ell'}{k'} + \cdots \quad (3.1)$$

where the suffices α, β, γ denote the cartesian coordinates. The second derivative $\phi_{\alpha\beta}\binom{\ell\ell}{kk'}$ is the negative of the force constant and gives the force in the α-direction on the ion (ℓk) when the ion ($\ell' k'$) is displaced by a (small) unit displacement in the β-direction. For a perfect crystal with uniform translational symmetry, no residual stresses remain. Hence,

$$\sum_{\ell' k'} \phi_{\alpha\beta}\binom{\ell\ell'}{kk'} = 0.$$

The equation of motion of the (ℓk)th atom is then

$$M_k \ddot{u}_\alpha\binom{\ell}{k} = -\sum_{\ell' k' \beta} \phi_{\alpha\beta}\binom{\ell\ell'}{kk'} u_\beta\binom{\ell'}{k'} \quad (3.2)$$

This equation is solved in a perfect crystal by finding a solution in terms of plane waves, such that

$$u_\alpha\binom{\ell}{k} = \frac{1}{\sqrt{M_\kappa}} u_\alpha(k|\underset{\sim}{q}) \exp i\,(\underset{\sim}{q}\cdot\underset{\sim}{r}\binom{\ell}{k} - \omega(\underset{\sim}{q})t) \quad (3.3)$$

where,

$$\omega(\underset{\sim}{q})^2 u_\alpha(k|\underset{\sim}{q}) = \sum_{k'\beta} D_{\alpha\beta}\binom{\underset{\sim}{q}}{kk'} u_\beta(k'|q) \quad (3.4)$$

Here, the Fourier transformed dynamical matrix, $D_{\alpha\beta}\left(\genfrac{}{}{0pt}{}{\underset{\sim}{q}}{kk'}\right)$ is given by

$$D_{\alpha\beta}\binom{\underset{\sim}{q}}{kk'} = \frac{1}{\sqrt{M_k M_{k'}}} \sum_{\ell'} \phi_{\alpha\beta}\binom{\ell\ell'}{kk'} \exp iq\cdot(\underset{\sim}{r}\binom{\ell'}{k'} - \underset{\sim}{r}\binom{\ell}{k}). \quad (3.5)$$

We have then $3rN$ coupled equations of motion. The standard technique which is used in solving these equations involves making a transformation to a set of normal coordinates. The squares of the frequencies are the eigenvalues of the dynamical matrix 3.5. The displacements of the atoms in one of these normal modes labelled by (qj) corresponds to a wave-like displacements of atoms, so that if the system is vibrating in one of its normal modes, we can write

$$u_\alpha\binom{\ell}{k} = \frac{1}{\sqrt{rM_k}} \sum_{\underset{\sim}{q}j} \underset{\sim}{e}_\alpha(k|\underset{\sim}{q}j) \exp i(\underset{\sim}{q}\cdot\underset{\sim}{r}\binom{\ell}{k}) P\binom{q}{j} \quad (3.6)$$

where $P\binom{q}{j}$ are the normal coordinates and $\underset{\sim}{e}(k|qj)$ is the eigenvector of the normal mode (qj), where j runs from 1 to 3N and is used to distinguish between the 3N normal modes at $\underset{\sim}{q}$, and $\omega_j(\underset{\sim}{q})/2\pi$ is the frequency of the normal mode. The frequencies of the normal modes (eigenvalues) and their eigenvectors are determined by diagonalizing the dynamical matrix through a solution of the secular equation:

$$\det\left| \omega_j(\tilde{q})^2 \, \delta_{kk'} \delta_{\alpha\beta} - D_{\alpha\beta}\binom{q}{\tilde{kk'}} \right| = 0, \tag{3.7}$$

The eigenvectors are normalized such that

$$\sum_{k\alpha} \tilde{e}^*_\alpha(k|\tilde{q}j)\tilde{e}_\alpha(k|\tilde{q}j')/M_k = \delta_{jj'} \tag{3.8}$$

The thermal occupancy number of the phonons is given by

$$n(\tilde{q}j) = [\exp \beta \hbar\omega(\tilde{q}j) - 1]^{-1}, \tag{3.9}$$

where $\beta = 1/k_BT$.

The frequency spectrum $g_j(\omega)$ is known as the <u>phonon density of states</u> and is given by

$$g_j(\omega) = \frac{NV}{8\pi^3} \int d\tilde{q} \tag{3.10}$$

where, V is the volume of the unit cell and the integration extends over the volume of the q-space lying between the surfaces corresponding to and $\omega + d\omega$. The elementary volume in $\tilde{q}$ space is given by

$$d\tilde{q} = \frac{d\,S\,d\,\omega}{|\text{grad}_{\tilde{q}}\, \omega_{\tilde{q}j}|},$$

where dS is the elementary surface area, the vector grad_q, being perpendicular to the surface. Now,

$$g_j(\omega) = \frac{NV}{8\pi^3} \iint \frac{d\,S}{|\text{grad}_{\tilde{q}}\, \omega_{\tilde{q}j}|} \tag{3.11}$$

This is the quantity we need to determine specific heat and other thermodynamic properties of crystals (see Kieffer, this volume).

4. APPENDIX 2

FERROELECTRIC TRANSITIONS IN BARIUM TITANATE: AN EXAMPLE

This example of possible phase transitions in cubic $BaTiO_3$ based on the theory of Devonshire (1949), similar to Landau's theory is from Boccara (1976).

The irreducible representation which characterizes the transition does not always determine the symmetry group of the ordered phase. If the dimension of the irreducible representation is greater than 1, the order parameter is not entirely defined by the quadratic invariant, and we have to examine the invariants of higher degree. Let us see how we can construct a phenomenological theory of the ferroelectric properties of barium titanate, whose cubic structure is represented in Figure 28.

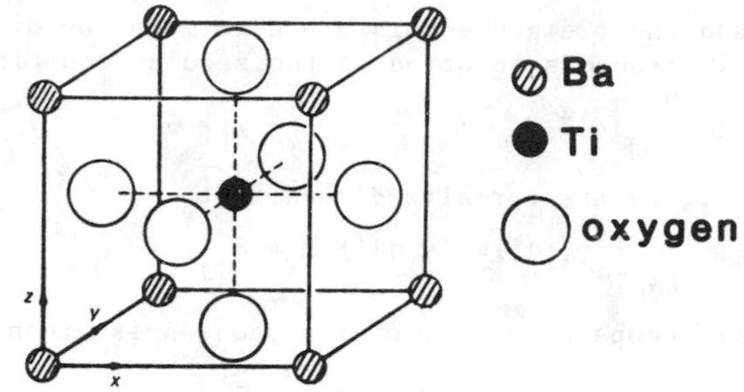

Figure 28. Crystal structure of cubic barium titanate. Ba at cube corners, Ti at body center and O on face-centers.

Since the ferroelectric phase is characterized by the existence of a spontaneous polarization $\underset{\sim}{P}$, we have to study the vectorial representation of the group of the cube. It is easy to confirm that this representation is irreducible, which is due to the fact that there exists only one quadratic invariant, $P_1^2 + P_2^2 + P_3^2$. P_1, P_2, P_3 are the components of $\underset{\sim}{P}$ in an axial system, which is parallel to the three axes of the cubic lattice. The direction of $\underset{\sim}{P}$, which determines the symmetry group of the ferroelectric phase is defined by a solution of the equation of minimization of the free energy $F(T; P_1, P_2, P_3)$. If the transition from the paraelectric phase to the ferroelectric phase is of second order, and if one assumes the free energy function to be an analytic function of the components of vector $\underset{\sim}{P}$, it is sufficient to consider the terms of the lowest degree. Knowing the symmetry properties of the crystal structure of barium titanate, we have

$$\Delta F = F(T; P_1, P_2, P_3) - F(T; 0, 0, 0)$$

$$= 1/2\ a\ (P_1^2 + P_2^2 + P_3^2) + 1/4\ b\ (P_1^4 + P_2^4 + P_3^4)$$

$$+ 1/2\ c\ (P_1^2 P_2^2 + P_2^2 P_3^2 + P_3^2 P_1^2),$$

where a, b, and c are functions of temperature. From this equation, we deduce the following equations of minimization:

$$\frac{\partial F}{\partial P_1} = a\ P_1 + bP_1^3 + c\ P_1\ (P_2^2 + P_3^2) = 0$$

$$\frac{\partial F}{\partial P_2} = a\ P_2 + b\ P_2^3 + c\ P_2(P_3^2 + P_1^2) = 0$$

$$\frac{\partial F}{\partial P_3} = a\ P_3 + b\ P_3^3 + c\ P_3(P_1^2 + P_2^2) = 0$$

This system of equations possesses several solutions shown in the table below.

Solutions	Symmetry of the Phase	ΔF	Condition of Stability
$P_1 = P_2 = P_3 = 0$	Cubic	0	$a > o$
$P_1 = P_2 = 0$, $P_3^2 = -\frac{a}{b}$	Tetragonal	$-\frac{1}{4}\frac{a^2}{b}$	$a < o$, $b > o$, $c > b$
$P_1 = 0$, $P_2^2 = P_3^2 = -\frac{a}{b+c}$	Orthorhombic	$-\frac{1}{2}\frac{a^2}{b+c}$	unstable for any value of b and c
$P_1^2 = P_2^2 = P_3^2 = -\frac{a}{b+2c}$	Rhombohedral	$-\frac{3}{4}\frac{a^2}{b+2c}$	$a < o$, $b > o$, $-\frac{b}{2} < c < b$

The stability conditions have been obtained by examining the matrix of elements

$$\frac{\partial^2 F}{\partial P_i \, \partial P_j} \qquad (i, j = 1, 2, 3)$$

is positive definite.

According to the relative values of b and c, one may go from the paraelectic state of cubic symmetry to a ferroelectric phase with tetragonal or rhombohedral symmetry through a second order transition. The passage to a ferroelectric state with orthorhombic symmetry is forbidden.

Barium titanate shows the following transitions:

cubic paraelectric → tetragonal ferroelectric at 130°C,
tetragonal ferroelectric → orthorhombic ferroelectric at 0°C.
orthorhombic ferroelectric → rhombohedral ferroelectric at -90°C.

The successive point symmetries are O_h, C_{4v}, C_{2v} and C_{3v}. All the transitions are of first order.

ACKNOWLEDGMENTS

I am greatly indebted to Prof. M.J. Brown, University of Washington, Seattle, Dr. S.W. Kieffer, U.S. Geological Survey, Flagstaff, Arizona, Prof. A. Navrotsky, Arizona State University, Tempe, and Dr. K.R. Rao, Neutron Physics Division, Bhabha Atomic Research Centre, Trombay, India for a critical review of an earlier version of this paper, which resulted in a considerable improvement of the presentation. This research has been supported by the NSF grant EAR 82-06526.

SUGGESTED READING

Ashcroft, N.W. and N.D. Mermin (1976) Solid State Physics. Saunders College, Philadelphia, PA, 826 pp.

Ausloos, M. and R.J. Elliott, eds. (1983) Magnetic Phase Transition. Springer Series in Solid State Sciences, v. 48, Springer-Verlag, New York, 269 pp.

Birman, J.L. (1984) Theory of Crystal Space Groups and Lattice Dynamics: infra-red and Raman optical processes in insulating crystals. Springer-Verlag, New York, 538 pp.

Bruce, A.D. and R.A. Cowley (1981) Structural Phase Transitions. Taylor and Francis, Ltd., London, 326 pp.

Cochran, W. (1973) Dynamics of Atoms in Crystals. Arnold, London, 145 pp.

Donovan, B. and J.F. Angress (1971) Lattice Vibrations. Chapman and Hall, London, 190 pp.

Ghatak, A.K. and L.S. Kothari (1972) An Introduction to Lattice Dynamics. Addison Wesley, New York, 234 pp.

Lovesey, S.W. and T. Springer, eds. (1977) Dynamics of Solids and Liquids by Neutron Scattering. Topics in Current Physics, v. 3, Springer-Verlag, New York, 379 pp.

Maraduddin, A.A., E.W. Montroll, G.H. Weiss and I.P. Ipatova (1971) Theory of Lattice Dynamics in the Harmonic Approximation (2nd ed.). Academic Press, New York, 708 pp.

Müller, K.A. and H. Thomas, eds. (1981) Structural Phase Transitions I. Springer-Verlag, New York, 190 pp.

Rao, C.N.R. and K.J. Rao (1978) Phase Transitions in Solids: An Approach to the Study of the Chemistry and Physics of Solids. McGraw-Hill, New York, 330 pp.

Reissland, J.A. (1973). The Physics of Phonons. John Wiley, New York, 319 pp.

Samuelson, E.J., E. Anderson, and J. Feder, eds. (1971) Structural Phase Transitions and Soft Modes. Universitetsforlaget, Oslo, 422 pp.

Venkataraman, G., L.A. Feldkamp and V.C. Sahni (1975) Dynamics of Perfect Crystals. MIT Press, Cambridge, MA, 517 pp.

Willis, B.T.M. and A.W. Pryor (1975). Thermal Vibrations in Crystallography, Cambridge University Press, Cambridge, England, 280 pp.

REFERENCES

Anderson, P.W. (1960) Качественные соображения относительно статистики фазового перехода в сегнетоэлектриках типа $BaTiO_3$ in Fizika Dielektrikov. G.I. Skanavi, ed., Akademiia Nauk SSSR, Moscow.

Andrews, T. (1869) On the continuity of the gaseous and liquid states of matter. Phil. Trans. Roy. Soc. London 159, 575-590.

Arndt, U.W. and B.T.M. Willis (1966) Single Crystal Diffractometry. Cambridge University Press, Cambridge.

Ausloos, M. and R.J. Elliot, eds. (1983) Magnetic Phase Transitions. Springer-Verlag, New York.

Axe, J.D. and G. Shirane (1970) Study of α-β quartz phase transition by inelastic neutron scattering. Phys. Rev. B 1, 342-348.

Bachheimer, J.P. and G. Dolino (1975) Measurement of the order parameter of α-quartz by second-harmonic generation of light. Phys. Rev. B 11, 3195-3205.

Barron, T.H.K., C.C. Huang and A. Pasternak (1976) Interatomic forces and lattice dynamics of α-quartz. J. Physics C, Solid State Phys. 9, 3925-3940.

Bilz, H. and W. Kress (1979) Phonon Dispersion Relations in Insulators. Springer-Verlag, New York.

Boccara, N. (1976) Symmètrie Briseés. Theorie des Transitions avec Paramètre d'Ordre. Hermann, Paris.

Born, M. and K. Huang (1954) The Dynamical Theory of Crystal Lattices. Clarendon Press, Oxford.

------ and Th. von Kármán (1912) Über Schwingungen in Raumgittern. Physikalische Zeitschrift 13, 297-309.

Boysen, H., B. Dorner, F. Frey and H. Grimm (1980) Dynamic structure determination of two interacting modes at the M-point in α- and β-quartz by inelastic neutron scattering. J. Phys. C, Solid State Phys. 13, 6127-6146.

Brockhouse, B.N. (1962). Interatomic forces from neutron scattering. J. Phys. Soc. Japan, Suppl. 17, BII, 363-367.

------, T. Arase, G. Caglioti, K.R. Rao and A.B. Woods (1962) Crystal dynamics of lead. I. Dispersion curves at 100 K. Phys. Rev. 128, 1099-1111.

Bruce, A.D. and R.A. Cowley (1981) Structural Phase Transitions. Taylor and Francis, London.

Cochran, W. (1960) Crystal stability and theory of ferroelectricity. Advances in Physics, 9, 387-423.

------ (1971) Lattice dynamics of ionic and covalent crystals. C. R. C. Critical Reviews in Solid State Sci. 2, 1-83.

------ and R.A. Cowley (1967) Phonons in perfect crystals. In Handbuch der Physik, S. Flugge, ed., v. XXV/2a. Light and Matter la, p. 59-156. Springer-Verlag, Berlin.

Cowley, R.A., W.J.L. Buyers and G. Dolling (1969) Relationship of normal modes of vibration of strontium titanate and its antiferroelectric phase transition at 110 K. Solid State Comm. I, 181-184.

Debye, P. (1912) Zur Theorie der spezifischen Wärmen. Annalen der Physik 39, 789.

Devonshire, A.F. (1949). Theory of barium titanate, Part I. Phil. Mag. 40, 1040-1063.

Dolino, G. and J.P. Bachheimer (1977) Fundamental and second-harmonic light scattering by the α-β coexistence state of quartz. Phys. Status Solidi (a) 41, 674-677.

------. ------, F. Gervais and A.F. Wright (1983) La transition α-β du quartz: le point sur quelques problèmes actuels: transition ordre-désordre ou displacive, comportement thermodynamique. Bull. Mineral. 106, 267-285.

------, ------, and C. Zeyen (1983) Observation of an intermediate phase near the α-β transition of quartz by heat capacity and neutron scattering measurements. Solid State Comm. 45, 295.

Donovan, B. and J.F. Angress (1971) Lattice Vibrations. Chapman and Hall, London.

Dorner, B. (1981) Investigations of structural phase transformations by inelastic neutron scattering. In: Structural Phase Transitions I. K.A. Müller and H. Thomas, eds., Springer-Verlag, New York, pp. 93-130.

Elcombe, M.E. (1967) Some aspects of lattice dynamics of quartz. Proc. Phys. Soc. London 91, 947-958.

Fleury, P.A. (1969) Paramagnetic spin waves and correlation functions in NiF_2. Phys. Rev. 180, 591-593.

------, J.F. Scott and J.M. Worlock (1968) Soft phonon modes and the 110 K phase transition in $SrTiO_3$. Phys. Rev. Letters 21, 16-19.

Guggenheim, E.A. (1945) The principle of corresponding states. J. Chem. Phys. 13, 253-261.

Höchli, U.T. and J.F. Scott (1971) Displacement parameter, soft-mode frequency and fluctuations in quartz (below its α-β phase transition). Phys. Rev. Letters 26, 1627-1629.

Iizumi, M., J.D. Axe, G. Shirane and K. Shimaoka (1977). Structural phase transformation in K_2SeO_4. Phys. Rev. B 15, 4392-4411.

Kellerman, E.W. (1940) Theory of the vibrations of the sodium chloride lattice. Phil. Trans. Royal Soc. 238, 513-548.

Kittel, C. (1971) An Introduction to Solid State Physics, 4th ed. John Wiley and Sons, New York.

Krishnan, R.S. (1947) Thermal scattering of light in diamond. Nature 159, 740-741.

Landau, L.D. (1937) On the theory of phase transitions. Zhurnal eksperimental'noi i teoreticheskoi fiziki, 7, 627. [English translation in "Collected Papers of L.D. Landau" (1965), p. 193-216, Pergamon Press, Oxford.]

------, and E.M. Lifshitz (1969) Statistical Physics, 2nd ed. Pergamon Press, Oxford.

Liebau, F. and H. Böhm (1982) On the co-existence of structurally different regions in the low-high quartz and other displacive phase transformations. Acta Crystallog. A38, 252-256.

Lovesey, S.W. and T. Springer, eds. (1977) Dynamics of Solids and Liquids by Neutron Scattering. Springer-Verlag, New York.

Megaw, H.D. (1973) Crystal Structures: a Working Approach. Saunders, Philadelphia.

Müller, K.A. and W. Berlinger (1971) Static critical exponents at structural phase transitions. Phys. Rev. Letters 26, 13-16.

------ and H. Thomas, eds. (1981) Structural Phase Transitions I. Springer-Verlag, New York.

Raman, C.V. and T.M.K. Nedungadi (1940) The α-β transformation of quartz. Nature 145, 147.

Rao, C.N.R. and K.J. Rao (1978) Phase Transitions in Solids, an approach to the study of the chemistry and physics of solids. McGraw-Hill International Book Co., New York.

Riste, T., E.J. Samuelson, and K. Otnes (1971) Critical neutron scattering from $SrTiO_3$. In: Structural Phase Transitions and Soft Modes, E.J. Samuelson, E. Andersen and J. Feder, eds., Universitetsforlaget, Oslo, p. 396-407.

Saksena, B.D. (1940) Analysis of the Raman and infra-red spectra of α-quartz. Proc. Indian Acad. Sci. 12, 93-138.

Salje, E. (1985). Thermodynamics of sodium feldspar: Order-parameter treatment and strain induced coupling effects. Physics and Chemistry of Minerals (in press).

Scott, J.F. (1974) Soft mode spectroscopy: experimental studies of structural phase transition. Rev. Modern Physics 46-83.

Shapiro, S.M. and H.Z. Cummins (1968) Critical opalescence in quartz. Phys. Rev. Letters 21, 1578-1582.

------, R.W. Gammon and H.Z. Cummins (1966) Brillouin scattering spectra of crystalline quartz, fused quartz and glass. Applied Phys. Letters 9, 157-159.

------, D.C. O'Shea and H.Z. Cummins (1967) Raman scattering study of the alpha-beta phase transition in quartz. Phys. Rev. Letters 19, 361-364.

Shirane, G. (1974) Neutron scattering studies of structural phase transitions at Brookhaven. Rev. Modern Physics 46, 437-449.

Shirane, G. and Y. Yamada (1969) Lattice dynamical study of the 110°K phase transition in $SrTiO_3$. Phys. Rev. 177, 858-863.

Stanley, H.E. (1971) Introduction to Phase Transitions and Critical Phenomena. Clarendon Press, Oxford.

Unoki, H. and T. Sakudo (1967) Electron spin resonance of Fe^{3+} in $SrTiO_3$ with special reference to the 110°K phase transition. J. Phys. Soc. Japan 23, 546-552.

Van Tendeloo, G., J. Van Landuyt and S. Amelinckx (1976) The α-β phase transition in quartz and $AlPO_4$ as studied by electron microscopy and diffraction. Phys. Status Solidi (2) 33, 723-737.

Wallis, R.F., ed. (1965) Lattice Dynamics. Proc. Int'l Conf., Copenhagen, Denmark, August, 1963. Pergamon Press, Oxford.

Weidner, D.J., J.D. Bass and M.T. Vaughan (1982) The effect of crystal structure and composition on elastic properties of silicates. In: High Pressure Research in Geophysics, S. Akimoto and M. Manghnani, eds., Center for Academic Publications, Japan, Tokyo.

Willis, B.T.M. and A.W. Pryor (1975). Thermal Vibrations in Crystallography. Cambridge University Press.

Woods, A.D.B., W. Cochran and B.N. Brockhouse (1960) Lattice dynamics of alkali halide crystals. Phys. Rev. 119, 980-999.

Wright, A.F. and M.S. Lehmann (1981) The structure of quartz at 25 and 590°C determined by neutron diffraction. J. Solid State Chem. 36, 371-380.

Yakolev, I.A., L.F. Mikheeva and T.S. Velichkina (1956) The molecular scattering of light and the α-β transformation in quartz. Soviet Phys. Crystallog. 1, 91-98.

Young, R.A. (1962) Mechanism of the phase transition in quartz. U.S. Air Force, Office of Scientific Research, Report N2569.

Chapter 5. J. Desmond C. McConnell

SYMMETRY ASPECTS of ORDER-DISORDER and the APPLICATION of LANDAU THEORY

1. INTRODUCTION

This chapter provides a formal but simple introduction to the application of symmetry principles to order-disorder problems. There are two areas in which it is essential to apply such considerations. The first of these is the symmetry of short range order above the order-disorder transition temperature. In this chapter a new symmetry treatment of short range order will be used to help explain the nature of short range order in minerals and to explain why the observed configurational entropy due to disorder just above the transition temperature is often much less than its maximum value. Our second purpose is to give a simple account of the use of Landau theory to describe the characteristics of certain order-disorder transformations. Landau theory, as it was originally presented, dealt with symmetry aspects of the development of long range order and was primarily concerned with changes in translational symmetry in alloys where there was a single average atom at each lattice point in the high temperature structure. We will also be concerned with the situation where there may be many ordering atoms per unit cell, and we will allow symmetry to act on a set of special or general equivalent positions within the unit cell, a situation frequently observed in minerals.

It is not possible to present the relevant symmetry arguments in this chapter without using group theory, since the whole of the application of Landau theory is based on this approach. Accordingly, some space has been devoted to explaining exactly how group representations are to be used in describing order-disorder phenomena. See the Appendix for a brief introduction. For those who find the related group theoretical principles difficult to follow, it is useful to note that group tables can be looked up and used in much the same way as one currently determines permissible modes of vibration for a molecule using the same symmetry data (see McMillan, this volume). The application of symmetry theory to order-disorder problems considerably simplifies the problems and also provides a great deal of new insight.

In the first of three sections in this chapter, symmetry properties are used to define the nature of short range order behavior in a single crystal above the order-disorder transformation temperature. Currently this is particularly important in the context of defining the thermodynamics of disordered mineral structures. The present treatment of short range order is based on the fact that the behavior of short range order must be described in terms of multi-dimensional representations.

This problem of dimensionality in group representations can be illustrated relatively simply in terms of representations of the point group 6/mmm associated with the loss of the three-fold axis, and this topic will be discussed in detail in this chapter. Where a three-fold axis is lost, two one-dimensional representations must be combined to form an irreducible two-dimensional representation because we are free to choose two linearly independent subgroup symmetries which are respectively even and odd in relation to the horizontal diad axis. The group table for point group 6/mmm as set out in the Appendix illustrates this point, and the representation E_{2g} of this table may be used to discuss the related problem in short range ordering in cordierite as described later in the present chapter with the necessary diagrams. In such short range order problems, multi-dimensional character is always present, and it is impossible to discuss this form of disorder at all meaningfully without reference to the dimensionality aspect of the symmetry description. We will attempt to illustrate the main features of this dimensionality problem here without deriving it rigorously from group theory where it is well understood.

In the second section of the chapter, Landau theory (introduced by Ghose, this volume) is used to study the character of an ordering transformation with falling temperature. Landau theory utilizes the irreducible representations of the symmetry group of the high temperature, disordered structure since it assumed that the symmetry of the low temperature structure is a subgroup of the high temperature symmetry in some sense. The theory is very much concerned with the nature of this group-subgroup relationship, and it is only strictly relevant where the symmetry of the low temperature, ordered structure is contained within the high temperature symmetry group as a specific irreducible representation. Landau theory cannot deal at all with transformations in which an arbitrary symmetry relationship exists between the high and low temperature structures. While it is possible to use it to demonstrate that a certain transformation may be of second order, it is not possible to prove that this is the case. On the other hand, it is usually possible to show that a chosen transformation is of first order by examining the related irreducible representation.

In Landau theory the change in free energy associated with the appearance of the low symmetry structure is described in terms of a power series expansion in the long range order parameter, as explained by Carpenter (this volume). In this expansion, the existence of a third order term precludes the existence of a second order transformation. Mathematical operations on the representations themselves are used to determine whether or not certain of these expansion terms can contribute to the free energy by considering whether or not there exists an invariant of the corresponding order. The operations associated with establishing the existence of such invariants have been set out in the Appendix where they are used to study the character of the hexagonal to orthorhombic transformation in cordierite.

The final section of this chapter is concerned with case studies of order-disorder transformations in minerals. We will examine both thermodynamic and kinetic aspects of such transformations, since many mineral transformations of order-disorder are very sluggish, and it is impossible to deal with their transformation behavior by using thermodynamic criteria alone. Fortunately, it is also possible to use symmetry arguments to unravel certain kinetic aspects of sluggish transformation behavior. This is possible where the system concerned is in metastable equilibrium, as is the case in the low temperature ordering of cordierite and potassium feldspar. The examples chosen to illustrate these topics start with the rather simple case of the transformation C2/m to $C\bar{1}$ in potassium feldspar. This transformation satisfies the symmetry conditions for possible second order behavior on Landau criteria. A study of the first order transformation from hexagonal to orthorhombic symmetry in cordierite provides an opportunity to study both the thermodynamic and kinetic aspects of a first order transformation from a rigorous symmetry point of view.

The chapter ends with a brief survey of the important symmetry aspects of the origin of incommensurate structures as observed in the intermediate plagioclase feldspars and in the refractory mineral mullite. Incommensurate behavior is very much concerned with short range order effects.

At this point it is convenient to provide a brief guide to existing literature on the application of symmetry to the problems discussed in this chapter. To the best of the author's knowledge there exists in the literature no simple treatment of the symmetry properties of short range order, although there is a voluminous literature on purely experimental aspects of this subject. An excellent account of experimental studies of short range order in Cu_3Au is provided by Bardhan and Cohen (1976), and an introduction to the symmetry of diffuse disorder scattering is provided by deFontaine (1975). There are many excellent treatments of the application of the Landau theory of phase transformations. Perhaps the most useful is that due to Birman (1966), but the reader may also consult Lyubarski (1960) and Chapter 14 of Landau and Lifshitz (1959) which contains a useful summary. The original papers of Lifshitz (1942a,b) may still be read with benefit.

In the study of the examples of the application of symmetry criteria to order-disorder in minerals in the final section of this chapter, much of what has been written here is quite new, and there are no references dealing specifically with the symmetry criteria as they are used here. Both point group and space group irreducible representations are available in Bradley and Cracknell (1972). The relevant theory of space group representations is provided by Bradley and Cracknell (1972) and Birman (1974). Finally, the symmetry and structural aspects of incommensurate structures are discussed in McConnell and Heine (1984).

2. SHORT RANGE ORDER

We start this section by defining terms. The concept of long range order is well understood, and it implies that, if we choose a particular atomic position in the unit cell with local coordinates x,y,z and note that, at one point in the single crystal, this site tends to be occupied by an A atom with some probability, pA, then this site will be similarly occupied with the same probability by an A atom throughout the single crystal. Here we assume, in the case of the development of a superlattice, that we define the local atomic position in terms of the new, large cell. This fixed probability for A occupation in relation to the average or completely disordered value may be used to define a long range order parameter. This is usually defined in such a way that its numerical value becomes zero for complete disorder.

Above the transformation temperature for disordering, long range order does not exist, but it is still possible that, if we find an A atom on any site, the probability of its neighboring site being occupied by B will be higher than that dictated by a purely random distribution. The short range order parameter (SROP) is therefore defined in terms of chosen pairs of sites, and the departure of pair probabilities (pAB) from the purely random value for the pair probability (2.xA.yB). In pAB we sum over probabilities with first A and then B on the origin site, and we obtain the average pair probability from the atom fractions of A and B (x and y): SROP = 1 - pAB/2xAyB.

Having defined the concept of short range order, we now proceed to demonstrate how the short range order behavior is affected by symmetry. First we note that the short range order pattern must be entirely compatible with the symmetry of the high temperature structure. Formally it must transform according to irreducible representations of the high symmetry group. This allows us to separate the ordering observed into its symmetry components and to assert that these components are linearly independent and need not interact directly with one another. We may use this principle to discuss the exact nature of the short range order in certain mineral systems such as adularia and orthoclase which show characteristic short range order scattering, as explained later in this chapter.

In order to show in outline how short range order theory may be developed from the ordering associated with point group symmetry alone, we will choose a simple example which has been used before in a rather similar context (McConnell, 1983). The example has the advantage that it is simple. It deals with the ordering of two A atoms and two B atoms on a set of general equivalent positions in the space group Pmm2, as illustrated in Figure 1. If we choose one or the other of the ordering schemes for A and B atoms illustrated in Figure 1 and then allow it to exist over the crystal as a whole, we create examples of long range order. Note that the ordering schemes which are set out in Figure 1 are very simply related to the corresponding irreducible representations in the table within the figure. Thus we may identify the character 1 with the existence of an A on the corresponding site, and -1 with a B atom. This method of using symmetry is dealt with further in the Appendix. Thus each of the ordering schemes corresponds to one of the three irreducible representations of the point group mm2. Note that the first representation in the table within Figure 1 corresponds to the identity, which implies that we cannot distinguish among the four items. This representation describes a situation of complete disorder. We note that the long range ordering schemes which we have developed from the point group representations cannot interact with one another since the product of any pair of representations fails to contain the identity representation. This is a general group theoretical rule and relates to the fact that, in interaction, final representations other than the identity representation can be ignored (McConnell, 1983). Note also that each of the representations in Figure 1 is one dimensional. Having established long range ordering patterns, we want next to derive the corresponding representations for short range order. These are more difficult since they are no longer simple one-dimensional representations. They are, however, based on the simple point group representations of Figure 1.

In building up the short range order, we modulate basis functions or ordering schemes as given in Figure 1 with functions of the form exp ik.R

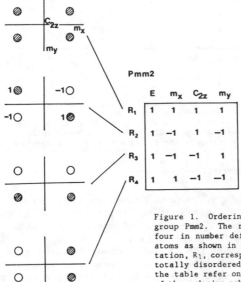

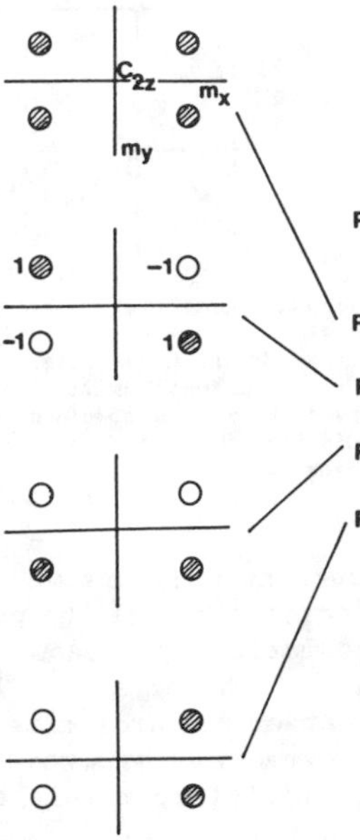

Pmm2

	E	m_x	C_{2z}	m_y
R_1	1	1	1	1
R_2	1	−1	1	−1
R_3	1	−1	−1	1
R_4	1	1	−1	−1

Figure 1. Ordering schemes for possible subgroups of space group Pmm2. The representation of the vector 000 which are four in number define possible ordering schemes for A and B atoms as shown in the diagram. Note that the first representation, R_1, corresponds to the average state which is also the totally disordered state, and that the characters 1 and -1 in the table refer only to the symmetry and not to the magnitude of the ordering schemes.

associated with the translation group of the lattice. This has already been explained in McConnell (1983). The modulation functions are associated with translations of the lattice other than the primary lattice repeats; they are best studied in k-space where they may be defined very easily in terms of the Brillouin zone, as illustrated in Figure 2 and in McConnell (1971, 1983).

To explain the multi-dimensional character of the representations for short range order, it is necessary to utilize a reciprocal or k-space representation and to consider the points in k-space which are equivalent by symmetry. Each of the point group ordering schemes which we have studied above relates to the origin vector 000 of the reciprocal lattice. This symmetry point is unique and hence we may avoid any reference to the translations of the lattice in developing representations. This and similar symmetry points have a special significance in the Landau theory of phase transformations, since they are either at the zone origin or on the zone boundary and are related by symmetry to points which are identical with them, rather than simply equivalent. If we choose a vector in k-space other than at a symmetry point, it is normally not unique and must be taken in conjunction with symmetry-related vectors in building up a representation.

The suite of vectors which are equivalent by symmetry is described as a star, and the number of vectors in the star is one factor which determines the dimensionality of the representation. In group theory, an important theorem proves that representations built up from the complete suite of vectors in a star are irreducible. Since short range order scattering occurs at points in reciprocal space other than the symmetry points, it must always involve multi-dimensional representations. On the

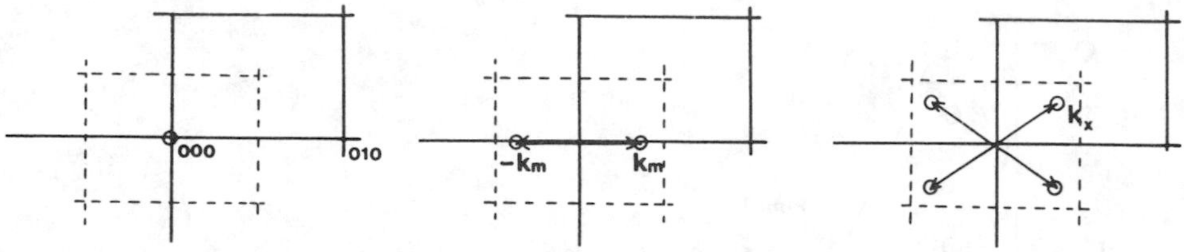

Figure 2. Reciprocal or k-space diagrams for the short range order components in space group Pmm2. In the first diagram the illustrated vector 000 is unique, since points to which it is related by symmetry are identical points of the reciprocal lattice. In the second diagram, the vector k_m is related by symmetry to the vector $-k_m$, and the associated functions are linearly independent of one another. The representations for k_m and $-k_m$ must be combined to form a two-dimensional representation which is irreducible. Finally, the last diagram of the set illustrates the situation with regard to a general vector k_x. This is related by symmetry to three other vectors, and representation functions for all four must be combined to form an irreducible four-dimensional representation.

other hand, the star of the symmetry point 000 contains only one vector. For this reason if we choose a one-dimensional representation of the point group of the crystal (vector 000), it automatically becomes a one-dimensional representation of the space group of the crystal. This point is illustrated by the fact that the transformation of cordierite can be described entirely in terms of the point group symmetry 6/mmm which is equivalent to the group for the vector 000 in the high temperature cordierite space group P6/mcc.

To illustrate these principles further we now choose a vector for ordering other than at a symmetry point 000 of Pmm2, as illustrated in Figure 2. If we choose a vector which lies in the mirror plane, such as the vector k_m of Figure 2, then symmetry requires that it be taken with its partner under the point group symmetry and, from equivalent representations of the two vectors, we build a full irreducible representation. Theory requires that this representation should be two dimensional and irreducible and that two ordering basis functions must be used to describe the local order induced by this representation. The two basis functions (or ordering schemes) for this representation are just two of the point group representations from the table in Figure 1. They have the property that they have the same symmetry in the small group of k_m. This group has the symmetry of the vector k_m and contains only the identity and the mirror as symmetry elements. It is easy to show that there are just two inequivalent irreducible representations of dimension two which can be formed from the pair of vectors k_m and $-k_m$ as chosen in this example. This means that the short range order may selectively involve two out of four possible ordering schemes. We will use this principle later to define short range order in potassium feldspar as observed, for example, in adularia (McConnell, 1971).

Finally, we choose an entirely arbitrary vector in the k-space representation of space group Pmm2. A vector such as k_x of Figure 2 is repeated four times by the point group symmetry, and theory now demands that the functions defined by these four vectors should transform through one another as the components of a four-dimensional irreducible representation. In this case there is only one symmetrically distinct irreducible representation associated with the suite of four vectors. The basis functions for ordering comprise all four of the ordering schemes shown in Figure 1. Since each of these is one dimensional, the final irreducible representation is four dimensional.

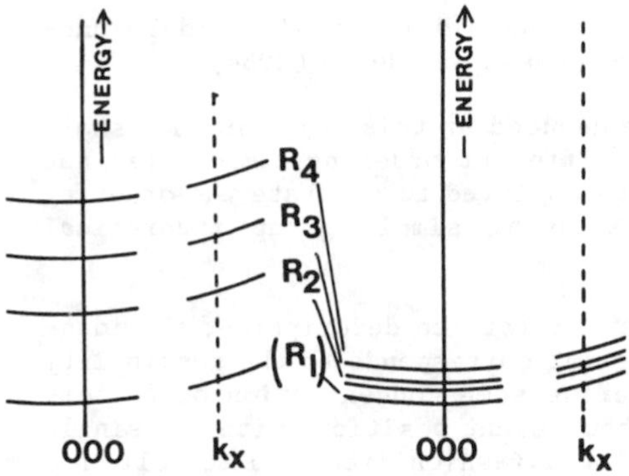

Figure 3. In short range order for vectors such as k_x of Figure 2, we have to consider the inclusion of all three of the ordering schemes R_2-R_4 of Figure 1 in a single band. In the lowest ordering energy band of Figure 3, which we have associated primarily with R_2 (P2), the amount of other ordering schemes included depends on the energy spacing of the bands. Thus in the first of the diagrams where the several bands are well separated in energy, the R_2 band will be fairly pure. On the other hand, if the ordering schemes R_2-R_4 are approximately equal in energy, then the R_2 band will contain an appreciable amount of all the ordering schemes. This simple argument implies that, in certain short range order situations, the local order will include all the ordering schemes and will vary between them from point to point in the crystal. At the same time the entropy will be very high and will approximate to the maximum possible. Where ordering bands are widely separated, configurational entropy will be much less than the maximum possible.

We are now in a position to take advantage of the piece of theory which we have discussed. The theory implies that vectors other than symmetry point vectors, such as 000, must define multi-dimensional irreducible representations. Since the short range order must transform according to these multi-dimensional representations, it itself must be defined in terms of suitable basis functions which are orthogonal. Further, if we select any vector or suite of vectors other than for a symmetry point such as 000 we must consider the possible admixture of different ordering schemes as components of these multi-dimensional representations. Since, at the same time, energy criteria require that the partners or components of such a multi-dimensional irreducible representation should be degenerate, i.e., they should be equal in energy, we discover that there is indeed a great deal of structure in the short range order pattern. If we were to assume that our simple order-disorder model in Pmm2 transforms with falling temperature to an ordered structure with space group P2, then above the transformation temperature we would expect to find that the multi-dimensional basis functions associated with short range order would include not only the basis function P2, as illustrated in Figure 1, but also would contain other possible ordering schemes chosen from the set in Figure 1 according to the symmetry characteristics of the relevant star which we chose. The proportion of the different ordering schemes to be included depends not only on symmetry but also on the relative energies to be associated with the different ordering schemes.

We may readily incorporate these findings in an extension of the simple band model for order-disorder presented earlier by the author (McConnell, 1978a). In that model, ordering behavior was described in terms of the variation of the ordering energy on a single energy surface throughout reciprocal space. If we now consider that there are several energy bands, that they have the same symmetry at some chosen point in k-space, and that they are widely spaced in energy, there will be little admixture of ordering schemes. On the other hand, if the ordering schemes differ little in free energy (Fig. 3), the lowest band is likely to contain substantial amounts of all four of the ordering schemes. So far in discussing short range order we have ignored the translations or modulation component of the order as defined by the value of, say, k_x. As explained previously (McConnell, 1983), the modulation introduces no new detail but merely produces a cosinusoidal modulation. Since ordering is a scalar function, we must deal with real functions, i.e., cosinusoidal and sinusoidal functions, as explained at some length in McConnell and Heine (1984).

The important result which we have deduced at this point is that short range order normally contains several different ordering components, not just one, and that symmetry rules can be employed to separate the ordering into linearly independent components using simple group theoretical rules.

At this point we may note that, in the lattice description, the identity representation of the point group mm2 corresponds to the possibility of spinodal behavior, i.e., we may define some concentration of A atoms which is uniform over all the general equivalent positions within a single unit cell but which changes in a modulated fashion from cell to cell on a spinodal wavelength. It is gratifying at this point in the development of the overall theory to find that spinodal effects are taken care of automatically.

It is very important to point out here that the combination of different ordering basis functions within the multi-dimensional representations has important implications in the lowering of the free energy of the system in its disordered state. Here we may use arguments which have previously been employed by Heine and McConnell (1984) and McConnell and Heine (1984). In brief, these imply that the combination of different ordering basis functions in the short range order modulation pattern may introduce extra energy-lowering terms due to interaction. In normal short range order behavior above Tc, these energy-lowering effects are usually insufficient to produce any dramatic effects, but nevertheless they must always be present to some extent. Later in this chapter we will employ this theory of interaction between component structures in the short range order pattern to explain the transformation behavior in potassium feldspar. This phase has strong short range order in a metastable, low temperature, ordered state. What we wish to assert at this point is simply that group theory can be employed to establish the symmetry of, and hence (by implication) the nature of the short range ordering in problems of direct interest in mineral thermodynamics.

So far we have discussed the existence of multi-dimensional representations which must be employed in describing short range order, but we have been concerned neither with their form nor how they are derived in practice. We now turn to examining examples of ordering for multi-dimensional representations. We carry out this exercise in order to be able to deal in practical terms with the transformation behavior in cordierite. Since this phase transforms from hexagonal to orthorhombic symmetry with the loss of a three-fold axis, it follows that the point group irreducible representation appropriate to the symmetry change is itself two-dimensional. By preparing and examining two appropriate ordering basis functions which may be used to satisfy this irreducible representation, a great deal of practical insight into the theory used in this chapter may be obtained. Of course in the case we consider, the two-dimensional representation is a point group representation, whereas the multi-dimensional representations which we have discussed for short range order are space group representations. However, the basis functions which we are concerned with are exactly equivalent in both cases.

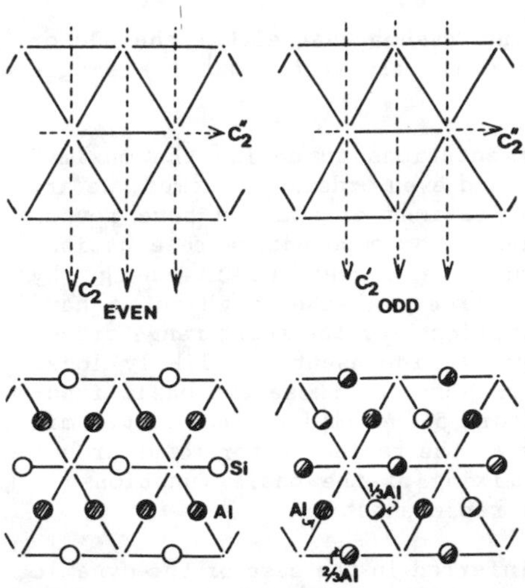

Figure 4. Even- and odd-ordering schemes associated with the representation E_{2g} of the group for the vector 000 of space group P6/mcc (high cordierite). In high cordierite three tetrahedral sites (T1) are equivalent to three others by the horizontal mirror plane m_z in representation E_{2g}. Since a three-fold axis is lost in this subgroup symmetry, the three sites can be described by both odd- and even-ordering schemes in this two-dimensional representation. The two ordering schemes are necessarily degenerate under hexagonal symmetry. A horizontal diad axis exists in the plane of the three sites and passes through one of them. While the even-ordering scheme is simply the scheme observed in orthorhombic low cordierite, the odd scheme requires that the site on the diad axis should be totally disordered, i.e., 1/3 Si and 2/3 Al. The remaining two sites have opposite ordering character, and one of them in the limit can be totally occupied by aluminum.

The irreducible representations of the point group 6/mmm are set out in Table 1 of the Appendix. Among the two-dimensional representations shown there is one, E_{2g}, which contains the representation of the subgroup mmm. It corresponds to the appearance of the low temperature orthorhombic structure. It is associated with a second symmetry component which corresponds to a monoclinic subgroup. As we have already noted, the two-dimensional character of this representation depends on the fact that it is associated with the loss of a three-fold axis. And, as we have also noted, the point group 6/mmm is isomorphic to the group for the vector 000 of the space group P6/mcc for high cordierite.

We may use the representation E_{2g} of this group table to study possible ordering schemes for ordering Si and Al on tetrahedral sites of type T_1 in the high temperature, disordered structure of cordierite. Here Si and Al atoms lie on six-fold special equivalent positions (6,f,222). Since the character for the horizontal mirror plane (m_z) is even (+1), it is possible to separate these six positions into two equivalent sets of three which are populated in some way by Si and Al atoms. In Figure 4 we have set out the symmetry conditions which govern the two orthogonal basis functions for ordering on these three tetrahedral sites in high cordierite according to the chosen two-dimensional irreducible representation. The scheme which is even in relation to the horizontal diad axis (which passes through one of the tetrahedral sites as indicated) is illustrated in Figure 4a. Clearly complete order is possible by placing one Si and two Al atoms on these three sites. We now turn to ordering on the basis of the second irreducible representation of the pair.

Here we are compelled to use the horizontal diad as an odd operator, i.e., one which turns 1 into -1. Now we find that we have no choice but to leave the site on the horizontal diad axis completely disordered and to define the ordering on the other two sites such that one is the negative of the other. The characters for this ordering scheme are therefore 0, 1, -1, and the maximal value of the deviation from the mean state is one-third. Maximum ordering on the odd scheme involves the site occupancy as in Figure 4b. For convenience here we may assume that either the odd or the even ordering scheme is developed uniformly throughout the single crystal.

Having used the irreducible representations to define the maximal order on these three sites for both odd and even ordering schemes, we are now in a position to discuss the implications of what we have found. Since in the high temperature state these two component representations are not only orthogonal, but also must be degenerate in energy by symmetry, we have to conclude that the ordering schemes which we have derived must both be present as basis functions for the short range order. It is convenient to use a simple scheme to represent the likely local behavior in the single crystal in the presence of these two basis functions alone. This is illustrated in Figure 5 and implies that we may mix the representations in any way we choose as the radius vector moves around the limiting circle. This arbitrary mixing of the basis functions is characteristic of such a two-dimensional representation.

The situation here is exactly that inferred in the case of the dynamic Jahn-Teller effect as described and discussed in McConnell and Heine (1982). In that case the two-dimensional behavior was associated with displacements. At any temperature the actual magnitude of the radius vector in Figure 5 is determined by the thermodynamic condition that the free energy should be a minimum. This condition is established in the usual way by the energy and entropy conditions associated with the ordering, and we expect to find that the radius vector increases in magnitude with falling temperature.

We have now established what we mean by a multi-dimensional representation. Within such a representation we are permitted to mix basis functions, and we may infer without further ado that, in the high temperature, disordered state, we have a similar multi-dimensional short range ordering pattern with basis functions chosen, as we have already described, from these and other point group irreducible representations. Clearly, if the degree of short range order is less than maximal, there must be a great deal of structure in the local disorder pattern, since definite phase relationships are implied between the basis functions for single multi-dimensional irreducible representations (McConnell and Heine, 1984). At this stage we may anticipate one of the important results used later, namely, that there is a maximum ordering possible in the odd ordering scheme which we have defined for cordierite. In fact, for maximum ordering this scheme has an entropy of approximately half of that of the completely disordered state. In some sense the transition in cordierite is conditioned by the fact that when maximum order is achieved in the odd ordering component, it is no longer possible to maintain the necessary degeneracy between the two ordering schemes with falling temperature. This basically is the mechanism through which the first order transformation in cordierite operates.

We complete this discussion of symmetry aspects of short range order by discussing the example of ordering in Cu_3Au. The disordered structure is cubic face-centered, and ordering at the transformation temperature produces a primitive cubic structure with one site for Au as illustrated in Figure 6. The irreducible representation of the high temperature space

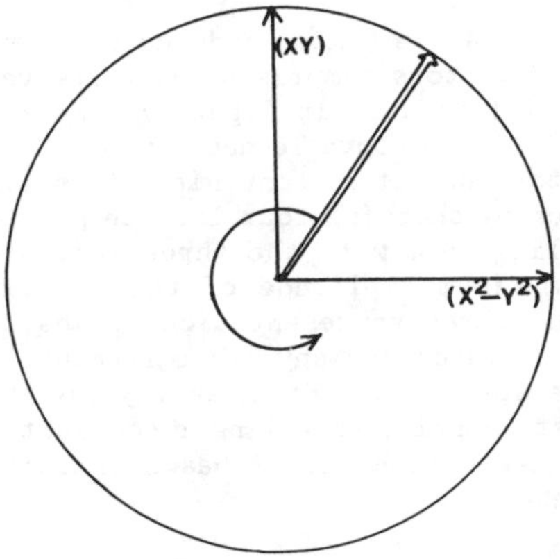

Figure 5. In this diagram we ilustrate the concept of a two-dimensional ordering representation. The radius vector can circulate at will around a circle and thus mix arbitrarily the odd (xy) and even (x^2-y^2) ordering schemes from point to point in the crystal.

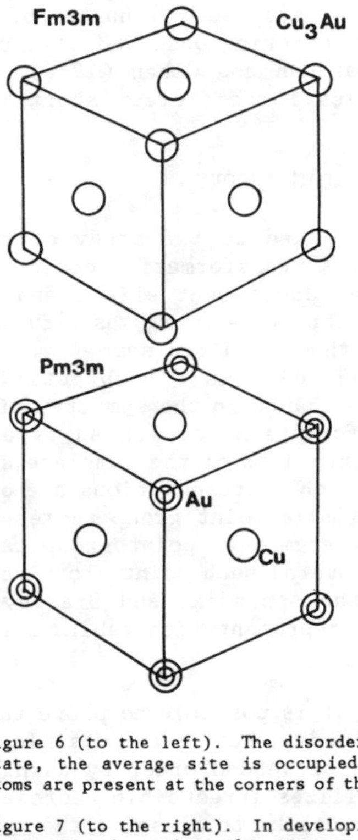

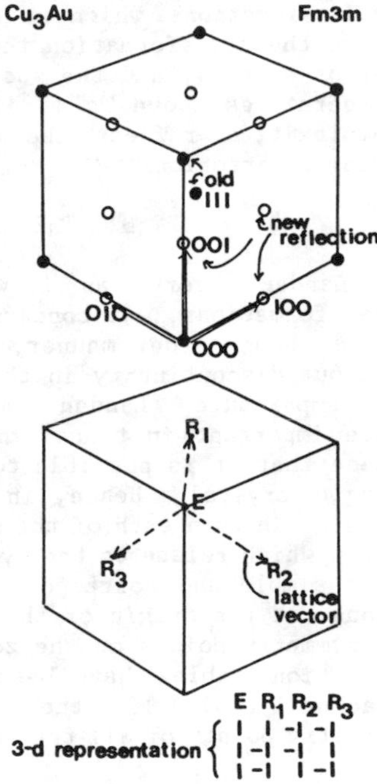

Figure 6 (to the left). The disordered and ordered structures of Cu_3Au. In the disordered state, the average site is occupied by 1/4 Au and 3/4 Cu. In the ordered structure, gold atoms are present at the corners of the large cell.

Figure 7 (to the right). In developing the ordered structure of Cu_3Au, new scattering vectors appear in reciprocal space, as shown in the first diagram. These three vectors require that the ordered structure (Pm3m) be described within the high symmetry space group (Fm3m) in terms of a three-dimensional representation. In the second diagram we show, in real space, the four lattice points of the disordered structure which can be used to define a representation for the three-dimensional short range ordering above the transformation temperature.

group (Fm3m) associated with the low temperature (Pm3m) ordered structure is three-dimensional, since the suite of vectors comprises the three vectors, 0,0,1/2; 0,1/2,0; 1/2,0,0 — as illustrated in Figure 7. We know immediately that the short range order in Cu_3Au above Tc must involve this multi-dimensional irreducible representation. It is convenient to define a basis function for this representation by choosing four lattice points, as illustrated in Figure 7, and labelling them with the three suites of characters as given. Note that the maximum amplitude of the ordering functions is 1/4 and that the sum of the three representations gives the low temperature ordering scheme. Without involving any other irreducible representations of the high temperature space group Fm3m, we may say that the local short range order pattern must comprise some linear combination of these ordering representations, at least as a set of basis functions for the full short range ordering scheme.

We may also note that the ordering vector in this case migrates over the surface of a sphere. As in the example of cordierite, the transformation in Cu_3Au appears to occur, with falling temperature, where the limit of ordering is achieved for certain combinations of the three ordering basis functions which we have chosen. The short range order in Cu_3Au above the transformation temperature has been studied by a number of different authors, and the short range order scattering observed at several temperatures above Tc is illustrated by Bardhan and Cohen (1976). Its complexities reflect the many possibilities for different short range order interactions.

3. THE APPLICATION OF LANDAU THEORY

Landau theory, as it was originally applied to the study of phase transformations, was concerned to show that a transformation could occur in a second order manner, i.e., without a latent heat effect and thus without discontinuity in the slope of the Gibbs free energy as a function of temperature. Landau demonstrated that the so-called symmetry points were important in transformation theory, since it is only by utilizing these that it is possible to make a discrete change in the symmetry of the single crystal. Hence, in studying transformations of Landau type, we examine in turn each of the subgroup representations of the complete space group which relate to the symmetry points. Such representations are often very simple and correspond frequently to simple point group representations (at the origin of the zone) or simply augmented point group tables at symmetry points on the zone boundary. Several such point group representation tables have been presented in the Appendix, and Bradley and Cracknell (1972) list the details of all the representation tables for all symmetry points of all the space groups.

In the process of applying Landau theory it is possible to prove that a given phase transformation is not of second order, but it is possible only to infer that a given transformation may be of second order by using the theory. As already noted, Landau theory utilizes irreducible representations of the high temperature space group and seeks to classify the corresponding transformation behavior by seeking for invariants in the equivalent of a power series expansion of the free energy associated specifically with the appearance of the chosen irreducible representation, i.e., the chosen ordering or other transformation scheme. Phenomenological aspects of the application of Landau theory are discussed in this volume by Carpenter. Here we will be concerned only with showing how Landau theory may be applied in detail. The existence of a second order transformation implies that the free energy expansion should not contain third order terms. Specifically, this can be checked by proving that the symmetrized cube of the irreducible representation does not contain the identity representation.

The fact that we cannot usually prove that a given transformation is of second order, in spite of the fact that a symmetrized cubic invariant does not exist, follows from the fact that it is usually impossible to study all the possible invariants of high order. For example, in one of the examples dealt with by Carpenter (this volume), a fourth order invariant associated with a negative rather than a positive coefficient leads to a first order transformation. In general, we do not know enough about the detailed interactions in the crystal to be certain about the magnitudes and signs of all the coefficients in the free energy expansion.

It is also necessary to prove that the antisymmetrized square of the chosen irreducible representation does not contain the representation of a polar vector. Both this operation, and the operation of taking the symmetrized cube on an irreducible representation are illustrated in the Appendix. It is important to note here that they are applied to the full matrices in the representation and not simply to the characters (traces) of the matrices. In many of the cases in which we are interested here, the chosen irreducible representation is real and one-dimensional. In this case the possibility of a second order transformation can usually be established from inspection. This can be readily illustrated in the case of the monoclinic to triclinic transformation associated with Al and Si ordering in potassium feldspar. In this case the volume per lattice point does not change.

This transformation may be described as a zone origin (000) transformation, and consequently the representations needed are simply those of the point group 2/m. These are listed in the Appendix and correspond to three possibilities: the retention of the mirror plane, the diad axis, or the center of symmetry. Taking the second representation which relates to the retention of the center of symmetry and applying the formulae for symmetrized cube and anti-symmetrized square from the Appendix, it is easy to show that neither invariant exists, and consequently, this transformation may be of second order.

It is equally easy to examine the nature of the transformation from $C\bar{1}$ to $I\bar{1}$ in the case of anorthite and the Ca-rich plagioclase feldspars. The relevant irreducible representations are given in the Appendix. In this case the point group symmetry remains the same, and the unit cell doubles in volume. This doubling is associated with the character -1 for the original translation vector parallel to c in the disordered structure. This transformation may be described as a zone boundary transformation. Since the representations of the doubled cell are one-dimensional, it is easy to show that neither the symmetrized cube nor the antisymmetrized square exist, and therefore the transformation may be second order in character. We note here that this transformation from $C\bar{1} \rightarrow I\bar{1}$, producing a doubled c axis, can occur in either of two ways, since it is necessary to choose between the two different symmetry centers present in the disordered structure. As Carpenter (this volume) has pointed out, the conditions for a second order transformation also require that the coefficient of the fourth order term in the Landau expansion of the Gibbs free energy should be positive rather than negative.

Finally, in this section we deal briefly with the symmetry analysis in situations where the ordering atoms in the unit cell lie on special equivalent positions. In this case not all of the irreducible representations of the point group are relevant, and it is possible to deal with the representations of a smaller (factor) group instead of those of the full group. Thus in the factor group certain of the symmetry elements may be combined, because all of the symmetry possibilities are not realized. A factor group description is the correct description for the set of atoms in special equivalent positions, because certain symmetry elements act to turn them into themselves. Note that the order of the factor group is equal to the number of special equivalent positions. It is easy to illustrate this principle by referring back to the case of ordering in Cu_3Au. Since in this case there is only one atom per lattice point in the high temperature, disordered structure, the factor group is of order one, and we may ignore the point group symmetry entirely. Here the average atom in the high structure lies on all the point symmetry elements at once. It is for this reason that the relevant representations can be constructed, as in Figure 7, from the translation group elements alone. In Figure 7 we were content to write out representations in terms of the translations between the four chosen lattice points, i.e., E, R_1, R_2, R_3, and to ignore the point group symmetry operations. We have used classical Landau theory here to determine certain important characteristics of transformations. In describing certain details of transformation behavior, particularly close to transition temperatures, it is now known that Landau theory does not provide an exact description of the thermodynamic behavior, but we are not concerned with such detail in the present chapter.

4. APPLICATION OF SYMMETRY CRITERIA TO TRANSFORMATION MECHANISMS

In this section, symmetry principles will be used to elucidate some problems of transformation and transformation mechanisms as observed in common mineral systems. First of all, we turn to a discussion of the nature of the symmetry constraints which act during the transformation of potassium feldspar and are responsible for the characteristic short range order scattering observed, for example, in adularia. It has already been pointed out that the mechanism of the order-disorder transformation in K-feldspar under equilibrium conditions is extremely simple from the point of view of the application of equilibrium Landau transformation theory. In order to explain the observed additional diffraction effects in adularia and orthoclase, it is necessary to assume that the transformation is sluggish and that the mechanism of transformation operates at some temperature below the ideal equilibrium temperature defined by thermodynamics. This is certainly a true interpretation in the case of ordering in adularia.

Transformation in the metastable temperature range below Tc permits a modulation mechanism for the transformation. The observed intensity and its position in reciprocal space can readily be interpreted in terms of

K-FELDSPAR

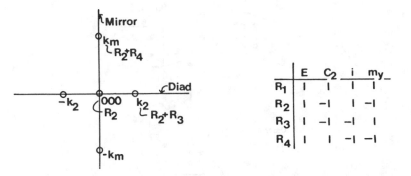

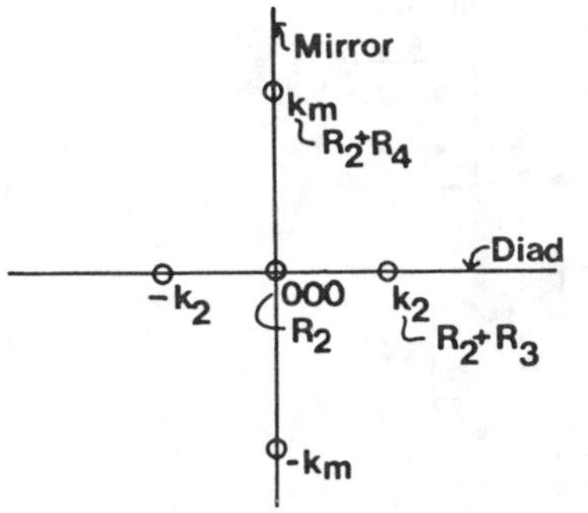

	E	C_2	i	m_y
R_1	1	1	1	1
R_2	1	−1	1	−1
R_3	1	−1	−1	1
R_4	1	1	−1	−1

Figure 8. Details of the several different stars for short range ordering in potassium feldspar. The vector 000 is associated with a single ordering scheme (R_2). Ordering associated with the vectors k_2 and $-k_2$ is two-dimensional and uses basis functions R_2 and R_3. Short-range ordering associated with k_m is similarly two-dimensional and based on R_2 and R_4. Finally, short range order associated with a general vector k_x involves all four ordering schemes and is four dimensional.

our analysis of short range ordering effects in Section 2 above. We have already pointed out that ordering other than that associated with the symmetry points requires the existence of multi-dimensional ordering, i.e., the combination of ordering belonging to different basis functions or ordering schemes. Details of the intense short range order x-ray scattering in adularia have been presented previously (McConnell, 1971). This scattering comprises relatively sharp streaks through the Bragg maxima both normal to the mirror plane and in one direction in the mirror plane. In Figure 8 the symmetry associated with points on these high intensity streaks has been defined, and the basis functions to be used for each star have been set down.

In one case the symmetry involves the diad axis and in the other the mirror plane. If we assume, as seems quite reasonable, that one component of the short range order involves the retention of the symmetry center as in microcline, then we may identify all the multi-dimensional irreducible representations associated with the observed short range order intensity. For k_2 parallel to the diad axis, the basis functions must comprise representations R_2 and R_3 of 2/m (see the Appendix). For k_m lying in the mirror plane, the basis functions are necessarily R_2 and R_4 of the same group table. The sharpness of the diffraction streaks indicates that the general representation of dimension four, made up of all the basis functions, is not at all favored energetically.

In Figure 9 the nature of the short range order pattern is illustrated for both k_2 and k_m. In each case, the combination of ordering schemes is clearly beneficial in reducing the free energy of the modulated structure. At this point we may note that the configurational entropy of short range ordered K-feldspar is likely to be extremely small, since a very high degree of order is obviously established throughout the modulated structure (McConnell, 1971). The same general principles have been employed recently to explain the stability of modulated incommensurate structures

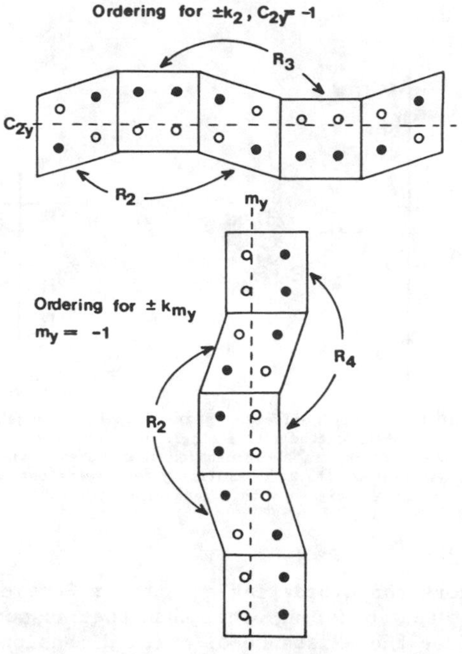

Figure 9. The actual short range order in low temperature monoclinic K-feldspar is based almost entirely on the stars k_2 and k_m. It is possible to use the data from Figure 8 to construct the ordering patterns for these two stars. Note that in the two diagrams the ordering schemes chosen clearly minimize the energy of the system. Inherently, these ordering schemes are capable of producing a very substantial reduction in the configurational entropy of the system.

where, once more, the existence of paired interacting structures, associated with multi-dimensional representations has been demonstrated (McConnell and Heine, 1984), and very low values of the configurational entropy are anticipated.

Having demonstrated that an analysis of short range order patterns in symmetry terms is a powerful analytical tool, we turn now to the more difficult problem of unraveling the transformation behavior of cordierite under irreversible conditions. As explained earlier, the short range order in cordierite in the hexagonal state must be based, at least, on the two short range order basis functions which are associated with the two-dimensional irreducible representation of the point group 6/mmm. This is associated with the loss of three-fold symmetry and the appearance of an orthorhombic structure. Further, as we have already been at pains to point out, short range order representations of even higher dimensionality will certainly be involved for low symmetry positions in k-space in relation to the high temperature (hexagonal) structure.

The first point which should be made here is simply that the condition of degeneracy imposed by the point group symmetry 6/mmm may no longer hold when we consider points in k-space other than 000. Thus, if we move off 000 along the a* axis, the three-fold symmetry becomes irrelevant, and the

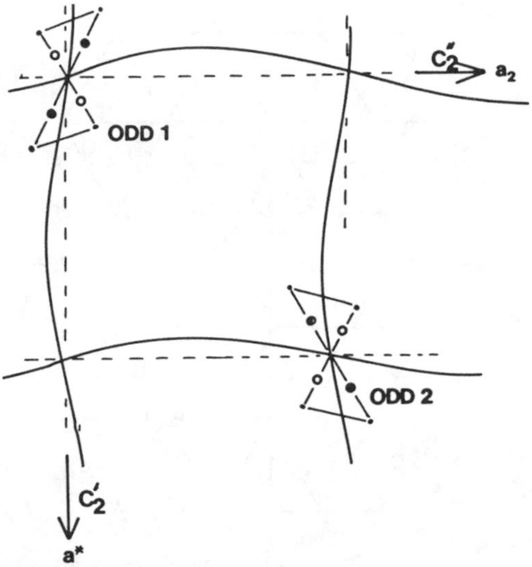

Figure 10. Illustration of the interaction of two orthogonal transverse wave systems in the plane 0001 of high cordierite to produce locally the two possible versions of the odd-ordered scheme for representation E_{2g}.

odd and even degenerate basis functions which we defined for the point group representation are no longer degenerate.

In general, this dictates that one or the other of these new representations may be of lower energy intrinsically, or may be involved in an energy-lowering interaction. In short, the crystal may employ one or the other of these basis functions to produce a modulated structure with a substantial reduction in free energy during ordering under metastable conditions well below Tc. Experimental data due to Putnis (1980a,b) have demonstrated that, in the earliest stages of the transformation of cordierite under irreversible conditions, the modulated structure has essentially transverse wave character. This may be interpreted quite simply to mean that there is an energy-lowering interaction between the odd component of the ordering scheme, which we have already defined, and transverse acoustical waves. The situation is in fact closely akin to that observed in the transformation behavior of K-feldspar. In the early-stage modulated structure studied by Putnis, two orthogonal transverse waves were observed. This implies that the full multi-dimensional short range order representation is four-dimensional and it is necessary to assume, quite correctly in the long wavelength limit, that the wave vectors for the two orthogonal transverse wave systems normal to the c axis are degenerate with one another.

At this stage in the transformation process, the crystal is still properly described as hexagonal but modulated. The four-dimensional irreducible representation which we have just described is responsible for a two-dimensional distribution of the odd-ordered component and its obverse in a pattern rather similar to that described for the ordered structure of the complex nitrite phase in McConnell and Heine (1982) or that inferred for short range order in K-feldspar (McConnell, 1971). This transverse wave-modulated structure is illustrated in Figure 10.

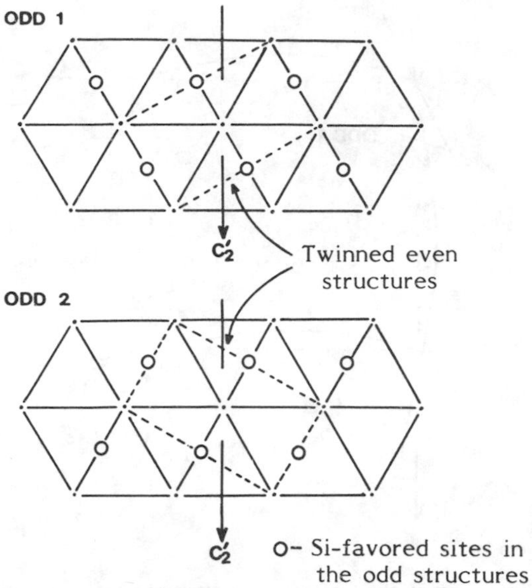

Figure 11. In the later stages of the ordering transformation in cordierite the odd-ordering scheme is made over to the even scheme, which is ultimately of very low energy and entropy. The diagrams illustrate the development of two twin-related, even-ordered schemes from the two odd-ordered alteratives of Figure 10. Note that the Si-favored sites in the odd-ordered variant become totally Si in this process, and that the ordering corresponds to a monotonic transfer of Si atoms to these sites. At this stage in the transformation process the crystal ceases to be hexagonal.

Earlier in this chapter we discussed the fact that even if the odd-ordered component in cordierite was developed uniformly throughout the single crystal with maximal possible order, the configurational entropy would still correspond to approximately half that for the totally disordered state. For the earliest stages of the metastable ordering process, as studied by Putnis, the disorder entropy determined by calorimetry by Carpenter et al. (1983) corresponds closely to this value of one-half, implying that the disordered phase rapidly achieves this odd-ordered state in the initial stages of equilibration (Putnis, 1980a,b; Putnis and Bish, 1983). This is very strong evidence that the present analysis, which bears on the nature of the early short range order, is correct.

It remains to discuss the subsequent ordering behavior of the cordierite which is clearly driven by the necessity to reduce the configurational enthalpy (and therefore entropy) of the odd-ordered state still further. How this is achieved locally by modification of the two alternatives in the odd-ordered modulation may readily be followed from study of Figure 11. This shows that the distribution of favored Si sites in both versions of the odd structure may readily be utilized to generate even-ordered structures by the transfer of Si to the Si-favored sites in the odd-ordered structure. This operation has, in addition, the effect of generating the even structure in two twin-orientated relationships.

Needless to say, this appearance of twinning corresponds to the loss of hexagonal symmetry. The appearance of twin boundaries at this stage in the transformation process in cordierite has been observed experimentally by Putnis (1980a,b). Further, the linear change in enthalpy as a function of log time, observed in this ordering process by Carpenter et al. (1983), may readily be explained by the simple monotonic process of Si transfer between the primary odd-ordered structure and the even-ordered structure.

Finally in this survey of short range order effects, we will comment briefly on the development of incommensurate structures in mineral solid solutions such as the plagioclases, mullite and nepheline (McConnell, 1978b; 1981a,b; 1983). Incommensurate structures may readily be explained in terms of the multi-dimensional irreducible representations which we have already used to study short range order. In the incommensurate structures, based on such favored non-origin scattering vectors, there is a strong energy-lowering interaction between the component structures in the ordering, such that the incommensurate structure is actually more stable than either normally ordered structure. The effect is rather easily explained in the case of mullite where the component structures relate to the ordering of Al and Si atoms on the one hand, and the ordering of oxygen vacancies on the other. The nature of the short range order in incommensurate structures and the application of Landau theory to a study of their stability have been the subjects of a number of papers recently (Heine and McConnell, 1981, 1984).

In conclusion, we wish to emphasize that the use of rigorous symmetry arguments based on group theoretical considerations can be an extremely powerful tool in the study of all manner of short range ordering phenomena and that all the problems which we have discussed here have a great deal in common from the fundamental group theoretical point of view.

APPENDIX

Some notes on the application of group representations in the study of ordering and in the application of Landau theory.

In order to describe ordering in symmetry terms, it is necessary to combine a function which describes the average or disordered state with some function which describes the subgroup symmetry associated with the ordering. Thus in Figure 1 we may utilize the fully symmetrical representation to describe the average state, i.e., all four atoms are labelled with the same label (+1) which stands for (A+B)/2. To create an ordered distribution, we must distribute and add the difference functions (A-B)/2 and -(A-B)/2 over the four sites according to the symmetry chosen. In this way any ordered distribution is defined as the sum of some function with full symmetry and a function with the symmetry of the chosen difference function. In a one-dimensional irreducible representation, which contains only characters 1 and -1, the characters accurately describe the subgroup symmetry but not the amplitude of the functions concerned.

In a two-dimensional representation symmetry operations are represented by two-by-two matrices which again determine the exact symmetry properties of possible subgroups without dictating the amplitude of the

Table 1. The representations of point group 6/mmm (D_{6h}).

D_{6h}	E	$2C_6$	$2C_3$	C_2	$3C_2'$	$3C_2''$	i	$2S_3$	$2S_6$	σ_h	$3\sigma_d$	$3\sigma_v$
A_{1g}	1	1	1	1	1	1	1	1	1	1	1	1
A_{2g}	1	1	1	1	−1	−1	1	1	1	1	−1	−1
B_{1g}	1	−1	1	−1	1	−1	1	−1	1	−1	1	−1
B_{2g}	1	−1	1	−1	−1	1	1	−1	1	−1	−1	1
E_{1g}	2	1	−1	−2	0	0	2	1	−1	−2	0	0
E_{2g}	2	−1	−1	2	0	0	2	−1	−1	2	0	0
A_{1u}	1	1	1	1	1	1	−1	−1	−1	−1	−1	−1
A_{2u}	1	1	1	1	−1	−1	−1	−1	−1	−1	1	1
B_{1u}	1	−1	1	−1	1	−1	−1	1	−1	1	−1	1
B_{2u}	1	−1	1	−1	−1	1	−1	1	−1	1	1	−1
E_{1u}	2	1	−1	−2	0	0	−2	−1	1	2	0	0
E_{2u}	2	−1	−1	2	0	0	−2	1	1	−2	0	0

functions concerned. The character associated with a given symmetry operation in a two- or multi-dimensional irreducible representation corresponds to the trace of the corresponding matrix, and certain operations on these representations can be performed using character data only. Landau tests for the presence of an invariant require that we utilize the full matrices rather than the characters, however.

The representations of the point group 6/mmm are given in Table 1. This group is isomorphic (identical) to that of the vector 000 for the high temperature space group for hexagonal cordierite (P6/mcc). The transformation to the low cordierite, which is orthorhombic with space group Cccm is represented by the two-dimensional representation E_{2g} in this table. Matrices associated with the first three symmetry elements listed are repeated in the remaining nine elements of symmetry so it is necessary only to use three representation matrices in carrying out Landau type expansions. Three matrices for this purpose are set out below.

$$E:\ \begin{bmatrix}1 & 0\\ 0 & 1\end{bmatrix} \qquad C3^{+}:\ \begin{bmatrix}-\frac{1}{2} & -\frac{1}{2}\sqrt{3}\\ \frac{1}{2}\sqrt{3} & -\frac{1}{2}\end{bmatrix} \qquad C2:\ \begin{bmatrix}1 & 0\\ 0 & -1\end{bmatrix}$$

$$\chi = 2 \qquad\qquad \chi = -1 \qquad\qquad \chi = 0$$

The use of Landau expansions is illustrated in what follows. First, the method of calculating the symmetrized cube of a given representation, for example E_{2g}, is to use the formula:

$$[\chi R]^3 = \frac{1}{6}\{[\chi(R)]^3 + 3\chi(R)\chi(R^2) + 2\chi(R^3)\} . \tag{1}$$

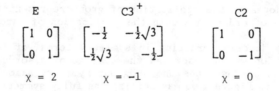

In this expression the term $[\chi(R)]^3$ corresponds simply to taking the cube of the character χ for the original matrix. The term $\chi(R^2)$ corresponds to finding the character for the product of the matrix with itself, i.e., the character of the corresponding operation (R) repeated twice. In the same way the term $\chi(R^3)$ corresponds to the character of the cube of the matrix, i.e., the character of the operation (R) repeated three times. An exactly similar formalism is used for calculating the antisymmetrized square of the chosen representation using the same set of matrices:

$$\{\chi R\}^2 = \tfrac{1}{2}\{[\chi(R)]^2 - \chi(R^2)\}\,. \qquad (2)$$

The calculation set out step by step below demonstrates that the representation E_{2g} for the vector 000 of space group P2/mcc yields an invariant in the symmetrized cube, and the representation of a non-polar vector in the antisymmetrized square. We can be certain that the corresponding transformation will be of first order.

$$E_{2g}\ \{\chi R\}_3 = \tfrac{1}{6}\{[\chi R]^3 + 3\chi R\cdot\chi(R^2) + 2\chi(R^3)\}$$

$$E = \begin{bmatrix} 1 & \\ & 1 \end{bmatrix} = \tfrac{1}{6}\{8 + (6\times 2) + 4\} = 4$$

$$C_3^+ = \begin{bmatrix} -\frac{1}{2} & -\frac{\sqrt{3}}{2} \\ +\frac{\sqrt{3}}{2} & -\frac{1}{2} \end{bmatrix} = \tfrac{1}{6}\{-1 + (-3\times -1) + 4\} = 1$$

$$C_2' = \begin{bmatrix} 1 & \\ & -1 \end{bmatrix} = \tfrac{1}{6}\{0 + (0\times 2) + 0\} = 0$$

	E	C_3^+	C_2'	
	4	1	0	
×	1	1	1	
	4	1×2	0	= 6/3 = $2A_{1g}$

$$E_{2g}\ \{\chi R\}_2 = \tfrac{1}{2}\{[\chi R]^2 - \chi(R^2)\}$$

$$\left.\begin{array}{lll} E & \tfrac{1}{2}\{4 - 2\} = & 1 \\ C_3^+ & \tfrac{1}{2}\{1 - -1\} = & 1 \\ C_2' & \tfrac{1}{2}\{0 - 2\} = & -1 \end{array}\right\} = A_{2g}$$

In Tables 2 and 3 below are set out the representations for the vector 000 of space groups C2/m and for the changes in translational symmetry in going from space group $C\bar{1}$ to $I\bar{1}$ in the plagioclase feldspars.

Table 2. Representations for vector 000 of C2/m

	E	2	$\bar{1}$	m	
R_1	1	1	1	1	2/m
R_2	1	-1	1	-1	$\bar{1}$
R_3	1	-1	-1	1	m
R_4	1	1	-1	-1	2

Table 3. Representations for $C\bar{1}$ to $I\bar{1}$

	E/0	i/0	E/c	i/c	
R_1	1	1	1	1	$C\bar{1}$
R_2	1	-1	1	-1	$C\bar{1}$
R_3	1	-1	-1	1	$I\bar{1}$
R_4	1	1	-1	-1	$I\bar{1}$

REFERENCES

Bardhan, P. and J.B. Cohen (1976) A structural study of the alloy Cu_3Au above its critical temperature. Acta Crystallogr. A32 597-614.

Birman, J.L. (1966) Simplified theory of symmetry change in second-order phase transitions: application to V_3Si. Phys. Rev. Lett. 17, 1216-1219.

__________ (1974) Theory of Crystal Space Groups and Infra-Red and Raman Lattice Processes of Insulating Crystals. In: Handbuch der Physik, XXV/2b, S. Flugge, ed., Springer-Verlag, Berlin.

Bradley, C.J. and A.P. Cracknell (1972) The Mathematical Theory of Symmetry in Solids. Clarendon Press, Oxford.

Carpenter, M.A., A. Putnis, A. Navrotsky and J.D.C. McConnell (1983) Entropy effects associated with Al/Si ordering in anhydrous magnesian cordierite. Geochem. Cosmochem. Acta 47, 899-906.

de Fontaine, D. (1975) k-space symmetry rules for order-disorder reactions. Acta Met. 23, 553-571.

Heine, V. and J.D.C. McConnell (1981) Origin of modulated incommensurate phases in insulators. Phys. Rev. Lett. 46, 1092-1095.

__________ and __________ (1984) The origin of incommensurate structures in insulators. J. Phys., C: Solid State Phys. 17, 1199-1220.

Landau, L.D. and E.M. Lifshitz (1958) Statistical Physics. Addison Wesley, Reading, Massachusetts.

Lifshitz, E.M. (1942a) On the theory of phase transitions of second order. I. Changes of the elementary cell of a crystal in phase transitions of the second order. J. Physics (Moscow) 6, 61-74.

__________ (19742b) On the theory of phase transitions of second order. II. Phase transitions of the second order in alloys. J. Physics (Moscow) 6, 251-263.

Lyubarski, G.Ya. (1960) Application of Group Theory in Physics. Pergamon Press, New York, Chapter VIII.

McConnell, J.D.C. (1971) Electron optical study of phase transformations. Mineral. Mag. 38, 1-20.

__________ (1978a) K-space symmetry rules and their application to ordering behavior in non-stoichiometric (metal enriched) chalcopyrite. Phys. Chem. Minerals 2, 253-265.

__________ (1978b) The intermediate plagioclase feldspars: an example of a structural resonance. Z. Kristallogr. 147, 45-62.

__________ (1981a) Electron optical study of modulated mineral solid solutions. Bull. Minéral. 104, 231-235.

__________ (1981b) Time-temperature study of the intensity of satellite reflections in nepheline. Am. Mineral. 66, 990-996.

__________ (1983) A review of structural resonance and the nature of long-range interactions in modulated mineral structures. Am. Mineral. 68, 1-10.

__________ and V. Heine (1982) Origin of incommensurate structure in the cooperative Jahn-Teller system $K_2PbCu(NO_2)_6$. J. Phys. C, Solid State Phys. 15, 2387-2402.

__________ and __________ (1984) An aid to the structural analysis of incommensurate phases. Acta Crystallogr. A40, 473-482.

Putnis, A. (1980a) Order-modulated structures and the thermodynamics of cordierite reactions. Nature 287, 128-131.

__________ (1980b) The distortion index in anhydrous Mg-cordierite. Contrib. Mineral. Petrol. 74, 135-141.

__________ and D.L. Bish (1983) The mechanism and kinetics of Al,Si ordering in Mg cordierite. Am. Mineral. 68, 60-65.

Chapter 6. Michael A. Carpenter

ORDER-DISORDER TRANSFORMATIONS in MINERAL SOLID SOLUTIONS

1. INTRODUCTION

When a structure is subject to cation substitution, its macroscopic thermodynamic properties are usually modified. The changes may be understood in terms of structural strains, electrostatic interactions between ions, bonding variations, etc., and, depending on the magnitude of these microscopic contributions, the structure may undergo a phase transformation. In the simplest treatments of solid solutions, a tendency to unmix and a tendency to order are regarded as opposite extremes which result from either positive or negative interactions. Thus, for a regular solution comprised of equal proportions of A and B atoms, the total energy of interaction between first nearest neighbors varies with $\varepsilon AB - \frac{1}{2}\varepsilon AA - \frac{1}{2}\varepsilon BB$, where εAB, εAA, εBB can be defined as the energies of A-B, A-A, B-B interactions (with negative values). If this sum is zero ($\varepsilon AB = \frac{1}{2}[\varepsilon AA + \varepsilon BB]$), the solid solution behaves as an ideal mixture with no macroscopic excess properties. If it is positive, phase separation will be favored as temperature is reduced, but, if it is negative, ordering of A and B atoms onto discrete sublattices will occur instead. In reality, mineral solid solutions frequently show tendencies both to order and to unmix in the same system; e.g., Figure 1: ilmenite-hematite (after Burton, 1984); jadeite-augite (Carpenter, 1978, 1980); calcite-magnesite (Goldsmith and Heard, 1961; Reeder, 1981, 1983; Goldsmith, 1983; and see Goldsmith, 1983, for analogous carbonate systems); albite-anorthite (schematic after: Smith, 1972, 1974, 1983; Carpenter, 1981; Ribbe, 1983a,b; Grove et al., 1983; Carpenter and McConnell, 1984). These two processes are not, therefore, mutually exclusive but can occur in a cooperative or interdependent manner, as is commonly found in metallic systems (Soffa and Laughlin, 1982).

That the macroscopic properties of solid solutions arise from the sum of microscopic interactions is obvious enough, but it is clear that physically realistic models cannot depend only on nearest neighbor effects. In order to reproduce the topology of the phase diagrams for some binary metallic alloys, using pairwise atomic interactions, at least nearest and next nearest neighbor effects must be included (Richards, 1971; Richards and Cahn, 1971; Inden, 1974; Büth and Inden, 1982). As a further sophistication, interactions within clusters of atoms instead of between pairs of atoms may be considered by the cluster variation method (CVM) of Kikuchi (1951) (and see Kikuchi and Sato, 1974; de Fontaine, 1979). Burton and Kikuchi (1984a,b) and Burton (1984) have successfully applied the latter approach to the systems calcite-magnesite and ilmenite-hematite. These more elaborate models do not yield thermodynamic expressions for immediate general use, however, and, until they are more widely applied, we are still left with the questions of why and how complex mineral solid solutions behave as they do.

The purpose of this chapter is to present an overview of the subsolidus behavior of mineral solid solutions which undergo discrete order-disorder transformations. In the absence of definitive atomistic models, it is still possible to outline the general forms of the principal

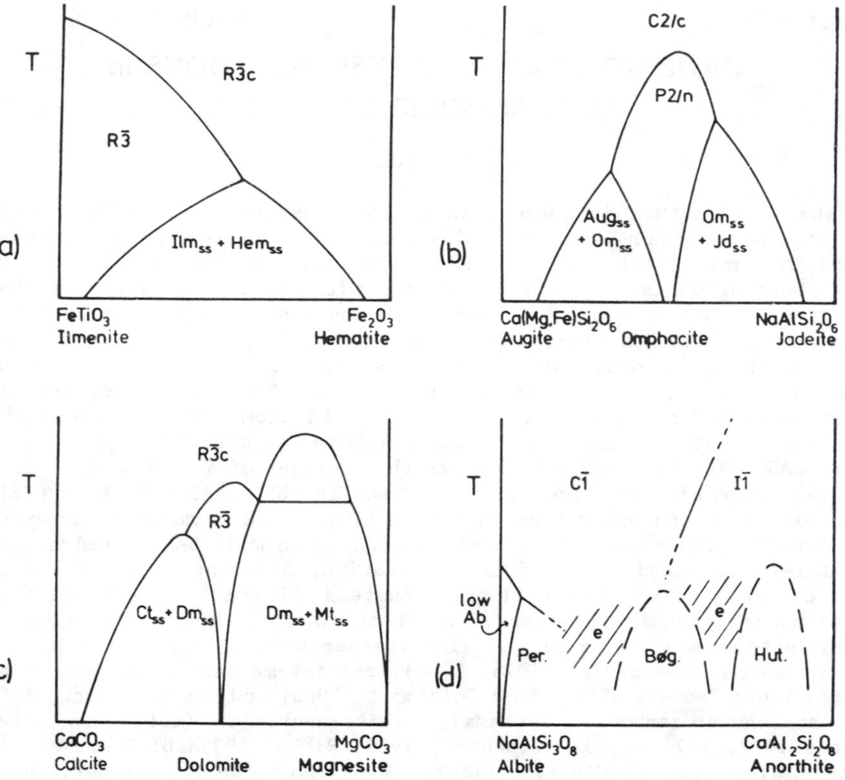

Figure 1. Schematic phase relations of solid solutions which undergo both ordering and exsolution reactions (ignoring melting). (a) Ilmenite-hematite; after Burton (1984). (b) Jadeite-augite; from Carpenter (1980). (c) Calcite-magnesite; after Goldsmith and Heard (1961), Reeder (1981), Merkel and Blencoe (1982), Goldsmith (1983). (d) Plagioclase feldspars: only the high ⇄ low transformation in pure albite and the C1̄ ⇄ I1̄ transformation line have been determined experimentally (Goldsmith and Jenkins, 1985; Carpenter and McConnell, 1984). The positions of stability fields for "e" plagioclase and their relationship with the three miscibility gaps are not known and therefore have been left undefined. Per = peristerite gap; Bøg = Bøggild gap; Hut = Huttenlocher gap. The displacive P1̄ ⇄ I1̄ transformation is not shown.

thermodynamic properties, enthalpy, entropy and volume as they vary with pressure, temperature and composition through an ordering transformation. A more general and phenomenological approach serves at least to define the macroscopic constraints on the way in which a given solid solution can evolve. It may also serve as a reference against which the results of microscopic theories can be compared.

The chapter is divided into three main sections. In the first, changes in physical properties associated with order-disorder transformations are described. The second section deals with examples of Al,Si ordering transformations for which some thermodynamic data are available (plagioclase and cordierite). Similar data for other mineral systems are also briefly reviewed. In the final section, a number of solid solutions are described in which both ordering and exsolution can occur at low temperatures. Quantitative analysis of individual systems and computational methods have been deliberately excluded in order to concentrate on a broader view. In addition, only discrete cation order-disorder transformations are considered; for example, non-convergent cation ordering is treated elsewhere by Navrotsky (this volume). The rigorous constraints of symmetry are applied to some of the examples described by McConnell (this volume).

Subsolidus processes may be seen as the result of short range and long range interactions between atoms or ions. In some solid solutions this combination may generate a delicate balance of forces such that a small change in temperature or composition makes ordering marginally more favorable than exsolution, or vice versa. The complex series of processes that can then occur are subject to thermodynamic and mechanistic constraints. We must therefore be interested not only in predicting equilibrium states but also in determining how those states may or may not be achieved. Thus if any general themes in this chapter can be isolated, they relate firstly to the relationship between equilibrium thermodynamics and the mechanisms of phase transformations, secondly to the role of short range ordering in making a link between microscopic and macroscopic properties, and finally to the paucity of hard data for cation ordering transformations in minerals.

No attempt has been made to produce a thorough bibliography. Reference is made throughout to appropriate reviews where they are available, and emphasis is placed on recent work. A more extensive selection of papers from the non-mineralogical literature is included, however.

2. PRINCIPAL THERMODYNAMIC PARAMETERS THROUGH AN ORDER-DISORDER TRANSFORMATION

2.1 Thermodynamic aspects

In the absence of physically realistic methods for calculating the thermodynamic properties of minerals through order-disorder transformations, it is necessary to consider, if only qualitatively, the general form which the principal parameters might have. In particular, we wish to know the possible forms of free energy (G), enthalpy (H), entropy (S) and volume (V) as functions of temperature (T) and pressure (P), and, in the case of solid solutions, as functions of compositions (X). To predict the form of these properties it is necessary first to consider what kinds of transformations, in the thermodynamic sense, are likely in minerals.

A convenient classification of phase transformations in the present context is that of Ehrenfest (1933) (and see Pippard, 1957; Sato, 1970; Parsonage and Staveley, 1978; Rao and Rao, 1978; Thompson and Perkins, 1981). The order of a transformation is defined in terms of the lowest derivative of free energy with respect to temperature and/or pressure which is a discontinuous function. From the restricted equilibrium conditions,

$$dG = VdP - SdT$$

the first and second derivatives are well known:

$$(\partial G/\partial P)_T = V \ , \quad (\partial G/\partial T)_P = -S \ , \quad (\partial^2 G/\partial P^2)_T = (\partial V/\partial P)_T = -V\beta \ ,$$

$$(\partial^2G/\partial T^2)_P = -(\partial S/\partial T)_P = -C_p/T \ , \ (\partial^2G/\partial P\partial T) = (\partial V/\partial T)_P = V\alpha \ ,$$

where Cp is heat capacity, α is isobaric thermal expansion and β is isothermal compressibility. Thus a first order transformation has discontinuities in H, S, V at the equilibrium transformation temperature and pressure (Tc, Pc); whereas, a second order transformation has discontinuities in Cp, α and β but is continuous in H, S, V through Tc, Pc. The forms of these parameters are shown in Figure 2, columns I and II. Landau and Lifshitz (1958) provided a further basis for this approach by describing transformations in terms of a long range order parameter (η_ℓ), which goes continuously to zero as T increases to Tc in a second order transformation, and discontinuously to zero in a first order transformation (Fig. 2).

In practice, this classical approach breaks down at temperatures close to Tc, and, for example, truly second order transformations with discrete and measurable values of ΔCp appear to be rare. Instead, some transformations (including atomic ordering and atomic displacements) have a λ-shaped anomaly in the heat capacity at Tc, and these are commonly described as λ transformations (Fig. 2, columns III and IV) (see reviews by Rao and Rao, 1978; Parsonage and Staveley, 1978; Thompson and Perkins, 1981). The origin of this departure lies in the fact that in classical theories pairwise interactions between atoms are replaced by a "mean field," which is equivalent to assuming that local structure is controlled by long range interactions and that only long range ordering can occur (Tanner and Leamy, 1974; Rao and Rao, 1978). Because of short range interactions, however, critical fluctuations or short range ordering can occur near Tc, with the development of pretransition phenomena above Tc (see Cook et al., 1977). Parsonage and Staveley (1978, p. 856) have suggested that λ-shaped anomalies in Cp are usually found for ordering transitions in which the interactions are short range, whereas, more classical heat capacity effects (Fig. 2, column II) indicate that long range interactions are important.

Lambda transformations can be first or second order in the Erhenfest sense of having discontinuous or continuous changes of H, V and S. For example, the $\alpha \rightarrow \beta$ transformation in quartz has a strong λ-shaped Cp anomaly with small discontinuities in H and V at Tc (Ghiorso et al., 1979; Dolino and Bachheimer, 1982). Recent structural studies have revealed the presence of an intermediate (incommensurate) phase over a range of ~1.5 degrees between the α (low) and β (high) phases (Gouhara et al., 1983a,b; Dolino et al., 1983, 1984a,b). The largest heat capacity effect occurs at the transition from α to the intermediate phase, which is first order (Dolino et al., 1983, 1984b; Gouhara and Kato, 1984). In contrast, Cu-Zn ordering in β-brass gives a λ-shaped anomaly in Cp and has a continuous change of H through the transition, with a cusp of the form predicted for a second order transformation (see Sato, 1970; Thompson and Perkins, 1981). In Figure 2 a distinction has therefore been made between λ transformations with and without a first order break at Tc, although both would show pretransition effects (dynamical fluctuations or short range order), and the first order break could be very small.

In principle, the nature of a transformation under consideration could be identified on the basis of its heat capacity. Aside from the problem of measuring Cp close to Tc when it can vary rapidly, this still is not a practicable approach for cation order-disorder transformations in minerals. The required cation exchanges are far too slow to allow a dynamic equilibrium to be maintained on the time scale needed for heat capacity measurements. A classification based on Cp is useful only in so far as it provides a framework for comparison with different types of transformations and transformations in other materials. As will be discussed below, short range order can be an important factor in order-disorder processes in minerals, so that the variations of H and S shown in columns II, III and IV of Figure 2 are most likely. Gradations between these idealized forms are possible, depending on the relative importance of short and long range interactions, and the magnitude of any first order break at Tc. More detailed analyses, taking account of possible interactions between order parameters and volume or strain, for example, which can cause transformations which would otherwise be second order to become first order (Bruce and Cowley, 1981), are beyond the scope of the present review.

2.2 Mechanistic aspects

The thermodynamic character of an ordering transformation exerts some influence over the mechanisms by which it can actually proceed under equilibrium and non-equilibrium conditions (de Fontaine, 1975, 1979; Cook, 1976; Chen and Cohen, 1977, 1979). Information on mechanisms and kinetics at different temperatures below Tc can be of value in unravelling transformation properties, and, conversely, an understanding of the thermodynamic character may allow prediction of the transformation mechanisms. The arguments are illustrated here using the Landau and Lifshitz (1958) free energy expansion in terms of the long range order parameter (η_ℓ). (The Landau-Lifshitz theory of phase transformations is dealt with in more detail by de Fontaine, 1975, 1979; McConnell, 1978a and this volume; Bruce and Cowley, 1981.) For small values of η_ℓ,

$$G = A\eta_\ell^2 + B\eta_\ell^3 + C\eta_\ell^4 + \dots ,$$

where G is the free energy of ordering, $A = a(T-T_i)$, and A, B, C, Ti are constants. For situations where G is symmetrical with respect to the $\eta_\ell = 0$ axis (the energy of an ordered state with $+\eta_\ell$ order is the same as for $-\eta_\ell$), odd order terms in the expansion must be zero. When the transformation is second order: $Tc = T_i$, $B = 0$, C is positive and, at T > Tc, A is positive; at T = Tc, A = 0; at T < Tc, A is negative. The relationship between the free energy of ordering and the long range order parameter given by these conditions is shown in Figure 3a. At any temperature, the equilibrium value of η_ℓ ($\eta_{\ell,eqm}$) occurs at the minimum in G. η_ℓ increases smoothly from zero as T decreases from Tc, and at any $T \leqslant Tc$ there is a continuous pathway (continuously decreasing G) between $\eta_\ell = 0$ and $\eta_\ell = \eta_{\ell,eqm}$. A second order transformation is therefore characterized not only by a continuous increase in $\eta_{\ell,eqm}$ from zero, but also by the availability of a continuous reaction pathway for ordering to proceed at any temperature up to Tc.

For a first order transformation in which, by symmetry, the condition of $+\eta_\ell$ being energetically equivalent to $-\eta_\ell$ still holds, odd order terms

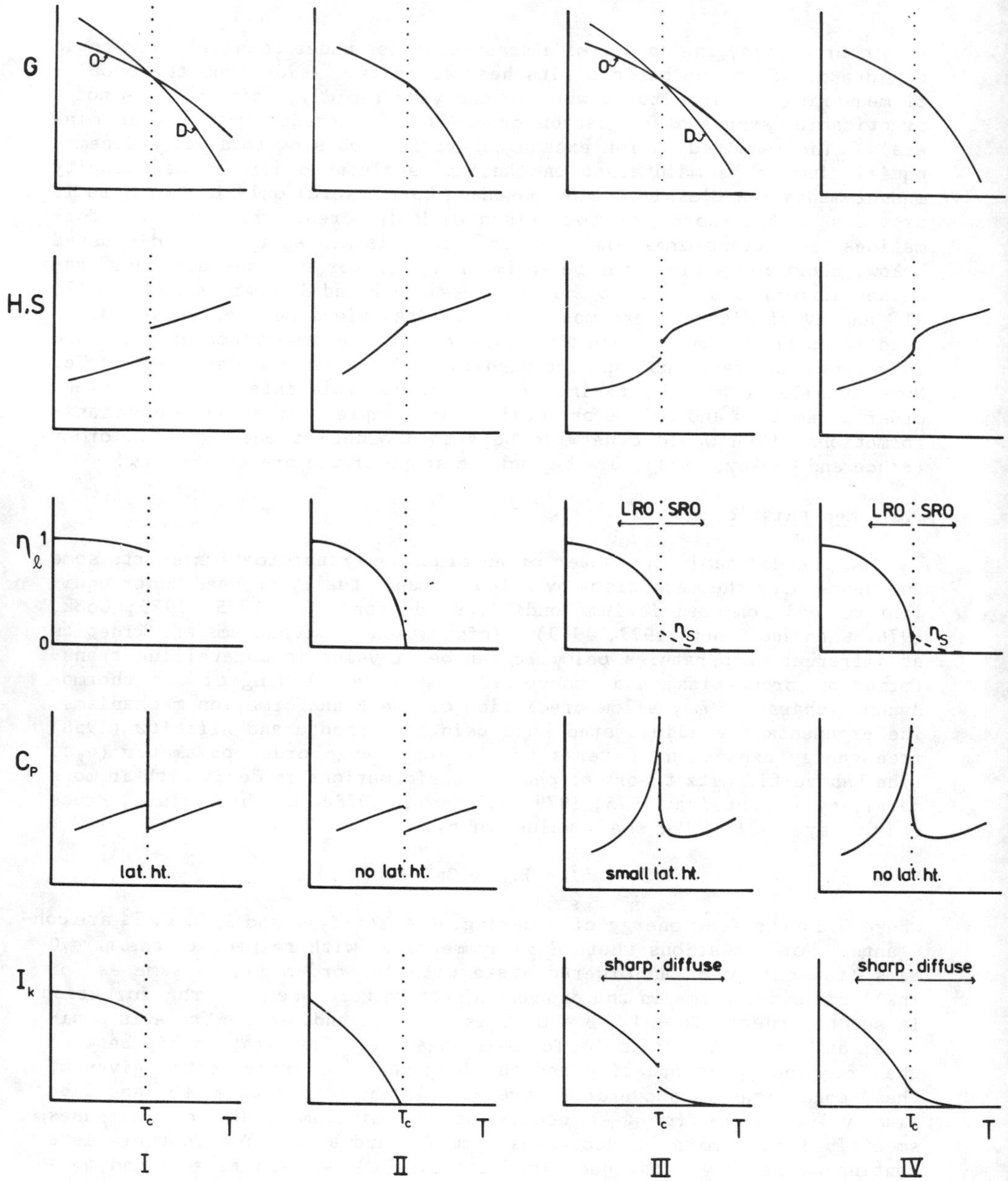

Figure 2. Schematic form of the principal thermodynamic parameters through a phase transformation at Tc (modified after Thompson and Perkins, 1981). Column I = first order; column II = second order; column III = λ transformation with a small first order break at Tc; column IV = λ transformation with no first order break. G = free energy, H = enthalpy, S = entropy, η_ℓ = long range order parameter, η_S = short range order parameter describing precursor ordering above Tc; Cp = specific heat; I_k = integrated intensity of a superlattice reflection. D = disordered state, O = ordered state. LRO = long range order, SRO = short range order. Volume is not shown, but must be continuous or discontinuous in some manner analogous to H and S.

in the expansion must still be zero. The equilibrium transformation temperature, Tc, is above the temperature T_i at which A goes to zero, however, and this may be described by A > 0, B = 0, C < 0 with a term, $E\eta_\ell^6$, added for stability, having E > 0. The resulting G - η_ℓ relations are shown in Figure 3b. In this case, at T = Tc, $\eta_{\ell,eqm}$ is greater than zero and there is an energy barrier between η_ℓ = 0 and $\eta_\ell = \eta_{\ell,eqm}$. A nucleation event is now required if the equilibrium ordered structure is to form from a disordered crystal. This energy barrier persists as T decreases but, at some lower temperature, T_i, it may disappear. At $T \leqslant T_i$ the disordered structure is unstable with respect to small fluctuations in order and a continuous reaction pathway between η_ℓ = 0 and $\eta_\ell = \eta_{\ell,eqm}$ becomes available. A first order transformation is therefore characterized not only by a break in $\eta_{\ell,eqm}$ at T = Tc but also by the need for a nucleation and growth mechanism when ordering occurs at T_i < T < Tc. By analogy with spinodal decomposition, the continuous ordering at T < T_i is sometimes known as spinodal ordering (de Fontaine, 1975; Tanner and Laughlin, 1975; Chen et al., 1977).

A final case is illustrated in Figure 3c. This represents first order transformations in which the odd order terms in the expansion are not equal to zero; at T = Tc, A > 0, B < 0, C > 0. Available ordering mechanisms are the same as described for the case illustrated in Figure 3b. The addition of short range ordering effects does not materially affect the mechanisms outlined for either first or second order transformations.

2.3 Physical observations

As has already been pointed, thermodynamic data for discrete cation order-disorder transformations in minerals are rather restricted. Given the kinetic problems, dynamic experiments involving direct measurements of structural changes at elevated temperatures and pressures are not really feasible. Most observations of physical properties have therefore been made on samples quenched after long annealing times at high temperatures and pressures.

The order parameter, η_ℓ, as defined in Landau theory, is a macroscopic property and can, in principle, be determined by measuring other macroscopic properties. The most direct information relating to η_ℓ is carried in the intensities (I_k) of sharp superlattice reflections, with $I_k \propto \eta_\ell^2$ (Sato, 1970; Cook et al., 1977; Bruce and Cowley, 1981). Similarly, the intensity of diffuse reflections around superlattice positions from samples held just above Tc carries information on the nature and extent of short range order (or dynamical fluctuations) (de Fontaine, 1975, 1979; Cook, 1975; Bardhan and Chen, 1976; Cook et al., 1977; Chen and Cohen, 1979). For transformations with a first order break (columns I and III, Fig. 2) the discontinuity in η_ℓ at Tc should be detectable experimentally as a discontinuity in I_k with T. Short range order, giving rise to diffuse superlattice reflections above Tc, and variations of I_k with T below Tc may also be observed. For the case of thermodynamically continuous transformations (Fig. 2, columns II and IV) I_k should be continuous through Tc, again with the possibility of short range order and diffuse diffraction maxima above Tc. In a classical second order transformation (Fig. 2, column II) I_k goes smoothly to zero at Tc. Quantitative relationships between η_ℓ, I_k and T remain to be explored for cation order-

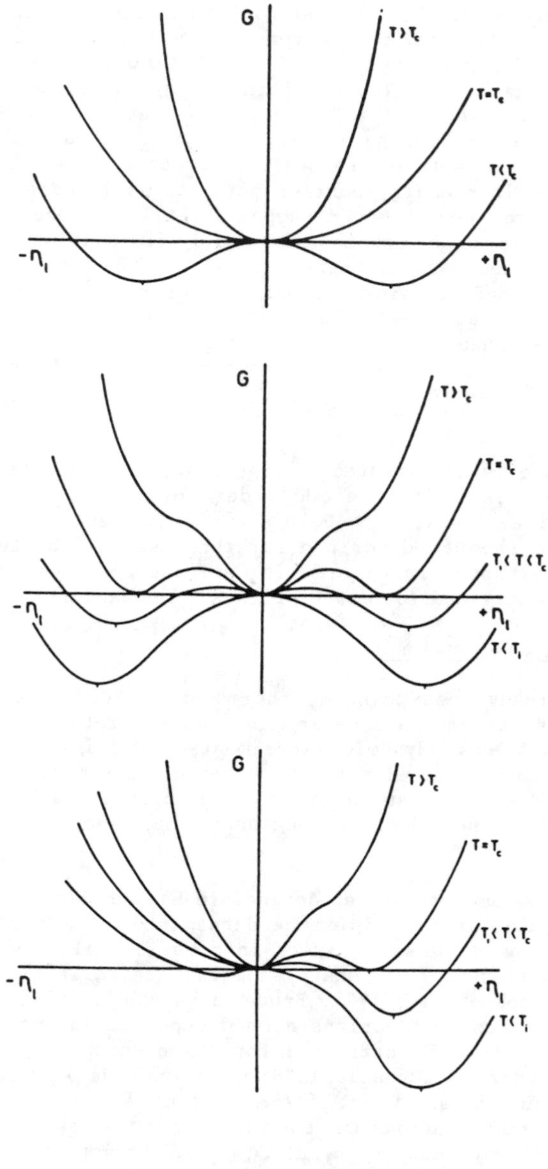

Figure 3. Schematic relationship between free energy (G) and long range order (η_ℓ), based on the Landau-Lifshitz free energy expansion. The equilibrium value of η_ℓ at each temperature is indicated by a small vertical line. (a) Second order; the equilibrium value of η_ℓ goes continuously to zero as T increases to Tc. A continuous mechanism of ordering is available at any T $\leqslant$ Tc. (b) First order, symmetrical; at Tc there is a jump in the equilibrium value of η_ℓ from 0 to $\eta_{\ell,eqm}$. Nucleation is required to form an ordered phase from its ordered host at $T_i < T < Tc$. Below T_i a disordered crystal would be unstable to ordering fluctuations and a continuous ordering mechanism becomes possible (spinodal ordering). (c) First order, asymmetric; mechanisms of ordering as in (b).

ing processes in minerals, but for both atomic ordering and displacive transformations in a variety of non-mineralogical systems I_k has been observed (in neutron diffraction studies) to vary with

$$(T - Tc)^{2\beta}, \text{ where } \beta \simeq 1/3 \text{ [i.e., } \eta_\ell \propto (T - Tc)^{\sim 1/3}]$$

(Als-Nielsen and Dietrich, 1967; Riste et al., 1971; Durand, 1982). Classical theories predict $\eta_\ell \propto (T - Tc)^{\frac{1}{2}}$ and hence, because $I_k \propto \eta_\ell^2$, a linear variation of I_k with T. Intensity measurements of this sort have to be made on crystals held at the appropriate temperature and not on quenched samples. In those transformations which do not give rise to superlattice reflections, the order parameter may be related to lattice distortions and short range order may be more difficult to detect.

It should be noted that a number of order parameters can be defined or measured for a given phase transformation, each of which will refer to a different aspect of the structure and each of which may have its own precise temperature dependence. The driving or "primary" order parameter of thermodynamic consequence must involve the correct breaking of symmetry for the transformation as a whole (Bruce and Cowley, 1981). Coupling between order parameters is, however, also possible and can lead to a variety of transformation topologies. For information on this aspect of order parameter theory, readers are referred to Achiam and Imry (1975), Imry (1975), Gufan and Larin (1980), Bruce and Cowley (1981), Salje and Devarajan (1985).

Enthalpy changes associated with variations in order can be determined in a solution calorimeter by measuring the heats of solution of samples equilibrated at different temperatures and pressures, since for minerals the state of order is generally quenchable. This method can give some insight into details of the enthalpy changes with temperature, but only if a rather large number of samples, annealed at closely spaced temperatures spanning Tc, are investigated. It has the merit, however, of being quantitative. An example of what the measurement entails is shown in Figure 4. If the solution calorimeter is operating at ~973 K (e.g., using a lead borate flux), the difference in enthalpy (ΔH_{ord}) between a sample equilibrated at some high temperature and a sample equilibrated at some low T, as measured at 973 K, is obtained. In the absence of Cp data for the ordered and disordered phases it is necessary to assume that ΔCp effects between the high and low equilibration temperatures are small.

Changes in degree of order can be determined by standard crystallographic (x-ray and neutron) methods (Burnham, 1973). Unfortunately, these usually only give average site occupancies and do not provide the information on local correlations and fluctuations necessary for an accurate calculation of configurational entropy. Moreover, no quantitative treatments of diffuse diffracted intensity appear to have yet been made for cation ordering in minerals. In this context, further development of spectroscopic methods for extracting short range cation ordering data would be invaluable (Burnham, 1973; Hawthorne, 1983; McMillan, this volume; Burns, this volume).

Volume can be measured as a function of order and could be used, in principle, to follow the nature of a transformation. In particular, the

Table 1. A summary of data for selected cation order-disorder transformations in minerals.

Mineral	Ions involved in ordering	Symmetry of high T (disordered) structure	Symmetry of low T (ordered) structure	Thermodynamic character of transformation[1]
Cordierite $Mg_2Al_4Si_5O_{18}$	Aℓ,Si	P6/mcc	Cccm	1st order
Anorthite $CaAl_2Si_2O_8$	Aℓ,Si	$C\bar{1}$	$I\bar{1}$	2nd order
Bytownite (An_{72}) $Na_{0.28}Ca_{0.72}Al_{1.72}Si_{2.28}O_8$	Aℓ,Si	$C\bar{1}$	$I\bar{1}$	2nd order
Labradorite (An_{60}) $Na_{0.4}Ca_{0.6}Al_{1.6}Si_{2.4}O_8$	Aℓ,Si,Na,Ca	$C\bar{1}$	incommensurate "e" structure	2nd order
Andesine (An_{40}) $Na_{0.6}Ca_{0.4}Al_{1.4}Si_{2.6}O_8$	Aℓ,Si,Na,Ca	$C\bar{1}$	incommensurate "e" structure	2nd order
Albite $NaAlSi_3O_8$	Aℓ,Si	$C\bar{1}$	$C\bar{1}$ ("low albite" structure)	non-first order[5]
K-feldspar $KAlSi_3O_8$	Aℓ,Si	C2/m	$C\bar{1}$	2nd order
Sillimanite Al_2SiO_5	Aℓ,Si	Pbam	Pbnm	2nd order
Mullite $Al_1[Al_{1+2x}Si_{1-2x}]O_{5-x}$	Aℓ,Si,O,□[7]	Pbam	incommensurate structure	2nd order
Omphacite $\sim Na_{0.5}Ca_{0.5}Al_{0.5}Mg_{0.5}Si_2O_6$	Na,Ca,Mg,Aℓ	C2/c	P2/n	2nd order
Dolomite $Ca_{0.5}Mg_{0.5}CO_3$	Ca,Mg	$R\bar{3}c$	$R\bar{3}$	2nd order (Goldsmith & Heard, 1961; Reeder, 1981,1983: Goldsmith, 1983)
Ilmenite $FeTiO_3$	Fe^{2+},Ti	$R\bar{3}c$	$R\bar{3}$	2nd order

1 Second order character is indicated where it is possible; it has not, of course, been proved.

2 An estimated value for the entropy change due to the transformation ($\Delta S_{ord,est}$) has been obtained by setting $\Delta S_{ord} = \Delta H_{ord}/T_c$. A maximum configurational entropy change ($\Delta S_{ord,max}$) has been calculated assuming a change from perfectly ordered to randomly disordered cation distributions.

3 The value given for the enthalpy of ordering of cordierite is for a change from a metastable state of disorder (hexagonal) to a state of order (orthorhombic).

Table I, continued.

T_c(K)	ΔH_{ord}(kJ/mole)	$\Delta S_{ord,est}$[2] (J/mole.K)	$\Delta S_{ord,max}$[2] (J/mole.K)	ΔV_{ord} (Å^3/unit cell)
∿1723 (Smart & Glasser, 1977)	40.84±6.53[3] (Carpenter et al., 1983)	19.9-27.5	51.4	-
∿2000<T_c<∿2250 (by extrapolation; Carpenter & Ferry, 1984)	15.5±2.5 (by extrapolation, Carpenter et al., 1985)	5.8-9.0	23.1	-
∿1525 (Carpenter & McConnell, 1984)	7.91±1.09 (Carpenter et al., 1985)	4.5-5.9	27.7	0.65±0.5[4] (≈0.1%) (Carpenter et al., 1985)
? ∿1125±50 (Carpenter, unpublished data)	11.72±1.17 (Carpenter et al., 1985)	9.0-12.0	28.0	1.1±0.5[4] (≈0.2%) (Carpenter et al., 1985)
<1025 (Carpenter, unpublished data)	5.86±1.13 (Carpenter et al., 1985)	<∿6	27.1	1.0±0.5[4] (≈0.2%) (Carpenter et al., 1985)
950-1020 (Goldsmith & Jenkins, 1984, 1985)	∿12.6±1.3 [973K] (Holm & Kleppa, 1968; Newton et al., 1980; Blinova & Kiseleva 1982; Carpenter et al., 1985) 11.00±1.67 [323K] (Thompson et al., 1974) 10.88±1.26 [323K] (Waldbaum & Robie, 1971)	∿11.4-14.1 (using T_c = 985K and ΔH_{ord} = 12.6±1.3 kJ/mole)	18.7	2.8±0.4[4] (≈0.4%) (Kroll & Ribbe, 1983; Carpenter et al., 1985)
∿723 (Bambauer & Bernotat, 1982; Bernotat & Bambauer, 1982)	∿8.16±1.34 [323K] (Waldbaum & Robie, 1971)	9.3-13.3	18.7	1.9±0.3 (≈0.3%) (Waldbaum & Robie, 1971)
∿2010 (Greenwood, 1972; Holland & Carpenter, 1983)	∿12[6]	-	11.5	-
? ∿2075 for $Al[Al_{1.4}Si_{0.6}]O_{4.8}$	∿15[6]	-	14.3	-
∿1138 for ∿$Jd_{50}Aug_{50}$ (Carpenter, 1981)	8.28±1.46 (Wood et al., 1980)	6.0-8.6	11.5	0.1±0.3
1373<T_c<1423 (Reeder & Nakajima, 1982; Reeder & Wenk, 1983)	∿4[6] ∿6.1 from Bragg-Williams model if T_c = 1473K (Helgeson et al., 1978; Navrotsky & Loucks, 1977)	-	5.8	-
∿1673 (Burton, 1984)	∿9.6[6] ∿8 (Burton, 1984)	-	11.5	-

4 c = 7Å unit cell.

5 Salje (1985), Salje et al. (1985) and Goldsmith & Jenkins (1984,1985) have shown that high/low albite ordering is probably thermodynamically continuous and does not give a discrete transformation.

6 Some values of ΔH_{ord} have been guessed by setting $\Delta H_{ord} = (\Delta S_{ord,max}/2).T_c$.

7 Ordering in mullite involves vacancies (□).

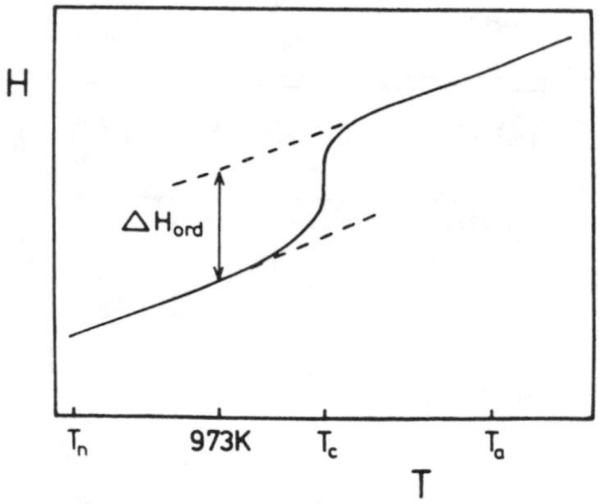

Figure 4. Schematic enthalpy changes with temperature through an ordering transformation with Tc > 973 K. ΔH_{ord} is the enthalpy change, measured by solution calorimetry at 973 K, say, between a natural sample equilibrated at T_n and the same sample annealed at some high temperature (T_a). Dashed lines indicate the enthalpy changes of the two samples due to lattice vibrational contributions to Cp.

observation of a discontinuity in V at T = Tc, or P = Pc, would be instructive, but the volume changes associated with ordering and disordering in minerals tend to be very small and, therefore, rather insensitive to varying degrees of order (Table 1). In addition, with small volume changes, the effects of impurities and changes in oxidation state may mask the effects of changes of ordering.

An alternative approach to the problem of how thermodynamic parameters vary through a transformation may be to consider their variation with composition. This is because it is usually easier to prepare an isothermal, isobaric suite of samples with different compositions than to equilibrate a single sample over a wide range of temperatures on both sides of Tc (see below). Study of the kinetics and mechanisms of ordering transformations at temperatures away from Tc, particularly using transmission electron microscopy, can also add a further dimension to the problem of defining the nature of order-disorder transformations (Tanner and Leamy, 1974; Heuer and Nord, 1976).

3. DISCRETE ORDER-DISORDER TRANSFORMATIONS IN MINERALS

Table 1 provides a summary of data available for some of the best known order-disorder transformations in minerals. It is immediately apparent that very few data are available for even the most basic properties, and not one system has been studied directly and systematically as a function of temperature, with the recent exception of albite (Salje et al., 1985; Goldsmith and Jenkins, 1984, 1985). There is actually far more information on the microstructures in these minerals which indicate that they have undergone phase transformations during their evolution in nature. In this section, two Al,Si order-disorder transformations will be compared; the $C\bar{1} \rightleftarrows I\bar{1}$ transformation in plagioclase feldspars has properties consistent with it being non-first order in character, while the $P6/mcc \rightleftarrows Cccm$ transformation in cordierite is first order. The symmetry properties of these same transformations are discussed by McConnell in the following chapter.

At temperatures up to its melting point, anorthite (An = $CaAl_2Si_2O_8$) has an ordered distribution of Al and Si on tetrahedral sites, and $I\bar{1}$ symmetry at high temperatures (Laves and Goldsmith, 1955; Laves et al., 1970; Bruno et al., 1976; Smith, 1974; Ribbe, 1983a). Its equilibrium order-disorder temperature would occur at a temperature well above the melting point but decreases rapidly as the composition changes by solid solution towards albite (Ab = $NaAlSi_3O_8$). At $Ab_{30}An_{70}$ Tc is ~1525 K (Carpenter and McConnell, 1984). Prolonged annealing experiments at different temperatures and on samples with different compositions (Carpenter and McConnell, 1984) suggested the following properties for the transformation: (a) As T increases up to Tc the superlattice reflections remain sharp but become increasingly weaker; (b) the actual transformation from the ordered ($I\bar{1}$, anorthite) structure to the disordered ($C\bar{1}$, high albite) structure is marked by an abrupt change from sharp to diffuse ordering reflections; (c) no evidence for a two phase field (which would be present if the transformation is first order) was detected between the $C\bar{1}$ and $I\bar{1}$ fields. The observations are thus consistent with, but not proof of non-first order character for the transformation. The ordering process, $C\bar{1} \rightarrow I\bar{1}$, results in the formation of antiphase domains, and these have been observed in many natural samples (reviewed by Heuer and Nord, 1976; Smith, 1974; Ribbe, 1983a). If the transformation is thermodynamically continuous (second or higher order in the Ehrenfest sense), a continuous ordering mechanism should be available at T = Tc. It has not been possible, however, to distinguish between a nucleation and growth mechanism and continuous ordering on the basis of antiphase domain distributions. Annealing experiments on anorthite at 1703 K give microstructures which are consistent with a continuous ordering mechanism (Kroll and Müller, 1980), although this is at T << Tc.

Limited enthalpy data have been obtained by solution calorimetry (Carpenter et al., 1985). The heat of solution for ordered crystals of 72 mol % An composition with $I\bar{1}$ structure is 69.71 ± 1.00 kJ/mole, and for the same crystals disordered ($C\bar{1}$) by annealing at 1620 K it is 61.80 ± 0.42 kJ/mole. The total enthalpy change between these samples due to ordering (ΔH_{ord}, as measured at 973 K, see Fig. 4) is therefore 7.91 ± 1.09 kJ/mole. A rough estimate of the entropy change can be made by assuming, as a first approximation, first order behavior and by then setting $\Delta S_{ord} \simeq \Delta H_{ord}/Tc = (7.19 \pm 1.09)/1525 = 4.5 - 5.9$ J/mole.K. This may be compared with the difference between a state of zero configurational entropy (complete order) and a state with random Al,Si distribution given by

$$-4R\left(\frac{2.28}{4}\ \ln\frac{2.28}{4} + \frac{1.72}{4}\ \ln\frac{1.72}{4}\right) = 22.7 \text{ J/mole.K} .$$

These estimates of the total enthalpy and entropy of ordering in plagioclases of 72 mol % An composition show that the enthalpy is a significant quantity in relation to the enthalpy changes associated with heterogeneous reactions involving feldspars. The low value of ΔS_{ord}, as compared with the maximum possible entropy of ordering, $\Delta S_{ord,max}$, suggests that the structure may not be fully ordered, or that the $C\bar{1}$ structure has substantial short range order, or both. This is, of course, a gross simplification in that the transformation probably is not first order and because the deviation in composition from the one-to-one Al:Si ratio of pure anorthite precludes perfect order under $I\bar{1}$ symmetry. There is, however, direct evidence of short range order above Tc, in the form of diffuse superlattice reflections in diffraction patterns, and it seems likely that this short range order makes some contribution to the entropy reduction relative to a random state of disorder.

The order-disorder transformation in Mg-cordierite ($Mg_2Al_4Si_5O_{18}$) also involves Al and Si on tetrahedral sites. The symmetry change is P6/mcc ⇄ Cccm (Gibbs, 1966; Meagher and Gibbs, 1977) and Tc is ~1725 K (Smart and Glasser, 1977). By symmetry arguments the transformation must be first order (see next chapter by McConnell). Cordierite crystals can be prepared with metastable Al,Si disorder by crystallization from glass (Schreyer and Schairer, 1961; Schreyer and Yoder, 1964). Al,Si ordering in these crystals proceeds at T < Tc via a distinctive series of structural states, as revealed by transmission electron microscopy (Putnis, 1980a,b; Putnis and Bish, 1983). After a few hours of isothermal annealing at temperatures of 1050-1300°C, the initially homogeneous, hexagonal crystals (Fig. 5a) develop a tweed-like texture of two mutually perpendicular modulations on a scale of a few hundred Ångströms (Fig. 5b,c). With further annealing these modulations coarsen (Fig. 5d,e), apparently continuously, until the crystals are orthorhombic and have optically visible lamellar twinning. At T = 1570-1670 K the kinetics of ordering are slightly different (Putnis and Bish, 1983). In this temperature range the order modulations develop but then seem to give way to coarse, homogeneous, untwinned regions, as if orthorhombic cordierite nucleates and grows at the expense of the modulated structure (Fig. 5f). Heat of solution measurements on crystals with a series of structural states from hexagonal to orthorhombic are consistent with a continuous ordering process at 1473 K because a plot of heat of solution against the logarithm of the annealing time is continuous (Fig. 5g). Measurements of the distortion from hexagonal symmetry on samples annealed for long times show that the final structural state is almost independent of temperature (at T < Tc), as would be expected for a truly first order transformation (Putnis, 1980b).

The evidence from experiments and from symmetry arguments is therefore consistent with a first order transformation in cordierite at ~1725 K. It is even possible to speculate that changes in the kinetics and mechanism of ordering at ~1573 K might correspond to the onset of continuous (spinodal) ordering as outlined in Figure 3b,c. In contrast with the $C\bar{1} \rightleftarrows I\bar{1}$ ordering transformation in calcic plagioclases, there is no direct diffraction evidence for short range ordering in the hexagonal crystals, because the ordering causes only a distortion of the lattice without the appearance of superlattice reflections. That short range order does develop, even before the order modulations can be detected, however, is clearly shown by changes in the infra-red spectra of crystals annealed for very short times (Putnis and Bish, 1983) and may also be indicated by changes in Raman spectra (McMillan et al., 1984). A rough estimate of the entropy change on going from hexagonal to orthorhombic structures may be made by using $\Delta S_{ord} \simeq \Delta H_{ord}/Tc$. ΔH_{ord}, in this case, is taken as the maximum enthalpy change observed between metastably disordered crystals and annealed, ordered crystals (40.84 ± 6.53 kJ/mole, Carpenter et al., 1983). This gives ΔS_{ord} = 19.9 - 27.5 J/mole.K, which may be compared with a maximum possible value (for random disorder to complete order and ignoring lattice vibrational contributions) of 51.4 J/mole.K (Navrotsky and Kleppa, 1973; Putnis, 1980a). Again the suggestion is that the nominally disordered structure has substantial local order, reducing its configura-

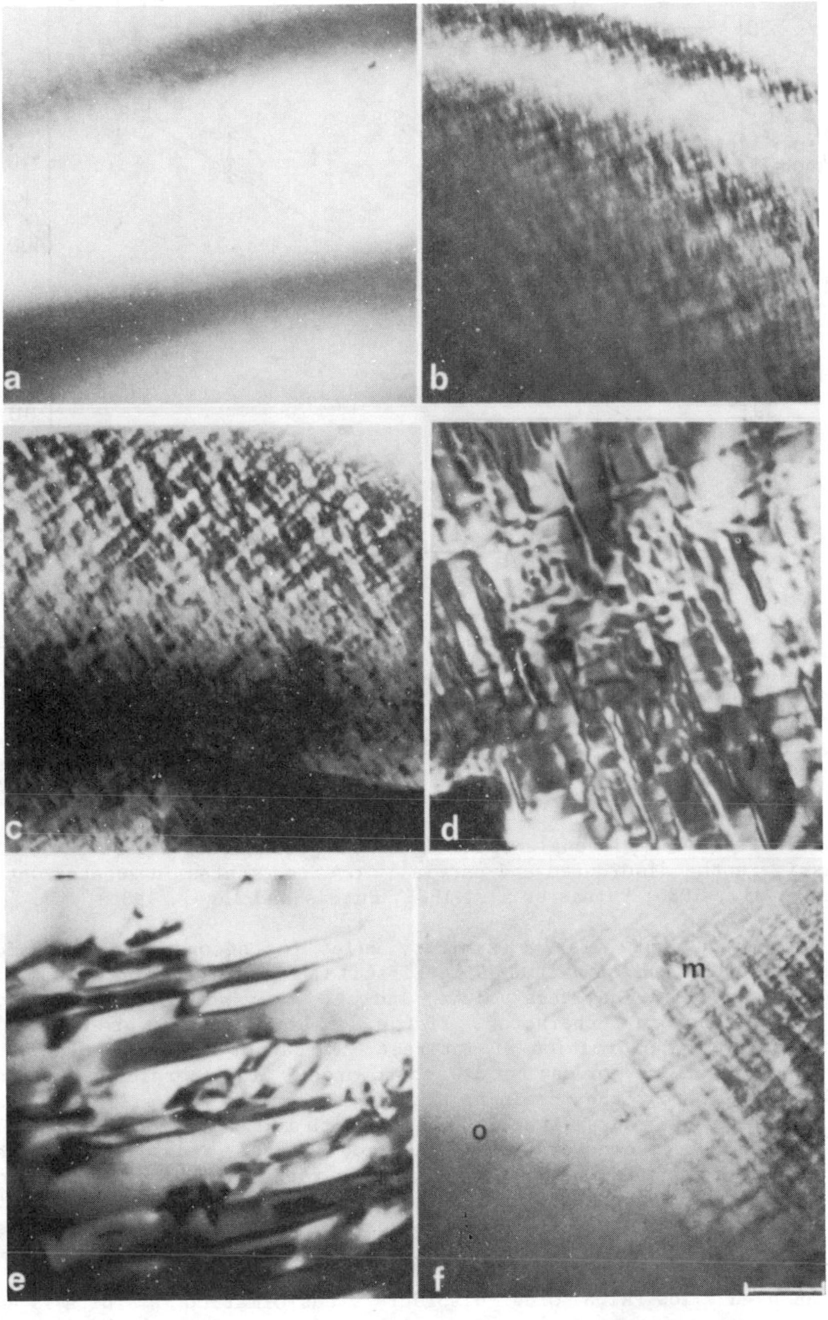

Figure 5. Legend next page.

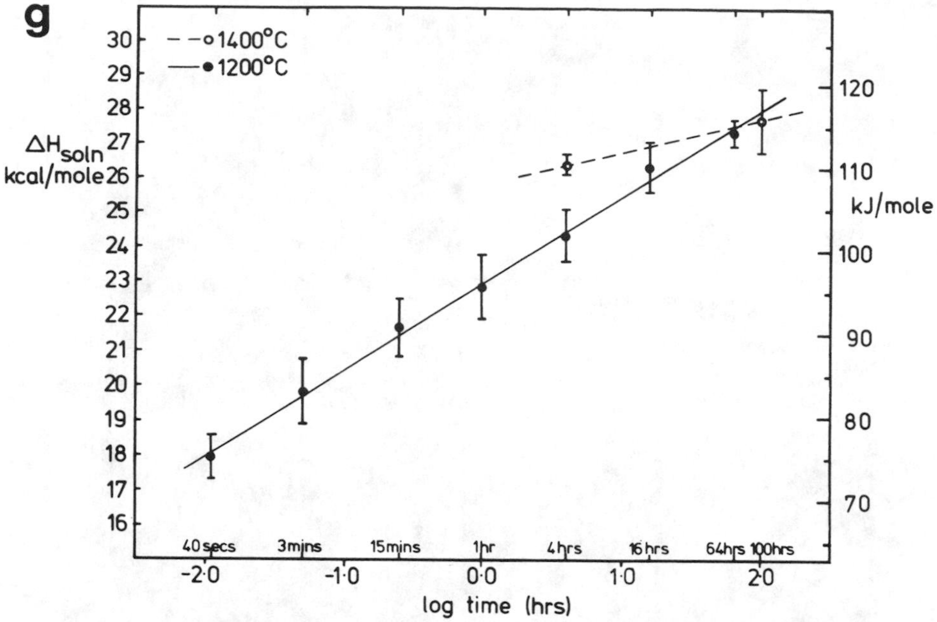

Figure 5. The ordering transformation in cordierite. (a-f) From Putnis (1980b) [see opposite page]. (a) Homogeneous, hexagonal cordierite. (b-e) Development and coarsening of distortion modulations due to ordering during isothermal annealing. (f) Growth of orthorhombic cordierite (o) within the modulated structure (m). The scale bar represents 0.2 μm. (g) Heat of solution (ΔH_{soln}) in lead borate at ~973 K against log (annealing time) for cordierite crystals prepared from glass at 1473 K and 1673 K. From Carpenter et al. (1983). At short times the crystals are hexagonal (disordered) and at long times they are orthorhombic (ordered). From the point of view of enthalpy, the overall transformation mechanism at 1473 K appears to be continuous.

tional entropy. Magic angle spinning n.m.r. techniques are being used to investigate the disordered and partially ordered states in greater detail (Fyfe et al., 1983; Putnis et al., 1985; Putnis and Angel, 1985).

The Landau and Lifshitz symmetry rules for second order transformations (Landau and Lifshitz, 1958; Lifshitz, 1942a,b; de Fontaine, 1975, 1979; McConnell, 1978a, this volume) indicate only which transformations may have second order character. If the symmetry criteria are not met by a particular transformation of interest, it cannot be second order. If they are met, it may or may not be second order. Thus cordierite can only be first order, but the remaining transformations in the silicate, oxide and carbonate minerals listed in Table 1, with the exception of high ⇄ low albite, could be second order. Actual experimental, mechanistic or thermodynamic evidence in each case, however, is rather limited. Second order behavior has been indicated in Table 1 wherever it is possible under the Landau and Lifshitz rules. The thermodynamic parameters could, of course, vary in the manner shown for a λ transformation.

The high ⇄ low (Al,Si order-disorder) transformation in albite is unusual in that it does not involve a symmetry change. Both the high and low albite structures have space group $C\bar{1}$ so that the transformation is not "symmetry breaking." Theoretical analysis, involving coupled order parameters (one order parameter relating to structural distortions and a second to the Al,Si ordering) indicates that the transformation is thermodynamically continuous (Salje, 1985; Salje et al., 1985), and the most recent experiments (Goldsmith and Jenkins, 1984, 1985) are consistent with this view. Exsolution features in the peristerite gap of plagioclase feldspars can also be interpreted in terms of non-first order behavior in albite (Carpenter, 1981, and see below). The C2/m $\rightleftharpoons$ C$\bar{1}$ (Al,Si ordering) transformation in K-feldspar, the C2/c $\rightleftharpoons$ P2/n (Na,Ca and Mg,Al ordering) transformation in omphacite, the Pbnm $\rightleftharpoons$ Pbam (Al,Si ordering) transformation in sillimanite and the R$\bar{3}$c $\rightleftharpoons$ R$\bar{3}$ transformations in dolomite (Ca,Mg ordering -- Goldsmith and Heard, 1961; Reeder, 1983; Goldsmith, 1983) and ilmenite (Fe^{2+},Ti^{4+} ordering) could all be second order. CVM calculations on the ilmenite and dolomite transformations predict strong short range order effects, with λ-shaped Cp anomalies, and are consistent with second order character (Burton and Kikuchi, 1984a). In dolomite a decrease in the intensities of ordering reflections and a decrease in order as Tc is approached from below has also been documented (Goldsmith and Heard, 1961; Reeder and Wenk, 1983). The Cd,Ca analogue of dolomite shows a similar steady decrease in the intensity of sharp ordering reflections (Goldsmith, 1972). There are other possible ordering transformations which have yet to be investigated; for example, P$\bar{4}$2m → P$\bar{1}$(?), Al,Si ordering in gehlenite (Louisnathan, 1971); Al,Si ordering in feldspathoids (e.g., Abbott, 1984; Merlino, 1984); Al,vacancy ordering in staurolite (Smith, 1968; Griffen and Ribbe, 1973).

The development of incommensurate ordered structures from disordered, high temperature states may also be thermodynamically continuous in minerals. Theoretical treatments of these disordered $\rightleftharpoons$ incommensurate transformations point, in general, to second order behavior at the initial stages of development of the incommensurate phase (McConnell, 1981a, 1983; Heine and McConnell, 1981, 1984). In cases where superlattice reflections have been monitored through a transformation, a continuous variation in intensity has been found (e.g., for $NaNO_2$, Durand, 1982). Short range ordering can occur above Tc (Durand, 1982; Durand et al., 1982). Continuous variations in thermodynamic properties should therefore be anticipated at Tc for the incommensurate ordering transformations (from disordered parent phases) in plagioclase feldspars, mullite, nepheline and yoderite (for yoderite structure see McKie and Radford, 1959; Fleet and Megaw, 1962; Higgins et al., 1982). Some intensity data have been published for the superlattice reflections of nepheline, and, although they are limited, they do show a steady linear decrease as Tc is approached from below (McConnell, 1981b).

Table 1 includes a summary of transformation temperature, enthalpy, entropy and volume data for ordering transformations in minerals. Some of the enthalpies of ordering come from solution calorimetric studies on ordered and disordered samples as outlined in Figure 4, and some are estimated. For systems where ΔH_{ord} and Tc values are known or can be extrapolated from data at different compositions, estimates of ΔS_{ord} have been made using the assumption that $\Delta S_{ord} \simeq \Delta H_{ord}/Tc$. In a surprising number of cases, the enthalpy of ordering is approximately equal to half the maximum possible change in configurational entropy multiplied by the transformation temperature [$\Delta H_{ord} \simeq (\Delta S_{ord,max}/2) \times Tc$]. This simple rule of thumb, which makes no attempt to include ΔCp effects associated with lattice vibrations, allows some unknown values to be guessed. For example, sillimanite has an equilibrium order-disorder temperature above ~1960 K (Holland and Carpenter, 1983), possibly in the region of ~2010 K (Greenwood, 1972). $\Delta S_{ord,max}/2$ = 5.8 J/mole.deg, giving $\Delta H_{ord} \simeq$ 11.7 kJ/mole, which may be compared with the value of ~16.7 kJ/mole estimated by Greenwood (1972), using a Bragg-Williams model of disordering. Navrotsky et al. (1973) suggested a higher value from calorimetric data, but their heat treated sillimanite samples may have been affected by slight partial melting (Holland and Carpenter, 1983). The value of $\Delta H_{ord} \simeq$ 4kJ/mole estimated in the same way for dolomite (Table 1) is similar to a Bragg-Williams value of 6.1 kJ/mole (using Tc = 1473 K, Navrotsky and Loucks, 1977; Helgeson et al., 1978) and to the enthalpy of formation of dolomite from calcite plus magnesite (~3.8 kJ/mole, Burton and Kikuchi, 1984a). For ilmenite, the estimated value of ΔH_{ord} = 9.6 kJ/mole compares with ~8 kJ/mole obtained by Burton (1984) in CVM calculations. Volume changes on disordering are all small, typically ~0.1% or less (Table 1).

4. ORDER-DISORDER AND SOLID SOLUTIONS: COMPOSITION AS A VARIABLE

4.1 Ordering

In the case of a transformation occurring within a solid solution, the derivatives of free energy with respect to composition (X) will be continuous or discontinuous in the same manner as the derivatives with respect to T and P. For a first order transformation, $\partial G/\partial X$ will be discontinuous. For a second order transformation, $\partial G/\partial X$ will be continuous but $\partial^2 G/\partial X^2$ will be discontinuous (Allen and Cahn, 1976a,b; Merkel and Blencoe, 1982). The form of the principal thermodynamic parameters as

Table 2. Some values for the maximum excess enthalpies of mixing observed for mineral solid solutions.

Solid solution	Maximum excess enthalpy of mixing
Jadeite-Diopside $NaAlSi_2O_6$ - $CaMgSi_2O_6$	~7.6 kJ/mole (Wood et al., 1980)
Enstatite-Ferrosilite $FeSiO_3$ - $MgSiO_3$	~1.0 kJ/mole (Chatillon-Colinet et al., 1983)
Diopside-CaTs $CaMgSi_2O_6$ - $CaAl_2SiO_6$	~7.1 kJ/mole (Newton et al., 1977)
Enstatite-diopside (C2/c structure) $Mg_2Si_2O_6$ - $CaMgSi_2O_6$	~7.3 kJ/mole (Newton et al., 1979)
Albite-orthoclase $NaAlSi_3O_8$ - $KAlSi_3O_8$	~6.1 kJ/mole (high structural state) (Hovis & Waldbaum, 1977) ~7.5 kJ/mole (low structural state) (Waldbaum & Robie, 1971)
Forsterite-Fayalite Mg_2SiO_4 - Fe_2SiO_4	~0 (Thierry et al., 1981) ~3.1 kJ/mole (Wood & Kleppa, 1981)
Åkermanite-gehlenite $Ca_2MgSi_2O_7$ - $Ca_2Al_2SiO_7$	~3.6 kJ/mole (Charlu et al., 1981)
Grossular-pyrope $Ca_3Al_2Si_3O_{12}$ - $Mg_3Al_2Si_3O_{12}$	~9.1 kJ/mole (Newton et al., 1977)

functions of composition should also be directly analogous to their form as functions of temperature, when crossing the boundary between the stability fields of ordered and disordered structures (Fig. 6a,b). In setting out these parameters schematically, however, it is necessary to include the normal effects of mixing as well. A set of mixing relations for first order, second order and λ transformations is shown in Figure 6c, in which, for simplicity, the ordered and disordered segments of the solid solution have been shown as ideal. The enthalpy of mixing for the solid solution as a whole then consists of two linear segments with a displacement due to ordering at composition Xc, etc. (Fig. 6c). Extrapolation of the mixing curves for the individual segments can be used to indicate relative energies for the ordered and disordered states at the pure end-member compositions. The effects of ordering could, of course, be superimposed on more complex mixing properties.

At high temperatures in the plagioclase solid solution, the $C\bar{1} \rightleftarrows I\bar{1}$ transformation occurs at intermediate compositions (Fig. 7a). If it is thermodynamically continuous, mixing properties of the form shown in Figure 6c, column II or IV, might be appropriate (Carpenter and McConnell, 1984; Carpenter and Ferry, 1984). Calorimetric data from heat treated natural samples are consistent with this simple picture, although they are not sufficiently precise to define the heat of mixing curve unequivocally (Fig. 7b). Unit cell volume data for the high structural state solid solution are also consistent with a thermodynamically continuous transformation (Carpenter et al., 1985).

Heat-treated plagioclases show the same variation in their diffraction patterns (from quenched samples) across the $C\bar{1} \rightleftarrows I\bar{1}$ transformation with changing composition, as has been discussed above for changing temperature (Carpenter et al., 1985). Samples with albite-rich compositions have no ordering reflections. Those closer to Xc, but still on the albite-rich side, have diffuse ordering reflections. A change from diffuse to sharp reflections occurs as Xc, again indicating a changeover from short range ordering to long range ordering. Exactly the same variations have been observed in diffraction patterns from natural sodium-rich pyroxenes which span the $C2/c \rightleftarrows P2/n$ boundary (Carpenter and Smith, 1981; Rossi et al., 1983). In neither the heat treated plagioclases (Carpenter et al., 1985) nor the high temperature sodic pyroxenes described by Carpenter and Smith (1981) was any evidence found for a two-phase region separating the stability fields of the ordered and disordered structures. Exactly the same behavior might be anticipated for the $R\bar{3}c \rightleftarrows R\bar{3}$ transformation in the ilmenite-hematite solid solution (Burton, 1984) and at high temperatures close to the dolomite order-disorder boundary in the calcite-magnesite system (Burton and Kikuchi, 1984a).

Thermodynamic formulations of the activity-composition (a-X) relations in these systems will need to take account of the discrete order-disorder transformations. Carpenter and Ferry (1984) have shown how a discrete order-disorder transformation can give deviations from ideal a-X behavior which are subject to quite different variations with temperature, relative to more normal non-ideal mixing (Fig. 8). Merkel and Blencoe (1982) have applied the constraint of non-first order properties for the $C2/m \rightleftarrows C\bar{1}$ displacive transformation on the mixing behavior of alkali feldspars. Similar constraints could also be included in models of

Figure 6. Legend with Figure 7.

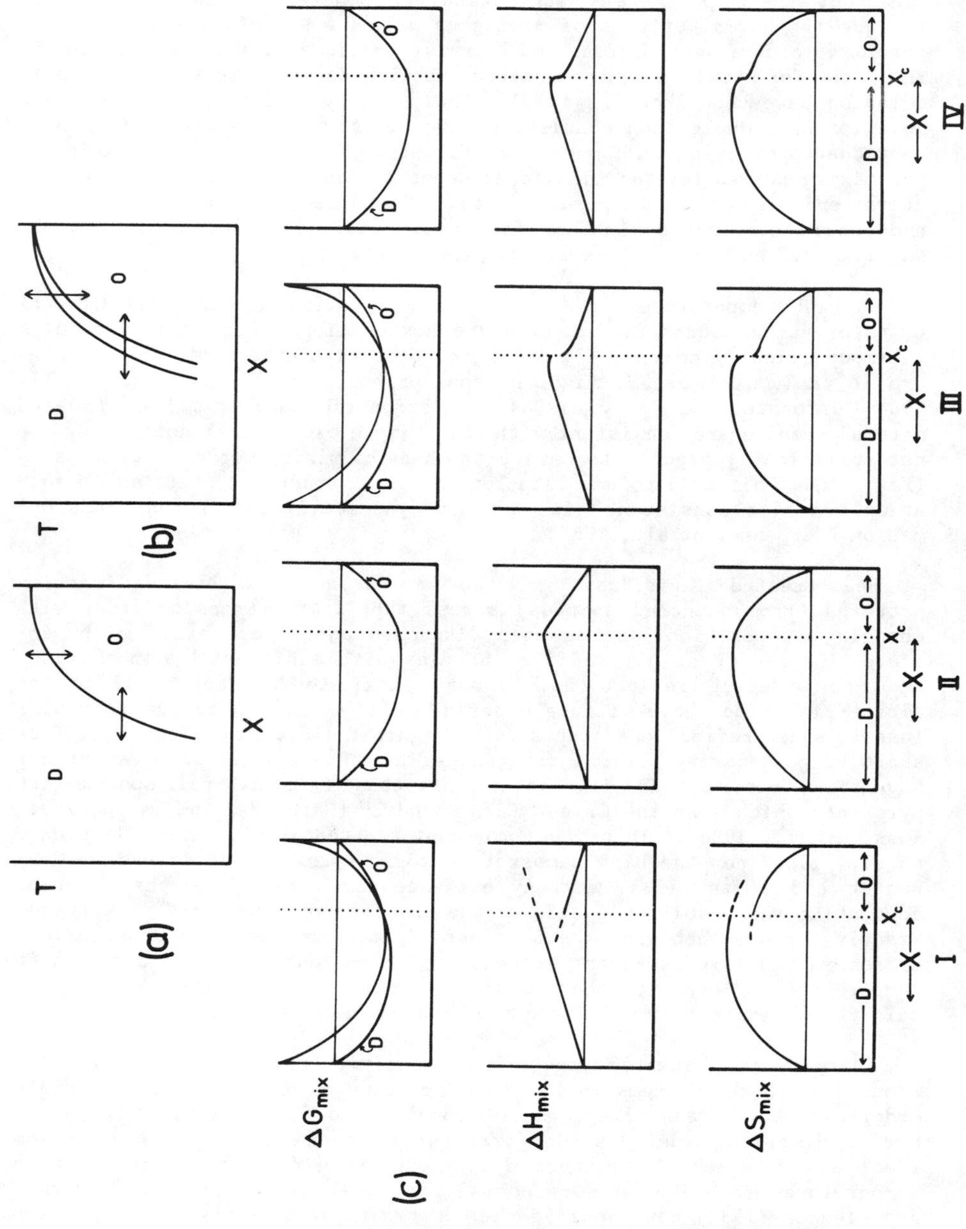

Figure 6. Schematic form of principal thermodynamic parameters as functions of composition in a solid solution. Changing composition (X) is analogous to changing temperature (T) in a second order transformation (a) and in a first order transformation (b). (c) The solid solution is shown as being made up of two ideal segments with the transformation at Xc. Column I = first order; column II = second order; column III = λ transformation with a small first order break at Tc; column IV = λ transformation with no first order break. ΔG_{mix} = free energy of mixing; ΔH_{mix} = enthalpy of mixing; ΔS_{mix} = enthalpy of mixing; ΔS_{mix} = entropy of mixing; O = ordered phase; D = disordered phase.

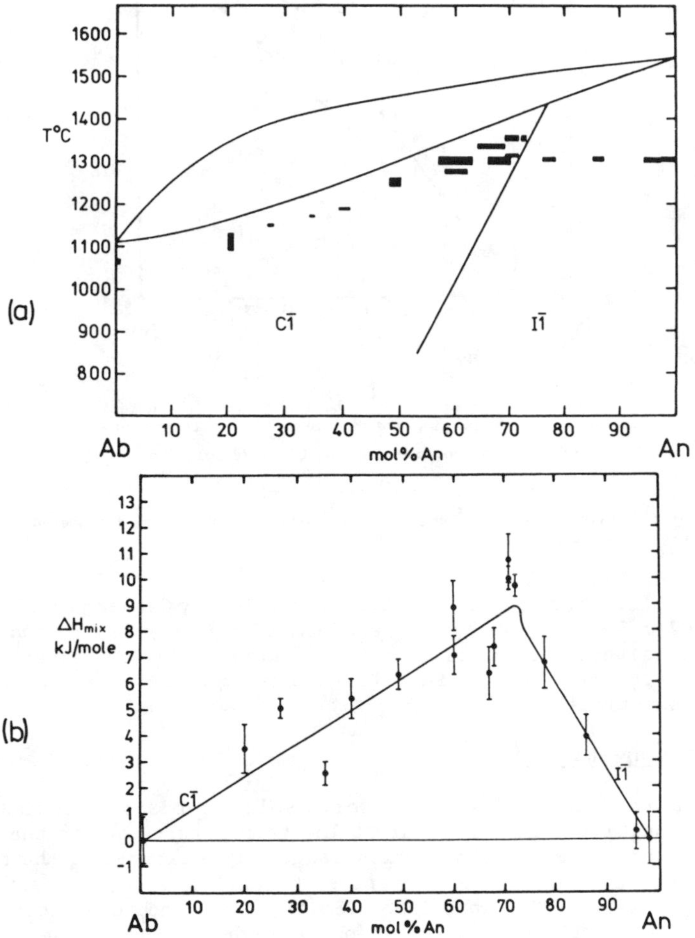

Figure 7. Enthalpy of mixing for high structural state plagioclase feldspars. (a) Range of compositions and annealing temperatures of samples used for solution calorimetry in relation to the position of the $C\bar{1} \rightleftarrows I\bar{1}$ line (Carpenter et al., 1985). (b) Enthalpy of mixing (ΔH_{mix}) for $C\bar{1}$ and $I\bar{1}$ samples with respect to end members having compositions $Ab_{99}An_1$ and Ab_2An_{98}. The data can be approximately described with two straight lines (ideal $C\bar{1}$ and $I\bar{1}$ solid solutions) and a continuous join at the composition where samples with $C\bar{1}$ structures give way to $I\bar{1}$ structures. Scatter in the values shown arises at least in part from different orthoclase and Fe contents of the natural samples. The data are from Carpenter et al. (1985) and are similar to data of Newton et al. (1980).

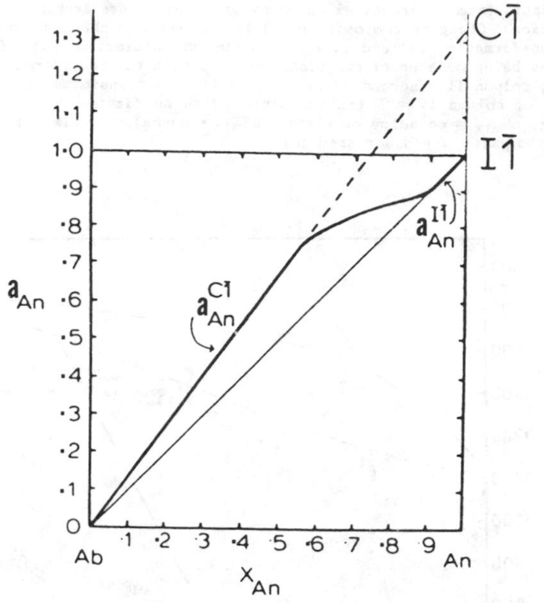

Figure 8. Schematic activity-composition relation showing the effect of a $C\bar{1} \rightleftarrows I\bar{1}$ order-disorder transformation in the plagioclase feldspars at high temperatures.

$a_{An}^{C\bar{1}}$ = activity of anorthite component in the C1 solid solution;

$a_{An}^{I\bar{1}}$ = activity of anorthite in the $I\bar{1}$ solid solution.

The $C\bar{1}$ solid solution is shown as ideal and extrapolates to a fictive disordered ($C\bar{1}$) anorthite at An_{100} (from Carpenter and Ferry, 1984).

pyroxene solid solutions of the type described by Grover (1980) to account for the $C2/c \rightleftarrows P2_1/c$ displacive transformation in pigeonite, and in models of the ilmenite-hematite solid solution (Rumble, 1977; Spencer and Lindsley, 1981) to allow for the Fe^{2+},Ti^{4+} ordering transformation at intermediate compositions.

4.2 Ordering and exsolution

In the event that a disordered solid solution mixes approximately ideally, a miscibility gap can arise at low temperatures due to the effect of ordering over a limited composition range. One example may be the peristerite gap in plagioclase feldspars (Orville, 1974; Carpenter, 1981; Maruyama et al., 1982, Ribbe 1983b; Smith, 1983; and references therein). At ~973 K, activity data for the sodium-rich portion of the plagioclase solid solution imply little, if any, deviation from ideality when the Al,Si distribution is largely disordered (Orville, 1972). Ordering in albite, however, leads to a reduction in free energy at the most anorthite-poor compositions. If the order-disorder transformation ($C\bar{1} \rightleftarrows I\bar{1}$) is first order, the G-X curves would cross as in Figure 9a. Exsolution in the resulting two-phase region (Fig. 9b) could then only occur by a nucleation and growth mechanism. If the transformation is thermodynamically continuous (Salje, 1985; Salje et al., 1985; Goldsmith and Jenkins,

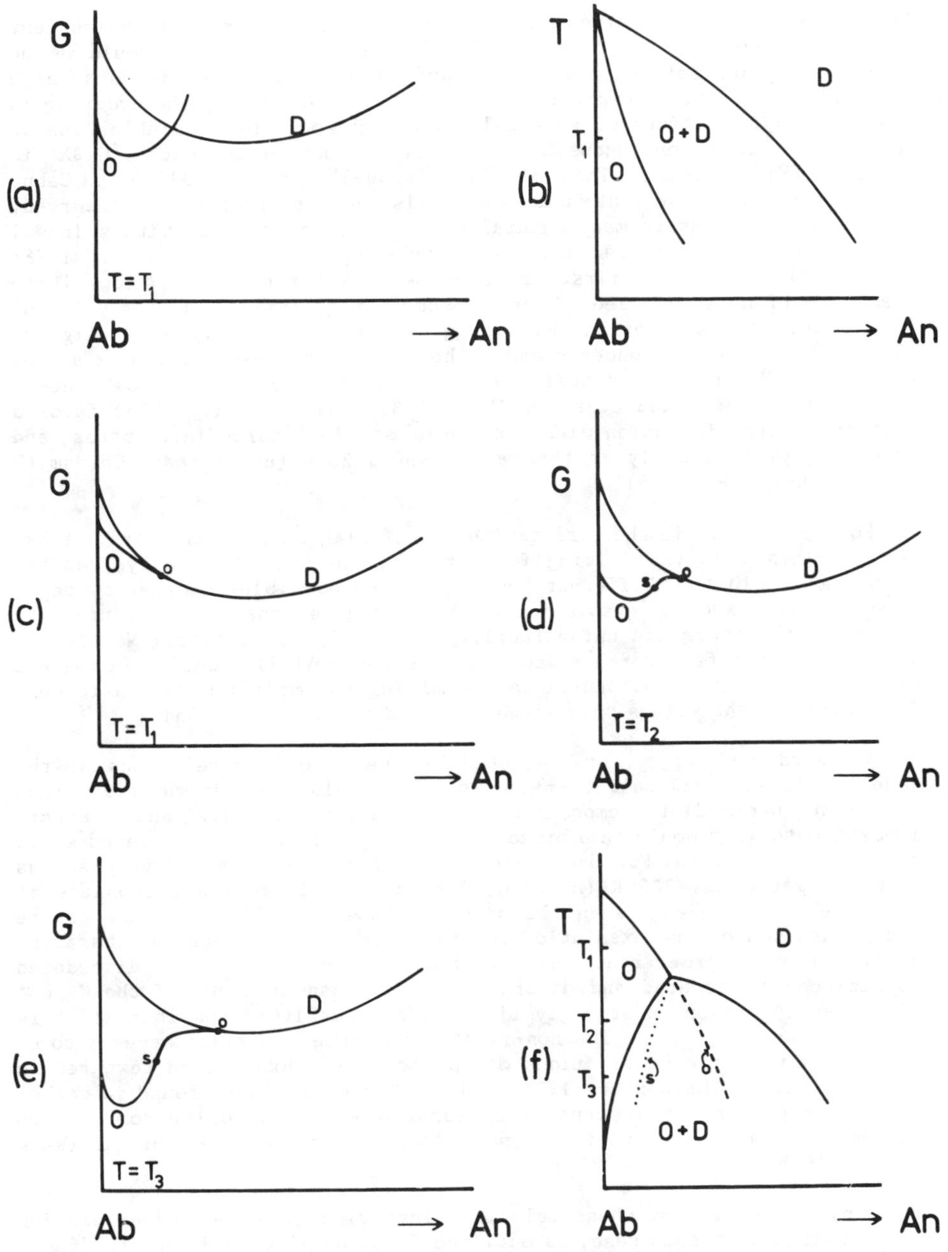

Figure 9. Schematic G-X and T-X relations for the peristerite miscibility gap in plagioclase felspars if the exsolution reaction is driven by ordering in albite. (a,b) First order transformation in albite. Exsolution in the two phase field can only occur by nucleation and growth. (c-f) Non-first order ($\partial G/\partial X$ continuous) transformation in albite. At low temperatures there is a composition range (between o and s) where $\partial^2 G/\partial X^2$ is negative, allowing spinodal decomposition into ordered (O) and disordered (D) components. Note that (a,c,d,e) are really projections of a three dimensional diagram of G, X, η_ℓ. After Carpenter (1981). High ⇄ low albite ordering probably does not occur by a discrete transformation (Salje, 1985; Salje et al., 1985), and should not, strictly speaking therefore, be shown as a line on the phase diagram.

1984, 1985), the G-X curves for the ordered and disordered phases must merge continuously (Fig. 9c-e). In the latter case there would be no two-phase region between the ordered and disordered phase fields at high temperatures. At low temperatures exsolution is driven by the ordering in albite, and a continuous, spinodal mechanism would be available because there would be a continuous G-curve with a range over which $\partial^2G/\partial X^2$ is negative (Fig. 9d,e,f). This is a "conditional" spinodal (Allen and Cahn, 1976a,b) in the sense that it does not exist unless ordering has occurred. Exsolution textures in many natural samples are consistent with spinodal decomposition into ordered and disordered components (Nord et al., 1978; Ribbe, 1983b), and non-first order properties for the high $\rightleftarrows$ low albite transformation are implied. There is some disagreement as to the value of Tc for this transformation, with suggestions that it may be as high as ~950 K (Raase, 1971; Senderov and Shchekina, 1976; Mason, 1979) or as low as ~850 K (McConnell and McKie, 1960; Orville, 1974). The most recent studies (Goldsmith and Jenkins, 1984, 1985; Salje et al., 1985) favor a continuous transformation with a range of stable intermediate states, and occurring predominantly in the range ~950-1020 K (at 18 kbar, Goldsmith and Jenkins, 1984, 1985).

For exsolution in the peristerite gap of plagioclase feldspars, it has been implied that the driving force for exsolution arises only from the ordering in albite. Different topologies are possible in other systems, however. For example, exsolution in the alkali feldspar solid solution is driven by the energetic unfavorability of simply substituting Na^+ for K^+ in the feldspar framework at low temperatures. Al,Si ordering in the end members is really coincidental to the mixing and modifies the basic configuration of the solvus only slightly (Yund and Tullis, 1983).

A third topology is represented by the stability relations of the jadeite-diopside and calcite-magnesite solid solutions, in which ordering occurs at intermediate compositions (omphacite or dolomite) and is superimposed onto what would be a broad miscibility gap in the disordered solid solution (Fig. 10). For disordered sodic pyroxenes the solvus peak has been estimated as <723 K (Holland, 1983); for disordered carbonates it would be much higher. When the ordered phases are involved, the nature and mechanisms of the exsolution reactions are dependent on the character of the order-disorder reactions. Spinodal decomposition of a disordered omphacite into ordered and disordered phases cannot occur if the C2/c $\rightleftarrows$ P2/n transformation is strongly first order, but it is possible if it is second order (Fig. 10). The non-first order interpretation appears to be consistent with the observations of spinodal type exsolution textures in sodic pyroxenes (Carpenter, 1978, 1980). Microstructures found in carbonates are consistent in general appearance with spinodal decomposition (Reeder, 1981), but there is now some doubt as to their real origin (Wenk et al., 1983).

Many other sets of phase relations involving ordering and exsolution are possible. In each case, as outlined first by Allen and Cahn (1976a,b) and Allen (1977) for the system Fe-Al, thermodynamic and mechanistic arguments can be used to predict behavior at low temperatures.

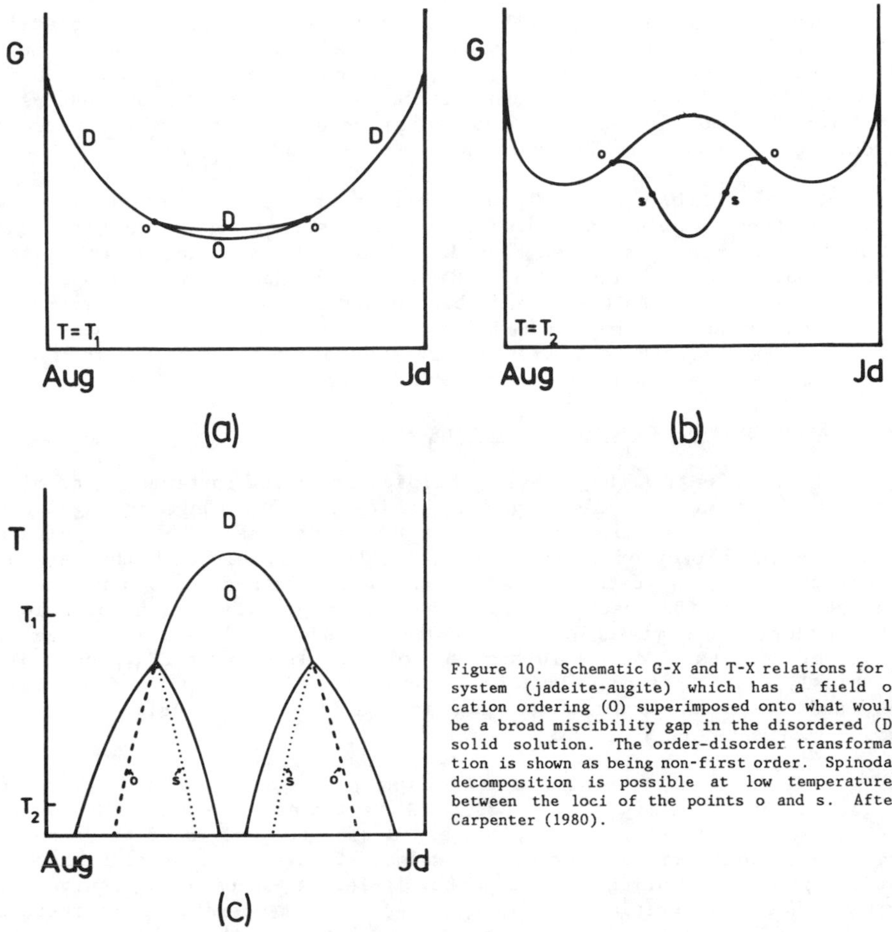

Figure 10. Schematic G-X and T-X relations for a system (jadeite-augite) which has a field of cation ordering (O) superimposed onto what would be a broad miscibility gap in the disordered (D) solid solution. The order-disorder transformation is shown as being non-first order. Spinodal decomposition is possible at low temperatures between the loci of the points o and s. After Carpenter (1980).

4.3 Development of incommensurate structures

A further possibility for the way in which a mineral system might respond to changes in temperature and composition is to develop an incommensurate superstructure by ordering. Two examples are the intermediate (or 'e') plagioclase structure (see reviews by Smith, 1974; Ribbe, 1983a) and mullite (Cameron, 1977a,b; Ribbe, 1980; Tokonami et al., 1980; Ylä-Jääski and Nissen, 1983). Both show a remarkable range of composition, reflecting the flexibility of incommensurate ordering. They adapt to variation in stoichiometry (and perhaps to variations in temperature as well) by changing the repeat distance of the ordering periodicity in a way that is impossible for a commensurate ordered structure (Grove, 1977; Kitamura and Morimoto, 1977, 1984).

The stability of an incommensurate phase is largely due to the interactions between two different ordering schemes or component structures (McConnell, 1978b, 1981a,b, 1983; Heine and McConnell, 1981, 1984). A solid solution in which, because of changing composition, more than one ordering scheme or structural modification becomes possible is then a good candidate for incommensurate behavior. In the plagioclase solid solution there is a continuous variation of the Na:Ca and Al:Si ratios, and in mullite the Al:Si and oxygen:vacancy ratios also both change with composition. These pairs of substitutions are considered to be responsible for the two interactive component structures of the incommensurate ordering (McConnell, 1978b, 1981a; Heine and McConnell, 1981).

Both the Al_2SiO_5-iota Al_2O_3 and albite-anorthite solid solutions show spinodal-type exsolution features between structures with different ordering schemes (reviewed for plagioclases by McLaren, 1974; Smith, 1974; Champness and Lorimer, 1976; Ribbe, 1983b; for mullite see Cameron, 1976; Wenk, 1983). As has already been discussed, non-first order transformation properties are favored for the early stages of incommensurate ordering. Relationships between ordering and exsolution of the form outlined for peristerites might, therefore, be appropriate.

4.4 Enthalpies of ordering and mixing

Maximum values of the excess enthalpies of mixing for a number of mineral solid solutions are shown in Table 2. They range in magnitude between 0 and 10 kJ/mole and are thus not very dissimilar in magnitude from the enthalpies of ordering given in Table 1, although they are of opposite sign. Any detailed comparison should be made on a per site or per atom rather than per mole basis, but the implication is nevertheless that ordering and mixing can contribute an essentially similar range of enthalpy effects. Cordierite has an exceptionally large ΔH_{ord} per mole (Table 1), but the formula unit has more atoms than most of the other cases, and the value obtained does not relate to an equilibrium change in order (Carpenter et al., 1983).

All the excess enthalpies of mixing which have been measured for solid solutions have turned out to be positive (Newton et al., 1981). Navrotsky (1971) and Newton et al. (1980) suggested that the origin of these positive deviations can be understood in terms of structural strains associated with substituting cations with different sizes onto equivalent sites. When the contributions of ordering at low temperatures in crystals of intermediate composition are included, however, the excess tends to zero (as in jadeite-diopside, Wood et al., 1980) or even becomes slightly negative (as in enstatite-ferrosilite, Chatillon-Colinet et al., 1983). Thus the observation of positive excess enthalpies of mixing for high temperature solid solutions does not preclude the possibility of ordering at low temperatures. Data for the mixing properties of plagioclase feldspars (Newton et al., 1980) have not been included in Table 2 because the deviation from ideality can be attributed to ordering at the anorthite end (Carpenter and McConnell, 1984). The excess enthalpy of mixing for $C\bar{1}$ (disordered) plagioclases could be close to zero.

Excess volumes of mixing are typically very small and are comparable in magnitude to the volume changes associated with ordering (Newton and Wood, 1980).

5. CONCLUSION

It is, of course, a truism to state that physically realistic models of the thermodynamic properties of minerals must take account of both ordering and mixing behavior. The intention of this chapter has been to show, however, that both effects are very closely related in energy, origin and development. They are manifestations of only one phenomenon, namely, the balance of short range and long range interactions between atoms. This more unified view of solid state processes has recently been emphasized for metallic systems (de Fontaine, 1975; Sofa and Laughlin, 1982). The net result of these interactions would be difficult to predict, a priori, for minerals but is such that at low temperatures a given solid solution may (a) order -- either at intermediate or at end-member compositions, (b) unmix, (c) order and unmix, or (d) produce an incommensurate structure. More importantly, all these processes can occur in a single system. Solid solutions which do nothing (e.g, forsterite-fayalite, ?) are exceptional.

Given this range of possibilities for any system, what constraints can be applied to predict or calculate the properties needed for general petrological applications? The most common and tractable approach is to use convenient mathematical expressions simply to describe measured properties. These are tested by their ability to reproduce experimentally derived equilibrium phase diagrams, and, for many purposes, it is unnecessary then to inquire as to the physical origin of the properties. Whether the correct activities, free energies or any other properties that are required are generated for conditions beyond the P,T range of actual, measured data will always be uncertain, however.

Alternatively, models can be produced which are based on a physical picture of the interactions between atoms in a structure. In general, these do not produce data of sufficient accuracy to be of immediate petrological consequence but are more revealing of why a solid solution behaves in a particular way. For example, regular solution or Bragg-Williams models can only present ordering and exsolution processes as opposite extremes because they are based only on the interactions of first nearest neighbors. It follows that other interaction energies are important and must be included in models which are intending to reproduce the topologies of real, complex systems. In effect, the most important missing ingredient is short range order, and the new approach of Burton and Kikuchi (1984a,b) and Burton (1984) using the cluster variation method holds great promise, because it includes both short range and long range ordering contributions.

Landau theory follows a more phenomenological approach such that the actual structures or atomic interactions need not be specified. It can be used to generate a complete set of thermodynamic properties (G, H, S, Cp, as functions of T) if the constants in the free energy expansion can be determined (Salje, 1985; Salje et al., 1985). The overall constraint in ordering systems, however, is symmetry. While an analysis of symmetry properties does not necessarily lead to a quantitative assessment of entropy or enthalpy, it does allow prediction of both the nature and mechanisms of phase transformations.

With regard to future research, perhaps the most pressing problem is to characterize the nature of disordered structures. From three independent lines of evidence (the brief review of thermodynamic data presented here, phase equilibrium data, Holland, 1980, Holland and Powell (pers. comm.), and comparisons of calorimetric and phase equilibrium data for solid solutions, Newton et al., 1981), many mineral systems do not appear to have truly random distributions of cations. The effect on configurational entropies could be most severe, with nominally disordered minerals having, perhaps, only half their maximum possible configurational entropy. Finally, a detailed study of the order-disorder transformation at equilibrium through Tc for any one mineral would be of enormous value in confirming or refuting many of the subjective judgments with which we are at present faced when treating cation ordering processes in natural materials.

ACKNOWLEDGMENTS

Critical comments on the original manuscript by A. Navrotsky, E. Salje, J.R. Goldsmith, T.J. Holland and J.D.C. McConnell are gratefully acknowledged. This paper is Cambridge Earth Sciences No. ES 562.

6. A WORKED EXAMPLE

The effect of a discrete order-disorder transformation on the activities of components in a solid solution.

An ideal binary (A-B) solid solution has no excess mixing properties, and the activity of component A can be written as $a_A = \gamma_A x_A$, where x_A is the mole fraction of component A and γ_A is the activity coefficient of component A which equals 1. Non-ideal solid solutions have activities (and activity coefficients) which vary with temperature. The simplest model for deriving this temperature dependence is the regular solution model, in which there is an excess enthalpy of mixing given by $\Delta H_{mix} = W_H x_A x_B$, W_H being a constant, and an ideal entropy of mixing. For this restricted case, the temperature dependence of γ_A is given by $RT \ell n\, \gamma_A = W_H(1 - x_A)^2$. The regular solution can be extended to temperature dependent excess mixing properties ($\Delta G_{mix} = W_G x_A x_B$, $W_G = W_H - TW_S$), to multi-component solutions, and to asymetric solutions. [B.J. Wood and D.G. Fraser (1977) Elementary Thermodynamics for Geologists, Oxford Univ.

Press.] In all these examples, however, the excess properties are due to simple mixing of two end-member phases which have the same essential structure.

The temperature dependence of ordering processes is rather different, and if a discrete order-disorder transformation occurs at some intermediate composition in a solid solution, such that the end-member structures have different symmetries (one phase ordered and the other disordered), then the temperature dependence of a_A, a_B, γ_A, γ_B will also be rather different. It is necessary to take account specifically of the ordering as well as of the straight mixing. Take, for example, the case of the high structural state plagioclase feldspar solid solution, with the simplifying assumptions that the $C\bar{1}$ solid solutions (Ab-rich compositions) and the $I\bar{1}$ solid solutions (An-rich compositions) are related by a discrete order-disorder transformation, $C\bar{1} \rightleftarrows I\bar{1}$, at an intermediate composition (Carpenter and Ferry, 1984).

The activity-composition relations for a_{An} in this simple model of the high structural state plagioclase solid solution are shown in Figure 8. The activity of anorthite in the $C\bar{1}$ solid solution is $a_{An}^{C\bar{1}}$ and the activity of anorthite in the $I\bar{1}$ solid solution is $a_{An}^{I\bar{1}}$. Both are referred to ordered $I\bar{1}$ anorthite as a standard state.

For the $C\bar{1}$ solid solution

$$\bar{G}_{An}^{C\bar{1}} = \bar{G}_{An}^{\circ I\bar{1}} + RT \ln a_{An}^{C\bar{1}} , \qquad (1)$$

and for the $I\bar{1}$ solid solution:

$$\bar{G}_{An}^{I\bar{1}} = \bar{G}_{An}^{\circ I\bar{1}} + RT \ln a_{An}^{I\bar{1}} , \qquad (2)$$

where $\bar{G}_{An}^{C\bar{1}}$ and $\bar{G}_{An}^{I\bar{1}}$ are the partial molar free energies of anorthite in the $C\bar{1}$ and $I\bar{1}$ solid solutions, respectively, $\bar{G}^{\circ I\bar{1}}_{An}$ refers to pure anorthite, R is the gas constant, and T is the temperature.

Subtracting Equation 2 from Equation 1 gives

$$\bar{G}^{C\bar{1}}_{An} - \bar{G}^{I\bar{1}}_{An} = 0 + RT \ln a^{I\bar{1}}_{An} - RT \ln a^{I\bar{1}}_{An} . \tag{3}$$

Now $a^{I\bar{1}}_{An} = \gamma^{I\bar{1}}_{An} \cdot x^{I\bar{1}}_{An}$, $a^{C\bar{1}}_{An} = \gamma^{C\bar{1}}_{An} \cdot x^{C\bar{1}}_{An}$, and for pure anorthite

$a^{I\bar{1}}_{An} = 1$, $a^{C\bar{1}}_{An} = \gamma^{C\bar{1}}_{An}$, so Equation 3 reduces to:

$$\bar{G}^{C\bar{1}}_{An_{100}} - \bar{G}^{I\bar{1}}_{An_{100}} = RT \ln \gamma^{C\bar{1}}_{An} . \tag{4}$$

$G^{C\bar{1}}_{An_{100}} - G^{I\bar{1}}_{An_{100}}$ is actually the energy of the disordering reaction ($I\bar{1} \rightarrow C\bar{1}$) in pure anorthite, ΔG_{ord}. It has been assumed that the $C\bar{1}$ solid solution is ideal (i.e., $\gamma^{C\bar{1}}_{An_{100}} = \gamma^{C\bar{1}}_{An_{30}} = \gamma^{C\bar{1}}_{An_{20}}$, etc.), so that a simple relationship between $\gamma^{C\bar{1}}_{An}$ and T has now been derived, where $\gamma^{C\bar{1}}_{An}$ depends only on ΔG_{ord} and T.

To obtain values for $\gamma^{C\bar{1}}_{An}$, it is necessary to assume a model for the order-disorder process in anorthite. If we treat the $C\bar{1} \rightleftarrows I\bar{1}$ transformation as being first order in character, involving a change from complete order to complete disorder at a transformation temperature Tc $\simeq$ 2000 K with an enthalpy change $\Delta H_{ord} \simeq 15.5$ kJ/mole, ΔG_{ord} can be obtained for all temperatures. Using $\Delta G_{ord} = 0 = \Delta H_{ord} - Tc\Delta S_{ord}$ to obtain a value for the entropy change $\Delta S_{ord} \simeq 7.8$ J/mole.K, leaves $\Delta G_{ord} = \Delta H_{ord} - T\Delta S_{ord}$. Thus at 1400, 1000, 600°C, ΔG_{ord} = 2450, 5570, 8690 J/mole, respectively, and $\gamma^{C\bar{1}}_{An}$ = 1.2, 1.7, 3.3, respectively.

Of course, in detail such a model is not numerically very realistic, but it does at least lead to the identification of real constraints on the overall mixing properties. Above Tc (2000 K) $\gamma_{An}^{C\bar{1}}$ would be 1, while, as T decreases from 2000 K, $\gamma_{An}^{C\bar{1}}$ increases from 1. Further elaboration to include real values for ΔG_{ord}, the probable second order character of the $C\bar{1} \rightleftarrows I\bar{1}$ transformation, and non-ideality in the $C\bar{1}$ and $I\bar{1}$ solid solution segments could give more accurate activities which take into account the discrete ordering transformation in the solid solution. The same general approach could be used to identify constraints on mixing behavior for solid solutions with different types of transformations.

REFERENCES

Abott, R.N., Jr. (1984) $KAlSiO_4$ stuffed derivatives of tridymite: phase relationships. Am. Mineral. 69, 449-457.

Achiam, Y. and Imry, Y. (1975) Phase transitions in systems with a coupling to a non-ordering parameter. Phys. Rev. B12, 2768-2776.

Allen, S.M. and Cahn, J.W. (1976a) On tricritical points resulting from the intersection of lines of higher-order transitions with spinodals. Scripta Met. 10, 451-454.

__________ and Cahn, J.W. (1976b) Mechanisms of phase transformations within the miscibility gap of Fe-rich Fe-Al alloys. Acta Met. 24, 425-437.

__________ (1977) Phase separation of Fe-Al alloys with Fe_3Al order. Phil. Mag. 36, 181-192.

Als-Nielson, J. and Dietrich, O.W. (1967) Long-range order and critical scattering of neutrons below the transition temperature in β-brass. Phys. Rev. 153, 717-721.

Bambauer, H.U. and Bernotat, W.H. (1982) The microcline/sanidine transformation isograd in metamorphic regions, I: Composition and structural state of alkali feldspars from granitoid rocks of two N-S traverses across the Aar Massif and Gottard <<Massif>> Swiss Alps. Schweiz. Mineral. Petrogr. Mitt. 62, 185-230.

Bardhan, P. and Cohen, J.B. (1976) A structural study of the alloy Cu_3Au above its critical temperature. Acta Crystallogr. A32, 597-614.

Bernotat, W.H. and Bambauer, H.U. (1982) The microcline/sanidine transformation isograd in metamorphic regions. II. The region of Lepontine metamorphism, Central Swiss Alps. Schweiz. Mineral. Petrogr. Mitt. 62, 231-244.

Blinova, G.K. and Kiseleva, I.A. (1982) Calorimetric study of the structural transitions in plagioclase. Geochem. Int'l 19, 74-80.

Burnham, C.W. (1973) Order-disorder relationships in some rock-forming silicate minerals. Earth Sci. Rev. 9, 313-338.

Bruno, E., Chiari, G. and Fachinelli, A. (1976) Anorthite quenched from 1530°C. I. Structure refinement. Acta Crystallogr. B32, 3270-3280.

Burton, B. (1984) Thermodynamic analysis of the system Fe_2O_3-$FeTiO_3$. Phys. Chem. Minerals 11, 132-139.

__________ and Kikuchi, R. (1984a) Thermodynamic analysis of the system $CaCO_3$-$MgCO_3$ in the tetrahedron approximation of the cluster variation method. Am. Mineral. 69, 165-175.

__________ and Kikuchi, R. (1984b) The antiferromagnetic-paramagnetic transition in α Fe_2O_3 in the single prism approximation of the cluster variation method. Phys. Chem. Minerals 11, 125-131.

Buth, J. and Inden, G. (1982) Ordering and segregation reactions in f.c.c. binary alloys. Acta Met. 30, 213-224.

Cameron, W.E. (1976) Exsolution in 'stoichiometric' mullite. Nature 264, 736-738.

__________ (1977a) Mullite: a substituted alumina. Am. Mineral. 62, 747-755.

__________ (1977b) Composition and cell dimensions of mullite. Am. Ceram. Soc. Bull. 56, 1003-1007, and 1011.

Carpenter, M.A. (1978) Kinetic control of ordering and exsolution in omphacite. Contrib. Mineral. Petrol. 67, 17-24.

__________ (1980) Mechanisms of exsolution in sodic pyroxenes. Contrib. Mineral. Petrol. 71, 289-300.

__________ (1981) A "conditional spinodal" within the peristerite miscibility gap of plagioclase feldspars. Am. Mineral. 66, 553-560.

__________ and Ferry, J.M. (1984) Constraints on the thermodynamic mixing properties of plagioclase feldspars. Contrib. Mineral. Petrol. 87, 138-148.

__________ and McConnell, J.D.C. (1984) Experimental delineation of the $C\bar{1} \rightleftharpoons I\bar{1}$ transformation in intermediate plagioclase feldspars. Am. Mineral. 69, 112-121.

__________, McConnell, J.D.C. and Navrotsky, A. (1985) Enthalpies of ordering in the plagioclase feldspar solid solution. Geochim. Cosmochim. Acta (in press).

__________, Putnis, A., Navrotsky, A. and McConnell, J.D.C. (1983) Enthalpy effects associated with Al/Si ordering in anhydrous magnesian cordierite. Geochim. Cosmochim. Acta 47, 899-906.

__________ and Smith, D.C. (1981) Solid solution and cation ordering limits in high temperature sodic pyroxenes from the Nybö eclogite pod, Norway. Mineral. Mag. 44, 37-44.

Champness, P.E. and Lorimer, G.W. (1976) Exsolution in silicates. In "Electron Microscopy in Mineralogy," H.-R. Wenk, ed. Springer-Verlag, Berlin, p. 174-204.

Charlu, T.V., Newton, R.C. and Kleppa, O.J. (1981) Thermochemistry of synthetic $Ca_2Al_2SiO_7$ (gehlenite) - $Ca_2MgSi_2O_7$ (åkermanite) melilites. Geochim. Cosmochim. Acta 45, 1609-1617.

Chatillon-Colinet, C., Newton, R.C., Perkins III, D. and Kleppa, O.J. (1983) Thermochemistry of $(Fe^{2+},Mg)\ SiO_3$ orthopyroxene. Geochim. Cosmochim. Acta 47, 1597-1603.

Chen, H. and Cohen, J.B. (1977) A comparison of experiment and the theory of continuous ordering. J. Physique C7, 314-327.

__________ and Cohen, J.B. (1979) Measurements of the ordering instability in binary alloys. Acta Met. 27, 603-611.

__________, Cohen, J.B. and Ghosh, R. (1977) Spinodal ordering in Cu_3Au. J. Phys. Chem. Solids 38, 855-857.

Cook, H.E. (1975) On the problem of fluctuations near first-order phase transformations in solids. J. Appl. Crystallogr. 8, 132-140.

__________ (1976) Continuous transformations. Materials Sci. Eng. 25, 127-134.

__________, Suezawa, M., Kajitani, T. and Rivaud, L. (1977) Pretransition phenomena. J. Physique C7, 430-439.

Dolino, G. and Bachheimer, J.P. (1982) Effect of the α - β transition on mechanical properties of quartz. Ferroelectrics 43, 77-86.

__________, Bachheimer, J.P., Berge, B. and Zeyen, C.M.E. (1974a) Incommensurate phase of quartz: I. Elastic neutron scattering. J. Physique 45, 361-371.

__________, Bachheimer, J.P., Berge, B., Zeyen, C.M.E., Van Tendeloo, G., Van Landuyt, J. and Amelinkx, S. (1984) Incommensurate phase of quartz: III. Study of the coexistence state between the incommensurate and the α-phases by neutron scattering and electron microscopy. J. Physique 45, 901-912.

__________, Bachheimer, J.P. and Zeyen, C.M.E. (1983) Observation of an intermediate phase near the α - β transition of quartz by heat capacity and neutron scattering measurements. Solid State Commun. 45, 295-299.

Durand, D. (1982) "Statique et dynamique de la phase incommensurate du nitrite du sodium." Thesis, Univ. Paris-Sud, Centre d'Orsay, France.

__________, Denoyer, F., Lambert, M., Bernard, L. and Currat, R. (1982) Etude par rayons X et par neutrons de la phase incommensurable du nitrite de sodium. J. Physique 43, 149-154.

Ehrenfest, P. (1933) Phasenumwandlungen im ueblichen und erweiterten Sinn, classifiziert nach den entsprechenden singularitaeten des thermodynamischen potentiales. Proc. Sec. Sciences, Koninklijke Akad. Wetenschappen Te Amsterdam 36, 153-157.

Fleet, S.G. and Megaw, H.D. (1962) The crystal structure of yoderite. Acta Crystallogr. 15, 721-728.

de Fontaine, D. (1975) k-space symmetry rules for order-disorder reactions. Acta Met. 23, 553-571.

__________ (1979) Configurational thermodynamics of solid solutions. Solid State Phys. 34, 73-274.

Fyfe, C.A., Gobbi, G.C., Klinowski, J., Putnis, A. and Thomas, J.M. (1983) Characterization of local atomic environments and quantitative determination of changes in site occupancies during the formation of ordered synthetic cordierite by ^{29}Si and ^{17}Al magic-angle spinning n.m.r. spectroscopy. J. Chem. Soc., Chem. Commun. 1983, 556-558.

Ghiorso, M.A., Carmichael, I.S.E. and Moret, L.K. (1979) Inverted high-temperature quartz. Contrib. Mineral. Petrol. 68, 307-323.

Gibbs, G.V. (1966) The polymorphism of cordierite I: The crystal structure of low cordierite. Am. Mineral. 51, 1069-1081.

Goldsmith, J.R. (1972) Cadmium dolomite and the system $CdCo_3$-$MgCo_3$. J. Geol. 80, 617-626.

__________ (1983) Phase relations of rhombohedral carbonates. In "Carbonates: Mineralogy and Chemistry," R.J. Reeder, ed. Reviews in Mineralogy 11, 49-76.

__________ and Heard, A.C. (1961) Subsolidus relations in the system $CaCo_3$-$MgCO_3$. J. Geol. 69, 45-74.

__________ and Jenkins, D.M. (1984) The high-low albite relations revealed by hydrothermal melting and reversal of degree of order. EOS, Trans. Am. Geophys. Union 65, 307.

__________ and Jenkins, D.M. (1985) The high-low albite relations revealed by hydrothermal melting and reversal of order at high pressures. Am. Mineral. (in press).

Gouhara, K., Ying Hao Li, and Kato, N. (1983a) Observation of satellite reflections in the intermediate phase of quartz. J. Phys. Soc. Japan 52, 3697-3699.

__________, Ying Hao Li, and Kato, N. (1983b) Studies on the α - β transition of quartz by means of in-situ X-ray topography. J. Phys. Soc. Japan 52, 3821-3828.

__________ and Kato, N. (1984) The study of α - β phase transition of quartz by means of in-situ X-ray topography and fine beam Laue photography. Acta Crystallogr. A40, suppl., C-326.

Greenwood, H.J. (1972) Al^{IV}-Si^{IV} disorder in sillimanite and its effect on phase relations of the aluminum silicate minerals. Geol. Soc. Am. Mem. 132, 553-571.

Griffen, D.T. and Ribbe, P.H. (1973) The crystal chemistry of staurolite. Am. J. Sci. 273A, 479-495.

Grove, T.L. (1977) A periodic antiphase model for the intermediate plagioclases (An_{33} to An_{75}). Am. Mineral. 62, 932-941.

__________, Ferry, J.M. and Spear, F.S. (1983) Phase transitions and decomposition relations in calcic plagioclase. Am. Mineral. 68, 41-59.

Grover, J.E. (1980) Thermodynamics of pyroxenes. In "Pyroxenes," C.T. Prewitt, ed. Reviews in Mineralogy 7, 341-417.

Gufan, Yu.M. and Larin, E.S. (1980) Theory of phase transitions described by two order parameters. Soviet Phys. Solid State 22, 270-275.

Hawthorne, F.C. (1983) Quantitative charaterization of site occupancies in minerals. Am. Mineral. 68, 287-306.

Heine, V. and McConnell, J.D.C. (1981) Origin of modulated incommensurate phases in insulators. Phys. Rev. Lett. 46, 1092-1095.

__________ and McConnell, J.D.C. (1984) The origin of incommensurate structures in alloys. J. Phys. C17, 1199-1220.

Helgeson, H.C., Delaney, J.M., Nesbitt, H.W. and Bird, D.K. (1978) Summary and critique of the thermodynamic properties of rock forming minerals. Am. J. Sci. 278A, 229p.

Heuer, A.H. and Nord, G.L. (1976) Polymorphic phase transitions in minerals. In "Electron Microscopy in Mineralogy," H.-R. Wenk, ed. Springer-Verlag, Berlin, p. 274-303.

Higgins, J.B., Ribbe, P.H. and Nakajima, Y. (1982) An ordering model for the commensurate antiphase structure of yoderite. Am. Mineral. 67, 76-84.

Holland, T.J.B. (1980) The reaction albite = jadeite + quartz determined experimentally in the range 600-1200°C. Am. Mineral. 65, 129-134.

__________ (1983) The experimental determination of activities in disordered and short range ordered jadeitic pyroxenes. Contrib. Mineral. Petrol. 82, 214-220.

__________ and Carpenter, M.A. (1983) Order/disorder and melting in sillimanite. Terra Cognita 3, 162.

Holm, J.L. and Kleppa, O.J. (1968) Thermodynamics of the disordering process in albite. Am. Mineral. 53, 123-133.

Hovis, G.L. and Waldbaum, D.R. (1977) A solution calorimetric investigation of K-Na mixing in a sanidine-analbite ion-exchange series. Am. Mineral. 62, 680-686.

Imry, Y. (1975) On the statistical mechanics of order parameters. J. Phys. C8, 567-577.

Inden, G. (1974) Ordering and segregation reactions in B.C.C. binary alloys. Acta Met. 22, 945-951.

Kikuchi, R. (1951) A theory of cooperative phenomena. Phys. Rev. 81, 988-1003.

__________ and Sato, H. (1974) Characteristics of superlattice formation in alloys of face centred cubic structures. Acta Met. 22, 1099-1112.

Kitamura, M. and Morimoto, N. (1977) The superstructure of plagioclase feldspars. Phys. Chem. Minerals 1, 199-212.

__________ and Morimoto, N. (1984) The modulated structure of the intermediate plagioclases and its change with composition. In "Feldspars and Feldspathoids," W.L. Brown, ed. NATO Advanced Study Inst., Ser. C, 137, p. 95-119. Reidel, Dordrecht, Holland.

Kroll, H. and Müller, W.F. (1980) X-ray and electron optical investigation of synthetic high-temperature plagioclases. Phys. Chem. Minerals 5, 255-277.

__________ and Ribbe, P.H. (1983) Lattice parameters, composition and Al,Si order in alkali feldspars. In "Feldspar Mineralogy," P.H. Ribbe, ed. Reviews in Mineralogy 2 (2nd ed.), 57-99.

Landau, L.D. and Lifshitz, E.M. (1958) "Statistical Physics." Addison Wesley, Reading, Massachusetts.

Laves, F., Czank, M. and Schulz, H. (1970) The temperature dependence of the reflection intensities of anorthite ($CaAl_2Si_2O_8$) and the corresponding formation of domains. Schweiz. Mineral. Petrogr. Mitt. 50, 519-525.

Laves, F. and Goldsmith, J.R. (1955) The effect of temperature and composition on the Al-Si distribution in anorthite. Z. Kristallogr. 106, 227-235.

Lifshitz, E.M. (1942a) On the theory of phase transformations of the second order. I. Changes of the elementary cell of a crystal in phase transitions of the second order. J. Phys. (Moscow) 6, 61-74.

__________ (1942b) On the theory of phase transitions of the second order. II. Phase transitions of the second order in alloys. J. Phys. (Moscow) 6, 251-263.

Louisnathan, S.J. (1971) Refinement of the crystal structure of a natural gehlenite, $Ca_2Al(AlSi)_2O_7$. Can. Mineral. 10, 822-837.

Maruyama, S., Liou, J.G. and Suzuki, K. (1982) The peristerite gap in low grade metamorphic rocks. Contrib. Mineral. Petrol. 81, 268-276.

Mason, R.A. (1979) The ordering behaviour of albite in aqueous solutions at 1 kbar. Contrib. Mineral. Petrol. 68, 269-273.

McConnell, J.D.C. (1978a) K-space symmetry rules and their application to ordering behavior in non-stoichiometric (metal enriched) chalcopyrite. Phys. Chem. Minerals 2, 253-265.

__________ (1978b) The intermediate plagioclase feldspars: an example of a structural resonance. Z. Kristallogr. 147, 45-62.

__________ (1981b) Electron-optical study of modulated mineral solid solutions. Bull. Minéral. 104, 231-235.

__________ (1981b) Time-temperature study of the intensity of satellite reflections in nepheline. Am. Mineral. 66, 990-996.

__________ (1983) A review of structural resonance and the nature of long-range interactions in modulated mineral structures. Am. Mineral. 68, 1-10.

__________ and McKie, D. (1960) The kinetics of the ordering process in triclinic $NaAlSi_3O_8$. Mineral. Mag. 32, 436-454.

McKie, D. and Radford, A.J. (1959) Yoderite, a new hydrous magnesium iron alumino-silicate from Mautia Hill, Tanganyika. Mineral. Mag. 32, 282-307.

McLaren, A.C. (1974) Transmission electron microscopy of the feldspars. In "The Feldspars," W.S. MacKenzie and J. Zussman, eds. Manchester University Press, Manchester, England, p. 378-424.

McMillan, P., Putnis, A. and Carpenter, M.A. (1984) A Raman spectroscopic study of Al/Si ordering in synthetic Mg-cordierite. Phys. Chem. Minerals 10, 256-260.

Meagher, E.P. and Gibbs, G.V. (1977) The polymorphism of cordierite: II. The crystal structure of indialite. Can. Mineral. 15, 43-49.

Merkel, G.A. and Blencoe, J.G. (1982) Thermodynamic procedures for treating the monoclinic/triclinic inversion as a high-order phase transition in equations of state for binary analbite-sanidine feldspars. In "Advances in Phys. Geochem.", S.K. Saxena, ed., 2, 243-284

Merlino, S. (1984) Feldspathoids: their average and real structures. In "Feldspars and Feldspathoids," W.L. Brown, ed. NATO Adv. Study Inst., ser. C, 137, p. 435-470. Reidel, Dordrecht, Holland.

Navrotsky, A. (1971) The intracrystalline cation distribution and the thermodynamics of solid solution formation in the system $FeSiO_3$-$MgSiO_3$. Am. Mineral. 56, 201-211.

__________ and Kleppa, O.J. (1973) Estimate of enthalpies of formation and fusion of cordierite. J. Am. Ceram. Soc. 56, 198-199.

__________ and Loucks, D. (1977) Calculation of subsolidus phase relations in carbonates and pyroxenes. Phys. Chem. Minerals 1, 109-127.

__________, Newton, R.C. and Kleppa, O.J. (1973) Sillimanite-disordering enthalpy by calorimetry. Geochim. Cosmochim. Acta 37, 2497-2508.

Newton, R.C., Charlu, T.V., Anderson, P.A.M. and Kleppa, O.J. (1979) Thermochemistry of synthetic clinopyroxenes on the join $CaMgSi_2O_6$-$Mg_2Si_2O_6$. Geochim. Cosmochim. Acta 43, 55-60.

__________, Charlu, T.V. and Kleppa, O.J. (1977) Thermochemistry of high pressure garnets and clinopyroxenes in the system CaO-MgO-Al_2O_3-SiO_2. Geochim. Cosmochim. Acta 41, 369-377.

__________, Charlu, T.V. and Kleppa, O.J. (1980) Thermochemistry of the high structural state plagioclases. Geochim. Cosmochim. Acta 44, 933-941.

__________ and Wood, B.J. (1980) Volume behavior of silicate solid solutions. Am. Mineral. 65, 733-745.

__________, Wood, B.J. and Kleppa, O.J. (1981) Thermochemistry of silicate solid solutions. Bull. Minéral. 104, 162-171.

Nord, G.L., Jr., Hammerstrom, J. and Zen, E.-An. (1978) Zoned plagioclase and peristerite formation in phyllites from south western Massachusetts. Am. Mineral. 63, 947-955.

Orville, P.M. (1972) Plagioclase cation exchange equilibria with aqueous chloride solution: results at 700°C and 2000 bars in the presence of quartz. Am. J. Sci. 272, 234-272.

__________ (1974) The "peristerite gap" as an equilibrium between ordered albite and disordered plagioclase solid solution. Bull. Soc. franc. Minéral. Cristallogr. 97, 386-392.

Parsonage, N.G. and Staveley, L.A.K. (1978) "Disorder in Crystals." Oxford University Press, Oxford, England.

Pippard, A.B. (1957) "The Elements of Classical Thermodynamics." Cambridge University Press, Cambridge, England.

Putnis, A. (1980a) Order-modulated structures and the thermodynamics of cordierite reactions. Nature 287, 128-131.

__________ (1980b) The distortion index in anhydrous Mg-cordierite. Contrib. Mineral. Petrol. 74, 135-141.

__________ and Angel, R.J. (1985) Al,Si ordering in cordierite using "magic angle spinning" NMR. II: Models of Al,Si order from NMR data. Phys. Chem. Minerals, submitted.

__________ and Bish, D.L. (1983) The mechanism and kinetics of Al,Si ordering in Mg cordierites. Am. Mineral. 68, 60-65.

__________, Fyfe, C.A. and Gobbi, G.C. (1985) Al,Si ordering in cordierite using "magic angle spinning" NMR. I: Si^{29} spectra of synthetic cordierites. Phys. Chem. Minerals, submitted.

Raase, P. (1971) Zur Synthese und Stabilität der Albit-Modificationen. Tschermaks Mineral. Petrogr. Mitt. 16, 136-155.

Rao, C.N.R. and Rao, K.J. (1978). "Phase Transitions in Solids." McGraw-Hill, New York.

Reeder, R.J. (1981) Electron optical investigation of sedimentary dolomites. Contrib. Mineral. Petrol. 76, 148-157.

__________ (1983) Crystal chemistry of the rhombohedral carbonates. In "Carbonates: Mineralogy and Chemistry," R.J. Reeder, ed. Reviews in Mineralogy 11, 1-47.

__________ and Nakajima, Y. (1982) The nature of ordering and ordering defects in dolomite. Phys. Chem. Minerals 8, 29-35.

__________ and Wenk, H.-R. (1983) Structure refinements of some thermally disordered dolomites. Am. Mineral. 68, 769-776.

Ribbe, P.H. (1980) Aluminum silicate polymorphs (and mullite). In "Orthosilicates," P.H. Ribbe, ed. Reviews in Mineralogy 5, 189-214.

__________ (1983a) Aluminum-silicon order in feldspars; domain textures and diffraction patterns. In "Feldspar Mineralogy," P.H. Ribbe, ed. Reviews in Mineralogy 2 (2nd ed.), 21-55.

__________ (1983b) Exsolution textures in ternary and plagioclase feldspars; interference colours. In "Feldspar Mineralogy," P.H. Ribbe, ed. Reviews in Mineralogy 2 (2nd ed.), 241-270.

Richards, M.J. (1971) "On the Effect of Second Neighbor Interactions in the Theory of Ordering and Clustering in Binary Alloys." D.Sc. Thesis, Massachusetts Inst. Technology, Cambridge, Massachusetts.

__________ and Cahn, J.W. (1971) Pairwise interactions and the ground state of ordered binary alloys. Acta Met. 19, 1263-1277.

Riste, T., Samuelson, E.J. and Otnes, K. (1971) Critical neutron scattering from $SrTiO_3$. In "Structural Phase Transitions and Soft Modes," E.J. Samuelson, E. Anderson, J. Feder, eds. Universitetsforlagt, Oslo, p. 395-408.

Rossi, G., Smith, D.C., Ungaretti, L. and Domeneghetti, M.C. (1983) Crystal-chemistry and cation ordering in the system diopside-jadeite: a detailed study by crystal structure refinement. Contrib. Mineral. Petrol. 83, 247-258.

Rumble, D., III (1977) Configurational entropy of magnetite-Ülvospinel$_{ss}$ and hematite-ilmenite$_{ss}$. Carnegie Inst. Wash. Yearbook 76, 581-584.

Sato, H. (1970) Order-disorder transformations. In "Physical Chemistry: an Advanced Treatise, vol. X, Solid State," W. Jost, ed. Ch. 10, p. 579-718.

Salje, E. (1985) Thermodynamics of sodium feldspar I: order parameter treatment and strain induced coupling effects. Phys. Chem. Minerals (in press).

__________, Devarajan, V. (1985) Phase transitions in systems with strain induced coupling between two order parameters. J. Phys. C. (in press).

__________, Kusholke, B., Wruck, B. and Kroll, H. (1985) Thermodynamics of sodium feldspar II: Experimental results and numerical calculations. Phys. Chem. Minerals (in press).

__________ and Viswanathan, K. (1976) The phase diagram calcite-aragonite as derived from the crystallographic properties. Contrib. Mineral. Petrol. 55, 55-67.

Schreyer, W. and Schairer, J.F. (1961) Compositions and structural states of anhydrous Mg-cordierites: a re-investigation of the central part of the system $MgO-Al_2O_3-SiO_2$. J. Petrol. 2, 324-406.

__________ and Yoder, H.S., Jr. (1964) The system Mg-cordierite-H_2O and related rocks. N. Jahrb. Mineral. Abh. 101, 271-342.

Senderov, E.E. and Schekina, T.I. (1976) Natural production conditions and stability of structural forms of albite. Geochem. Int'l 13, 99-112.

Smart, R.M. and Glasser, F.P. (1977) Stable cordierite solid solutions in the $MgO-Al_2O_3-SiO_2$ system: composition, polymorphism and thermal expansion. Science of Ceramics 9, 256-263.

Smith, J.V. (1968) The crystal structure of staurolite. Am. Mineral. 53, 1139-1155.

__________ (1972) Critical review of synthesis and occurrence of plagioclase feldspars and a possible phase diagram. J. Geol. 80, 505-525.

__________ (1974) "Feldspar Minerals, Vol. 1: Crystal Structure and Physical Properties." Springer-Verlag, Berlin.

__________ (1983) Phase equilibria of plagioclase. In "Feldspar Mineralogy," P.H. Ribbe, ed. Reviews in Mineralogy 2 (2nd ed.), 223-239.

Soffa, W.A. and Laughlin, D.E. (1982) Recent experimental studies of continuous transformations in alloys: an overview. In "Solid Phase Transformations," H.I. Aaronson and D.E. Laughlin, eds. Metallurgical Soc. A.I.M.E., 159-183.

Spencer, K.J. and Lindsley, D.H. (1981) A solution model for coexisting iron-titanium oxides. Am. Mineral. 66, 1189-1201.

Tanner, L.E. and Laughlin, D.E. (1975) Experimental observations of continuous transformations in binary alloys. Scripta Met. 9, 373-378.

__________ and Leamy, H.J. (1974) The microstructure of order-disorder transitions. In "Order-Disorder Transformations in Alloys," H. Warlimont, ed. Springer-Verlag, Berlin, p. 180-239.

Thierry, P., Chatillon-Colinet, C., Mathieu, J.C., Regnard, J.R. and Amossé, J. (1981) Thermodynamic properties of the forsterite-fayalite ($Mg_2SiO_4-Fe_2SiO_4$) solid solution. Determination of heat of formation. Phys. Chem. Minerals 7, 43-46.

Thompson, A.B. and Perkins, E.H. (1981) Lambda transitions in minerals. In "Thermodynamics of Minerals and Melts," A. Navrotsky, R.C. Newton and B.J. Wood, eds. Advances in Physical Geochemistry, vol. 1. Springer-Verlag, New York, p. 35-62.

Thompson, J.B., Jr., Waldbaum, D.R. and Hovis, G.L. (1974) Thermodynamic properties related to ordering in end-member alkali feldspars. In "The Feldspars," W.S. MacKenzie and J. Zussman, eds. Manchester University Press, Manchester, England, p. 218-248.

Tokonami, M., Nakajima, Y. and Morimoto, N. (1980) The diffraction aspect and a structural model of mullite, $Al(Al_{1+2x}Si_{1-2x})O_{5-x}$. Acta Crystallogr. A36, 270-276.

Waldbaum, D.R. and Robie, R.A. (1971) Calorimetric investigation of Na-K mixing and polymorphism in the alkali feldspars. Z. Kristallogr. 134, 381-420.

Wenk, H.-R. (1983) Mullite-sillimanite intergrowths from pelitic inclusions in Bergell tonalite. N. Jahrb. Mineral. Abh. 146, 1-14.

__________, Barber, D.J. and Reeder, R.J. (1983) Microstructures in carbonates. In "Carbonates: Mineralogy and Chemistry," R.J. Reeder, ed. Reviews in Mineralogy 11, 301-367.

Wood, B.J., Holland, T.J.B., Newton, R.C. and Kleppa, O.J. (1980) Thermochemistry of jadeite-diopside pyroxenes. Geochim. Cosmochim. Acta 44, 1363-1371.

__________ and Kleppa, O.J. (1981) Thermochemistry of forsterite-fayalite olivine solutions. Geochim. Cosmochim. Acta 45, 529-534.

Ylä-Jääski, J. and Nissen, H.-U. (1983) Investigation of superstructures in mullite by high resolution electron microscopy and electron diffraction. Phys. Chem. Minerals 10, 47-54.

Yund, R.A. and Tullis, J. (1983) Subsolidus phase relations in the alkali feldspars. In "Feldspar Mineralogy," P.H. Ribbe, ed. Reviews in Mineralogy 2 (2nd ed.), 141-176.

Chapter 7. Alexandra Navrotsky

CRYSTAL CHEMICAL CONSTRAINTS on the THERMOCHEMISTRY of MINERALS

1. INTRODUCTION

Phase equilibria in mineral systems are energetically very finely balanced phenomena. Near phase boundaries and invariant points in (P, T, X) space, the free energies of a number of phases are very similar and, very frequently, their enthalpies, entropies, and volumes are such that the ΔH, $T\Delta S$, and $P\Delta V$ terms are small and all of comparable importance in determining ΔG. Thus to understand, on a microscopic and structural scale, as well as in terms in bulk thermodynamic properties, why a given phase (and structure) is stable, one must consider the systematic trends and crystal chemical constraints which influence energy (enthalpy), entropy, and volume. The effects of temperature, pressure, and composition on crystal structures, while systematic data on volumes, thermal expansivities, and compressibilities are discussed by Hazen (this volume). This chapter will emphasize enthalpies and entropies, with discussion of enthalpies of formation of minerals from the binary oxides, entropies arising from both vibrational and configurational (disorder) contributions, systematics of coupled substitutions in silicates, some aspects of high pressure phase transitions (also treated by Jeanloz, this volume), and some aspects of solid solution thermodynamics (cf. the chapter on order-disorder phenomena by Carpenter, this volume).

A comment on the magnitudes of various energetic parameters is a good starting point. Relative to isolated atoms or ions, the total energy of a crystal is on the order of thousands of kilojoules (kJ). The heat of formation of a typical silicate from the elements is of the order of hundreds of kJ, while its formation from the oxides is of the order of tens of kilojoules. Polymorphic transitions and order-disorder reactions typically have energies of a few kJ or less. Thus understanding the energetics of solid-solid transitions from the point of view of the total lattice energy of the crystal (or the total energy of a molecular cluster) requires confidence in the latter at the level of a few tenths of a percent or better. It is for this reason that first principles calculations of absolute energetics generally offer qualitative insight into the relative energies of different possible phase assemblages rather than reliable numerical values for such energies (cf. Burnham, this volume). Thus semi-empirical and correlative approaches to systematizing and predicting the energies of mineral reactions continue to be useful and indeed necessary. Such approaches can best be based on considerations of crystal chemistry and bonding, with concepts such as bond length, relative atomic size, electronegativity, acid-base characteristics, and site preference energy playing useful roles. In the examples in this chapter, I will use such semiquantitative concepts freely since their quantification is not yet generally attainable.

2. FORMATION OF TERNARY COMPOUNDS FROM THE OXIDES

Consider the enthalpy, ΔH, of the reaction:

$$AO + B_aO_b = AB_aO_{1+b}, \qquad (1)$$

with A = Ca, Mg, Mn, Fe, Co, Ni and B = Si, W, C, S (see Figure 1).

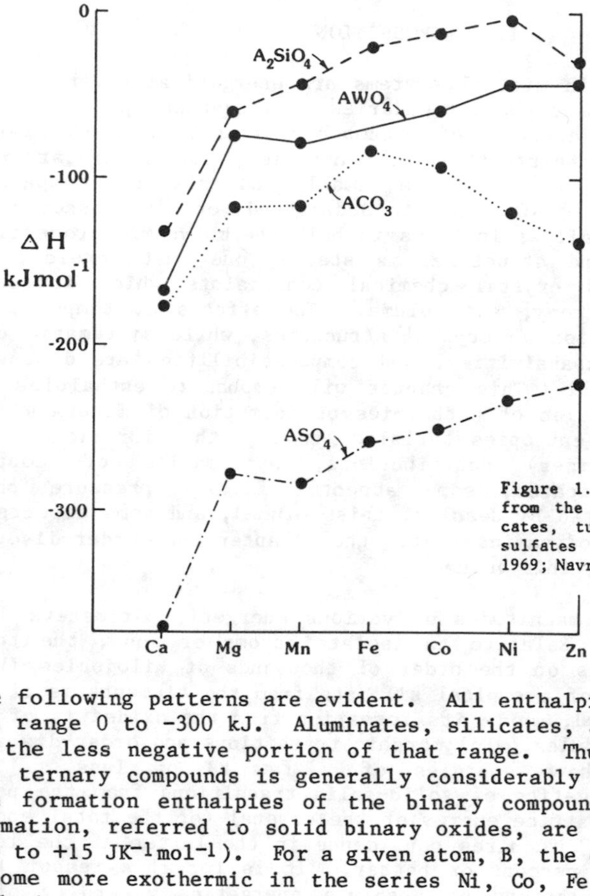

Figure 1. Enthalpies of formations from the oxides of series of silicates, tungstates, carbonates, and sulfates (Navrotsky and Kleppa, 1969; Navrotsky, 1974).

The following patterns are evident. All enthalpies of formation are in the range 0 to -300 kJ. Aluminates, silicates, and germanates also lie in the less negative portion of that range. Thus the stabilization of the ternary compounds is generally considerably less in magnitude than the formation enthalpies of the binary compounds. The entropies of formation, referred to solid binary oxides, are also quite small (from -15 to +15 $JK^{-1}mol^{-1}$). For a given atom, B, the enthalpies of formation become more exothermic in the series Ni, Co, Fe, Mn, Mg, Ca, that is, with increasing basicity of the oxide AO. For a given atom, A, the enthalpies of formation become more exothermic in the series Al, Si, Ge, W, C, S, that is, with increasing acidity of the oxide Al_2O_3, SiO_2, GeO_2, WO_3, CO_3, SO_3. Thus the most stable compounds form when the most complete transfer of oxide ion from base to acid occurs, and the ternary structures then contain well-defined covalently bonded anions (SiO_4, GeO_4, WO_4, CO_3, SO_4). For compounds between more similar binary oxides, e.g., $Al_2O_3 + SiO_2$, $CuO + Fe_2O_3$, $Fe_2O_3 + TiO_2$, the enthalpies of formation are much smaller in magnitude and in the above three cases are actually endothermic. Thus $3Al_3O_3 \cdot 2SiO_2$ (mullite), $CuFe_2O_4$ (spinel), and Fe_2TiO_5 (pseudobrookite) derive their high temperature stability from their entropies (configurational and/or vibrational) and at low temperature these compounds generally are metastable or decompose.

The terms "acid" and "base," as used loosely above, defy easy quantification. In the Lux-Flood sense (Lux, 1939; Flood and Førland, 1947), a basic oxide (e.g., Na_2O) is one which readily transfers its oxygens to the coordination sphere of an acid oxide (e.g., SiO_2) to form a complex anion (e.g., SiO_4^{2-}). Yet in many cases in the solid state complex anions are not easily defined, and the oxygens are bonded to a significant extent to several types of cations. Even in cases where a silicate anion or polymerized silicate or aluminosilicate framework can be defined topologically, the extent of interaction and perturbation of the anion or framework by the cation must be considered. In that context, approaches using ab initio molecular orbital calculations on small molecular clusters (Gibbs, 1982; Geisinger et al., 1985; Navrotsky et al., 1985) are useful in quantifying the observed trends. Alternately, various empirical acidity scales (defining the relative strength of interaction of a cation and an oxygen anion) may be used. Considerations from UV-visible spectra and oxidation-reduction equilibria lead to the optical basicity scale of Duffy (Duffy and Ingram, 1971). The ionic potential (formal charge divided by ionic radius) is another rough measure of the ability of a cation to compete for bonding to oxygen, as is the Pauling electronegativity.

If one considers mixing properties in the molten state, trends similar to those shown for crystalline ternary compounds are generally seen, though the heats of formation are generally less exothermic by 20-50% in the molten (or glassy) state compared to the crystals (Navrotsky et al., 1982).

Tardy (Tardy and Gartner, 1977; Tardy and Vieillard, 1977) proposed a scheme which parameterizes the free energies of formation of phosphates, carbonates, sulfates, and other minerals in terms of properties of corresponding aqueous ions. Such regularities are another manifestation of the relatively constant "acid-base" characteristics of the oxides involved, analogous to the trends discussed above. Indeed in organic chemistry, the application of "linear free energy relations," which parameterize the pK's of organic acids, the activation energies for nucleophilic substitutions, or other thermodynamic properties in terms of characteristics of various functional groups, is a well-established and useful approach (Lowry and Richardson, 1981). Analogous correlations, with parameters related to bonding and spectroscopic factors, will have increasing application to mineral reactions.

3. CHARGE-COUPLED SUBSTITUTIONS IN MINERALS, GLASSES, AND MELTS

Many minerals can be related to each other by the substitution of cations, either singly or as charged balanced pairs, for each other in specific crystallographic sites. A simple classification of sublattices on which such substitutions can occur would distinguish among the tetrahedral (T) sites normally occupied by Si and Al, the octahedral or distorted octahedral (M) sites occupied by Mg^{2+}, Fe^{2+} and other small cations, and the large sites (A) with 7-fold or higher coordination generally occupied by alkali and larger alkaline earth ions. Thus, some common substitutions may be written as follows:

ferromagnesian substitution ($Mg_M^{2+} \rightarrow Fe_M^{2+}$),

plagioclase substitution ($Na_A^+ + Si_T^{4+} \rightarrow Ca_A^{2+} + Al_T^{3+}$),

stuffed silica substitution ($Si_T^{4+} \rightarrow Al_T^{3+} + 1/n\ N_A^{n+}$),

Tschermak's substitution ($M_M^{2+} + Si_T^{4+} \rightarrow Al_M^{3+} + Al_T^{3+}$).

Such substitutions can occur in a variety of minerals; for example the stuffed silica substitution can be observed in framework silicates (stuffed tridymite, stuffed cristobalite, and stuffed β-quartz structures), in aluminosilicate glasses and melts, and in amphiboles, micas, and clays. From an energetic viewpoint, several questions can be formulated. For a given structure, how does the energy of a given type of substitution depend on the nature of the substituting cation? Can this variation be explained in terms of crystal chemical and bonding factors? For different structures, do analogous substitutions have similar energetics? Can such trends be used to predict thermodynamic properties, especially of complex sheet and chain silicates? This section will summarize some preliminary answers to these questions. Much experimental and theoretical work still needs to be done.

A series of aluminosilicate glasses SiO_2-$M_{1/n}^{n+}AlO_2$ (M = alkali or alkaline earth) have been studied by high temperature solution calorimetry (Navrotsky et al., 1982; Hervig and Navrotsky, 1984; Roy and Navrotsky, 1984). The stuffed-silica type of substitution introduces a metal cation into the interstices of a tetrahedral framework, with charge compensation and a nominally complete degree of polymerization. For aluminosilicate glasses, both x-ray scattering and Raman spectroscopy suggest that, when Al substitutes in a charge-coupled fashion for Si, an aluminosilicate framework is maintained although there may be changes in regularity (increased bond angle and ring size variation) with Al content. This perturbation of the framework also depends strongly on the nature of the charge balancing cation, greater disturbance occurring with ions of smaller size and larger charge. The heat of solution data are shown in Figure 2. The process of dissolving an aluminosilicate glass in molten lead borate to form a dilute (<1 wt %) solution consists of breaking up an aluminosilicate framework structure into isolated species, presumably SiO_4 and AlO_4 tetrahedra and alkali and alkaline-earth cationic species dissolved in a borate matrix. Thus the enthalpy of solution may be considered a measure of the strength of bonding in the aluminosilicate glass, at least in a relative sense when comparing various compositions. Three points are evident. First, the enthalpies of solution generally become more endothermic as $M_{1/n}^{n+}AlO_2$ is substituted for SiO_2, for $x < 0.5$ (see Navrotsky et al., 1985). This increase becomes more pronounced with decreasing field strength (or increasing basicity) of the cation, e.g., in the series Mg, Ca, Sr, Pb, Ba, Li, Na, K, Rb, Cs. Second, the enthalpy of solution relations tend to curve back toward more exothermic values, with a maximum near $x = 0.5$. This maximum reflects an exothermic enthalpy of mixing for the reaction:

$$xM_{1/n}^{n+}AlO_2 + (1-x)SiO_2 = M_{x/n}^{n+}Al_xSi_{1-x}O_2. \qquad (2)$$

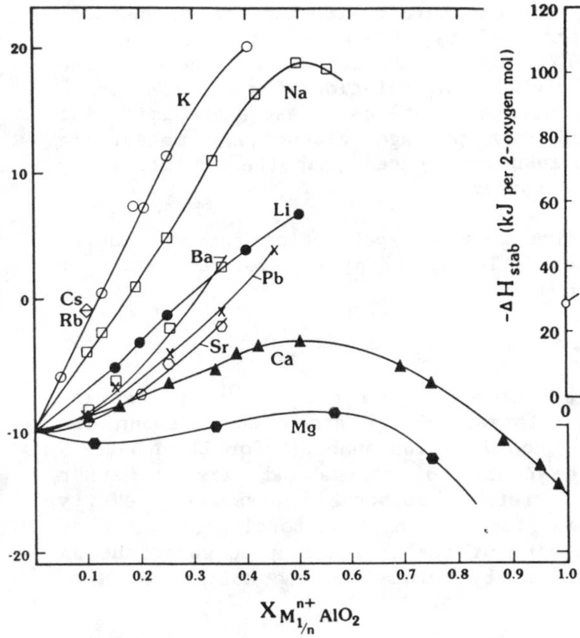

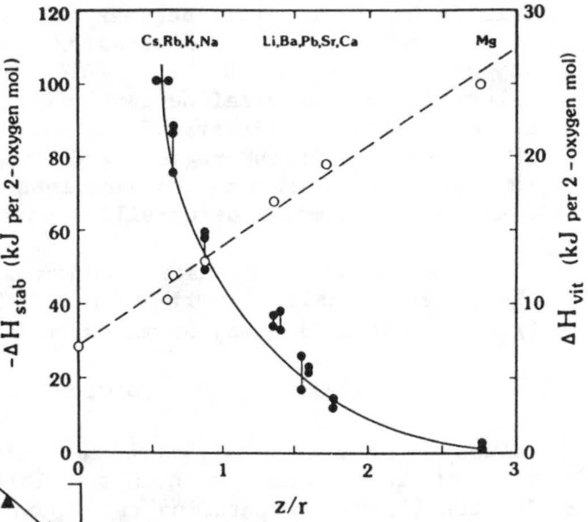

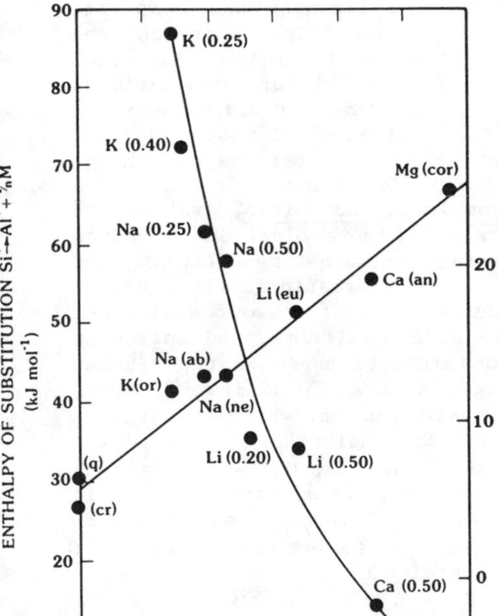

Figure 2 (above, left). Enthalpies of solution of glasses AlO_2-SiO_2 in molten $2PbO \cdot B_2O_3$ at 973 K (Navrotsky et al., 1982; Roy and Navrotsky, 1984; Hervig and Navrotsky, 1984).

Figure 3 (to the left). Enthalpy stabilization of glasses (Eqn. 3) plotted against ionic potential (z/r) of cation M (left-hand scale and curve). Enthalpy of vitrification (right-hand scale and line). From Roy and Navrotsky (1984); Navrotsky et al. (1985).

Figure 4 (above, right). Enthalpy stabilization (left-hand scale and curve) and enthalpy of vitrification (right-hand scale and line) plotted against average tetrahedral bond length pertubation parameter Δ(TO). Same references as Figure 3.

This enthalpy of mixing becomes more exothermic with increasing basicity of the metal oxide or decreasing ability of the M cation to bond to oxygen. Third, at $0 < x < 0.4$, pronounced curvature occurs in the relations for the alkaline-earths while the relations for the alkalis are approximately linear. This curvature implies a less exothermic heat of mixing in this region and may presage glass-glass metastable immiscibility analogous to but less pronounced than the immiscibility seen in binary metal oxide-silica systems.

The calorimetric data confirm regular systematics for the entire alkali and alkaline earth series. The enthalpy of the $Si^{4+} \rightarrow Al^{3+} + 1/nM^{n+}$ substitution may be measured by

$$\Delta H(stab) = [-\Delta H_{sol}(SiO_2) + \Delta H_{sol}(M_{x/n}Al_x\ Si_{1-x}O_2)]/x. \quad (3)$$

Values of ΔH(stab) are plotted in Figure 3 against the ionic potential, z/r, of the cation, with z the formal charge and r the Shannon and Prewitt (1969) ionic radius using coordination number 6 for Li and Mg, 8 for the other ions. Though the choice of charge and size is rather arbitrary, the ionic potential offers a reasonable parameter, easily calculated for all ions, for comparing the relative bonding strength of different cations. The stabilization of the charge coupled substitution in aluminosilicate glasses is inversely related to the ability of M to bond to oxygen.

Ab initio molecular orbital calculations on aluminosilicate clusters chosen to model the above systems have recently been completed (Geisinger et al., 1985; Navrotsky et al., 1985). This approach (see Gibbs, 1982) chooses small molecular clusters to model the local environments of atoms within a solid. The methods of computational quantum chemistry are used to solve the Schrödinger Equation (see the introduction to this volume) and find eigenvalues for physical parameters such as bond lengths, bond angles, total energies and charge distributions. Different levels of approximation, ranging from empirical or semi-empirical to almost completely _ab initio_, are used to describe the wave functions. In general, the more electrons involved (heavier atoms, larger molecular clusters), the more time-consuming and expensive the computer calculations and the more drastic is the level of approximation that must be used. Fortunately, for framework silicates, the light elements Si, Al, and O are major constituents, and molecules containing these atoms and hydrogens to terminate bonds in the cluster can be calculated fairly rigorously (e.g., at the STO-3G level, Gibbs, 1982). Thus in the problem above, the basic cluster, which contains an SiOAl linkage (bridging oxygen), is an H_6AlSiO_7 molecule. The bridging oxygen is then approached by a metal atom M (Li, Na, Mg, Al) at a bond length characteristic of the MO bond in minerals. This results in clusters of the type $M(OH)_3-H_6T_2O_7$ and $M(H_2O)_5-H_6T_2O_7$ for which STO-3G molecular orbital calculations result in the following conclusions. When an octahedrally or tetrahedrally coordinated cation cluster (M = Li, Na, Mg) is attached to the bridging oxygen of an SiOSi or AlOSi linkage, the TO bonds are perturbed such that the TOT angle narrows and the bridging TO bonds lengthen slightly. This perturbation increases in the order Na, Li, Mg. The bridging AlO bond is lengthened more than the bridging SiO bond. The effects of tetrahedrally and octahedrally coordinated M atoms (at their "normal" MO distances) are generally comparable. Calculated and experimental SiO and AlO bond lengths are in good agreement for ordered framework aluminosilicate crystals containing Li, Mg, and Na.

The thermochemical data for glasses along the joins SiO_2-$M^{n+}_{1/n}$ AlO_2 can be correlated with the changes (both calculated and observed) in TO bond lengths mentioned above. The average perturbation of the bridging TO bond length is calculated as follows. The AlO and SiO distances in the ordered crystals for those oxygens bonded to two tetrahedrally bonded framework cations plus at least one other nonframework cation can be compared to the "ideal" TO bond lengths of 1.712 (AlO) and 1.581 (SiO) taken as the average of SiO and AlO bond lengths in framework aluminosilicates for which the oxygen is only 2-coordinate. Thus

$$\Delta(\text{AlO}) = (\text{AlO})_{\text{obs}} - 1.712, \quad (4)$$

$$\Delta(\text{SiO}) = (\text{SiO})_{\text{obs}} - 1.581, \quad (5)$$

and, at a composition with Al/(Al+Si) = x and Si/(Al+Si) = 1-x,

$$\Delta(\text{TO})_{\text{av}} = x\,\Delta(\text{AlO}) + (1-x)\,\Delta(\text{SiO}), \quad (6)$$

The resulting plot (Fig. 4) shows strong correlation between ΔH(stab) and Δ(TO). Thus one can parameterize the perturbation of the aluminosilicate framework using a bond length perturbation parameter consistent with molecular orbital calculations on a covalently bonded cluster or by using an essentially ionic field strength parameter, z/r.

The enthalpy difference between crystal and glass (enthalpy of vitrification, ΔH(vit), shows a linear correlation with z/r or Δ(TO) (see Figs. 3 and 4). If one considers silica (quartz and glass) as a framework structure with the interstices occupied by a cation (i.e., a vacancy) of infinitely weak ability to perturb the bridging oxygen, then the point for SiO_2 with z/r = 0 and Δ(TO) = 0 lies on the same trend as the other data. The heat of vitrification increases with increasing z/r or Δ(TO), indicating that increasing perturbation of the aluminosilicate framework not only decreases the stability of the glass with respect to mixing properties in the amorphous state but also with respect to the crystalline state. Similar trends probably hold for the molten state as for the glass, but discussion of the relations among glass and melt properties is beyond the scope of this chapter.

To compare the same substitution in different crystalline and glassy systems, Table 1 lists the energetics of a number of substitutions, calculated from heat of solution data in a manner analogous to that for the enthalpy stabilization of aluminosilicate glasses discussed above. The table reveals several regularities. For the substitution, $Si^{4+}_T \rightarrow Al^{3+}_T + 1/nM^{n+}_A$, the enthalpy becomes less

Table 1

Enthalpies of Substitution Reactions in Silicates

Substitution and Mineral Pair[a]	ΔH ($kJ\ mol^{-1}$)
$Si^{4+}{}_T \rightarrow Al^{3+}{}_T + Cs^+{}_A$	
SiO_2(glass)-$Cs_{0.1}Al_{0.1}Si_{0.9}O_2$(glass)	-112[b]
$Si^{4+}{}_T \rightarrow Al^{3+}{}_T + Rb^+{}_A$	
SiO_2(glass)-$Rb_{0.1}Al_{0.1}Si_{0.9}O_2$(glass)	-115[b]
SiO_2(quartz)-$Rb_{0.25}Al_{0.25}Si_{0.75}O_2$(microcline)	-128[b]
$Si^{4+}{}_T \rightarrow Al^{3+}{}_T + K^+{}_A$	
SiO_2(glass)-$K_{0.25}Al_{0.25}Si_{0.75}O_2$(glass)	- 88[b,c]
SiO_2(quartz)-$K_{0.25}Al_{0.25}Si_{0.75}O_2$(high sanidine)	-100[c,d]
SiO_2(quartz)-$K_{0.33}Al_{0.33}S_{0.67}O_2$(leucite)	- 82[d]
SiO_2(quartz)-$K_{0.5}Al_{0.5}Si_{0.5}O_2$(high kalsilite)	- 78[d]
$Si^{4+}{}_T \rightarrow Al^{3+}{}_T + Na^+{}_A$	
SiO_2(glass)-$Na_{0.25}Al_{0.25}Si_{0.75}O_2$(glass)	- 61[b,c]
SiO_2(quartz)-$Na_{0.25}Al_{0.25}Si_{0.75}O_2$(high albite)	- 85[b,c]
SiO_2(quartz)-$Na_{0.5}Al_{0.5}Si_{0.5}O_2$(nepheline)	- 70[b,c]
$Ca_2Mg_5Si_8O_{22}F_2$(F-tremolite)-$NaCa_2Mg_5AlSi_7O_{22}F_2$ (F-edenite)	- 79[f]
$Si^{4+}{}_T \rightarrow Al^{3+}{}_T + Li^+{}_A$ (or M)	
SiO_2(glass)-$Li_{0.25}Al_{0.25}Si_{0.75}O_2$(glass)	- 37[b]
SiO_2(quartz)-$Li_{0.5}Al_{0.5}Si_{0.5}O_2$(β-eucryptite)	- 41[b]
$Ge^{4+}{}_T \rightarrow Al^{3+}{}_T + Na^+{}_A$	
GeO_2(glass)-$Na_{0.25}Al_{0.25}Ge_{0.75}O_2$(albite)	-105[g,h]
GeO_2(quartz)-$Na_{0.25}Al_{0.25}Ge_{0.75}O_2$(albite)	- 85[g,h]
$Si^{4+}{}_T \rightarrow Al^{3+}{}_T + 1/2Mg^{2+}{}_A$	
SiO_2(glass)-$Mg_{0.222}Al_{0.444}Si_{0.556}O_2$(glass)	~ 0[b,i]
SiO_2(quartz)-$Mg_{0.222}Al_{0.444}Si_{0.556}O_2$(β-quartz)	+ 3.1[e,i]
$Si^{4+}{}_T \rightarrow Al^{3+}{}_T + 1/2Mg^{2+}{}_M$	
SiO_2(quartz)-$Mg_{0.222}Al_{0.444}Si_{0.556}O_2$(cordierite)	- 25 to - 36[e,i,j]
$Si^{4+}{}_T \rightarrow Al^{3+}{}_T + 1/2Ca^{2+}{}_A$	
SiO_2(glass)-$Ca_{0.25}Al_{0.5}Si_{0.5}O_2$(glass)	- 14[b,e]
SiO_2(quartz)-$Ca_{0.25}Al_{0.5}Si_{0.5}O_2$(anorthite)	- 38[b,e]
$Si^{4+}{}_T \rightarrow Al^{3+}{}_T + 1/2Sr^{2+}{}_A$	
SiO_2(glass)-$Sr_{0.125}Al_{0.25}Si_{0.75}O_2$(glass)	- 22[b]

Continued next page

Substitution and Mineral Pair[a]	ΔH (kJ mol^{-1})
$Si^{4+}{}_T \rightarrow Al^{3+}{}_T + 1/2Ba^{2+}{}_A$	
SiO_2(glass)-$Ba_{0.125}Al_{0.25}Si_{0.75}O_2$(glass)	- 33[b]
$Si^{4+}{}_T \rightarrow Ge^{4+}{}_T$	
SiO_2(glass)-GeO_2(glass)	- 25[e,h]
SiO_2(quartz)-GeO_2(quartz)	- 15[e,h]
$Na_{0.25}Al_{0.15}Si_{0.75}O_2$(glass)-$Na_{0.25}Al_{0.25}Ge_{0.75}O_2$(glass)	- 17[e,g]
$Na_{0.25}Al_{0.25}Si_{0.75}O_2$(albite)-$Na_{0.25}Al_{0.25}Ge_{0.75}O_2$(albite)	- 18[e,g]
$Si^{4+}{}_M \rightarrow Ge^{4+}{}_M$	
SiO_2(stishovite)-GeO_2(rutile)	- 89[h,k]
$Mg^{2+}{}_M \rightarrow Fe^{2+}{}_M$	
$MgSi_{0.5}O_2$(forsterite)-$FeSi_{0.5}O_2$(fayalite)	- 26[d]
$MgSiO_3$(enstatite)-$FeSiO_3$(ferrosilite)	- 31[d]
$MgCaSi_2O_6$(diopside)-$FeCaSi_2O_6$(hedenbergite)	- 49[d,l]
$Mg^{2+}{}_A \rightarrow Fe^{2+}{}_A$	
$MgAl_{0.67}SiO_4$(garnet)-$FeAl_{0.67}SiO_4$(garnet)	14[d,l]

a. Subscripts refer to sites, T = tetrahedral framework site A = large (8-10 coordinate) non-framework site, M = smaller (6 coordinate) non-framework site
b. Roy and Navrotsky, 1984
c. Hervig and Navrotsky, 1984
d. Robie et al., 1978
e. Navrotsky et al., 1982
f. Graham and Navrotsky, in prep.
g. Capobianco and Navrotsky, 1982
h. Navrotsky, 1971
i. Carpenter et al., 1983
j. depending on degree of Al,Si order in cordierite
k. Akaogi and Navrotsky, 1984b
l. O'Neill and Navrotsky, unpublished

exothermic with increasing z/r of the cation for crystals as well as for glasses, with values of (negative) 110-120 kJ for Cs and Rb, 80-100 kJ for K, 60-85 kJ for Na, near 40 kJ for Li, and substantially less exothermic values for the alkaline earths. For the alkalis, values generally overlap for glasses and crystals and span a range of about ±10% around of the average value. Thus, to within an accuracy of about ±10%, it appears that the enthalpy of the coupled silica-stuffing substitution is constant in different long range environments (e.g., for Na; glass, feldspar, nepheline, amphibole). This suggests that such substitutions are controlled energetically primarily by the local environment, and that as long as the framework (T) and interstitial (A) sites remain reasonably similar, the energy of the substitution remains rather constant. For the alkaline earths (Mg and Ca), the difference between crystal and glass becomes more pronounced (although Al,Si order in anorthite and cordierite may be a complicating feature). The substitution which puts Mg into octahedral (M) sites in cordierite is energetically much more favorable than that which places it into large (A) interstitial sites in the stuffed β-quartz structure or in glass. The effect of different coordination is seen in comparing the Mg^{2+}-Fe^{2+} substitution in olivine and pyroxene with that in garnet, with the

the vibrations of the rather small Cd^{2+} ion in the large central site of the perovskite (Neil et al., 1971). Similar arguments can be made for $MgSiO_3$ perovskite, and the ilmenite-perovskite transition in mantle substitution in 8-fold coordination much less favorable than that in 6-fold. This reflects the relative instability, due to ligand field effects, of Fe^{2+} in 8-fold coordination and argues for a low stability for phases like $FeSiO_3$ perovskite, (see section 4.5 and Burns, this volume). Indeed, irregularities or surprises in thermochemical trends may be clues to differences in local coordination, and conversely.

On the basis of the above, several predictions can be made, which need verification by direct thermochemical measurements. (1) The substitutions $Si^{4+}_T \rightarrow Al^{3+}_T + M^+_A$ (M = Cs, Rb, K, Na, Li) should have enthalpies comparable to those given in Table 1 for a variety of framework silicates, amphiboles, micas and clay minerals. (2) For alkaline earths such predictions will be less reliable, both because the overall stabilization is less and because the variation in individual cation coordination (M^{2+} in A or M sites) can be greater. (3) For clays and micas with K (or Rb or Cs) in much larger sites than Na or Li, these systematics may need to be modified. (4) The energetics of homovalent substitution (Si→ Ge, Mg→ Fe, Mn, Co, Ni, Zn, etc.) will depend strongly on the coordination number (4, 6, or 8) the cation assumes. The salient point is that a general pattern of systematic trends does exist and can be used to predict, with due caution, unknown thermodynamic properties.

4. PHASE TRANSITIONS AT HIGH PRESSURE AND TEMPERATURE

4.1 General considerations

As temperature increases, a phase transition will occur when the positive $T\Delta S$ term compensates for less favorable energetics. Thus entropy is the driving force for the transition. Thus ΔS for the transition must be positive but the volume change may be positive or negative (e.g., a positive ΔV for the melting of NaCl, a negative ΔV for the melting of H_2O). When pressure increases, a phase transition occurs when the negative $P\Delta V$ term overcomes the initially unfavorable free energy. At a phase boundary with equilibrium at a given P,T,

$$0 = \Delta G\ (P,T) = \Delta G°(T) + \int_{1\ atm}^{P} \Delta V\ dP$$

$$= \Delta H°(T) - T\Delta S°(T) + \int_{1\ atm}^{P} \Delta V\ dP. \qquad (6)$$

Thus, the high pressure phase has to be the phase of lower volume but the entropy of the transition from low to high pressure phase can be positive or negative.

Since the P-T slope of a univariant phase transition is given by the Clausius-Clapeyron equation,

$$(dP/dT)_{equil} = \Delta S/\Delta V, \qquad (7)$$

a negative dP/dT results when ΔS and ΔV have opposite signs.

An approximate limit can be placed on how large $\Delta H°$ can be for a phase transition to occur. If the commonly accessible pressure range is 0-150 GPa, and the maximum volume contraction in a solid=solid phase transition is on the order of -10 cc mol^{-1}, the maximum value of $\Delta G° \sim P\Delta V°$ is 1.5 x 10^3 cc GPa or 152 kJ. Since for such a transition, $\Delta S°$ is likely to be small, this represents the maximum value of $\Delta H°$ also. Similarly, for a phase transition with increasing temperature at constant P, $\Delta S°$ must be positive and is unlikely to be greater than 20 JK^{-1} mol^{-1} for a solid-solid transition. If the highest accessible temperature is on the order of 3000 K, then the maximum value of $\Delta H°$ would be 60 kJ. Thus phases energetically unstable by greater than about 150 kJ are unlikely to be stabilized by either pressure or temperature, and, in most cases (see below), $\Delta H°$ for a phase transition is in fact much smaller than these limits.

The sections below will deal with some of the systematics of the thermochemistry of phase transitions relevant to the mineralogy of the earth's mantle. A complementary approach to similar problems is given by Jeanloz (this volume).

4.2 Entropies of high pressure phase transitions

The following arguments present evidence that positive entropies of transition and negative dP/dT slopes can be expected to become more frequent for transitions occurring in the 15-100 GPa region (Navrotsky, 1980).

The common mineral structures at and near atmospheric pressure (silica and its derivatives, feldspar, spinel, olivine, pyroxene, amphibole) represent a relatively loose packing of metal atoms and aluminosilicate units, with densities less than those of mixtures of corresponding binary oxides. Transitions among these phases involve, to a first approximation, a rearrangement of the packing of these isolated units, with a decrease in the empty space between them in the higher pressure polymorphs. The only changes in cation coordination occurring in the 0-7.5 GPa range are those for aluminum (4-fold to 6-fold) and in the distorted coordination polyhedra of large alkali and alkaline earths. Under these conditions, the entropy of a phase correlates well with its molar volume, and ΔS and ΔV have the same sign. At higher pressures, new phases with higher density are produced by increasing the coordination number of other cations (6-coordinated silicon in stishovite and silicate ilmenites and perovskites, 8- and 12-coordinated divalent ions in garnets and perovskites). The overall increase in density is achieved by a denser and more symmetrical packing of polyhedra, but is accompanied by an increase in nearest neighbor bond distances. Therefore with increasing coordination number and bond length, the individual MO bonds weaken. These changes can lead to a higher entropy for the denser polymorph, which has been found experimentally for a number of phases. The high standard entropy of pyrope garnet may be related to the vibration of Mg in dodecahedral sites. The high entropy of $CdTiO_3$ and $CdSnO_3$ perovskites, resulting in a negative dP/dT for the ilmenite-perovskite transition, may arise from silicates is expected to have a negative P-T slope. The positive $\Delta S°$ for the NaCl → CsCl transition in halides with small radius ratio is another example of this phenomenon (see section 4.3), as is the virtually zero entropy change for the ZnO transition (zincite → rocksalt) (Davies and Navrotsky, 1981), for which the volume change is -20%. The directional nature of the bonding in ZnO (and in diamond-related structures in general) results in high Debye temperatures and rather low standard entropies. The magnitude of $\Delta S°$ for the zincite → rocksalt transition then may reflect this low standard entropy of the zincite phase rather than an anomalously high vibrational entropy of the rocksalt phase. Analogous behavior might be seen in silicates for transitions from structures with well-defined covalently bonded SiO_4 tetrahedra at low pressure to structures with octahedrally coordinated Si forming part of a structure of high symmetry at high pressure. Stishovite, the rutile polymorph of SiO_2, may be considered as an intermediate case. It has octahedrally coordinated silicon but a structure of relatively low symmetry, well defined SiO_6 octahedra, and considerable covalency. The coesite-stishovite transition has a small positive P-T slope. The hypothetical transition of SiO_2 (stishovite) to the highly symmetrical fluorite structure is, by all the above arguments, a likely candidate for a negative P-T slope.

The trend toward higher entropies for high pressure silicates with increasing cation coordination number can be related to systematic changes in the vibrational spectrum. Specifically, several factors are important. (1) The high pressure phase often has higher symmetry and a smaller primitive unit cell than the low-pressure phase. Thus the low frequency acoustic modes comprise a greater fraction of the total number of degrees of freedom and increase $S°_{298}$. (2) High frequency tetrahedral SiO stretching modes (near 800-1100 cm^{-1}) are transformed into mid-frequency lattice modes (400-600 cm^{-1}) associated with lattice vibrations involving bonds of the SiO_6 octahedra. (3) Low frequency modes involving vibrations associated with large distorted MO_8 groups (in garnet and perovskite) may occur. The overall effect is to compress the vibrational spectrum into a lower range of frequencies and, indeed, to make the high pressure phase deviate less from a Debye solid, leading to a generally increased entropy. Because of the interplay of complex factors, crossovers in heat capacities as a function of temperature are frequently observed when comparing phases of quite different structures.

Other factors may also increase the entropy of high pressure phases. O'Keeffe and Bovin (1979) have suggested that $MgSiO_3$ perovskite may become a solid electrolyte (oxide ion conductor) at high temperature, in analogy to the behavior of the fluoride $NaMgF_3$. The result is an increase in the enthalpy and entropy of the conducting solid phase, and a decrease in its enthalpy and entropy of fusion. Possible solid electrolyte behavior would further increase the entropy of silicate perovskite phases at high temperature, as well as affecting their physical properties. Such effects are also likely in phases with the fluorite structure. Configurational disorder would also raise the entropy of a high pressure phase. $MgGeO_3$ and $ZnGeO_3$ ilmenites have been reported to undergo transitions to a corundum, i.e., disordered ilmenite, phase (Ito and Matsui, 1979). The likely large configurational entropy of this disordered phase will almost certainly result in a negative P-T slope for that transition.

Pressure affects the electronic band structure of materials. When metal-metal distances shorten, band gaps generally narrow. Semiconductor-metal transitions have been inferred at very high pressure in oxides, including Fe_2O_3, Cr_2O_3, TiO_2, NiO, and others (Kawai et al., 1977). A number of semiconductor-metal transitions are known at atmospheric pressure. These include the solid-solid transitions in NiS, Sn, and VO_2 and the solid-liquid transitions observed in the melting of Si, Ge, and III-V semiconductors. In these cases, the metallic phase is invariably the high-temperature phase, implying a positive $\Delta S°$, consistent with the contribution from electron delocalization. When collapse of a covalent structure framework occurs, this positive $\Delta S°$ is accompanied by a large negative $\Delta V°$, and dP/dT is strongly negative. When the structural rearrangement consists only of a small crystallographic distortion, as in NiS and VO_2, the volume change is very small. For VO_2 Pintchovski et al. (1978) have argued that changes in the band structure are responsible for about two-thirds of the $\Delta H°$ and $\Delta S°$ of the transition, with the lattice distortion accounting for the remainder.

Bukowinski and Hauser (1980) have suggested that a change in the band structure occurs at high pressure for many alkali and alkaline earth elements which drops a band of "d" symmetry below the Fermi level, causing the metals to exhibit increasing transition-metal character. This will change the chemical behavior of these elements and perhaps permit their concentration as chalcophiles in the core. For such a transition, which need not be first order, the overall volume change is negative and the overall entropy change is expected to be positive, reflecting the greater density of states of the d band. Thus, for these various insulator-metal and metal-metal transitions, negative P-T slopes appear probable. In particular, if the earth's core contains an Fe alloy with substantial amounts of light elements (e.g., H, O, C, and/or Si), then the distribution equilibria of these elements between crystalline silicate phases and the molten outer core will involve a change from a predominantly ionic environment to a predominantly metallic one. Thus, at the high temperatures involved, one can expect a large positive $T\Delta S$ term to play a major role.

4.3 Stability of rocksalt, cesium chloride, and nickel arsenide structures

Crystal chemical and thermodynamic systematics can be used to assess the possible energetics of transformation of (Fe,Mg)O from the rocksalt (B1) structure to either the cesium chloride (B2) or nickel arsenide (B8) structure types (Navrotsky and Davies, 1981). Such reactions may be possibilities for the lower mantle.

The transition from NaCl (octahedral coordination) to CsCl (cubic coordination) is well known in halides. This transformation also occurs in BaO and CaO and in BaS, although BaO transforms to a slightly distorted form of the CsCl structure. The radius ratio seems to determine the volume change (see Fig. 5). Thus ΔV for the transition is not constant, but ranges from near -15% (CsCl, KF) to about -3% (NaCl). Jamieson (1977) suggested that MgO in the rocksalt and cesium chloride structure would have virtually the same volume, making that transition highly unlikely.

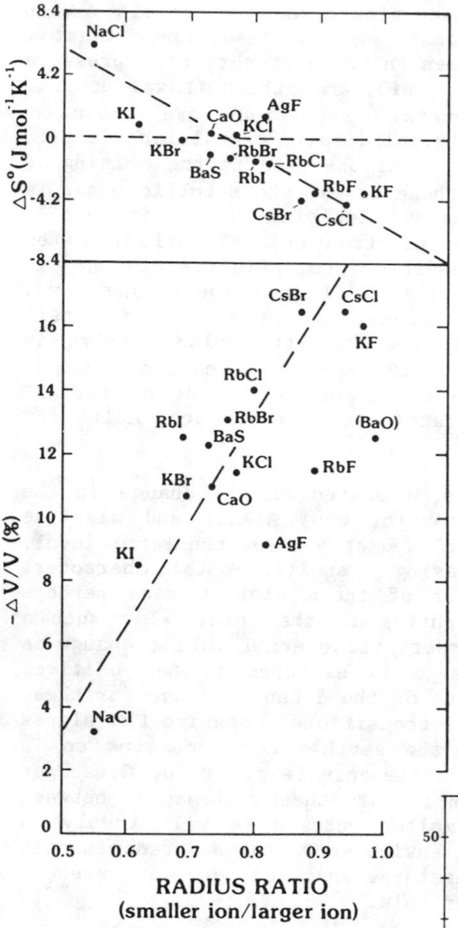

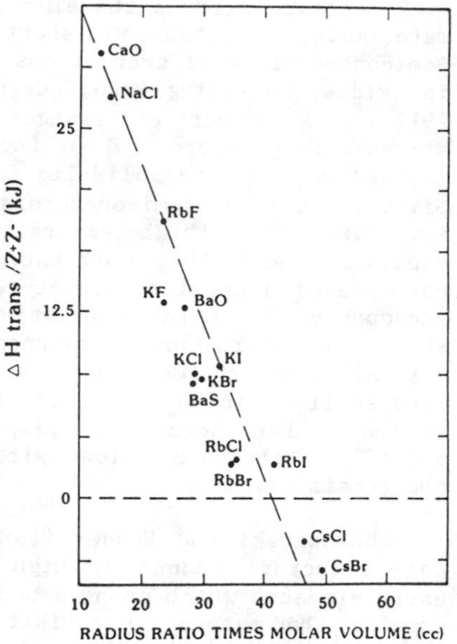

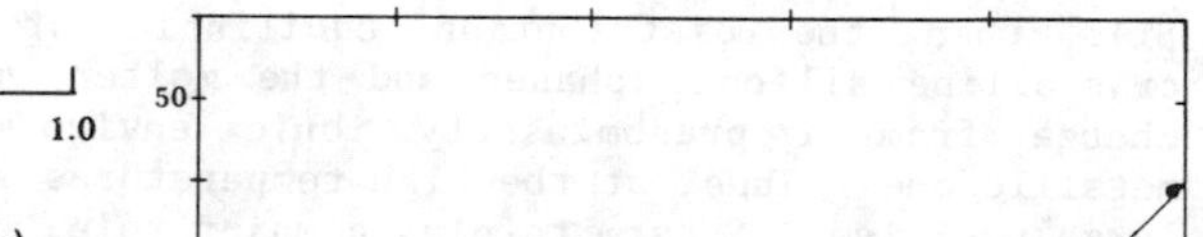

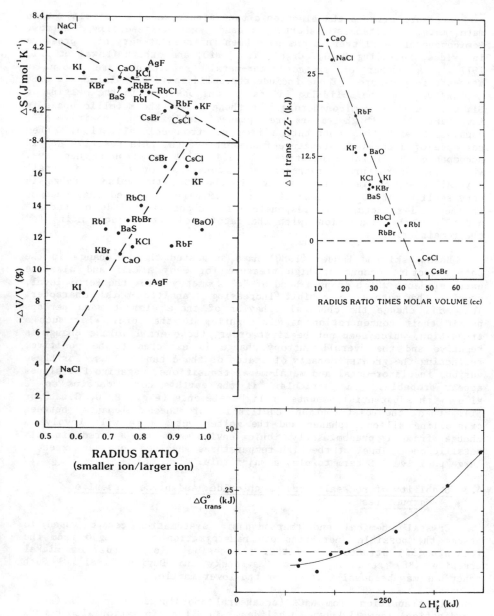

Figure 5 (upper right). Volume and entropy change for the rocksalt to cesium chloride transformation plotted against radius ratio (Davies and Navrotsky, 1981).

Figure 6 (upper left). Calculated enthalpy of the rocksalt to cesium chloride structure divided by the product of the ionic charges plotted against the product of the radius ratio and the molar volume of the rocksalt phase (Davies and Navrotsky, 1981).

Figure 7. Standard free energy of the rocksalt to nickel arsenide transition plotted against the standard enthalpy of formation of rocksalt phase at 298 K (Davies and Navrotsky, 1981).

For transition-metal oxides, the CsCl structure would be destabilized both by the predicted small volume change (analogous to MgO) and by the loss of crystal field stabilization energy (see Burns, this volume). This would make a transition to the CsCl structure unlikely for MgO, FeO, CoO, and NiO.

Bassett et al. (1969) suggested that the entropy change of the NaCl → CsCl transition is strongly correlated with the percentage volume change, and thus with radius ratio. Indeed, their correlation shows that for halides with small cations $\Delta S°$ for the NaCl → CsCl transition is positive, while for halides with cations and anions of comparable size it is negative (see Fig. 5).

Bassett's entropy correlation and the free energies of transition from high-pressure and solid solubility data may be used to estimate the enthalpies of the NaCl → CsCl transition. These are shown in Figure 6 as a function of the product of the radius ratio and molar volume of the rocksalt phase. This product, rather than the radius ratio alone, is used to plot the data in order to account for the observation that the enthalpy (and free energy and pressure) for the transition generally increases in magnitude from bromides to chlorides to oxides. This is presumably related to the larger magnitude of the lattice energy in materials having a smaller interatomic separation, which leads to a larger difference in energy between polymorphs. BaO, BaS, and CaO fall on the same linear trend if their estimated enthalpies of transition are divided by four, the product of the ionic charges. The data suggest an enthalpy of transition of greater than 200 kJ mol^{-1} for MgO, which, coupled with the small (or maybe even positive) ΔV (Jamieson, 1977), further supports the contention that for MgO the CsCl phase is not likely to exist.

The transition rocksalt → NiAs occurs at atmospheric pressure in MnSe and MnTe. The NiAs structure occurs in transition metal chalcogenides, while the rocksalt structure is limited to the alkaline-earth chalcogenides and to oxides and halides.

Can an analogous transition occur in oxides at high pressure? Compared to the rocksalt structure, the NiAs structure represents a smaller degree of ionicity and a larger contribution of metal-metal bonding. For a series of MX oxides or chalcogenides, this decrease in ionicity, uncompensated by any strong covalency as seen in tetrahedral compounds, can be expected to lead to a diminished stability of the compound relative to the elements. Therefore one may expect the NiAs structure to become more competitive, relative to the rocksalt structure, as the enthalpy of formation becomes less negative, since both reflect a relative decrease in ionicity and an increase in the contributions of electron delocalization (Navrotsky and Davies, 1981).

Figure 7 shows the available data for the standard free energy, $\Delta G°$(trans), for the rocksalt → NiAs transformation plotted against the standard enthalpy of formation, ΔH_f°, of the rocksalt phase. A smooth trend can be used to estimate the free energies of transition for the oxides. The results show that the NiAs structure appears plausible for FeO under lower mantle conditions, but would be possible for MgO, only at pressures near 300 GPa. Thus no phase transition in MgO appears likely in the mantle.

4.4 Relations among olivine, spinel and modified spinel phases

The structural basis of the stability of these three polymorphs has been discussed by several authors (Kamb, 1968; Tokonami et al., 1972; Sung and Burns, 1978; Akaogi et al., 1984; Burns, this volume; Price and Parker, 1984). Molar volumes, compressibilities, and thermal expansion coefficients increase in the order γ, β, α, as do vibrational entropies. Therefore the stability field of the β-phase, when it exists, is predicted to be limited to high temperatures, and negative entropies are predicted for the α-β, β-γ, and α-γ transitions, in accord with the observed positive P-T slopes of these transitions. Ligand field effects are predicted to stabilize the spinel more than the olivine, with the modified spinel presumably intermediate (see Burns, this volume). They may be the major reason for the much lower α-γ transition pressures for Co_2SiO_4, Fe_2SiO_4, and Ni_2SiO_4 than for Mg_2SiO_4 (Syono et al., 1971).

Recent calorimetric data (see Table 2) confirm this general picture but present some additional details. Firstly, ΔH° ($\alpha \rightarrow \gamma$) increases in the order Fe_2SiO_4, Ni_2SiO_4, Co_2SiO_4, Mg_2GeO_4, Mg_2SiO_4, while the pressure for the transition (near 1000 K) increases in the order Mg_2GeO_4, Ni_2SiO_4, Fe_2SiO_4, Co_2SiO_4, Mg_2SiO_4. Thus the internal energy is not the only important factor in determining variations in the transition pressure, because ΔS° and ΔV° for the transition are not constant. Indeed, ΔS° varies by almost a factor of four (compare Ni_2SiO_4 with Fe_2SiO_4 in Table 2).

The calorimetric data show that β-Co_2SiO_4 and β-Mg_2SiO_4 are intermediate in enthalpy and entropy between olivine and spinel. This correlates with the fact that the β-phases, in terms of bond lengths and regularity of coordination are intermediate between the corresponding α and γ-phases. At present there is no convincing explanation why the β-phase occurs for Mg_2SiO_4, Co_2SiO_4 and Mn_2GeO_4 but not for Fe_2SiO_4, Ni_2SiO_4, Mn_2SiO_4, Zn_2SiO_4, or Mg_2GeO_4. Phases similar to the β-phase occur in the Ni_2SiO_4-$NiAl_2O_4$ system. A series of phases related to the spinel structure by different stacking sequences of slabs of the spinel structure (Horiuchi et al., 1982), and referred to as spinelloids, form at various compositions. They are less stable in enthalpy and larger in volume than the corresponding spinels, so entropy, (perhaps related to Ni,Al distributions) must play a dominant role in their stability (Akaogi and Navrotsky, 1984a). The situation may be further complicated by intergrowths of stacking sequences in $NiAl_2O_4$-Ni_2SiO_4 spinelloids seen by high resolution transmission electron microscopy (Davies and Akaogi (1983) though such intergrowths were not seen in β-Mg_2SiO_4 (P.K. Davies, pers. comm. 1984).

Vibrational calculations using data from IR and Raman spectra for α, β, and γ-Mg_2SiO_4 confirm the pattern of entropies inferred from a combination of calorimetry and high pressure synthesis (Akaogi et al., 1984). Thus β-Mg_2SiO_4 is a phase intermediate in lattice entropy, energy, and volume to α and γ.

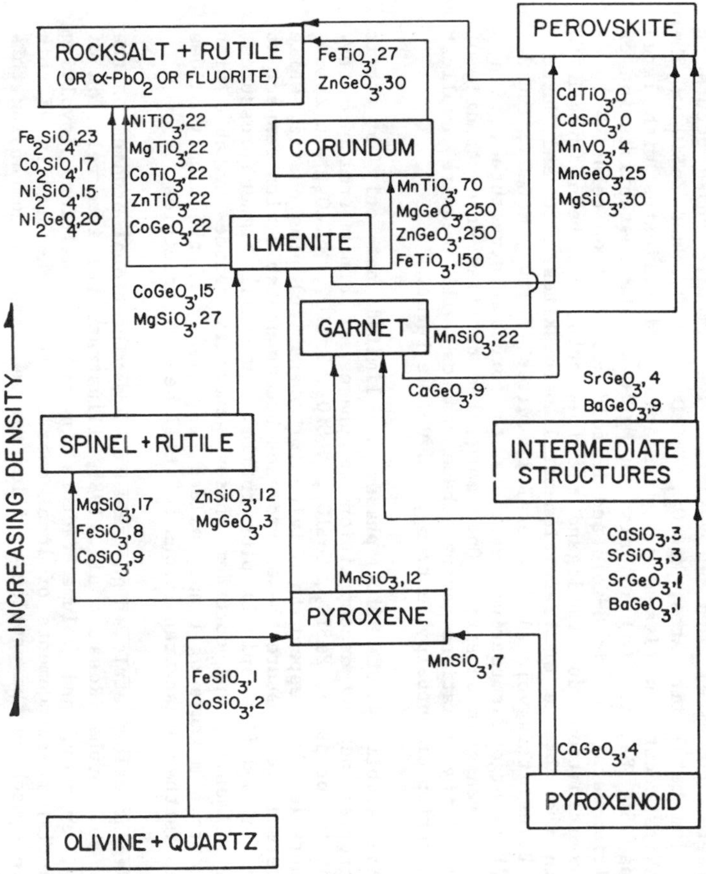

Figure 8. Schematic representation of observed phase transitions of ABO_3 and related compounds. Specific examples of each transition and their pressures in GPa are shown (Navrotsky, 1981).

Table 2

Thermochemical Parameters for α,β,γ Transitions in Silicates and Germanates

	$\Delta G°(T)$ ($J\ mol^{-1}$)	$\Delta H°_{1000}$ ($J\ mol^{-1}$)	$\Delta S°_{1000}$ ($J\ mol^{-1}\ K^{-1}$)	$\Delta V°_{298}$ ($cm^3\ mol^{-1}$)
$\alpha\rightarrow\gamma$				
Mg_2SiO_4 [a]	20417 (1000)	36777	-16.74	-4.13
Co_2SiO_4 [a]	24395 (1000)	11255	-13.14	-3.92
Fe_2SiO_4 [a]	20511 (1000)	2941	-17.57	-4.35
Ni_2SiO_4 [b]	11803 (1000)	5983	-5.82	-3.42
Mg_2GeO_4 [b]	0 (1086)	-12678	-11.67	-3.52
Fe_2GeO_4 [b]	-29706 (1300)			-4.07
Co_2GeO_4 [b]	-14644 (1473)			-3.89
Ni_2GeO_4 [b]	-34308 (1473)			-3.12
$\alpha\rightarrow\beta$				
Mg_2SiO_4 [a]	40417 (1000)			
Co_2SiO_4 [a]	18036 (1000)	29957	-10.46	-3.24
Fe_2SiO_4 [a]	27941 (1000)	8996	-9.04	-2.90
Mn_2GeO_4 [b]	15949 (1273)			-3.31
$\beta\rightarrow\gamma$				
Mg_2SiO_4 [a]	13100 (1000)	6820	-6.28	-0.89
Co_2SiO_4 [a]	6359 (1000)	2259	-4.10	-1.02
Fe_2SiO_4 [a]	-7430 (1000)	-13410	-5.98	-1.04

[a] summarized in Akaogi and Navrotsky (1984)
[b] from Navrotsky, 1983a,b; Navrotsky et al., 1979

4.5 Garnet, ilmenite, perovskite, and related dense phases

Figure 8 shows schematically the observed transitions in silicates, germanates, and titanates and gives approximate pressures at about 1273 K for specific transformations. The rich variety of transition sequences suggests that the energetic, entropic, and volumetric factors which stabilize one structure versus another are very finely balanced. The following generalizations can be made. Pyroxenoid-pyroxene stability relations appear to be determined largely by cation size; thus $MnSiO_3$ transforms first to a pyroxene structure, while the pyroxenoids of the larger cations (Ca, Ba, Sr) transform directly to much denser modifications. Garnet occurs relatively rarely for Al-free systems ($CaGeO_3$ and $MnSiO_3$ are examples), but that structure is greatly stabilized by the incorporation of Al_2O_3. The garnet structure, with formula $(MTO_3)_4$, can be written as having the structural formula $A_3M_2T_3O_{12}$, where A represents 8-coordinated sites, M octahedral sites, and T tetrahedral sites. Stabilization by aluminum then arises from the ability of Al to assume octahedral coordination at lower pressure than silicon, and the idealized aluminosilicate garnet formula is $3/4(MSiO_3)$-$1/4(Al_2O_3)$ or $M_3Al_2Si_3O_{12}$. At high pressure, solid solubility up to about 0.6 $M_4Si_4O_{12}$-0.4 $M_3Al_2Si_3O_{12}$, that is, 0.9 $MSiO_3$-0.1 Al_2O_3, is observed (Akimoto and Akaogi, 1977). However, since the garnet phase still contains tetrahedrally coordinated silicon, it is bounded at high pressure by phases containing all their silicon in octahedral coordination.

These high-pressure ternary phases are ilmenite, corundum (disordered ilmenite), and perovskite. The ilmenite phase occurs in transition metal titanates, $MgGeO_3$, $CoGeO_3$, $ZnGeO_3$, and $ZnSiO_3$. The perovskite occurs in titanates, germanates, and silicates with large cations (Sr, Ba, Ca, Cd) and also in $MgSiO_3$. The destabilization of transition-metal ions in 8-fold and 12-fold coordination relative to 6-fold coordination, due to ligand field effects, has been invoked to explain the absence of transition-metal perovskites (Ito and Matsui, 1979; Burns, this volume). Several ilmenites ($MnTiO_3$, $MgGeO_3$, $ZnGeO_3$, $FeTiO_3$) undergo disordering reactions to corundum structures, with a slight increase in density. Once more, the detailed understanding of the stability relations among these phases will require explicit consideration of enthalpy, entropy, and volume terms.

The stability of $MSiO_3$ phases is limited not only by their polymorphism but by decomposition to phases of other stoichiometry. Thus at low pressure, $FeSiO_3$, $CoSiO_3$, $NiSiO_3$, $ZnSiO_3$, $NiGeO_3$, and $ZnGeO_3$ are unstable with respect to olivine (or phenacite) plus quartz, while at intermediate pressure these pyroxenes become unstable with respect to spinel (or modified spinel) plus rutile. At still higher pressures, spinel, ilmenite, and corundum phases generally become unstable with respect to a mixture of binary oxides, especially if the binary oxides undergo further phase transitions to still denser structures.

Recent calorimetric, high pressure, and vibrational studies of the $CaGeO_3$ polymorphs (Ross et al., 1985) illustrate the interrelation of energy, entropy, and volume factors in phase transitions involving substantial rearrangements of local coordination. The details of this study are used as the problem or illustrative example accompanying this chapter.

5. CONFIGURATIONAL ENTROPY IN SOLID STATE REACTIONS

Many phases are stable at high temperature because of their higher entropy which more than compensates for their higher energy relative to lower temperature phase assemblages. The melting of a solid and the vaporization of a solid or liquid are obvious examples of such entropy-driven reactions, but in many instances involving only crystalline phases, entropy effects are equally important. Three major factors contribute to the entropy of a phase: lattice vibrations, positional disorder, and magnetic and electronic disorder. This section stresses the configurational entropy arising from partially disordered cation distributions in minerals.

5.1 General considerations

Consider a mineral with several (j) sublattices, each containing n_j moles of sites per mole of formula unit. If each sublattice is completely filled by only one kind of ion or completely empty, the mineral has no configurational entropy. If several (i) ions are distributed over the available sites such that their mole fraction on each sublattice is X_i, and if on each sublattice the various ions are distributed at random, then the configurational entropy is given by

$$\Delta S(\text{conf}) = -R \sum_j n_j \sum_i X_i \ln X_i . \qquad (8)$$

At low temperature ($T \rightarrow 0$ K) a completely ordered state is stable (though may be unattainable kinetically) while as temperature increases, the free energy of the phase can be lowered by partial disordering in which the enthalpy of disordering is balanced by the $T\Delta S$ term arising from ΔS(conf) above (modified by any vibrational entropy changes associated with disordering and by any short range order on individual sublattices). Thus in principle, if one could identify the dependence of the enthalpy of disordering on temperature, pressure, composition and on the extent of disorder, one could predict the equilibrium thermodynamic state. This dependence can be related to nearest, next nearest and more distant neighbor interactions on an atomic scale, and some of the complexity of the resulting behavior, leading to both first order and higher order phase transitions, is summarized in the chapter by Carpenter (this volume). In this section I present some systematics of the energetics of disordering reactions other than those involving Al and Si on tetrahedral sites which are discussed in Carpenter's chapter.

A useful distinction can be made between convergent and non-convergent disordering. In the former, the inequivalent sites in the ordered phase became crystallographically indistinguishable when occupied by a random distribution of cations in the totally disordered phase. Examples include Si,Al disorder in albite and Ca,Mg disorder in dolomite. In non-convergent disordering, the sites remain crystallographically distinct even when occupied by a random distribution of ions. Examples include octahedral-tetrahedral disorder in spinels, and M1-M2 disorder in the pseudobrookite, olivine, and pyroxene structures.

5.2 Spinels

The spinel unit cell is face-centered cubic, space group Fd3m. Since the anions occupy general positions, both the lattice parameter, "a", and a parameter, generally designated "u" and known in oxide spinels as the oxygen parameter, are required for a complete description. The cation-anion distances, R, are given by:

$$R_{tet} = a\left(\sqrt{3}(u - 1/8)\right), \tag{9}$$

$$R_{oct} = a(3u^2 - 2u + 3/8)^{1/2}. \tag{10}$$

Thus the octahedral and tetrahedral bond lengths can be used to determine the two structural parameters, a and u, and vice versa. A perfectly normal spinel is one in which the single A cation of the formula unit occupies the tetrahedral sites, and the two B cations the two equivalent octahedral sites. If the symbols () and [] denote tetrahedral and octahedral sites, respectively, then this distribution is $(A)[B_2]O_4$. The distribution $(B)[AB]O_4$, called inverse, is also possible, as are all distributions falling between these two extremes. One can define an additional parameter, x, the degree of inversion, which is the fraction of B ions occupying tetrahedral sites. Thus x may vary between 0 for the perfectly normal case and 1 for the perfectly inverse case. Of special note is the completely random arrangement, $(A_{1/3}B_{2/3})[A_{2/3}B_{4/3}]O_4$, with x = 0.667.

A thermodynamic treatment of cation distributions in simple spinels was presented by Callen et al. (1956) and by Navrotsky and Kleppa (1967). The configurational entropy is given, assuming completely random mixing of ions on each site by:

$$S_c = -R[x \ln x + (1-x) \ln (1-x) + x \ln (x/2) + (2-x) \ln (1-x/2)]. \tag{11}$$

If we write the changes in internal energy and volume on disordering as ΔU_D and ΔV_D, and the change in the nonconfigurational entropy as ΔS_D, then the change in free energy on disordering, ΔG_D, is:

$$\Delta G_D = \Delta U_D - T(\Delta S_C + \Delta S_D) + P\Delta V_D \tag{12}$$

At equilibrium

$$\partial \Delta G_D/\partial x = 0 \tag{13}$$

and

$$\ln[x^2(1-x)^{-1}(2-x)^{-1}] = -(RT)^{-1}(\partial \Delta U_D/\partial x) + R^{-1}(\partial \Delta S_D/\partial x) - P(RT)^{-1}(\partial \Delta V_D/\partial x). \tag{14}$$

ΔV_D is generally sufficiently small so that the last term in the above will be negligible at low or moderate pressures. The term inside the logarithm on the left hand side of Equation 14 is equivalent to an equilbrium constant, K, for the interchange reaction (Navrotsky and Kleppa, 1967).

$$(A) + (B) = (B) + [A], \tag{15}$$

thus:

$$K = \frac{[A](B)}{(A)[B]} = \frac{x^2}{(1-x)(2-x)} \cdot \tag{16}$$

Dunitz and Orgel (1957) and McClure (1957) showed that a number of the observed cation distributions in spinels could be explained by crystal field theory. However, in about half of the cases where this model can be applied, it does not predict the observed cation distribution correctly. Nor can it be extended easily to the many spinels (e.g., $MgAl_2O_4$, $ZnFe_2O_4$, Mg_2TiO_4) which do not contain ions with partially filled "d" shells. Price et al. (1982) used pseudopotential orbital radii to distinguish between normal and inverse structures. They make the point that this approach links concepts of size and of electronegativity, and make the cation distribution depend on bonding parameters similar to those which determine structure in general. Navrotsky and Kleppa (1967) used an essentially empirical approach. They assumed that ΔS_D was negligible, and that ΔU_D was proportional to the degree of disorder, x and to the difference in the "site preference energies", assumed constant for all spinels, of the two ions involved. Values for these site preference energies were derived from a small number of observed cation distributions and applied to predict x in other spinels.

Both Dunitz and Orgel and McClure dismissed the change in lattice energy with cation distribution as being relatively unimportant. However, Glidewell (1976) has shown that in many instances the change in the electrostatic part of the lattice energy with disordering is of much greater magnitude than any crystal field effect. For a crystal composed of point charge ions the internal energy, U, is given by

$$U = U_E + U_R + U_V, \tag{17}$$

where U_E is the electrostatic energy, U_R is an amalgam of other, noncoulombic, terms, including dipole-dipole and higher order "Van der Waals" attractions, and a repulsion term which is usually given an inverse power or exponential form. The sum of these terms is opposite in sign to, and typically about 10-20% of, U_E. U_V is the vibrational energy, which we approximate as being independent of a, u, and x.

The electrostatic term may be expanded into the form:

$$U_E = \frac{Ne^2M}{4\pi\varepsilon_0 a} = 1389\ \frac{M}{a}\ \text{kJ mol}^{-1}, \tag{18}$$

where M is the Madelung constant, which in this case will be a function of the charges on each ion, the cation distribution, x, and the oxygen parameter, u.

Both a and u are uniquely determined by a combination of the octahedral and tetrahedral bond lengths (Eqs. (9) and (10)). Hill et al. (1979) showed that, by using Shannon and Prewitt's (1969) ionic radii, 96.9% of the variation in a in a sample of 149 oxide spinels could be accounted for.

O'Neill and Navrotsky (1983) used this observation of constant bond lengths, R_{oct} and R_{tet}, to compute the change of a and u, and hence of the electrostatic energy, ΔU_E, with x. The data base of Hill et al. has been used to obtain a set of optimized bond lengths and ionic radii for the spinel structure which differ slightly from the radii of Shannon and Prewitt.

Constant ionic radii predict that a and u are linear functions of x. Substituting this into the equations for the electrostatic energy gives a quadratic dependence of ΔU_E on x.

Reasonable assumptions about the variation of the short range force terms with interatomic distance also suggest a quadratic dependence on x. A similar equation was derived by Urusov (1983) using a regular-solution approach to the mixing properties of ions on each sublattice.

Thus, the form to be considered for the change in enthalpy on disordering, ΔH_D, is also a quadratic one,

$$\Delta H_D = \alpha x + \beta x^2, \qquad (19)$$

in which α and β are generally expected to be of approximately equal magnitude and opposite sign. The coefficients α and β will depend mainly on the differences in radii of the two ions, but these coefficients must be determined empirically from observed cation distributions.

Then, using data on cation distributions for binary spinels and spinel solid solutions, values of α and β have been derived (O'Neill and Navrotsky, 1983; 1984). The model can be further simplified by noting that the values of β appear to be approximately constant for each spinel charge type, with values of -20 kJ mol^{-1} for 2-3 spinels and about -60 kJ mol^{-1} for 2-4 spinels. With constant β, the α term can be identified with the difference of site preference energies of ions A and B, analogous to the site preference energies used by Navrotsky and Kleppa (1967). A set of these revised site preference energies (O'Neill and Navrotsky, 1984) is given in Table 3. For transition metal ions having different distributions of "d" electrons among E_g and T_{2g} orbitals in octahedral and tetrahedral coordination, a temperature dependent term related to the difference in electronic entropy is included.

This model, with the interchange enthalpy depending on the degree of disorder, can predict a number of features, especially for spinel solid solutions, which are not predictable from constant interchange enthalpies. Deviations from Vegard's law in lattice parameters of solid solutions between a normal and an inverse spinel can be related to the change in cation distribution with composition. Immiscibility in such normal-inverse solid solutions can be predicted. It results from the change in cation distribution with compositions. Asymmetric solvi in systems such as Fe_3O_4-$FeAl_2O_4$, Fe_3O_4-$FeCr_2O_4$, and Fe_3O_4-Fe_2TiO_4 can be generated without invoking asymmetric Margules parameters for the solid solutions. Activity-composition relations in spinel solid solutions can be modelled. These various predictions are compared with experimental data by O'Neill and Navrotsky (1984).

Table 3
Octahedral Site Preference Energies in Spinels, from O'Neill and Navrotsky (1984)

Cation	ε_i (kJ)	σ_i (JK^{-1})
2+		
Cd	70[a]	0
Co	20	0
Cu	6	- 9.13
Fe	16	- 5.76
Mg	20	0
Mn	45	0
Ni	-28	- 9.13
Zn	53	0
3+		
Al	-36	0
Co L.S.	-375	0
Co H.S.	- 30	- 5.76
Cr	-160	- 9.13
Fe	0	0
Ga	- 4	0
Mn	- 95	- 9.13
Rh	-545	0
V	- 55	0
4+		
Ge	110	0
Si	150	0
Sn	0	0
Ti	50	0

[a]For each cation, the site preference is given by $E_i = \varepsilon_i - \sigma_i T$. The interchange enthalpy for ions A and B, $\alpha_{A-B} = E_A - E_B$. L.S. = low spin, H.S. = high spin

Inverse 2-4 spinels, e.g. titanates, form tetragonal phases at low temperature which have ordered distributions of M^{2+} and Ti^{4+} on octahedral sites (with tetrahedral sites occupied by only M^{2+}) (see Wechsler and Navrotsky, 1984 for discussion). Detailed calorimetric and structural study of Mg_2TiO_4 coupled with analysis of free energies in the $MgO-TiO_2$ system at high temperature has shed some light on the complexity of this order-disorder process (Wechsler and Navrotsky, 1984). The temperature of the tetragonal-cubic transition is 930 ± 20 K. The enthalpy and entropy change at the transition are an order of magnitude smaller than would be expected for complete disordering on octahedral sites. The tetragonal phase prepared at 773 K has a small amount of octahedral site disorder and the cubic material at 950-1200 K is inferred to have considerable short range order on octahedral sites. However at temperatures near 1600 K, the spinel appears to have close to its full configurational entropy, suggesting an essentially random octahedral Mg,Ti distribution. Thus, the change from long range order to short range order to essentially complete disorder occurs over a large temperature range and is not concentrated at or near the crystallographically observable symmetry change.

The disordering of Mg and Ti on octahedral sites is a convergent process, in contrast to non-convergent octahedral-tetrahedral disordering. Thus many of the highly cooperative phenomena discussed by

Carpenter and by McConnell (this volume) may apply to octahedral site order-disorder in Mg_2TiO_4 and the Mg_2TiO_4 order-disorder process may be analogous in some ways to that seen in cordierite (Carpenter et al., 1983).

5.3 Pseudobrookites

These materials of stoichiometry AB_2O_5 contain two nonequivalent octahedral sites, M1 and M2, in a 1:2 ratio. Thus their disordering equilibria may be described by a formalism identical to that developed for octahedral-tetrahedral disorder in spinels. Indeed many pseudobrookites are stable only at high temperature, decomposing to other phase assemblages (ilmenite + rutile or corundum + rutile) at low temperature, and indicating that they are entropy-stabilized phases. Thermochemical and structural data are summarized in Table 4. For $MgTi_2O_5$, the lattice parameters, enthalpy of formation, and cation distribution depend on quench temperatures. Within the uncertainties in the data, the dependence is described adequately by the simple equilibrium model with $\Delta H_{int} = 27 \pm 2$ kJ (Wechsler and Navrotsky, 1984). Because the cation distribution and hence the configurational entropy depend on temperature, an accurate description of the entropy of this phase must include a temperature-dependent configurational term and not simply a constant increment added to the entropy obtained by heat capacity measurements. In addition the rate of equilibration of the cation distribution is very slow at low temperature but rapid above ~1400 K. Thus it is very difficult to determine what portion of the enthalpy and heat capacity associated with changes in cation distribution has actually been measured in drop calorimetric studies of high temperature heat contents. To know this one must combine detailed calorimetric, kinetic, and structural information. Such kinetic complexity may be encountered quite generally in order-disorder reactions in oxides (pseudobrookites, spinels, pyroxenes, ilmenites) and is very relevant to issues such as closure temperatures for proposed geothermometers. The defect chemistry, especially of Fe-bearing phases, can strongly affect kinetics.

5.4 Olivines, pyroxenes, pyroxenoids, amphiboles

These contain two or more nonequivalent M sites and, particularly for Fe-Mg solid solutions, order-disorder is an important feature. Because of their potential use as geothermometers and geospeedometers, Mg-Fe cation distributions have been studied extensively both by Mössbauer spectroscopy and by x-ray structural methods. The result is a wealth of information, some of it contradictory. Recent summaries include review articles by Ghose and Ganguly (1982) and by Hawthorne (1983). In addition, the distribution of other cation pairs, especially Mg and a transition metal (Mn, Co, Ni, Zn) has been studied. Models to systematize and compare the data typically must take into account at least two factors, namely an Mg,Fe interchange free energy (which may depend on temperature) and a description of the mixing parameters in the solid solutions arising from the differences in size and electronic properties of the cations. However, these two effects are strongly interrelated, both theoretically and in terms of strong correlations

Table 4
Thermochemical and Structural Data for Some Compounds with Pseudobrookite Structure

Compound	ΔH_f° (at T,K) kJ mol^{-1}	ΔG_f° (at T,K) kJ mol^{-1}	ΔS_f° (at T,K) kJ mol^{-1}	Cation Distribution x in $(A_{1-x}B_x)_{M1}(A_xB_{2-x})_{M2}O_5$	ΔH° (interchange) kJ mol^{-1}	Decomposition
$MgTi_2O_5$	-14.0 (973)[a]	-28.5 (1573)[a] 30.5 (1673)[a]	11.7 (1573)[a] 13.0 (1673)[a]	0.217 (973)[a] 0.395 (1773)[a] 0.316 (1773)[a]	27.2[a]	$MgTiO_3$ and TiO_2 at 400-500K[a] 500 K[a]
$CoTi_2O_5$		-16.3 (1573)[b]				$CoTiO_3$ and TiO_2 below ~ 823 K[b]
$FeTi_2O_5$		-18.0 (1573)[b]		0.28 (1673)[b]	38-54[b]	$FeTiO_3$ and TiO_2 below 1413 K[b]
Fe_2TiO_5	+ 7.9 (973)[b]	0 (838)[b]				Fe_2O_3 and TiO_2 below 838 K[b]
Ti_3O_5 (high)					-43[b]	low-Ti_3O_5 below ~ 500 K[b]
Al_2TiO_5		endothermic[b]				α-Al_2O_3 and TiO_2 below ~ 1273 K[b]
Ga_2TiO_5		endothermic[b]				β-Ga_2O_3 and TiO_2 below ~ 1273 K[b]
$NiTi_2O_5$	- 8.4 (1673)[b]					unstable relative to $NiTiO_3$ and TiO_2 at all T[b]
$MnTi_2O_5$	- 26.4 (1523)[b]					unstable relative to $MnTiO_3$ and TiO_2 at all T[b]

[a]Wechsler and Navrotsky, 1984
[b]From data summerized by Navrotsky, 1975
[c]Lind and Housley, 1972

among coefficients derived in various regression analyses. Thus empirical parameterizations derived by various investigators can be hard to compare, and parameters for a given system may depend significantly on the data set used. (Ghose and Ganguly, 1982). Nevertheless, a series of exchange parameters can be derived and are summarized in Table 5. Several generalizations can be made. In olivines, the tendency for Fe,Mg ordering is weak and may depend on oxygen fugacity (and Fe^{3+} and defect concentrations). The tendency for M,Mg (M = Ni,Co,Mn, Zn) ordering in olivines is much more pronounced. In pyroxenes, all M,Mg pairs show substantial ordering. It has been suggested (Ghose et al., 1975) that larger ions prefer the larger M2 site in both olivine and pyroxene but that specific covalent bonding interactions cause transition metals to favor M1 in olivine and M2 in pyroxene. For Fe,Mg partitioning, the similarity of $\Delta G°$ terms for pyroxenes and amphiboles having similar M site geometries suggests that local geometry rather than specific crystal structure is dominant in determining site preference in chain silicates.

6. SOLID SOLUTIONS

6.1 General considerations

The driving force for forming a solid solution is usually the configurational entropy of mixing. Destabilizing contributions to the enthalpy of mixing arise from unfavorable interactions between unlike species, related in part to differences in size, which lead to a tendency toward clustering and phase separation. Stabilizing contributions come from favorable interactions between unlike species which lead to a tendency toward ordering and compound formation. The complex behavior resulting from these two competing tendencies has been summarized by Carpenter (this volume). This section describes some of the systematics of solid solutions formation, primarily in "simpler" systems in which the tendency toward order is small.

A solid solution of given composition will only be stable if its free energy is less than that of an equivalent mechanical mixture of its components, or of any possible exsolution products. Figure 9 shows schematic free energy of mixing curves for a binary system. At high temperature, a single phase is stable over the whole composition range but at lower temperature, a region can develop where the system can lower its free energy by separating into two phases, whose compositions are given by points of common tangency to the free energy curve. Note that even when the system has phase-separated, the free energies of the two phases are represented by portions of a single free energy curve. The compositions of coexisting phases define a solvus or binodal (see Fig. 9) which terminates at high temperature in a critical point. In general the miscibility gap need not be symmetric about the 50 mole % composition, though for a regular solution (see below) it is. Within the region marked by the dotted curve (spinodal) in Figure 9, the single phase not only is unstable with respect to exsolution to phases of markedly different compositions, but also, it can lower its free energy by any small compositional fluctuation.

One can represent the heat of mixing of a binary solid solution by,

Table 5

M,Mg Exchange Free Energies for Olivines, Pyroxenes, and Amphiboles

Structure	Exchange Pair	$\Delta G^{\circ}_{exchange}$ (kJ mol^{-1})[a]
Olivine	Mg,Fe	-2.1[b]
	Mg,Mn	+17.2[c]
	Mg,Co	-17.2[c]
	Mg,Ni	-27.2[c]
	Mg,Zn	-7.1[c]
Orthopyroxene	Mg,Fe	+13.0 - 0.001T[d], +31.4[e]
	Mg,Mn	+38.9[c]
	Mg,Co	+10.9[c]
	Mg,Ni	-3.3[c]
	Mg,Zn	+25.5[c]
Cummingtomite (amphibole)	Mg,Fe	+10.1[d]

[a] for reaction Mg(M1) + M(M2) = Mg(M2) + M(M1)
[b] Brown and Prewitt, 1973
[c] Ghose et al., 1975
[d] Ghose and Ganguly, 1982
[e] $\Delta H^{\circ}_{exchange}$ from Chatillon-Colinet et al., 1983, calorimetric results

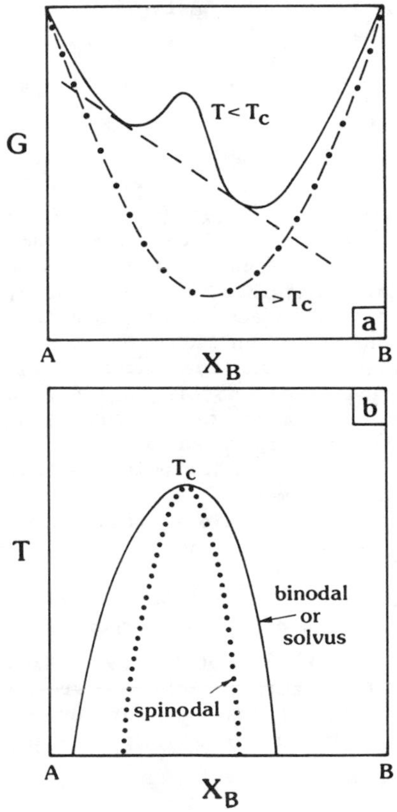

Figure 9. Schematic representation of (a) free energy of mixing and (b) solvus and spinodal in a binary system in which the end members are isostructural.

in order of increasing complexity, a simple regular solution, a subregular or two parameter Margules fit which leads to an asymmetric solvus, or a generalized polynomial in the mole fraction, X,

$$\text{Regular solution:} \quad \Delta H = WX(1-X). \tag{20}$$

$$\text{Subregular or Margules equation:} \quad \Delta H = X(1-X)[W_1X + W_2(1-X)]. \tag{21}$$

$$\text{General polynomial:} \quad \Delta H = X(1-X)(\sum_0^n W_n X^n). \tag{22}$$

One can generalize further by letting the coefficients W depend on temperature. The configurational entropy of mixing, assuming random substitution, is given by, per mole of species being mixed.

$$\Delta S = -R[X \ln X + (1-X) \ln (1-X)] \tag{23}$$

If ΔH is zero, and ΔS is given by Eq. (23), the solution is ideal, and one obtains thermodynamic activities in accord with Raoult's Law

$$a_i = X_i. \tag{24}$$

One can extend this definition of ideality carefully for statistically random mixing over more than one set of sites. Thus for the case of random mixing of Fe and Mg in olivine ideality would imply, since two moles of ions are being mixed per mole of $(Mg,Fe)_2SiO_4$

$$a_{Mg_2SiO_4} = X^2_{Mg_2SiO_4}, \tag{25}$$

but $a_{MgSi_{0.5}O_2} = X_{MgSi_{0.5}O_2}$. (26)

More complex expressions for the entropy of mixing arise when coupled substitutions occur on several sublattices, even when such substitutions are assumed random within each sublattice. Such expressions can be derived from statistical mechanics. As an example, Kerrick and Darken (1975) derived equations for the activities of albite and anorthite in plagioclase solid solutions based on three models (a) mixing of molecular-like $NaAlSi_3O_8$-$CaAl_2Si_2O_8$ units, (b) random mixing of Na and Ca on one sublattice and of Si and Al on another, and (c) an aluminum avoidance model in which Na and Ca mix at random but AlOAl linkages are avoided. The resulting equations are shown in Table 6. They will be discussed further in section 6.2. The major point is that each model is, in a sense, an ideal mixing model, with a random distribution of species under a specified set of conditions. Only the first model results in Raoult's Law ($\Delta\bar{s}_{Ab} = -R \ln X_{Ab}$, $\Delta\bar{h}_{Ab} = 0$, $a_{Ab} = X_{Ab}$, and $\Delta\bar{s}_{An} = -R \ln X_{An}$, $\Delta\bar{h}_{An} = 0$, $a_{An} = X_{An}$, $\Delta S = -R[X_{Ab} \ln X_{Ab} + X_{An} \ln X_{An}]$. The other two models give entropies of mixing larger than the first and result in negative deviations from Raoult's Law. Such negative deviations arise not from any heat of mixing terms but from a larger configurational entropy. Similar considerations hold for certain spinel solid solutions and for coupled Al substitutions in pyroxenes.

Table 6

Configurational Entropies for Different Albite-Anorthite Mixing Models, per mole $Na_xCa_{1-x}Al_{2-x}Si_{2+x}O_8$ (Kerrick and Darken, 1975)

Model	$\Delta\bar{s}_{Ab}$	$\Delta\bar{s}_{An}$
Mixing of Ab - and An-like molecular units	$-R\ln x$	$-R\ln(1-x)$
Random mixing of Na,Ca and of Al,Si	$-R\ln[x(2+x)^3(2-x)/27]$	$-R\ln[(1-x)(2+x)^2(2-x)^2/16]$
Al avoidance but random Na,Ca mixing	$-R\ln[x^2(2-x)]$	$-R\ln[(1-x)(2-x)^2/4]$

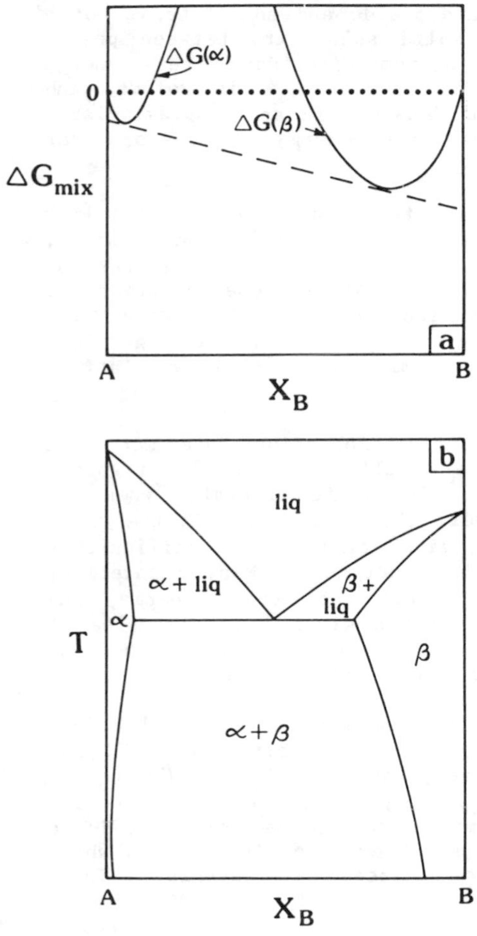

Figure 10. (a) Schematic free energy relations at a given temperature in the system A-B where A has the structure α and B the structure β. The two phases then follow different free energy curves as shown. (b) Schematic phase relations for this system.

When defining statistically ideal mixing involving several sublattices, one has to express very carefully the model and assumptions used. Statistically ideal mixing need not be the same as Raoult's Law. This has been a point of considerable confusion both in the literature and among students.

If one is interested in solid solubility among end members of different structures, the treatment above must be modified. The two end members no longer fall on the same free energy surface in P,T,X space; rather each phase defines its own surface. Complete solid solubility in general cannot occur, but the solubility at a given T and P is governed by free energy relations shown schematically in Figure 10. Since complete solid solubility is not possible, the solid regions terminate in a modified eutectic at high temperature, as shown. The thermodynamic properties of such a system can be described by sets of mixing parameters for each phase, plus free energy terms describing the transformaiton of each end member to the opposite structure or cryptomodification (e.g., Davies and Navrotsky, 1981).

Thus the solid solubility between diopside and enstatite is not a true solvus but represents terminal solid solubility between phases (orthopyroxene and clinopyroxene) having distinct free energy surfaces. Similarly, as discussed by Carpenter (this volume) the (apparently second order) transition between $C\bar{1}$ and $I\bar{1}$ plagioclases makes the use of single continuous free energy expressions for plagioclase solid solutions not strictly correct.

Four main factors on an atomic scale affect the range and stability of isostructural solid solutions. These are size difference, charge, covalency, and specific electronic factors. The first is usually the most important factor; it results in a strain energy which is necessarily larger the greater the difference in size. For a given size difference, it is often easier to put a smaller atom into a larger host structure than vice versa, resulting in asymmetric solubility relations and heats of mixing.

Theoretical models of solid solution formation have met with considerable success in alloy and alkali halide systems. A rigorous calculation of the partition function for ionic crystals, when the nearest neighbors of the cations (namely the anions) remain the same, but the next nearest neighbors change, gives rise to some difficulties in the calculation of pair interaction energies. Many theoretical models for ionic crystals calculate a form of strain energy, and sometimes allow for local relaxation about a given ion. Examples of such models, applied to ionic crystals include the work of Fancher and Barsch (1969), Urusov (1975), Driessens (1968), and many others.

However, for many complex crystals, the ions being mixed occupy only a small fraction of the total volume of the structure. If one compares solid solubility in CaO-MgO (very limited), $CaCO_3$-$MgCO_3$ (quite extensive) and $Ca_3Al_2Si_3O_{12}$-$Mg_3Al_2Si_3O_{12}$ (complete), one is drawn to the conclusion that greater ionic size mismatch can be tolerated by a structure in which the ions being mixed are embedded in a matrix which can itself change geometry slightly to absorb the strain. In a thermodynamic sense, the volume of a phase is a more convenient parameter for correlation than any individual bond length or radius. With these considerations in mind, Davies and Navrotsky (1983) developed a correlation between the excess free energy in a regular or subregular solution model and the volume mismatch for a variety of oxide, chalcogenide, halide, and silicate systems. The use of a volume mismatch, $\Delta V = (V_2-V_1)/V$ term rather than a bond mismatch enables one to consider simultaneously many diverse systems in which divalent ions are being mixed (oxides, chalcogenides, spinels, garnets, olivines, other silicates). The correlation gives

$$W = 100.8\ \Delta V - 0.4 \quad \text{kJ mol}^{-1}. \tag{27}$$

Within this correlation are points for cation mixing on 4-, 6-, and 8-coordinated sites and for anion mixing. Thus, once the effect of coordination number is included in the molar volume of the phases involved, it does not further explicitly affect the correlations.

The alkali halide systems as a group show much smaller positive deviations from ideality than the oxide and chalcogenide systems. This confirms the expectation that more highly charged ions mix less easily than ions of lower charge. A few points (e.g., Cr_2O_3-Al_2O_3, TiO_2-SnO_2) for trivalent and tetravelent ions also support this trend by showing more positive heats of mixing than predicted for divalent systems of comparable volume difference. These correlations, though empirical, allow one to predict the magnitudes of deviations from ideality in a wide range of simple ionic solid solutions.

6.2 The high plagioclases from several points of view

Ion exchange equilibria between high plagioclase and aqueous fluid near 1000 K permit the determination of activity-composition relations in high plagioclase, even though the high structural state is metastable under those conditions. Activities have been calculated by Orville (1972), by Saxena and Ribbe (1972), and by Seil and Blencoe (1979 and personal communication, 1982). These activites at 973 K are shown in Figure 11a. Orville chose to separate his activity plots (see Fig. 11a) into three regions. (a) $0 \leq X_{An} < \sim 0.5$, $\gamma_{An} = 1.27$, $\gamma_{Ab} = 1$ (Henry's Law for An, Raoult's Law for Ab), (b) $0.5 < X_{An} < \sim 0.9$, activity coefficients change with composition, and (c) $0.9 < X_{An} \leq 1$, $\gamma_{An} = 1$, $\gamma_{Ab} = 1.89$. Though Orville did not explicitly discuss the reason for this choice, the separation into two Henry's Law regions separated by an intermediate region inplies that at albite-rich compositions, the anorthite component is substituting into a rather constant albite-like environment, at anorthite-rich compositions an anorthite-like environment persists, and a fairly rapid change in structure and thermodynamic properties occurs at intermediate compositions which are, in fact, close to those for the $C\bar{1}$ - $I\bar{1}$ transition (Carpenter, this volume). Saxena and Ribbe fit Orville's data using a continuous polynomial function for $\ln\gamma$ (see Fig. 11a). Seil and Blencoe, using a more extensive ion exchange data set, fit a

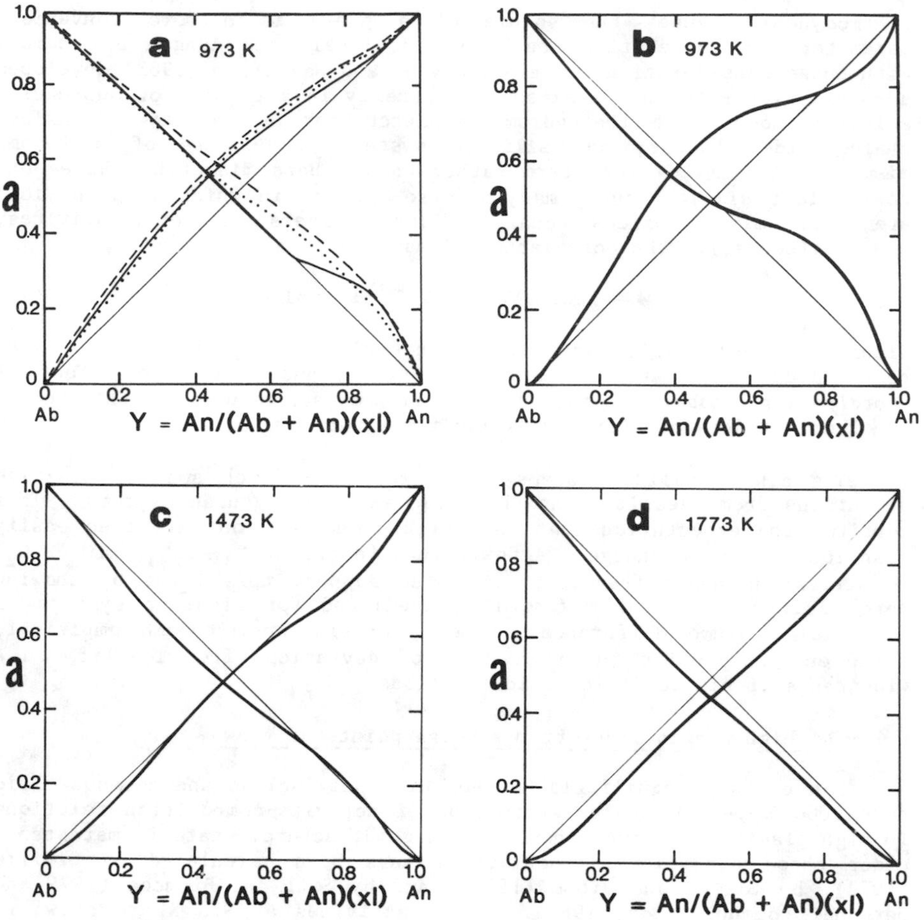

Figure 11. (a) Activity-composition relations in crystalline high plagioclase at 973 K from Orville (1972) (solid curves), Saxena and Ribbe (1972) (dotted curves), and Seil and Blencoe (1978) (dashed curves). (b) Activity-composition relations at 973 K calculated from the enthalpy of mixing of Newton et al. (1980) and the aluminum-avoidance entropy model. (c) Activity-composition relations at 1473 K, calculated as in (b). (d) Activity-composition relations at 1773 , calculated as in (b). Figure modified from Henry et al. (1982).

polythermal polybaric polynomial to describe activity coefficients as functions of composition, temperature, and pressure. These polynomials have no physical significance but represent a good statistical fit. Their use tacitly assumes that a single continuous function describes G(P,T,X) and that its derivatives (activities) also vary smoothly with composition.

Newton et al. (1980) reported enthalpies of mixing in high plagioclases, measured by oxide-melt solution calorimetry. They pointed out that the entropies of mixing calculated by combining the calorimetric data with the activity data of Orville (1972) were in quite good agreement with values predicted assuming complete aluminum avn high plagioclases, measured by oxide-melt solution calorimetry. They pointed out that the entropies of mixing calculated by combining the calorimetric data with the activity data of Orville (1972) were in quite good agreement with values predicted assuming complete aluminum avoidance in the crystals (Kerrick and Darken, 1975). Similar conclusions (Henry et al., 1982) can be reached if one uses the activity data calculated by Saxena and Ribbe (1972) and Seil and Blencoe (1978), and pers. comm. Newton et al. (1980) give the enthalpy of mixing for high plagioclase as:

$$\Delta H = X_{An}X_{Ab}(28225X_{An} + 8473X_{Ab}) \text{ J mol}^{-1} \qquad (26)$$

The resulting expressions for the partial molar enthalpies of mixing are

$$\Delta\bar{h}_{Ab} = X_{An}^2 (-11280 + 39505X_{An}) \text{ J mol}^{-1}, \qquad (27)$$

$$\Delta\bar{h}_{An} = X_{Ab}^2 (47978 - 39505X_{Ab}) \text{ J mol}^{-1}. \qquad (28)$$

Assuming complete aluminum avoidance, but random Ca-Na mixing, the partial molar entropies of mixing are given in Table 6. The activities in high plagioclase may be calculated as

$$RT \ln a_{Ab} = \Delta\bar{h}_{Ab} - T\Delta\bar{S}_{Ab} \qquad (29)$$

and analogously for a_{An}.

The activities at 973, 1473, and 1773 K, calculated from these equations, are shown in Figures 11b-d. At 973 K, the calculated activities are in fairly good agreement with those calculated from ion-exchange equilibria. The small differences can be attributed to errors associated with measurement of the heats of solution, to the small effect of pressure on the activity-composition relations, to possible slight differences in the degree of Al,Si order between plagioclase synthesized hydrothermally and that synthesized from a glass, to possible small departures of the entropy of mixing from that predicted by the aluminum avoidance model, especially in albite-rich plagioclase or to other inadequacies in the model. At 1473 K, the calculated activities in crystalline plagioclase approximate Raoult's Law but show slight negative deviations over part of the composition region. This behavior provides some justification for the use of a Raoult's Law approximation for high temperature plagioclase (e.g. Loomis, 1979). Such "pseudo-ideality" parallels that observed in pyroxenes along the $CaMgSi_2O_6$-$CaAl_2SiO_6$ join (Wood, 1979) and the $NaAlSi_2O_6$-$CaMgSi_2O_6$ join (Wood et al., 1980). The complex shape of the calculated activity curves at temperatures near 1473 K is caused by the competition of entropy and enthalpy terms and also by the fact that Newton et al. (1980) fit the enthalpy of mixing data by an expression whose coefficients cause $\Delta\bar{h}_{Ab}$ to become negative at albite-rich compositions.

The change from mainly positive calculated deviations from Raoult's Law at 800-1100 K, to approximate Raoultian behavior near 1500 K, and finally to negative deviations above 1750 K suggests that Raoult's Law need not be the high temperature limiting behavior. Though this may appear surprising, the reason for it is straightforward. The aluminum avoidance model gives an entropy of mixing different from and larger than the configurational entropy for one mole of mixture, $\Delta S = -R(X_{An} \ln X_{An} + X_{Ab} \ln X_{Ab})$, which forms the statistical basis for Raoult's Law. At low temperatures, the positive heats of mixing found by Newton et al. (1980) cause the overall free energy of mixing to be less negative than the Raoult's Law value, but at high temperature, where the $T\Delta S$ term dominates, the free energy of mixing becomes more negative than that predicted by Raoult's Law. Note that the above formulation ignores order-disorder and assumes that a single continuous expression describes the free energy of mixing.

Carpenter et al. (1985 and this volume) have reinterpreted the solution calorimetric data of Newton et al. (1980) on high structural state plagioclases and their own data on several series of plagioclases of different structural states. Stressing the importance of order-disorder and the $C\bar{1}$ - $I\bar{1}$ higher order phase transition, they suggest that the heat of solution data, rather than being represented by a single polynomial, consist of two distinct regions, each being almost straight lines suggestive of zero heat of mixing, one for $C\bar{1}$ (largely Al,Si disordered) and the other for the $I\bar{1}$ (largely Al,Si ordered) structures (see Fig. 7 in Carpenter, this volume) . The linear trend for $C\bar{1}$ can be extrapolated to pure anorthite to give a heat of solution for the hypothetical $C\bar{1}$ Al,Si disordered $CaAl_2Si_2O_8$. Similarly, the trend for $I\bar{1}$ can be extrapolated (though with greater uncertainty through a much longer composition range) to hypothetical $I\bar{1}$ $NaAlSi_3O_8$. The apparent positive heat of mixing between $C\bar{1}$ high albite and $I\bar{1}$ anorthite then arises from the choice of different structures as standard states. The activity-composition relations shown by Orville (1972) are generally consistent with this interpretation of two separate solid solution segments. This formalism could be extended to calculate activities if one knew the appropriate entropies of transition for $NaAlSi_3O_8$ $(C\bar{1} \rightarrow I\bar{1})$ and $CaAl_2Si_2O_8(I\bar{1} \rightarrow C\bar{1})$ and the entropies of mixing in the $I\bar{1}$ and $C\bar{1}$ solid solutions. These are not sufficiently well-constrained at present and are difficult to model because of likely short range order. The development of such models and the better characterization of structural states, as perhaps by new solid state magic angle spinning nuclear magnetic resonance techniques, represent fruitful areas for the future.

From the above discussion, several conclusions may be drawn. Even when calorimetric data and activity data can be measured for complex solid solutions, their interpretation depends on assumptions made about the appropriate form of equations used to fit the data. Whenever one chooses a set of equations, one tacitly defines standard states and assumes ideal mixing relative to some assumed structural models. Though activity data over a limited P,T,X range can often be fit equally well by a number of equations, extrapolations, especially to higher and lower temperatures, can be very model-dependent. A microscopically valid thermodynamic description still awaits development in both detailed structural information and models of ordering. On the other hand, for petrologic applications at 1273-1473 K, plagioclase activities are approximately given by Raoult's law, although the reasons for this appear to be complex balancing of enthalpy and entropy factors rather than any simple microscopic mixing behavior.

6.3 Mg_2SiO_4-Fe_2SiO_4 phase relations as a function of pressure

To illustrate the interplay of phase transitions and solid solution thermodynamics, a good example is the system Mg_2SiO_4-Fe_2SiO_4, in which olivine (α), modified spinel (β) and spinel (γ) phases must be considered at pressures up to about 20 GPa (Navrotsky and Akaogi, 1984).

Consider equilibrium, at a given P and T, where ferromagnesian olivine and spinel coexist. In the $(Fe_xMg_{1-x})Si_{0.5}O_2$ olivine solid solution, the chemical potential of the $FeSi_{0.5}O_2$ component may be written as:

$$\mu_{FeSi_{0.5}O_2(ol)} = \mu^\circ_{FeSi_{0.5}O_2(ol)} + P\bar{V}_{FeSi_{0.5}O_2(ol)} + RT \ln a_{FeSi_{0.5}O_2(ol)}. \quad (32)$$

In the spinel $(Fe_yMg_{1-y})SiO_{0.5}O_2$, the chemical potential of $FeSi_{0.5}O_2$ is

$$\mu_{FeSi_{0.5}O_2(sp)} = \mu^\circ_{FeSi_{0.5}O_2(ol)} + \Delta H^\circ_{FeSi_{0.5}O_2(ol\to sp)} - T\Delta S^\circ_{FeSi_{0.5}O_2(ol\to sp)} + P\bar{V}_{FeSi_{0.5}O_2(sp)} + RT \ln a_{FeSi_{0.5}O_2(sp)}. \quad (33)$$

At equilibrium at a given pressure and temperature

$\mu_{FeSi_{0.5}O_2(ol)} = \mu_{FeSi_{0.5}O_2(sp)}$. Therefore

$$0 = \Delta H^\circ_{FeSi_{0.5}O_2(ol\to sp)} - T\Delta S^\circ_{FeSi_{0.5}O_2(ol\to sp)} + P\Delta\bar{V}_{FeSi_{0.5}O_2(ol\to sp)} + RT \ln a_{FeSi_{0.5}O_2(sp)} - RT \ln a_{FeSi_{0.5}O_2(ol)} \quad (34)$$

An analogous set of equations can be written for the $MgSi_{0.5}O_2$ component. These equations can be solved for coexisting compositions, x and y, of olivine and spinel, providing one knows the following: the standard enthalpies and entropies of the olivine-spinel transition in the end-members, the volumes of the olivine and spinel solid solutions as functions of pressure, temperature, and composition and the activity-composition relations in the olivine and in the spinel solid solutions (relative to olivine end-members and spinel end members respectively) as functions of pressure, temperature, and composition. The enthalpies, entropies, and volumes of transition are taken from Table 2. In the present example, the partial molar volumes shall be assumed constant.

The activity-composition relations in the olivine and spinel at high T and P are also poorly known. Nishizawa and Akimoto (1973) have suggested that both olivine and spinel show positive deviations from ideality at high pressure, but numerical values of activity coefficients could not be determined. We shall assume that both olivine and spinel mix ideally, i.e. that $a_{FeSi_{0.5}O_2(ol)} = x$, $a_{MgSi_{0.5}O_2(ol)} = 1-x$, $a_{FeSi_{0.5}O_2(sp)} = y$, and $a_{MgSi_{0.5}O_2(sp)} = 1-y$. An alternate approach would be to take the regular solution parameters estimated from empirical correlations (Eqn. 29) discussed above. This would make a small difference in the widths of calculated binary loops. With the above simplifications, the equations for olivine-spinel equilibrium at (P,T) are

$$0 = 1/2\ \Delta H_{Fe_2SiO_4(ol\rightarrow sp)} - 1/2\ T\Delta S_{Fe_2SiO_4(ol\rightarrow sp)} + 1/2\ P\Delta V_{Fe_2SiO_4(ol\rightarrow sp)} + RT\ln y - RT\ln x, \tag{35}$$

and

$$0 = 1/2\ \Delta H_{Mg_2SiO_4(ol\rightarrow sp)} - 1/2\ T\Delta S_{Mg_2SiO_4(ol\rightarrow sp)} + 1/2\ P\Delta V_{Mg_2SiO_4(ol\rightarrow sp)} + RT\ln(1-y) - RT\ln(1-x) \tag{36}$$

At a given temperature Equations (35) and (36) can be solved at various pressures to give pairs (x,y) of coexisting olivine and spinel compositions which define a binary olivine-spinel loop in pressure-composition space.

However, the α-γ loop is metastable at Mg_2SiO_4-rich compositions since a field of stability of modified spinel (β) separates the fields of olivine and spinel (Kawada, 1977; Suito, 1977). Thus the Mg_2SiO_4-Fe_2SiO_4 phase diagram must include stability fields for α, α+β, β, β+γ, and γ. These can be obtained by calculating the α-γ, α-β, and β-γ loops and taking the stable portions of each to construct the phase diagram. The resulting Mg_2SiO_4-Fe_2SiO_4 phase diagram at 1273 K is shown in Figure 12. The calculated region of stability of β-phase is limited to Mg_2SiO_4-rich compositions. The calculated two phase field (α+β) covers a narrow pressure range (< 0.3 GPa) while the field (β+γ) is much wider, spanning on the order of 1.8 GPa at Mg/(Mg+Fe) = 0.9.

These calculated results may be compared to several experimental determinations of the Mg_2SiO_4-Fe_2SiO_4 phase relations at high pressure (Akimoto and Fujisawa, 1968; Ringwood and Major, 1970; Kawada, 1977). In the region containing α+γ the calculated boundaries fall among the experimentally determined ones (see Fig. 12). For more Mg-rich compositions, Ringwood and Major had real difficulty attaining equilibrium involving β-phase and concluded that β-Mg_2SiO_4 probably did not transform to the spinel at all. Kawada (1977) did obtain transformation of β-Mg_2SiO_4 to γ-Mg_2SiO_4, but only at pressures near 20 GPa. Nevertheless his Mg_2SiO_4-Fe_2SiO_4 diagram shows a topology very similar to that calculated. On the other hand, Suito (1977) obtained

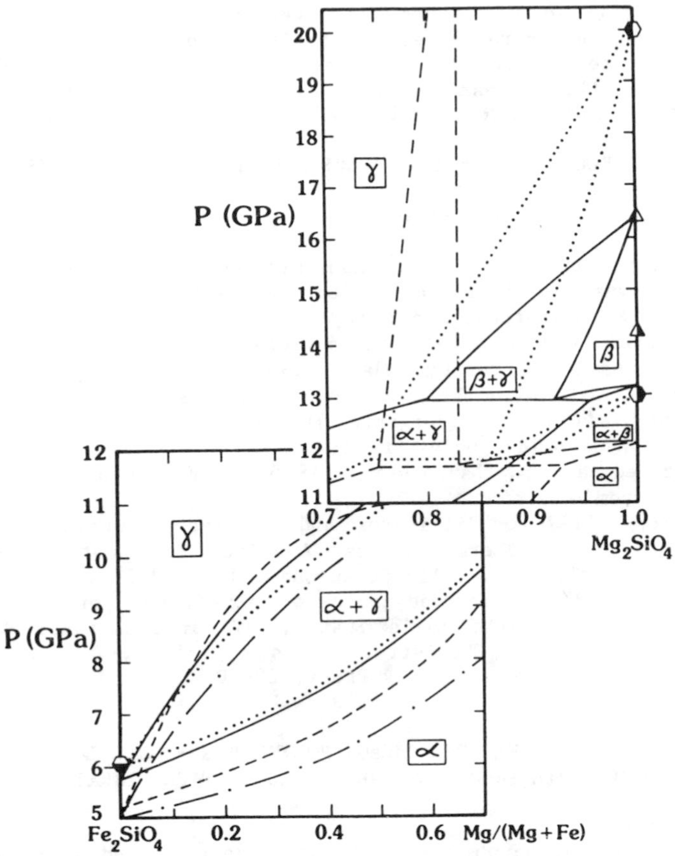

Figure 12. The system Mg_2SiO_4-Fe_2SiO_4 at 1273 K. Solid curves are boundaries calculated in this study, dashed boundaries are those of Ringwood and Major (1970), dotted boundaries are those of Kawada (1977), and dot-dashed boundaries are those of Akimoto and Fujisawa (1968). Hexagons represent transitions in end members determined by Kawada (1977); those half filled on bottom represent $\alpha \rightarrow \gamma$, half filled on right $\alpha \rightarrow \beta$, half filled on left $\beta \rightarrow \gamma$. Triangles represent data of Suito (1977); half filled on right represent $\alpha \rightarrow \beta$, half filled on left $\beta \rightarrow \gamma$. Note change in scale of inset. From Navrotsky and Akaogi (1984).

the β-γ Mg_2SiO_4 transformation at pressures near 16.5 GPa. Bearing in mind the uncertainties in attaining equilibrium, in pressure calibration, and in the thermochemical data, one must consider the overall agreement between calculated and experimental phase relations in Mg_2SiO_4-Fe_2SiO_4 to be satisfactory.

7. CONCLUSIONS

An approach to mineral thermodynamics has been described which makes maximum use of microscopic structural and bonding information both to constrain the forms of equations and to obtain, through semi-empirical correlations, systematic values of thermochemical parameters. Although such an approach can not always match polynomial curve-fitting in its accuracy of describing a particular set of data, its usefulness is predictive and systematic. Thermochemical properties for uncharacterized (or unknown) phases can be predicted and known properties can be extrapolated to conditions of more extreme pressure and temperature. Future developments will continue to combine structural and thermodynamic models. A special need for mineralogy is the further development of models for complex order-disorder phenomena.

8. BOOKS AND REVIEW PAPERS RELATED TO THIS CHAPTER

The following discuss systematics:

P.P. Mchedlow-Petrossyan, V.I. Babushkin and G.M. Matveyev. (1984) "Thermodynamics of Silicates", Springer-Verlag.

R.C. Newton, A. Navrotsky, and B.J. Wood, Eds. (1981) "Thermodynamics of Minerals and Melts", Springer-Verlag.

M. O'Keeffe and A. Navrotsky, Eds. (1981) "Structure and Bonding in Crystals, Vol. I and II", Academic Press.

W.E. Dasent (1982) "Inorganic Energetics, an Introduction, 2nd Edition", Cambridge Univ. Press.

H. Schmalzried and A. Navrotsky (1975) "Festkörperthermodynamik", Verlag Chemie.

A. Navrotsky (1976) "Silicates and Related Minerals: Solid State Chemistry and Thermodynamics Applied to Geothermometry and Geobarometry", Prog. Solid State Chem. 11, 203-264.

A. Navrotsky (1974) "Thermodynamics of Binary and Ternary Transition Metal Oxides in the Solid State", MTP International Reviews of Science, Inorganic Chemistry, Series 2, Vol. 5, D.W.A. Sharp, Ed. Butterworths-University Park Press, 29-70.

9. ILLUSTRATIVE EXAMPLE: THE $CaGeO_3$ POLYMORPHS - CALCULATIONS OF PHASE EQUILIBRIA FROM CALORIMETRY AND LATTICE VIBRATIONS

9.1 The problem

The $CaGeO_3$ polymorphs (wollastonite, garnet, and perovskite) are interesting analogues for lower mantle silicates. Recent calorimetric study (Ross et al. 1985) provides thermochemical data relevant to these phase transitions, see Table 7. Data on molar volume, thermal expansion, and bulk modules are given in Table 8. Infrared and Raman spectra are shown in Figures 13 (wollastonite = wo), 14 (garnet = gar), and 15 (perovskite = pv). The transitions wo→gar occurs at 0.8 GPa at 975 K, gar→per occurs at 6.2 GPa at 1273 K. Calculate the enthalpies and entropies of the wo→gar and gar→pv transitions, using the above thermochemical and high pressure data and construct the P-T diagram. Construct lattice vibrational models consistent with the above data and use then to calculate the entropies of the phase transitions. Discuss the relation of structure, vibrational spectra, and entropies.

9.2 Calculation from calorimetry and phase studies

From Table 7, for the reaction $CaGeO_3$ (wo→gar), $\Delta H^\circ_{975} = \Delta H^\circ_A - \Delta H^\circ_B = -1046$ J. ΔV°_{298} for the transition is -5.97 cc mol^{-1}. Neglecting thermal expansion and compressibility effects, $\Delta G^\circ_{975} = -P\Delta V = 0.8$ GPa x 5.97 = 4.776 GPa cc = 4839 J. Then $\Delta S^\circ = (\Delta H^\circ - \Delta G^\circ)\ T = -6.04$ JK^{-1}.

Table 7. Calorimetric Data for $CaGeO_3$ Polymorphs (Ross et al., 1985)

Step	Reaction	$\Delta H°(J\ mol^{-1})$
A	$CaGeO_3$ (wo, 975 K) → solution (975 K)	19351 ± 1054
B	$CaGeO_3$ (gar, 975 K) → solution (975 K)	20376 ± 908
C	$CaGeO_3$ (wo, 298 K) → $CaGeO_3$ (wo, 975 K)	80040 ± 2732
D	$CaGeO_3$ (gar, 298 K) → $CaGeO_3$ (gar, 975 K)	83852 ± 2820
E	$CaGeO_3$ (pv, 298 K) → $CaGeO_3$ (wo, 298 K)	-38522 ± 3000

a. Solution calorimetry in molten $2PbO{\bullet}B_2O_3$ at 975 K.
b. Transposed temperature drop calorimetry for measurement of heat contents.
c. Difference between first and second transposed temperature drop on same capsule. Value is corrected for small amounts of graphite, water, and wollastonite in original sample.

Table 8. Volume, Thermal Expansion, Bulk Modulus, and Acoustic Velocities for $CaGeO_3$ Polymorphs (Ross et al., 1985).

Compound	$V°_{298}$ ($cm^3\ mol^{-1}$)	α ($x10^5\ K^{-1}$)	K (GPa)	K'	V_ℓ (kms^{-1})	V_s (kms^{-1})
Wollastonite	42.39	1.8	82	4	6.01	3.31
Garnet	36.42	2.3	124	4	6.97	3.91
Perovskite	31.07	3.3	198	4	8.15	4.59

If one includes the dependence of volume on pressure and temperature, then one can use (Akaogi et al., 1984)

$$V(T,P) = V°_{298}(1 + \alpha(T-298)) \bullet (PK'/K + 1)^{-1/K'}) \qquad (37)$$

Using this in the VdP integral (and a microcomputer program) one gets $\Delta S° = -5.86\ JK^{-1}$. For the gar→pv transition, $\Delta H°_{298}$ can be found by a thermochemical cycle. Referring to Table 7, $\Delta H°_{298}$ (gar→pv) = $\Delta H_D - \Delta H_A + \Delta H_B - \Delta H_C - \Delta H_E = 44436$ J. At higher temperatures, ΔH(gar→pv) is not known accurately because the heat capacity of perovskite (which decomposes) has not been measured. Therefore we shall assume that $\Delta H°_{298} = \Delta H°_T$. If compressibility and thermal expansion are neglected as well, then $-P\Delta V°_{298}$ = 6.2 GPa x 5.35 cc = 33605 J and $\Delta S°$ = (44436 - 33605)/1273 = $+8.5\ JK^{-1}$. Taking $\int VdP$ according to Eq. (37), we get $\Delta S = +10.9\ JK^{-1}$.

The phase diagram may then be constructed according to the Clausius-Clapeyron equation, $dP/dT = \Delta S/\Delta V$. The results are shown as solid lines in Figure 16. The agreement with experiments is satisfactory. The wo→gar transitions has a positive dP/dT, the gar→pv transition has a negative slope. In the latter, the discrepancy between experimental and calculated slopes may be explained by uncertainty in the calorimetric data, in thermal expansion and compressibility, and in pressure calibration at high temperature as well as by neglect of the difference in heat capacity between garnet and perovskite. The negative P-T slope for the gar→pv transition is strongly supported by the calorimetric data.

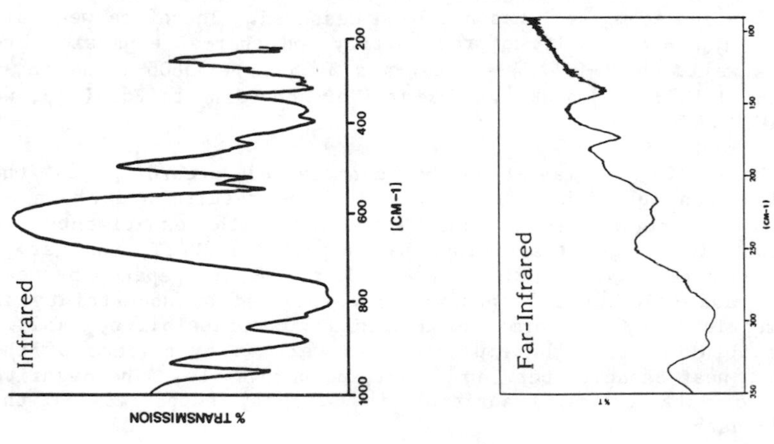
Infrared
% TRANSMISSION
1000
800
600
400
200
[CM-1]
Far-Infrared
T%
350
300
250
200
150
100
(CM-1)

13

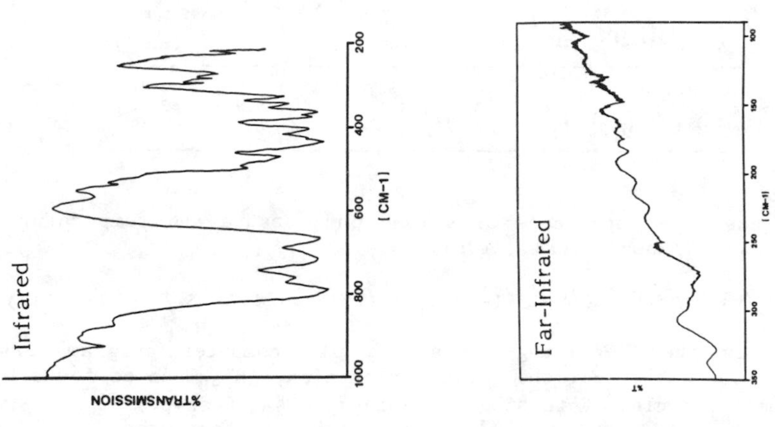
Infrared
%TRANSMISSION
1000
800
600
400
200
[CM-1]
Far-Infrared
T%
350
300
250
200
150
100
[CM-1]

14

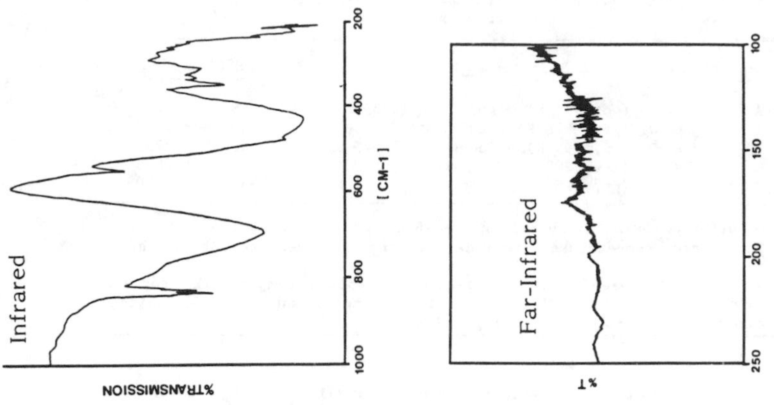
Infrared
%TRANSMISSION
1000
800
600
400
200
[CM-1]
Far-Infrared
T%
250
200
150
100
(CM-1)

15

Figures 13, 14, 15. Infrared and Raman spectra of wollastonite, garnet, and perovskite polymorphs of $CaGeO_3$.

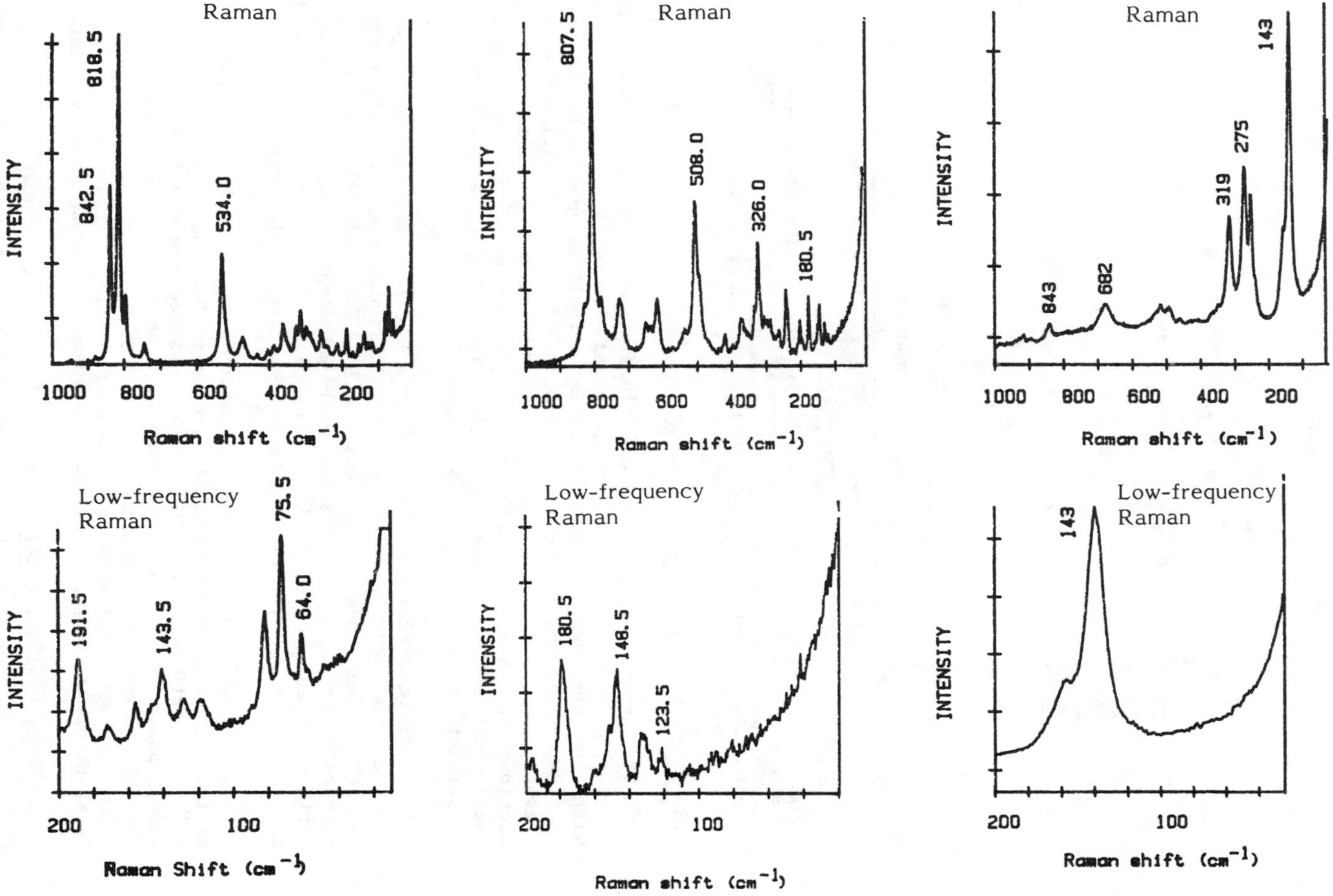

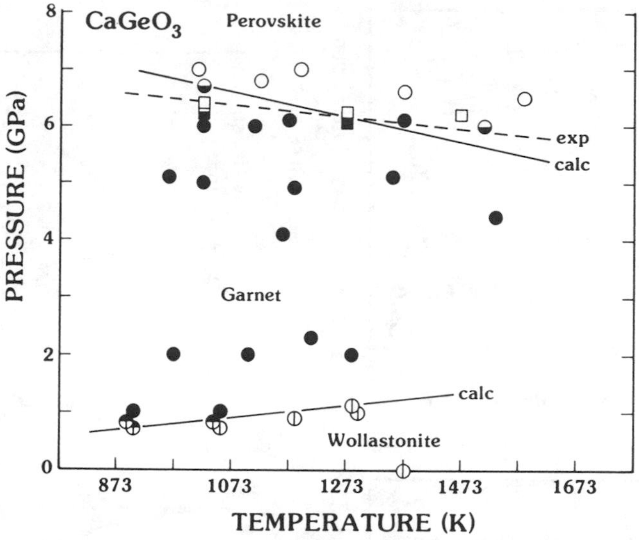

Figure 16. Phase equilibrium studies of $CaGeO_3$ polymorphism. All circles represent wollastonite as starting material; all squares represent perovskite as starting material. Open symbols represent perovskite as product, filled symbols represent garnet, open symbols with vertical line represent wollastonite. Calculated phase boundaries (see text) are shown by solid lines, experimental boundary by dashed line (Ross et al., 1985).

9.3 Vibrational calculations

The general approach to the vibrational modelling of the $CaGeO_3$ polymorphs is to develop a set of related models which are consistent with the observed spectra, acoustic velocities and crystallographic data, as described in Akaogi et al. (1984). Small differences exist between plausible models in the partitioning of modes into optic continua and Einstein oscillators. No one model is unique but the family of related models for each polymorph puts a reasonably tight constraint on the entropy of each phase.

A difficulty in applying Kieffer's model is the lack of dispersion data for the optic modes across the Brillouin zone (see Kieffer, this volume). For the present calculations we have chosen to exclude dispersion in all the models for each polymorph because behavior of the optic modes across the Brillouin zone is not known and there are no low-temperature heat capacity data to help constrain dispersion.

The model treats each vibration as harmonic and gives C_v. Anharmonicity has been included to convert C_v to C_p. Liebermann (1974) and Liebermann et al. (1977) have measured compressibilities of the $CaGeO_3$ polymorphs (Table 8). Thermal expansivity data do not exist so thermal expansion coefficients of $CaGeO_3$ (wo), $CaGeO_3$ (gar), and $CaGeO_3$ (pv) are approximated by those of $CaSiO_3$ wollastonite, $Ca_3Al_2Si_3O_{12}$ grossularite and $CaTiO_3$ perovskite, respectively (Table 8).

<u>$CaGeO_3$ (wollastonite):</u> $CaGeO_3$ (wo) is triclinic and belongs to space group $P\bar{1}$. The primitive unit cell contains twelve formula units. Thus there are 60 atoms and 180 degrees of freedom associated

with the cell. The unit cell volume is 844.70 x $10^{-24}cm^3$ and the radius of the Brillouin zone (approximated by a sphere) is 4.123 x 10^7 cm^{-1}.

$CaGeO_3$ (wo) has longitudinal and shear wave velocities of 6.01 and 3.31 kms^{-1}, respectively. The corresponding directionally-averaged acoustic velocities are u_1 = 3.17 kms^{-1}, u_2 = 3.48 kms^{-1} and u_3 = 6.01 kms^{-1}. These velocities characterize acoustic branches that reach 44, 49 and 84 cm^{-1} at the Brillouin zone boundary. The acoustic modes constitute 1.67% (3/180) of the total number of degrees of freedom.

Features of the $CaGeO_3$ wollastonite infrared and Raman spectra (Fig. 13) which are important for the lattice vibrational models are the low frequency cutoff determined from Raman spectroscopy at 65 cm^{-1}, and the two distinct bands of frequencies ranging from 64 to 540 cm^{-1} and 765 to 947 cm^{-1}. Models which are consistent with the combined crystallographic, acoustic velocity data and spectroscopic data are presented in Figure 17. All models have at least one continuum spanning 64 to 540 cm^{-1}. The high frequency part of the spectrum can be modelled with Einstein oscillators, continua or a combination of the two. 27% of the optic modes are assigned to the high frequency part of the spectrum, analogous to the mode partitioning for pyroxene chain structures. This distribution assumes that the high frequency modes are predominantly Ge-O stretching vibrations. A similar value for mode partitioning can be obtained by assuming a uniform distribution of optic modes and taking relative proportions of the frequency range spanned by the optic modes. For example, the total number of wavenumbers spanned by optic modes in the wollastonite spectrum is 660 cm^{-1} (475 + 185 cm^{-1}). The proportion of modes in the high frequency part of the spectrum is 185/660 = 28% of the modes. The models yield values for Cp and S° that all agree within 2% for the temperature range 300-1500 K.

<u>$CaGeO_3$ (garnet)</u>: $CaGeO_3$ (gar) is distorted from ideal cubic symmetry and belongs to the tetragonal space group $I4_1/a$. The primitive unit cell contains 16 formula units and hence there are 80 atoms and 240 degrees of freedom. The primitive unit cell volume is 969.565 x $10^{-24}cm^3$ and the radius of the Brillouin zone is 3.938 x 10^7 cm^{-1}.

Longitudinal and shear wave velocities for $CaGeO_3$ (gar) are 6.97 kms^{-1} and 3.91 kms^{-1}, respectively. The corresponding directionally-averaged acoustic velocities are u_1 = 3.75 kms^{-1}, u_2 = 4.11 kms^{-1} and u_3 = 6.97 kms^{-1}. These velocities characterize acoustic branches that reach 50, 55 and 93 cm^{-1} at the Brillouin zone boundary. The acoustic modes constitute 1.25% (3/240) of the total number of degrees of freedom.

Features of the $CaGeO_3$ (gar) spectra that are important for vibrational modelling are the low frequency cutoff of 100 cm^{-1} observed in far-infrared spectroscopy, bands extending from 100 to 575 cm^{-1} in the infrared spectrum and up to 616 cm^{-1} in the Raman spectrum, intense bands in the infrared spectrum at 675, 700 and 800 cm^{-1}, and a very intense band at 810 cm^{-1} in the Raman spectrum (Fig. 14). Models that are consistent with the crystallographic, acoustic velocity and spectroscopic data are summarized in Figure 17. The first three models have a continuum extending from 100 to 616 cm^{-1}. The high frequency part of the spectrum is modelled with Einstein oscillators, a second continuum or a combination of the two. 21% of the modes are assigned to this part of the spectrum, similar to Kieffer's distribution of modes for silicate garnets. Using a simple proportionation of modes as described for $CaGeO_3$ (wo), we would partition the same amount (135/651 = 21%) of the modes to the 675-810 cm^{-1} region of the spectrum.

Because it is not definitive where to separate the low and high frequency part of the spectrum, the fourth model shows a second way to partition the modes by choosing an upper cutoff limit of the lower continuum at 575 cm^{-1} and an upper continuum extending from 616 to 810 cm^{-1}. A higher percentage of modes, 28%, is assigned to the high frequency continuum since more modes are being included. The fifth model is a simple continuum which extends from 100 to 800 cm^{-1} and has no breaks between the low- and high-frequency part of the spectrum. This model gives lower values of S° below 500 K than the other models, but all models have entropies which agree within 3% above 500 K (Table 5). The heat capacities calculated from the models all agree within 2% in the temperature range 300-1500 K.

$CaGeO_3$ (perovskite): $CaGeO_3$ (pv) is pseudocubic and crystallizes in space group Pbnm. The primitive unit cell contains four formula units and hence there are 20 atoms and 60 degrees of vibrational freedom. The unit cell volume is 206.36 x 10^{-24} cm^3 and the radius of the Brillouin zone is 6.59 x 10^7 cm^{-1}.

The longitudinal and shear wave velocities for $CaGeO_3$ (pv) are 8.15 kms^{-1} and 4.59 kms^{-1}, respectively. The corresponding directionally-averaged acoustic velocities are u_1 = 4.40 kms^{-1}, u_2 = 4.82 kms^{-1} and u_3 = 8.15 kms^{-1}. These velocities characterize acoustic modes that reach 99, 108 and 182 cm^{-1} at the Brillouin zone boundary. The acoustic modes comprise 5% of the total degrees of freedom which is greater than for either $CaGeO_3$ (wo) or $CaGeO_3$ (gar).

Features of the $CaGeO_3$ (pv) spectra that are important for vibrational modelling are the lowest optic mode observed at 135 cm^{-1} in the far-infrared spectrum and the high frequency band at 700 cm^{-1} seen in both the Raman and infrared spectra. This band has a shoulder at 780 cm^{-1} in the infrared spectrum (Fig. 15). The peaks above 800 cm^{-1} observed in the Raman spectrum may be questionable, especially since the perovskite transforms under the laser beam. Models for $CaGeO_3$ (pv) that are consistent with the crystallographic, acoustic and spectroscopic data are presented in Figure 17. The simplest model is a single continuum extending from 135 to 700 cm^{-1}, the region in which the bulk of the optic modes are concentrated. Since there are few modes between 580 and 700 cm^{-1}, the 700 cm^{-1} band can alternately be treated separately as an individual Einstein oscillator or as a narrow continuum with 17% (90/545) of the modes assigned to it. This mode assignment is based on the simple proportionation described for $CaGeO_3$ (wo) and $GaGeO_3$ (gar). An advantage of this mode-partitioning scheme is that it provides approximations for relative mode partitioning that can be used for structure types which do not have tetrahedrally-coordinated cations.

$CaGeO_3$ - WOLLASTONITE

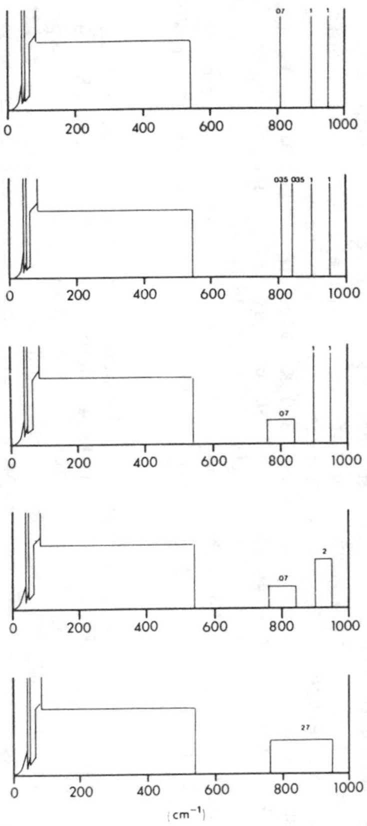

T[K]	Cp (J/mol·K)	S° (J/mol·K)
298	83.8	83.0
700	114.0	169.7
1000	119.8	211.5

T[K]	Cp	S°
298	83.6	82.9
700	114.0	169.5
1000	119.8	211.3

T[K]	Cp	S°
298	83.8	83.1
700	114.0	169.7
1000	119.8	211.5

T[K]	Cp	S°
298	83.8	83.0
700	113.9	169.6
1000	119.7	211.4

T[K]	Cp	S°
298	85.1	83.8
700	114.6	170.7
1000	120.1	212.7

$CaGeO_3$ - GARNET

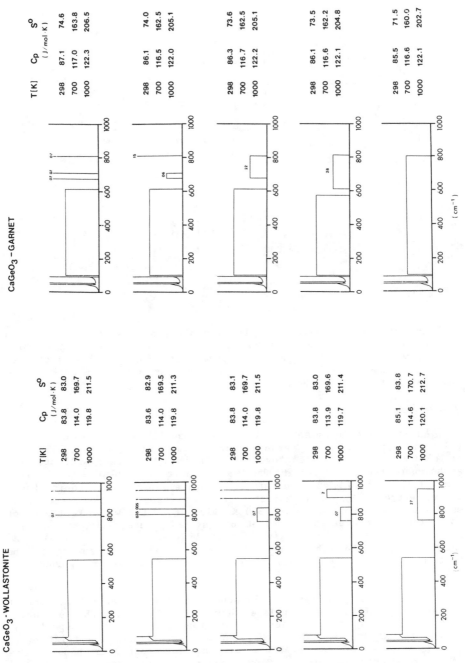

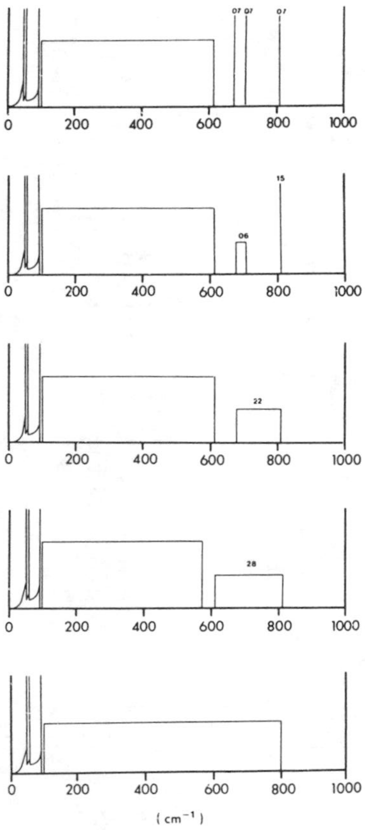

T[K]	Cp (J/mol·K)	S° (J/mol·K)
298	87.1	74.6
700	117.0	163.8
1000	122.3	206.5

T[K]	Cp	S°
298	86.1	74.0
700	116.5	162.5
1000	122.0	205.1

T[K]	Cp	S°
298	86.3	73.6
700	116.7	162.5
1000	122.2	205.1

T[K]	Cp	S°
298	86.1	73.5
700	116.6	162.2
1000	122.1	204.8

T[K]	Cp	S°
298	85.5	71.5
700	116.6	160.0
1000	122.1	202.7

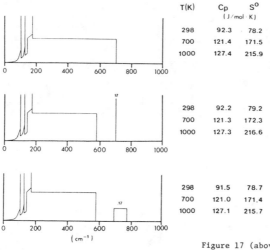

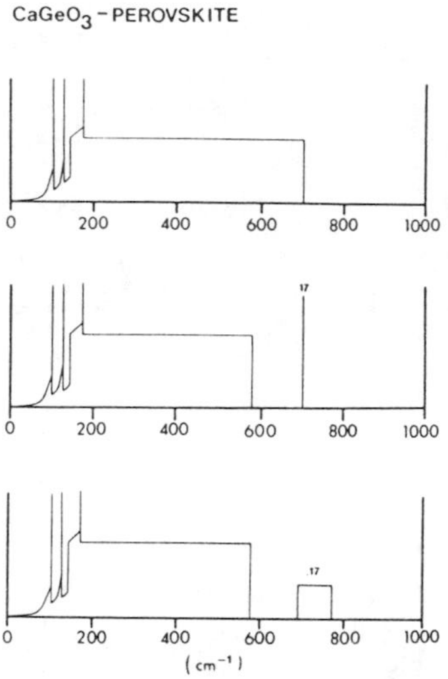

T(K)	Cp (J/mol·K)	S° (J/mol·K)
298	92.3	78.2
700	121.4	171.5
1000	127.4	215.9
298	92.2	79.2
700	121.3	172.3
1000	127.3	216.6
298	91.5	78.7
700	121.0	171.4
1000	127.1	215.7

Figure 17 (above & previous page). Vibrational models for $CaGeO_3$ polymorphs (Ross et al., 1985).

The simple continuum model is probably preferred because it does not introduce additional assumptions about mode partitioning, while still giving reasonable results. Indeed there are differences of less than 2% in the Cp and S° values predicted from all the models in the temperature range 300-1500 K.

Discussion

An average of each set of vibrational models for the $CaGeO_3$ polymorphs predict the Cp versus T curves shown in Figure 18. From 300 to 1500 K, Cp curves lie above each other in the order wollastonite-garnet-perovskite. Although the heat capacity of $CaGeO_3$ (wo) is less than that of $CaGeO_3$ (gar) in this temperature range, wollastonite has a higher entropy, S°_{298} than garnet. To understand why this occurs, we must study the Cp curves below 300 K (Fig. 18). Below 250 K, the Cp of wollastonite is greater than that of garnet. Integration of Cp/T to obtain S° will therefore give larger values for $CaGeO_3$ (wo) than $CaGeO_3$ (gar). The vibrational models also predict crossovers in the heat capacity curves of wollastonite and perovskite and garnet and perovskite.

Entropies of the $CaGeO_3$ polymorphs are plotted as a function of temperature in Figure 19. It is evident from this figure that entropies of transition are not constant as a function of temperature, especially, of course, at low T. The higher entropy of wollastonite relative to garnet is due to the presence of the low frequency band at 64 cm^{-1} compared to 100 cm^{-1}. Heat capacities and entropies are very sensitive to the presence of the lowest frequency optic mode; as the frequency of this mode decreases, Cp and S° increase.

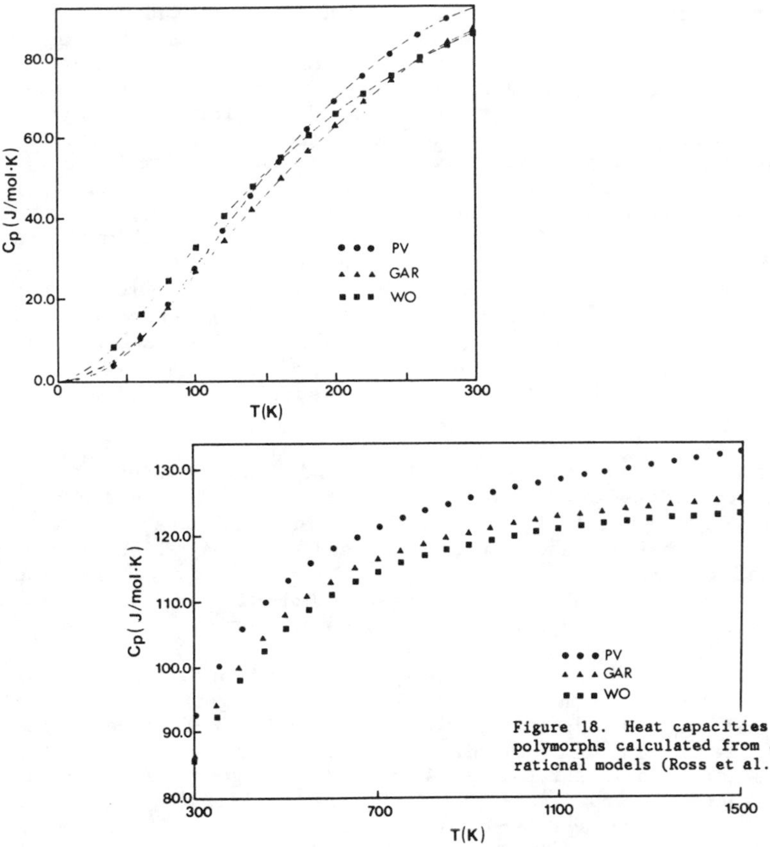

Figure 18. Heat capacities (Cp) of $CaGeO_3$ polymorphs calculated from average of vibrational models (Ross et al., 1985).

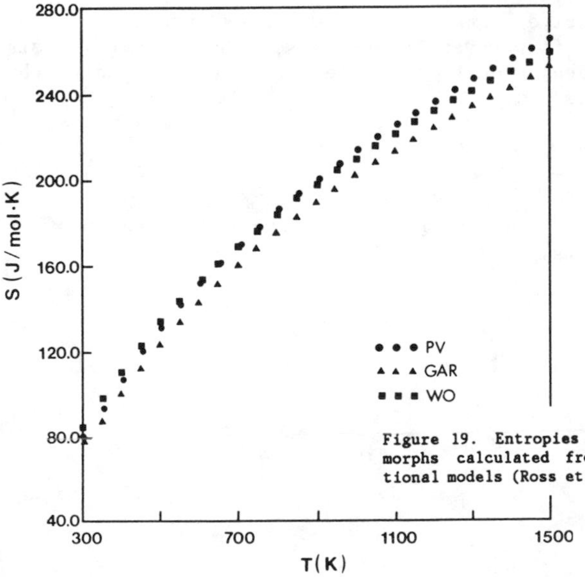

Figure 19. Entropies (S_T^o) of $CaGeO_3$ polymorphs calculated from average of vibrational models (Ross et al., 1985).

$CaGeO_3$ (pv) has a markedly higher Cp and S° than $CaGeO_3$ (gar). There are two major vibrational contributions to perovskite's higher entropy. First, the contribution from acoustic modes is greater for perovskite than for garnet; acoustic modes constitute 5% of the total degrees of freedom for $CaGeO_3$ (pv) and only 1.25% for $CaGeO_3$ (gar). Because the acoustic modes occur at low frequencies, a greater percentage of acoustic modes will increase Cp and S° significantly. Secondly, the overall distribution of optic modes affects the entropy. The optic modes in perovskite are concentrated at lower frequencies with the majority of modes below 700 cm^{-1}. This causes an increase in S°. A third contribution is the anharmonicity correction (Cv to Cp) which is greater for $CaGeO_3$ (pv) than $CaGeO_3$ (gar) because of perovskite's greater thermal expansion. This correction increases the entropy by 3% for $CaGeO_3$ (pv) and 1% for $CaGeO_3$ (gar) at 1000 K. The rather large thermal expansion typical of the perovskite structure may be related to changes in the tilting of octahedra and the resulting distortions from cubic symmetry.

The entropies of transition at 1000 K and 1300 K based on the vibrational calculations for the wo→gar and gar→pv transitions are -6.8 J $mol^{-1}K^{-1}$ and +12.7 J $mol^{-1}K^{-1}$, respectively. These compare well with $\Delta S°$ based on the combination of calorimetry and phase equilibria data. The vibrational calculations predict a slightly positive dP/dT for the wo→gar transition and a definitely negative dP/dT for the gar→pv transition which are consistent with the phase equilibria.

Crystal structure analyses of $CaGeO_3$ garnet and perovskite reveal that crystal chemical characteristics of the garnet-perovskite transition in $CaGeO_3$ are consistent with the negative P-T slope of the garnet perovskite transition. Coordination numbers increase across the transition: 6 and 8 to 12 for Ca, and 4 and 6 to 6 for Ge. Average Ca-O distances increase from 2.29 Å (^{vi}Ca-O) and 2.47 Å (^{viii}Ca-O) to 2.65 Å (^{xii}Ca-O), and average Ge-O bond lengths increase slightly from 1.78 Å (^{iv}Ge-O) and 1.92 Å (^{vi}Ge-O) to 1.89 Å (^{vi}Ge-O). The vibrational calculations give quantitative support to these considerations. The longer and weaker M-O bonds in perovskite are the crystal chemical basis for the changes in vibrational spectra discussed above which lead to the higher entropy of perovskite.

ACKNOWLEDGMENTS

Much of the work leading to the conclusions summarized here was supported by the National Science Foundation and the U.S. Department of Energy over a number of years.

REFERENCES

Akaogi, M. and Navrotsky, A. (1984a) Calorimetric study of the stability of spinelloids in the system $NiAl_2O_4$-Ni_2SiO_4. Phys. Chem. Minerals 10, 166-172.

_____ and _____ (1984b) The quartz-coesite-stishovite transformations: new calorimetric measurements and calculation of phase diagrams. Phys. Earth Planet. Interiors 36, 124-134.

_____ Ross, N.L., McMillan, P. and Navrotsky, A. (1984) The Mg_2SiO_4 polymorphs (olivine, modified spinel, and spinel) - thermodynamic properties from oxide melt calorimetry, phase relations, and models of lattice vibrations. Am. Mineral. 69, 499-512.

Akimoto, S. and Fujisawa, H. (1968) Olivine-spinel solid solution equilibria in the system Mg_2SiO_4-Fe_2SiO_4. J. Geophys. Res. 73, 1467-1479.

_____ and Akaogi, M. (1977) Pyroxene-garnet solid solution equilibria in the systems $Mg_4Si_4O_{12}$-$Mg_3Al_2Si_3O_{12}$ and $Fe_4Si_4O_{12}$-$Fe_3Al_2Si_3O_{12}$ at high pressures and temperatures. Phys. Earth Planet. Interior 15, 90-106.

Bassett, W.A., Takahashi, T. and Campbell, J.K. (1969) Volume changes for the B1-B2 phase transformations in three potassium halides at room temperature. Trans. Amer. Crystallogr. Assoc. 5, 93-103.

Brown, G.E. and Prewitt, C.J. (1973) High temperature crystal chemistry of hortosnolite. Am. Mineral. 58, 577-587.

Bukowinski, M.S.T. and Hauser, J. (1980) Pressure induced metallization of SrO and BaO: theoretical estimate of transition pressures. Geophys. Res. Lett. 7 (9), 689-692.

Callen, H.B., Harrison, S.E. and Kriessman, C.J. (1956) Cation distributions in ferrospinels. Phys. Rev. 103, 851-856.

Capobianco, C. and Navrotsky, A. (1982) Calorimetric evidence for ideal mixing of silicon and germanium in glasses and crystals of sodium feldspar composition. Am. Mineral. 67, 718-724.

Carpenter, M.A., McConnell, J.D.C. and Navrotsky, A. (1985) Enthalpies of ordering in the plagioclase feldspar solid solution. Geochim. Cosmochim. Acta (in press).

_____ Navrotsky, A. and McConnell, J.D.C. (1983) Enthalpy effects associated with Al/Si ordering in anhydrous Mg-cordierite. Geochim. Cosmochim. Acta 47, 899-906.

Chatillon-Colinet, C., Newton, R.C., and Kleppa, O.J. (1983) Thermochemistry of $(Fe^{2+},Mg)SiO_3$ orthopyroxene. Geochim. Cosmochim. Acta 47, 1597-1603.

Davies, P.K. and Akaogi, M. (1983) Phase intergrowths in spinelloids. Nature 305, 788-790.

_____ and Navrotsky, A. (1983) Quantitative correlations of deviations from ideality in binary and pseudo-binary solid solutions. J. Solid State Chem. 46, 1-22.

_____ and _____ (1981) Thermodynamics of solid solution formation in the systems NiO-MgO and NiO-ZnO. J. Solid State Chem. 38, 264-276.

Driessens, F.C.M. (1968) Thermodynamics and defect chemistry of some oxide solid solutions I. nearest-neighbor interactions and the effect of substitutional disorder. Ber. Bunsengesell. Phys. Chem. 82, 754-764.

Duffy, J.A. and Ingram, M.D. (1971) Establishment of an optical scale for lewis basicity in inorganic oxyacids, molten salts, and glasses. J. Am. Chem. Soc. 93, 6448-6454.

Dunitz, J.D. and Orgel, L.E. (1957) Electronic properties of transition metal oxides II. J. Phys. Chem. Solids 3, 318-323.

Fancher, D.L. and Barsch, G.R. (1969) Lattice theory of alkali halide solid solutions - I. heat of formation. J. Phys. Chem. Solids 30, 2503-2516.

Flood, H. and Forland, T. (1947) The acidic and basic properties of oxides. Acta Chem. Scand. 1, 592-604.

Geisinger, K.L., Gibbs, G.V. and Navrotsky, A. (1985) A molecular orbital study of bond length and angle variation in framework silicates. Phys. Chem. Minerals 11, 266-283.

Ghose, S., Wan, C., Okamura, F.P., Ohashi, H. and Weidner, J.R. (1975) Site preference and crystal chemistry of transition metal ions in pyroxenes and olivines. Acta Crystallogr. 31A, 576 (abstr.).

_____ and Ganguly, J. (1982) Mg-Fe order-disorder in ferromagnesian silicates. In "Advances in Physical Geochemistry", Vol. II. Saxena, ed., p. 3-100, Springer-Verlag, N.Y.

Gibbs, G.V. (1982) Molecules as models for bonding in solids. Am. Mineral. 67, 421-450.

Glidewell, C. (1976) Cation distribution in spinels. Inorg. Chim. Acta 19, L45-48.

Hawthorne, F. C. (1983) The crystal chemistry of the amphiboles. Canadian Mineral. 21, 173-480.

Henry, D.J., Navrotsky, A. and Zimmermann, H.D. (1982) Thermodynamics of plagioclase-melt equilibria in the system albite-anorthite-diopside. Geochim. Cosmochim, Acta 46, 381-391.

Hervig, R.L. and Navrotsky, A. (1984) Thermochemical study of glasses in the system $NaAlSi_3O_8$-$KAlSi_3O_8$-Si_4O_8, and the join $Na_{1.6}Al_{1.6}Si_{2.4}O_8$ - $K_{1.6}Al_{1.6}Si_{2.4}O_8$. Geochim. Cosmochim. Acta 48, 513-522.

Hill, R.J., Craig, J.R. and Gibbs, G.V. (1979) Systematics of the spinel structure type. Phys. Chem. Minerals 4, 317-340.

Horiuchi, H., Akaogi, M., and Sawamoto, H. (1982) Crystal structure studies on spinel-related phases, spinelloids: implications to olivine-spinel transformations and systematics. In "High Pressure Research in Geophysics", S. Akimoto, and M. H. Manghnani, eds., Center for Academic Publications, Tokyo, 391-404.

Ito, E. and Matsui, Y. (1979) High-pressure transformation in silicates, germanates and titanates with ABO_3 stoichiometry. Phys. Chem. Minerals 4, 265-273.

Jamieson, J.C. (1977) Phase relations in rutile-tyle compounds. In "High Pressure Research, Application to Geophysics", M. H. Manghnani, and S. Akimoto, eds., p. 209-218, Academic Press, N.Y.

Kamb, B. (1968) Structural basis of the olivine-spinel stability relation. Am. Mineral 53, 1439-1455.

Kawada, K. (1977) The system Mg_2SiO_4-Fe_2SiO_4 at high pressures and temperatures and the earth's interior. Ph.D. Thesis, Univ. of Tokyo, Japan.

Kawai, N., Togaya, M. and Mishima, O. (1977) Metallic transitions of oxides, H_2O, and hydrogen. In "High Pressure Research, Application to Geophysics", M. H. Manghnani, and S. Akimoto, eds., p. 267-280, Academic Press. N.Y.

Kerrick, D. M. and Darken, L.S. (1975) Statistical thermodynamic models for ideal oxide and silicata solid solutions, with application to high-temperature plagioclase. Geochim. Cosmochim. Acta 39, 1431-1442.

Liebermann, R.C. (1974) Elasticity of pyroxene-garnet and pyroxene-ilmenite phase transformations in germanates. Phys. Earth Planet. Interiors 8, 361-374.

_____ Jones, L.E.A. and Ringwood, A.E. (1977) Elasticity of aluminate, titanate, stannate and germanate compounds with the perovskite structure. Phys. Earth Planet. Interiors 14, 165-178.

Lind, M.D. and Housley, R.M. (1972) Crystallization studies of lunar igneous rocks. crystal structure of synthetic armalcolite. Science 175, 521-523.

Loomis, T.P. (1979) An empirical model for plagioclase equilibrium in hydrous melts. Geochem. Cosmochim. Acta 43, 1753-1759.

Lowry, T.H. and Richardson, K.S. (1981) "Mechanism and Theory in Organic Chemistry", 2nd ed., p. 130-144, Harper & Row, N.Y.

Lux, H. (1939) "Acids" and "bases" in a fused salt bath; the determination of oxygen-ion concentration. Z. Electrochem. Soc. 45, 303-309.

McClure, D.S. (1957) The distribution of transition metal cations in spinels. J. Phys. Chem. Solids 3, 311-317.

Navrotsky, A. (1971) Thermodynamics of formation of the silicates and germanates of some divalent transition metals and of magnesium. J. Inorg. Nucl. Chem. 33, 4035-4050.

_____ (1973a) Ni_2SiO_4 - enthalpy of the olivine-spinel transition by solution calorimetry at 713°. Earth Planet. Sci. Lett. 19, 471-475.

_____ (1973b) Enthalpy of the olivine-spinel transition in magnesium orthogermanate and the thermodynamics of olivine-spinel-phenacite stability relations. In "Phase Transition-1973, Proc. Conf. on Phase Transitions and their Applications in Materials Science, University Park, PA, May 23-25, 1973", L.E. Cross, ed., p. 393-398, Pergamon Press, N.Y.

_____ (1974) Thermodynamics of binary and ternary transition metal oxides in the solid state. "MTP International Reviews of Science, Inorganic Chemistry", Ser. 2, Vol. 5, D.W.A. Sharp, ed. Butterworths-University Park Press, Baltimore, MD, 29-70.

_____ (1975) Thermodynamics of formation of some compounds with the pseudobrookite structure and of the $FeTi_2O_5$-Ti_3O_5 solid solution series. Am. Mineral. 60, 249-256.

_____ (1980) Lower mantle phase transitions may generally have negative pressure - temperature slopes. Geophys. Res. Lett. 7, 709-711.

_____ (1981) Energetics of phase transitions in AX, ABO_3, and AB_2O_4 compounds. In "Structure and Bonding in Crystals". M. O'Keeffe and A. Navrotsky, eds. Academic Press, N.Y., Vol. II, 71-93.

_____ and Akaogi, M. (1984) α-β-γ phase relations in Fe_2SiO_4-Mg_2SiO_4 and Co_2SiO_4-Mg_2SiO_4: calculation from thermochemical data and geophysical applications. J. Geophys. Res. 89, 10135-10140.

_____ and Davies, P.K. (1981) Cesium chloride versus nickel arsenide as possible structures for (Mg,Fe)O in the lower mantle. J. Geophys. Res. 86, 3689-3694.

_____ Geisinger, K.L., McMillan, P. and Gibbs, G.V. (1985) The tetrahedral framework in glasses and melts - inferences from molecular orbital calculations and implications for structure thermodynamics, and physical properties. Phys. Chem. Minerals 11, 284-298.

_____ and Kleppa, O.J. (1969) Enthalpies of formation of some tungstates MWO_4". Inorg. Chem. 8, 756-758.

_____ and _____ (1967) The thermodynamics of cation distributions in simple spinels. J. Inorg. Nucl. Chem. 29, 2701-2714.

_____ Peraudeau, G., McMillan, P. and Coutures, J.P. (1982) A thermochemical study of glasses along the joins silica-calcium aluminate and silica-sodium aluminate. Geochim. Cosmochim. Acta 46, 2039-2047.

_____ Pintchovski, F.S. and Akimoto, S. (1979) Calorimetric study of the stability of high pressure phases in the systems CoO-SiO_2 and "FeO"-SiO_2 and calculation of phase diagrams in MO-SiO_2 systems. Phys. Earth Planet. Interiors 19, 275-292.

Neil, J.M., Navrotsky, A., and Kleppa, O.J. (1971) The enthalpy of the ilmenite-perovskite transformation in cadmium titanate. Inorg. Chem. 10, 2076-2077.

Newton, R.C., Charlu, T.V. and Kleppa, O.J. (1980) Thermochemistry of the high structural state plagioclases. Geochim. Cosmochim. Acta 44, 933-941.

Nishizawa, O. and Akimoto, S. (1973) Partitioning of magnesium and iron between olivine and spinel and between pyroxene and spinel. Contrib. Mineral. Petrol. 41, 217-230.

O'Keeffe, M. and Bovin, J.-O. (1979) Solid electrolyte behavior of $NaMgF_3$: geophysical implications. Science 206, 599-600.

O'Neill, H.St.C. and Navrotsky, A. (1983) Simple spinels: crystallographic parameters, cation radii, lattice energies, and cation distributions. Am. Mineral. 68, 181-194.

____ and ____ (1984) Cation distributions and thermodynamic properties of binary spinel solid solutions. Am. Mineral. 69, 733-755.

Orville, P.M. (1972) Plagioclase cation exchange equilibria with aqueous chloride solution: results at 700°C and 2000 bars in the presence of quartz. Am. J. Science 272, 234-272.

Pintchovski, F., Glaunsinger, W.S. and Navrotsky, A. (1978) Experimental study of the electronic and lattice contribution to the VO_2 transition. J. Phys. Chem. Solids 39, 941-949.

Price, G.D. and Parker, S.C. (1984) Computer simulations of the structural and physical properties of the olivine and spinel polymorphs of Mg_2SiO_4. Phys. Chem. Minerals 10, 209-216.

Price, G.D., Price, S.L., and Burdett, J.K. (1982) The factors influencing cation site preference in spinels, a new Mendeleyevian approach. Phys. Chem. Minerals 8, 69-82.

Ringwood, A.E. and Major, A. (1970) The system Mg_2SiO_4-Fe_2SiO_4 at high pressures and temperatures. Phys. Earth Planet. Interiors 3, 89-108.

Robie, R.A., Hemingway, B.S. and Fisher, J.R. (1978) Thermodynamic properties of minerals and related substances at 298.15 K and 1 bar (10^6 pascals) pressure and at higher temperatures. U.S. Geol. Survey Bull. 1452.

Ross, N.L., Akaogi, M., Navrotsky, A., Susaki, J. and McMillan, P. (1985) Phase transitions among the $CaGeO_3$ polymorphs (wollastonite, garnet, and perovskite structures): Studies by high pressure synthesis, high temperature calorimetry, and vibrational spectroscopy and calculation. J. Geophys. Res. (submitted).

Roy, B.N. and Navrotsky, A. (1984) Thermochemistry of charge-coupled substitutions in silicate glasses: the systems $M^{n+}_{1/n}AlO_2$-SiO_2 (M = Li, Na, K, Rb, Cs, Mg, Ca, Sr, Ba, Pb). J. Am. Ceram. Soc. 67, 606-610.

Saxena, S. and Ribbe, P.H. (1972) Activity-composition relations in feldspars. Contrib. Mineral. Petrol. 37, 131-138.

Seil, M.K. and Blencoe, J.G. (1978) Activity-composition relations of $NaAlSi_3O_8$-$CaAl_2Si_2O_8$ feldspars at 2 kb, 600-800°C. Geol. Soc. Am. Abstrs. Progrs. 11, 513.

Shannon, R.D. and Prewitt, C.T. (1969) Effective ionic radii in oxides and fluorides. Acta Crystallogr. B25, 925-946.

Suito, K. (1977) Phase relations of pure Mg_2SiO_4 up to 200 kbars. In "High Pressure Research-Application to Geophysics" M. H. Manghnani, and S. Akimoto, eds. p. 255-266, Acad. Press. N.Y.

Sung, C.M. and Burns, R.G. (1978) Crystal structural features of the olivine → spinel transition. Phys. Chem. Minerals 2, 177-198.

Syono, Y., Tokonami, M. and Matsui, Y. (1971) Crystal field effect on the olivine-spinel transformation. Phys. Earth Planet. Interiors 4, 347-352.

Tardy, Y. and Vieillard, P. (1977) Relationships among gibbs free energies and enthalpies of formation of phosphates, oxides and aqueous ions. Contrib. Mineral. Petrol. 63, 75-88.

____ and Gartner, L. (1977) Relationships among gibbs energies of formation of sulfates, nitrates, carbonates, oxides and aqueous ions. Contrib. Mineral. Petrol. 63, 89-102.

Tokonami, M., Morimoto, N., Akimoto, S., Syono, Y. and Takeda, H. (1972) Stability relations netween olivine, spinel and modified spinel. Earth Planet. Sci. Lett. 14, 65-69.

Urusov, V.S. (1975) Energetic theory of miscibility gaps in mineral solid solutions. Fortschr. Mineral. 52, 141-150

______ (1983) Interaction of cations on octahedral and tetrahedral sites in simple spinels. Phys. Chem. Minerals 9, 1-5.

Wechsler, B.A. and Navrotsky, A. (1984) Thermodynamics and structural chemistry of compounds in the system MgO-TiO_2. J. Solid State Chem. 55, 165-180.

Wood, B.J., Holland, T.B., Newton, R.C. and Kleppa, O.J. (1980) Thermodynamics of jadeite-diopside pyroxenes. Geochim. Cosmochim. Acta 44, 1363-1371.

____ (1979) Activity-composition relations in $Ca(Mg,Fe)Si_2O_6$- $CaAl_2SiO_6$ clinopyroxene solid solutions. Am. J. Sci. 279, 854-875.

Chapter 8. Roger G. Burns

THERMODYNAMIC DATA from CRYSTAL FIELD SPECTRA

It is embarrassing that the pressure experiments show the dependence of Δ upon the distance R between the metal ion and the ligand to be fairly close to that given by the point charge model.

S. Sugano and S. Ohnishi, in: Material Science of the Earth's Interior (I. Sunagawa, ed., Terra Scientific Publ. Co., Tokyo, 1984, p. 174).

1. INTRODUCTION

This chapter describes how thermodynamic data for transition metal-bearing minerals may be derived from absorption bands in the visible and near-infrared regions of the electromagnetic spectrum. Such optical or crystal field spectra originate from electronic transitions between unfilled 3d atomic orbitals within the cations and provide crystal field stabilization energy (CFSE) data. These CFSE values are useful for comparing relative stabilities and geochemical behaviors of cations of Ti, V, Cr, Mn, Fe, Co, Ni, and Cu when they occur in similar oxidation states and coordination environments in minerals.

The connection between spectroscopy and thermodynamics was demonstrated by Orgel (1952) when he showed that spectroscopically-determined CFSE's contribute to thermodynamic data of transition metal compounds. For example, the heats of hydration of hexahydrated transition metal cations shown in Figure 1 fall on characteristic two-hump curves with maximum values for ions with three and eight 3d electrons. Orgel (1952) found that when CFSE's estimated from visible region absorption spectra of hydrated cations are deducted from the measured heats of hydration, the "corrected" values lie near smooth curves passing through experimental values for cations with zero, five and ten 3d electrons which acquire zero CFSE. Figure 1 also demonstrates that CFSE's constitute only a small fraction of the absolute heats of hydration, lattice energies, etc., but nevertheless their values are comparable to enthalpy changes in many chemical reactions. Orgel's (1952) observations also heralded the rediscovery of crystal field theory, which was used subsequently to explain various aspects of transition metal crystal chemistry and geochemistry (Burns, 1970a). Many of these early applications were based on data derived from the then newly acquired polarized absorption spectral measurements of transition metal-bearing minerals. In the past fifteen years there have been numerous ingeneous applications of crystal field theory to a variety of problems in the earth and planetary sciences, utilizing mineral spectroscopic data acquired at elevated temperatures and pressures. Some of these results are summarized in this chapter. Following a survey of branches of spectroscopy in which excitations of electrons play a role, elements of crystal field theory are described, because these concepts provide the simplest explanation of energy splittings of transition metal 3d orbitals in mineral structures. Available crystal field spectra of transition metal ions in periclase, corundum, garnet, olivine, pyroxene, and silicate spinel phases are then reviewed, providing data for con-

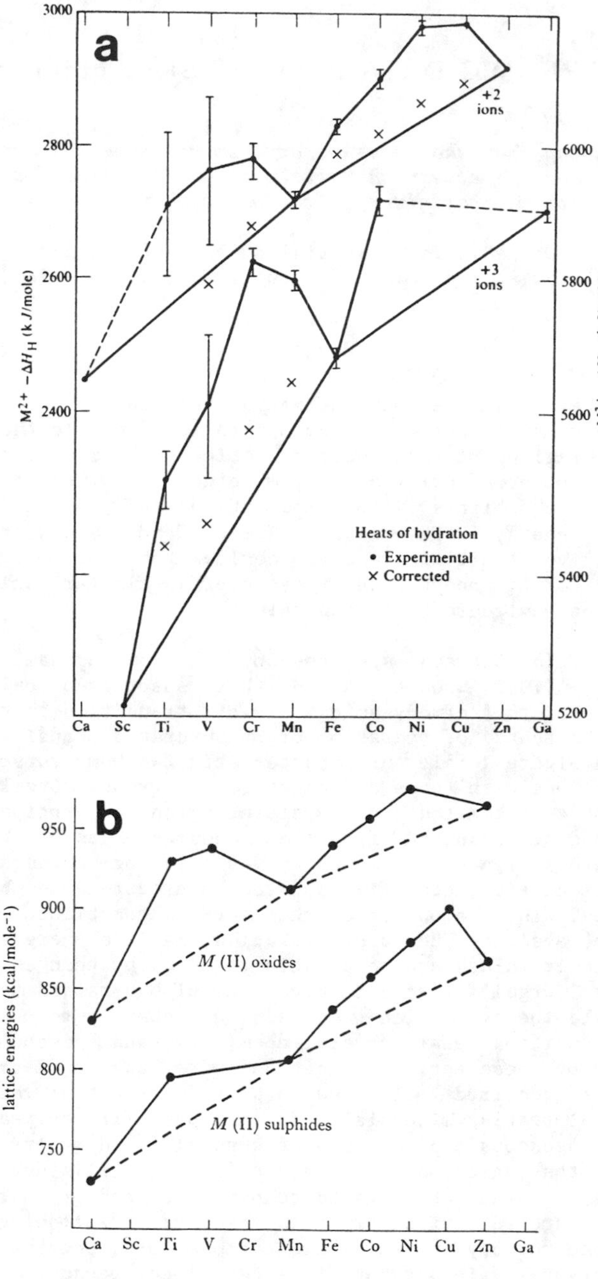

Figure 1. Variations of thermodynamic data for first series transition elements. (a) Hydration enthalpies of the divalent and trivalent ions (from George and McClure, 1959); (b) Lattice energies for oxides and sulfides of divalent ions (from Burns, 1970a).

structing 3d orbital energy level diagrams and estimating CFSE's for several of the cations in these minerals. Finally, some recent examples of the use of CFSE data in geochemical and geophysical problems are described.

2. TYPES OF ELECTRONIC SPECTRA

Optical, crystal field or d→d spectra represent one of many branches of spectroscopy involving excitations of electrons between split energy levels. A transition between two levels separated by energy ΔE takes place when

$$\Delta E = hc/\lambda, \quad (1)$$

where c and λ, respectively, are the velocity and wavelength of light (electromagnetic radiation), and h is Planck's constant. Different branches of spectroscopy are classified by the wavelengths of light or energies of photons required to induce transitions between various types of split energy levels. Examples of energy levels and branches of spectroscopy are shown in Figure 2 which also correlates some of the different wavelength scales and energy units commonly encountered in spectroscopy. Because wavelength is inversely proportional to energy (Eqn. 1), it is more convenient to use a wavenumber unit that is directly related to energy units when extracting data from spectra. In this chapter the wavenumber unit used is the reciprocal centimeter, cm^{-1}, which is related to the wavelength scale as follows:

$$1 \text{ micron} = 10^{-6} \text{ m} = 1000 \text{ nm} = 10000 \text{ Å} = 10000 \text{ cm}^{-1}. \quad (2)$$

Wavenumbers are sometimes expressed as kaysers, K, and 10000 cm^{-1} = 10kK. Since geochemical thermodynamic data are expressed conventionally in units of kJ $mole^{-1}$ (where 4.184 kJ = 1 k.cal), three convenient bench marks in optical spectroscopy are:

2000 nm = 5000 cm^{-1} = 59.8 kJ (14.29 k.cal) [near infrared]

1000 nm = 10000 cm^{-1} = 119.7 kJ (28.59 k.cal) [near infrared]

500 nm = 20000 cm^{-1} = 239.4 kJ (57.18 k.cal) [green light]

3. ORIGIN OF CRYSTAL FIELD SPECTRA

Electronic spectra of transition metal compounds in the visible region originate when electrons are excited by light between incompletely filled 3d orbital energy levels within the transition metal ion. The origin of the splitting of the 3d energy levels may be described by three different models: crystal field theory (CFT), ligand field theory (LFT), and molecular orbital theory (MOT) (Ballhausen, 1962; Burns, 1970a; Burdett, 1978). The CFT treats the transition metal ion in a crystalline environment as though it was subjected to purely electrostatic interactions with surrounding anions, which are approximated as point negative charges. The LFT introduces some empirical corrections, such as the Racah

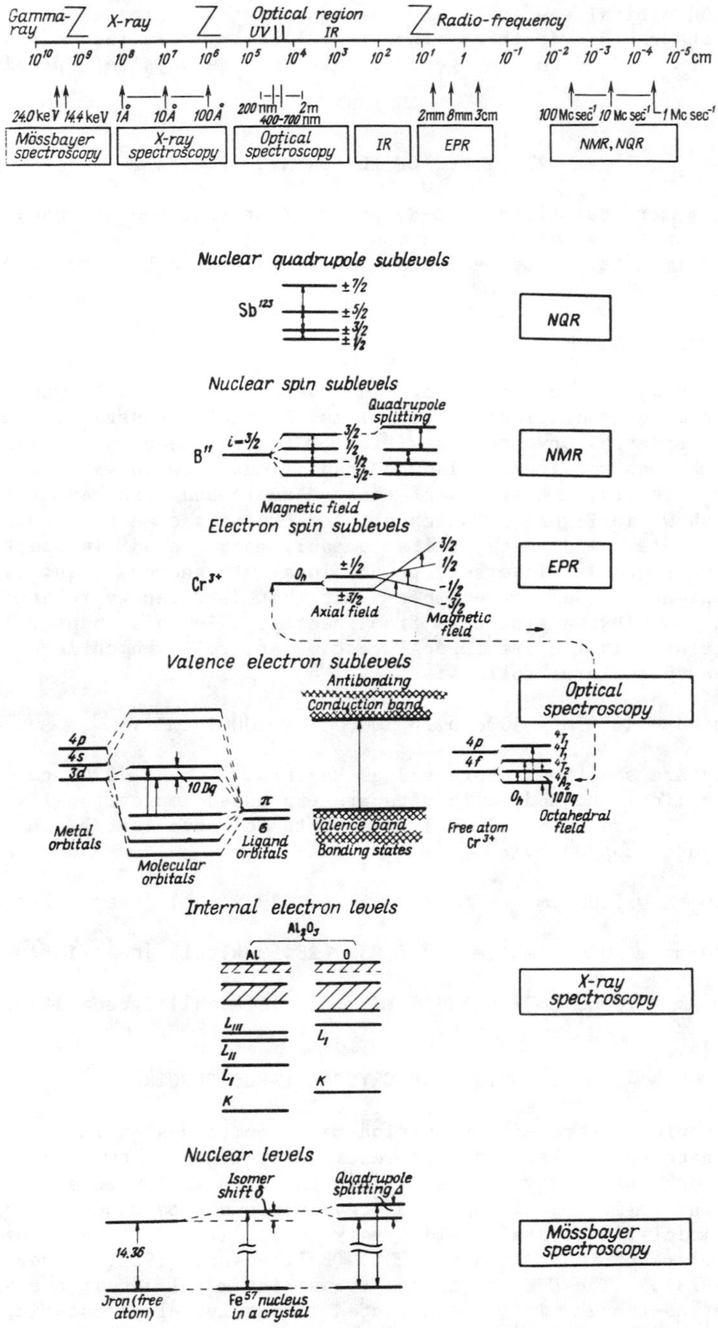

Figure 2. Branches of spectroscopy and related energy levels of solid phases (from Marfunin, 1979).

B and C parameters, to account for covalent bonding interactions of the cation with the coordinated anions or dipolar molecular species (e.g., H_2O, NH_3) which are referred to collectively as ligands. The MOT model focusses on orbital overlap and exchange interactions between the central transition element and its surrounding ligands. All three models make use of the symmetry properties of both the metal orbitals and the configuration of neighboring anions or ligands. The key feature of each model is that each transition metal has five 3d orbitals which are equivalent in a spherical environment such as a (hypothetical) gaseous free ion. However, this five-fold degeneracy is removed when the cation is surrounded by ligands in a crystal structure, the relative energies of the 3d orbitals depending on the symmetry of the ligand environment defining the geometry of the coordination site. Energy level diagrams based on the CFT and MOT models for a transition metal ion in octahedral coordination are compared in Figure 3. Note that both diagrams depict three orbitals of the t_{2g} group (consisting of d_{xy}, d_{yz}, and d_{zx}) to be more stable than the two orbitals of the e_g group (comprising $d_{x^2-y^2}$ and d_{z^2}) localized on the transition metal ion. The energy separation between the t_{2g} and e_g (or antibonding e_g*) levels is denoted by Δ_0 (or 10Dq) and is termed the crystal field splitting (or ligand field splitting) parameter. Δ_0 is obtained directly, or can be estimated, from spectra of transition metal-bearing phases in the visible and near infrared regions.

In the CFT model, the split 3d orbital energy levels are assumed to obey a "center of gravity" rule such that in octahedral (6-fold) coordination the three t_{2g} orbitals are lowered by 0.4 Δ_0 below, and the two e_g orbitals are raised by 0.6 Δ_0 above, the baricenter (corresponding to the energy levels for a cation in a spherically symmetric environment). Each electron in a t_{2g} orbital stabilizes a transition metal ion by $-0.4\ \Delta_0$, whereas every electron in an e_g orbital diminishes its stability by $+0.6\ \Delta_0$. The resultant net stabilization energy is called the crystal field stabilization energy (CFSE). The relative energies of the split 3d orbitals are reversed when transition metals occur in tetrahedral (4-fold), cubic (8-fold), or dodecahedral (12-fold) coordination sites (Fig. 4). Simple electrostatic calculations assuming identical cation, anion, and cation-anion distances for the three coordinations indicate relative crystal field splittings for octahedral (Δ_0), cubic (Δ_c), dodecahedral (Δ_d), and tetrahedral (Δ_t) coordinations to be

$$\Delta_o : \Delta_c : \Delta_d : \Delta_t = 1 : -8/9 : -1/2 : -4/9, \qquad (3)$$

where the negative signs indicate inverted energies of the two groups of split 3d orbitals. An important consequence of the much smaller tetrahedral and dodecahedral crystal field splitting parameters is that CFSE's of transition metal ions are significantly lower in 4- and 12-coordinated sites than in octahedral (6-coordinated) sites. This is expressed as an octahedral site preference energy, OSPE, which is defined by

$$\text{OSPE} = \text{CFSE (octahedral)} - \text{CFSE (tetrahedral)}. \qquad (4)$$

The OSPE parameter figured prominently in early interpretations of transition metal geochemistry (Burns, 1970a; Henderson, 1982). In Table 1 are summarized crystal field splitting and octahedral CFSE data for transition metal ions coordinated to H_2O in hexahydrate complexes, as well as OSPE data for cations in oxides.

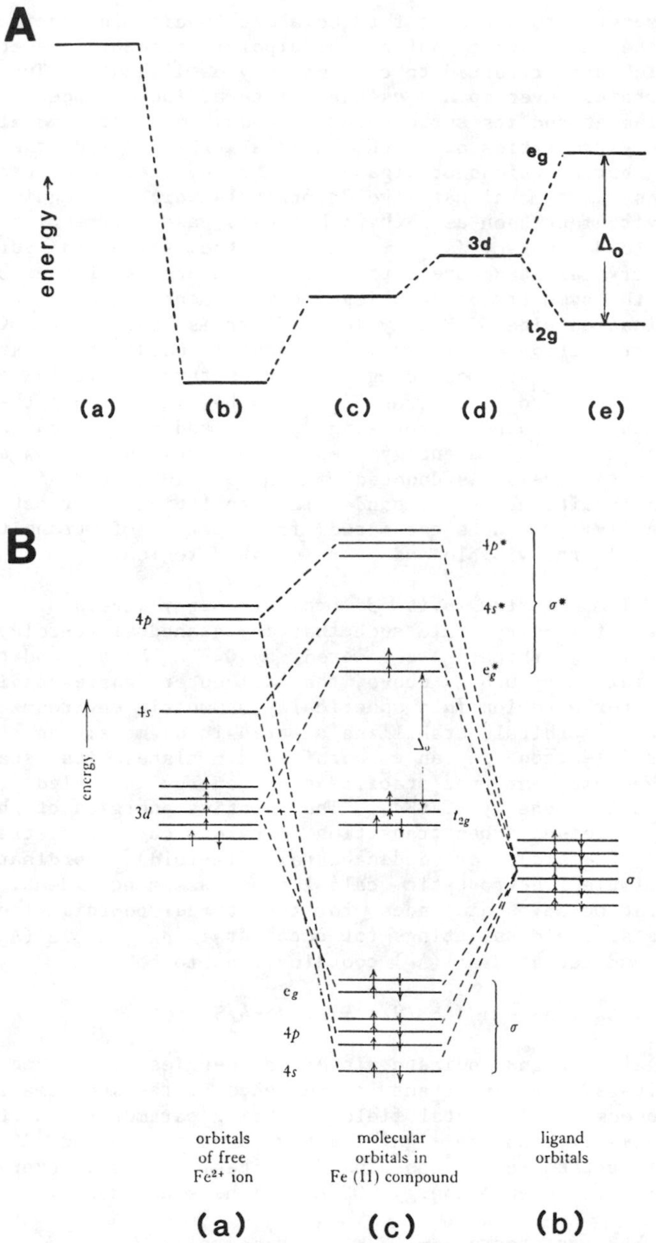

Figure 3. Relative energy levels of a transition metal 3d orbitals in octahedral coordination: (A) crystal field theory model: (a) energy levels of free ion; (b) electrostatic attraction between cation and anions; (c) repulsion between anions and electrons on cation other than those in 3d orbitals; (d) repulsion between anions and 3d electrons; (e) splitting of 3d orbital energy levels in an octahedral crystal field; (B) molecular orbital theory model: (a) energy levels of atomic orbitals of the free Fe^{2+} cation; (b) energy levels for the six L ligands before bonding; (c) molecular orbital energy levels for the octahedral $[ML_6]$ cluster

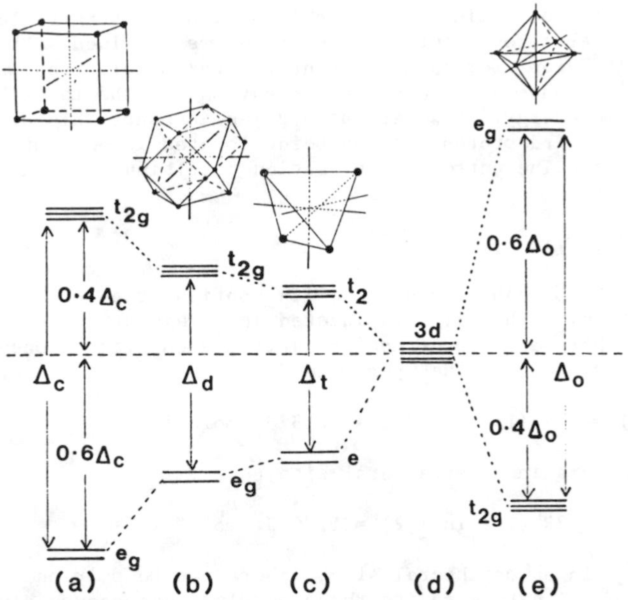

Figure 4. Crystal field splittings of transition metal 3d orbitals in (a) cubic (8-fold); (b) dodecahedral (12-fold); (c) tetrahedral (4-fold); (d) spherical; and (e) octahedral (6-fold) coordinations.

Table 1. Crystal Field Splittings (Δ_o) and Stabilization Energies in Octahedrally Coordinated Transition Metal Ions

Number of Electrons	Ion		Electronic Configuration	Δ_o (cm^{-1}) for Hexahydrated Cation	CFSE cm^{-1} (kJ. $mole^{-1}$)	OSPE (kJ. $mole^{-1}$)
1		Ti^{3+}	$(t_{2g})^1$	20300	0.4 Δ_o = 8120 (97.1)	28.9
2		V^{3+}	$(t_{2g})^2$	17700	0.8 Δ_o = 14160 (169.0)	53.6
3	V^{2+}		$(t_{2g})^3$	12600	1.2 Δ_o = 15120 (180.7)	-
3		Cr^{3+}	$(t_{2g})^3$	17400	1.2 Δ_o = 20880 (249.4)	157.3
4	Cr^{2+}		$(t_{2g})^3(e_g)^1$	13900	0.6 Δ_o = 8340 (99.6)	71.1
4		Mn^{3+}	$(t_{2g})^3(e_g)^1$	21000	0.6 Δ_o = 12600 (150.6)	95.4
5	Mn^{2+}		$(t_{2g})^3(e_g)^2$	7800	0	0
5		Fe^{3+}	$(t_{2g})^3(e_g)^2$	13700	0	0
6	Fe^{2+}		$(t_{2g})^4(e_g)^2$	10400	0.4 Δ_o = 4160 (49.8)	16.74
6		Co^{3+}	$(t_{2g})^6$	18600	2.4 Δ_o = 44640 (533.5)*	79.5
7	Co^{2+}		$(t_{2g})^5(e_g)^2$	9300	0.8 Δ_o = 7440 (89.1)	31.0
8	Ni^{2+}		$(t_{2g})^6(e_g)^2$	8500	1.2 Δ_o = 10200 (123.8)	86.2
9	Cu^{2+}		$(t_{2g})^6(e_g)^3$	12600	0.9 Δ_o = 11340 (90.4)	63.6

* This value for low-spin Co^{3+} is too high. It needs to be reduced by the electron pairing energy, which is approximately equal to Δ_o (i.e., 18600 cm^{-1} or 222.6 KJ. $mole^{-1}$).

Another thermodynamic property arising from different electronic configurations is electronic entropy, which may be obtained as follows (Wood, 1981). The Ti^{3+} ion, for example, contains one 3d electron which in octahedral coordination may be localized in any one of the three t_{2g} orbitals having the same energy. However, at a given instant, this single 3d electron has a one-third probability of being located in each of the d_{xy}, d_{yz} or d_{zx} orbitals. The entropy contribution resulting from this disorder is given by

$$S_{el} = -R \Sigma P_i \ln P_i, \quad (5)$$

where P_i is the probability of the ith configuration occurring and R is the gas constant. When Ti^{3+} is located in a regular (undistorted) octahedral site, there are three configurations each with a one-third probability of occurrence, so that per mole

$$S_{el}(oct) = -3R\ (1/3 \ln 1/3) = 9.13 \text{ J. mole}^{-1}. \quad (6a)$$

If the Ti^{3+} occurs in a tetrahedral site,

$$S_{el}(tet) = -2R\ (1/2 \ln 1/2) = 5.76 \text{ J. mole}^{-1}. \quad (6b)$$

For Cr^{3+} in an octahedral site, there can be only one unique electronic configuration because its three 3d electrons occupy singly each t_{2g} orbital. With probability of occurrence equal to one, the electronic configuration of Cr^{3+} is given by

$$S_{el}(oct) = -\ R\ (1 \ln 1) = 0. \quad (7)$$

The Ti^{3+} ion would also have zero S_{el} if its single 3d electron were located in a distorted octahedral site, because the 3-fold degeneracy of the t_{2g} orbitals is removed. This aspect of low symmetry coordination sites is discussed in the next section.

In addition to coordination number, other factors that affect the crystal field splitting and CFSE parameters include:

(1) The type of anion coordinated to the cation. Ligands can be arranged in order of increasing Δ_o, called the spectrochemical series, such that

$$I^- < Br^- < Cl^- < F^- < OH^- < CO_3^{2-} < O^{2-} < H_2O < NH_3 < HS^- < S^{2-} < CN^- \quad (8)$$

In transition metal-bearing minerals discussed in this chapter, oxygen is the ligand coordinated to cations in oxides and silicates. However, the oxygen bond-type varies from O^{2-} in simple oxides (periclase, corundum, spinel) to $Si\text{-}O^-$ (garnets, olivines, pyroxenes) to Si-O-Si (pyroxene M2 site).

(2) The cation-oxygen interatomic distance, R. For a point-charge model (i.e., oxygen anions are assumed to be point negative charges surrounding the central cation):

$$\Delta \propto 1/R^5. \quad (9)$$

Since the effect of rising pressure is to shorten interatomic distances, Δ increases so that CFSE's of most transition metal ions are raised at elevated pressures. Thus,

$$\Delta_P/\Delta_0 = (R_0/R_P)^5, \tag{10}$$

where Δ_P, Δ_0 and R_P, R_0 are crystal field splittings and average cation-oxygen distances at high pressures and 1 bar, respectively. Since $R = 1/V^3$ (V = volume),

$$\Delta \propto V^{-5/3} \text{ or } \ln \Delta \propto -5/3 \ln V. \tag{11}$$

Differentiation of this equation gives

$$\frac{d\Delta}{\Delta} = -\frac{5dV}{3V} = \frac{5d\rho}{3\rho} = \frac{5dP}{3K}, \tag{12}$$

where ρ is density and K is the bulk modulus or incompressibility of the coordination polyhedron. K is related to the coordination site compressibility, β, by

$$\beta = \frac{1}{K} = -\frac{VdP}{dV}. \tag{13}$$

Rewriting Equation 12 gives $$\frac{d\Delta}{dP} = \frac{5\Delta}{3K} \tag{14}$$

from which the site incompressibilities of transition metal-bearing minerals may be estimated.

Similarly, at elevated temperatures

$$\Delta_T/\Delta_0 = (V_0/V_T)^{5/3} = [1 + \alpha (T - T_0)]^{-5/3}, \tag{15}$$

where Δ_T, Δ_0 and V_T, V_0 are crystal field splittings and specific volumes at temperature T and 25°C (T_0), respectively, and α is the volume coefficient of thermal expansion.

(3) The type of cation. For example, values of Δ are generally higher for trivalent cations than for divalent cations. The data in Table 1 also show that cations such as Cr^{3+} and Ni^{2+} have relatively high CFSE, whereas Mn^{2+} and Fe^{3+} acquire zero CFSE because of their unique (high-spin) d^5 configurations. Cations with zero 3d electrons (e.g., Mg^{2+}, Ca^{2+}, Sc^{3+}, Ti^{4+}) and ten 3d electrons (e.g., Cu^+, Zn^{2+}, Ga^{3+}, Ge^{4+}) also have no CFSE.

4. LOW SYMMETRY ENVIRONMENTS IN DISTORTED COORDINATION SITES

We have seen how ligands forming high symmetry, regular octahedral, tetrahedral, dodecahedral and cubic environments about transition metal ions produce splittings of the five 3d orbitals into two energy levels with 3-fold and 2-fold degeneracies. Such coordination sites are highly idealized, but occur nevertheless in the periclase (octahedra), ideal perovskite (octahedra and dodecahedra), and spinel (tetrahedra) structures. The more important rock-forming oxide and silicate minerals, however, provide low symmetry environments, such as trigonally distorted octahedra in corundum and spinel structures, the rhombic distorted cubic site in garnets, and the very distorted 6-coordinated sites in the olivine, orthopyroxene and clinopyroxene structures. The occurrence of transition metal cations in these low symmetry sites leads to further resolution of the 3d orbitals into additional energy levels.

Examples of such splittings are shown in Figure 5. These energy level diagrams show that orbitals of the three-fold degenerate t_{2g} and two-fold degenerate e_g groups are resolved into additional energy levels when a transition element occurs in a low symmetry coordination site, and that the split energy levels obey "center of gravity" rules about the t_{2g} and e_g baricenters. As a result, additional absorption bands having different polarization dependencies (intensities) are observed when polarized light is transmitted through a non-cubic crystal. Often, the lower level splittings shown in Figure 5 are small and insufficiently sharp or intense to be detected in infrared spectral measurements. Such ground-state splittings are usually estimated or obtained from indirect measurements, including temperature dependencies of the Mössbauer quadrupole splitting parameter (Huggins, 1976).

To designate point symmetries of regular and distorted coordination sites in transition metal-bearing phases, spectroscopists conventionally use Schoenflies symbols rather than Hermann-Mauguin symbols more familiar to mineralogists. Thus, undistorted octahedra or tetrahedra have point symmetries O_h or T_d, respectively, while D_{4h} and C_{3v} designate tetragonally and trigonally distorted octahedra, for example. Other examples of Schoenflies point group symbols for coordination sites of minerals appear later.

Distorted coordination sites in oxide and silicate minerals have several important consequences in transition metal geochemistry. Not only do they contribute to the visual pleochroism of many minerals under the polarizing microscope, but low symmetry environments also stabilize unusual valencies of cations subject to the Jahn-Teller effect (e.g., Cr^{2+}, Mn^{3+}, Ti^{3+}) and induce preferred site occupancies in several ferromagnesian silicates. Examples are discussed later.

5. CRYSTAL FIELD STABILIZATION ENERGIES FROM OPTICAL SPECTRA

During the past 30 years, visible-region spectral measurements of numerous transition metal compounds and many natural and synthetic minerals hosting different cations have been made. The development of microscope spectrophotometers in the past decade has extended the spectral measurements to most rock-forming minerals in polarized light. The evolution of microscope absorption spectroscopy has paralleled developments of high pressure diamond anvil cells, with the result that pressure variations of mineral spectra are now made routinely. The stability of transparent diamonds to temperature also indicates the potential of using dia-

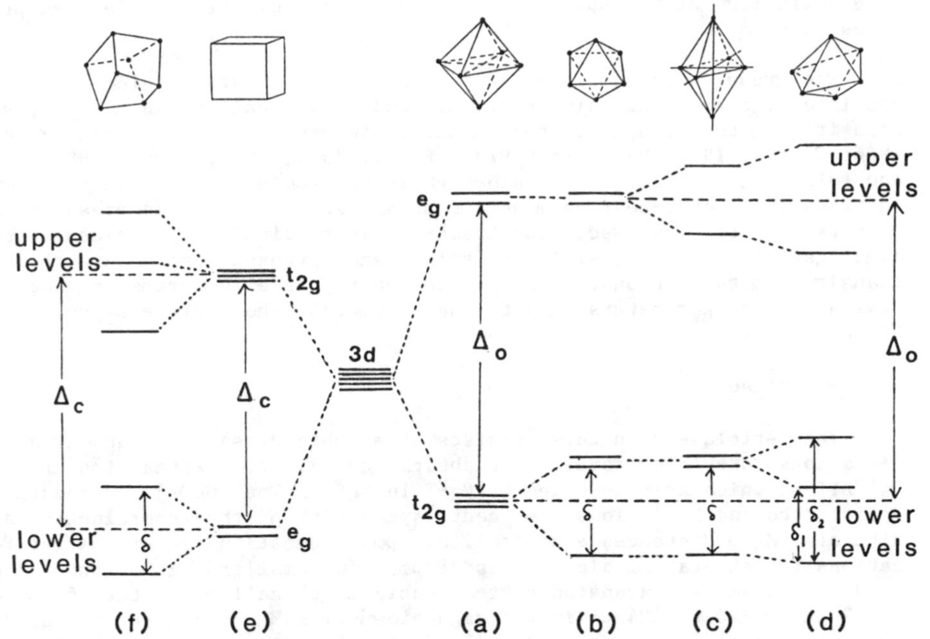

Figure 5. Relative energies of 3d orbital energy levels of a transition metal ion in low symmetry distorted sites: (a) regular octahedron (e.g., periclase); (b) trigonally distorted octahedron (e.g., corundum, spinel, approximate olivine M2 site); (c) tetragonally elongated octahedron (e.g., approximate olivine M1 site); (d) distorted 6-coordinated site (e.g., M1 and M2 sites of pyroxene and olivine); (e) regular cube; (f) distorted cube (e.g., triangular dodecahedral site of garnet).

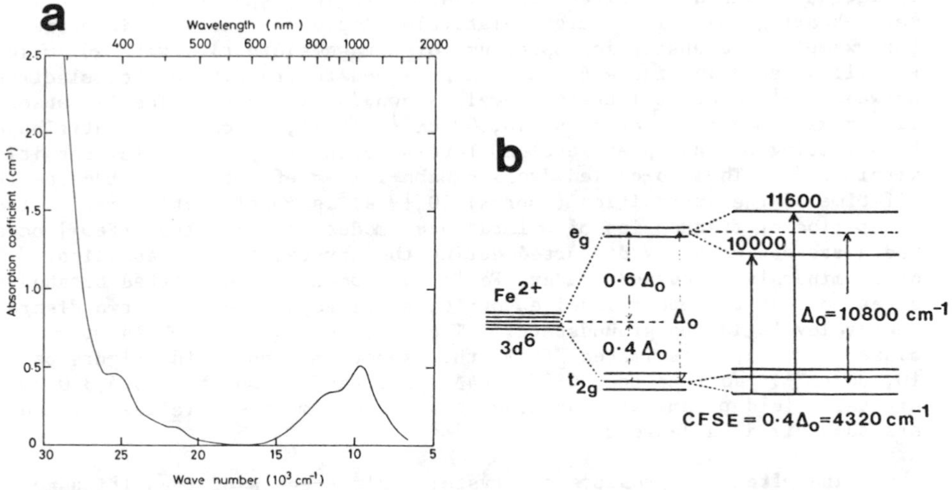

Figure 6. Crystal field spectrum and energy levels of Fe^{2+}-bearing periclase: (a) spectrum of magnesiowüstite ($Mg_{0.74}Fe_{0.26}$)O (from Goto et al., 1980); (b) energy level diagram of Fe^{2+} in magnesiowüstite.

mond cells for high temperature spectral measurements at elevated pressures.

The group theoretical basis for assigning crystal field spectra to specific electronic transitions within individual cations in a variety of transition metal compounds and minerals is described elsewhere (Figgis, 1966; Cotton, 1971; Marfunin, 1979; Burns, 1982, 1985; Lever, 1985). In the following sections, CFSE values obtained from spectral data of selected transition metal-bearing minerals at ambient and elevated pressures and temperatures are reviewed. The treatment is confined to periclase, corundum, garnet, olivine, silicate spinel and pyroxene structures hosting transition metal cations. Many of the energy level diagrams and derived crystal field parameters update the data published fifteen years ago (Burns, 1970a).

5.1 Periclase

The periclase structure consists of a cubic closest packed array of O^{2-} anions providing regular octahedral coordination sites (point group O_h) all of which are occupied by Mg^{2+} in MgO. The $[MgO_6]$ octahedra are undistorted and Mg^{2+} ions are centrosymmetric in the coordination site with all Mg-O distances equal to 210.6 pm. Substitution of larger Fe^{2+} cations (octahedral ionic radius 78 pm) for smaller Mg^{2+} ions (72 pm) leads to a small expansion of the cubic unit cell parameter from a = 421.2 pm in MgO to 424 pm in the magnesiowüstite $Mg_{0.75}Fe_{0.25}O$ (denoted as $Wü_{25}$). The Ni^{2+} (70 pm) and Co^{2+} (73.5 pm) cations are accommodated in transition metal-doped MgO crystals without much change in the unit cell parameters. Small trivalent cations such as Cr^{3+} (61.5 pm) and Ti^{3+} (67 pm), however, find themselves in comparatively large sites in MgO which would be expected to be manifested in relatively low crystal field splitting parameters compared to hexahydrated cations (cf. Table 1). Because the cubic periclase structure contains cations in regular centrosymmetric octahedral sites, the visible region spectra of transition metal-bearing periclases are relatively simple and easily interpreted. For example, the absorption spectrum of a magnesiowüstite with 26 mole % FeO illustrated in Figure 6a may be assigned to transitions of electrons between Fe^{2+} 3d orbital energy levels shown in Figure 6b. The two absorption maxima occurring at about 10,000 cm^{-1} and 11,600 cm^{-1} are attributed to splitting of the upper level e_g levels during the electronic transition within Fe^{2+}. This so-called dynamic Jahn-Teller effect occurs because the lifetime of the transition (approx. 10^{-13} s) is considerably smaller than the period of frequencies of vibrational modes so that the $[FeO_6]$ octahedra are not actually distorted during the crystal field transition. In other minerals discussed later, Fe^{2+} ions occur in distorted octahedral sites so that the upper level e_g orbitals are separated into two discrete energy levels in the groundstate. The value of Δ_o for Fe^{2+} in the periclase structure estimated from the spectrum shown in Figure 6a is 10,800 cm^{-1}, so that the CFSE ($0.4\Delta_o$ for Fe^{2+}) amounts to 4,320 cm^{-1}. Crystal field parameters for other transition metal-periclase structures are summarized in Table 2.

The effect of pressure on crystal field spectra of transition metal-doped periclases was first studied by Drickamer and coworkers (see Drickamer and Frank, 1973), and typical results are illustrated in Figure 7a.

Table 2. Crystal Field Parameters for Transition Metals in Periclase Structures

Oxide	$\Delta_o(cm^{-1})$	$CFSE(cm^{-1})$	References
$MgO : Fe^{2+}$	10800	4320	a-c
$MgO : Co^{2+}$	9000	7200	d-g
CoO	8500	6800	g
$MgO : Ni^{2+}$	8500	10200	h-k
NiO	8750	10500	l
$MgO : Cr^{3+}$	16200	19440	k,m
$MgO : Ti^{3+}$	11360	4544	k

a K.W. Blazey, J. Phys. Chem. Solids 38, 671 (1977)
b T. Goto, T.J. Ahrens, G.R. Rossman, and Y. Syono, Phys. Earth Planet. Interiors 22, 277 (1980)
c G. Smith, Phys. Status Solidi A, 61A, K191 (1980)
d D. Reinen, Monatsch. Chem. 96, 730 (1964)
e R. Pappalardo, D.L. Wood and R.C. Linares, Jr., J. Chem. Phys., 35, 2041 (1961)
f O. Schmitz-DuMont and C. Friebel, Monatsch. Chem. 98, 1583 (1967)
g G.W. Pratt, Jr., and R. Coelho, Phys. Rev. 116, 261 (1959)
h W. Low, Phys. Rev. 109, 247 (1958)
i O. Schmitz-DuMont, H. Gossling, and H. Brokopf, Z. anorg. allgem. Chemie 300, 159 (1959)
j R. Pappalardo, D.L. Wood and R.C. Linares, Jr., J. Chem. Phys. 35, 1460 (1961)
k H.G. Drickamer and C.W. Frank, Electronic Transitions and the High Pressure Chemistry and Physics of Solids. Chapman and Hall, (1973), p. 74
l G.R. Rossman, R.D. Shannon and R.K. Waring, J. Solid State Chem. 39, 277 (1981)
m W. Low, Phys. Rev. 105, 801 (1957)

Table 3. Crystal Field Parameters for Transition Metals Ions in Corundum Structures

Oxide	$\Delta_o(cm^{-1})$	$CFSE(cm^{-1})$	Reference
$Al_2O_3 : Ti^{3+}$	19500	7800	a,b
$Al_2O_3 : V^{3+}$	18730	14980	a,c
$Al_2O_3 : Cr^{3+}$	18150	21780	a
Cr_2O_3	16600	19920	c-f
$Al_2O_3 : Mn^{3+}$	19470	11680	a
$Al_2O_3 : Fe^{3+}$	14300	0	g-i
Fe_2O_3	14000	0	j
$Al_2O_3 : Co^{3+}$	18300	25620*	a,k
$Al_2O_3 : Ni^{3+}$	18000	14400*	a,k
$Al_2O_3 : Co^{2+}$	12300	9840	k
$Al_2O_3 : Ni^{2+}$	10700	12840	k

* Low-spin cations. These CFSE's are corrected for electron pairing energies, which are assumed to approximate the Δ_o values.
a D.S. McClure, J. Chem. Phys. 36, 2757 (1962)
b M.G. Townsend, Solid State Comm. 6, 81 (1968)
c H.U. Rahman and W.A. Runciman, J. Phys. C. 4, 1576 (1971)
d C.P. Poole, Jr., J. Phys. Chem. Solids 25, 1169 (1964)
e A. Neuhaus, Zeit. Krist. 113, 195 (1960)
f D. Reinen, Struct. Bonding 6, 30 (1969)
g G. Lehman and H. Harder, Amer. Mineral. 55, 98 (1970)
h J. Ferguson and P.E. Fielding, Chem. Phys. Lett. 10, 262 (1971); Australian J. Chem. 25, 1371 (1972)
i K. Eigenmann, K. Kurtz, and H. H. Gunthard, Helv. Phys. Acta 45, 452 (1972); Chem. Phys. Lett. 13, 58 (1971)
j R.V. Morris and H.V. Lauer, Lunar Planet. Science XIV, 524 (1983)
k R. Muller and H.H. Gunthard, J. Chem. Phys. 44, 365 (1966)

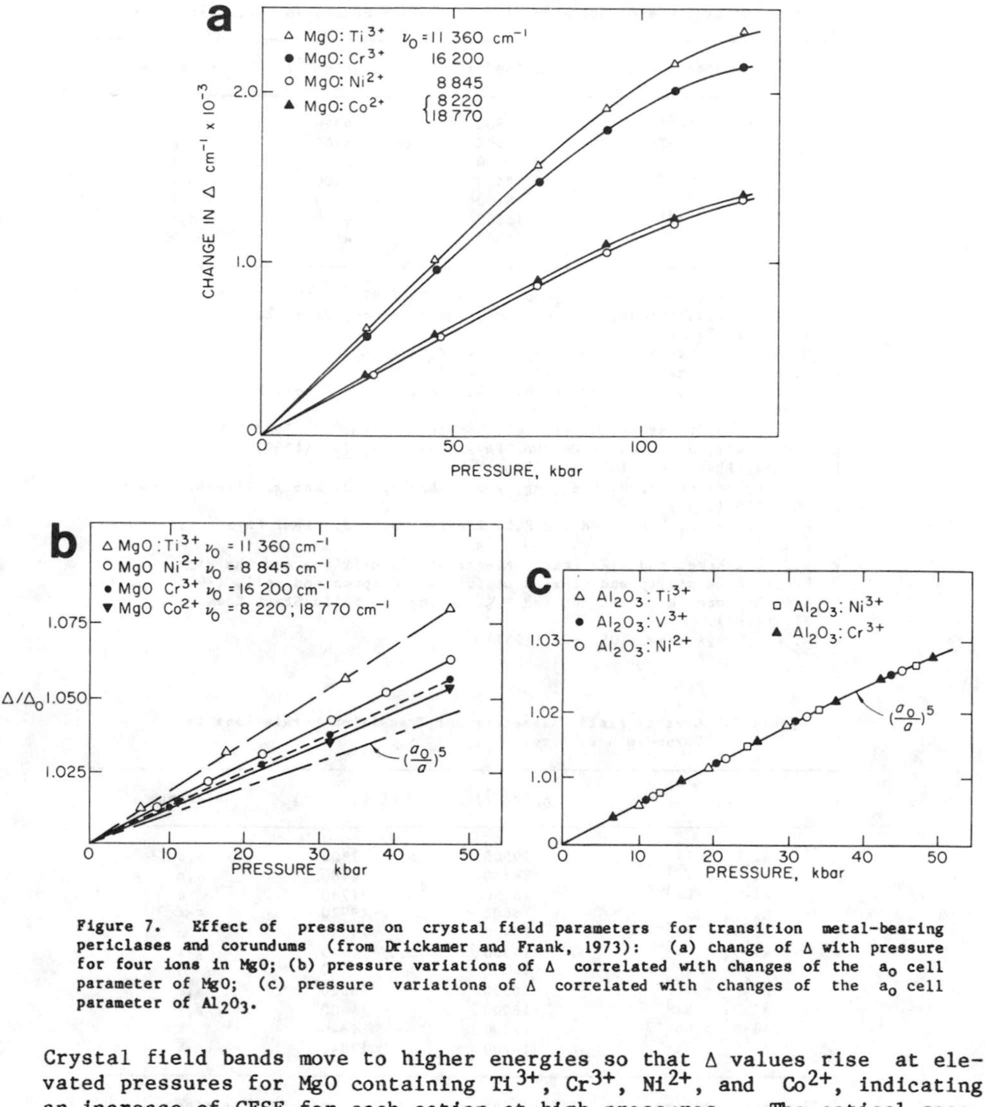

Figure 7. Effect of pressure on crystal field parameters for transition metal-bearing periclases and corundums (from Drickamer and Frank, 1973): (a) change of Δ with pressure for four ions in MgO; (b) pressure variations of Δ correlated with changes of the a_0 cell parameter of MgO; (c) pressure variations of Δ correlated with changes of the a_0 cell parameter of Al_2O_3.

Crystal field bands move to higher energies so that Δ values rise at elevated pressures for MgO containing Ti^{3+}, Cr^{3+}, Ni^{2+}, and Co^{2+}, indicating an increase of CFSE for each cation at high pressures. The optical spectra of Fe^{2+}-doped periclase and magnesiowüstites show similar pressure-induced shifts (Shankland et al., 1974). However, as shown in Figure 7b, the pressure shifts of Δ (and hence CFSE) deviate from the $\Delta \propto R^{-5}$ dependence predicted from a simple point charge model (Eqn. 9). In Figure 7b, the fractional changes in Δ at elevated pressures over the one atmosphere value, Δ_0, for several cations in MgO plotted against the change in lattice parameter (where a_0 = 421.2 pm at 1 atmosphere) are somewhat higher than the predicted values. As frames of reference, CFSE of some cations in MgO at 100 kbar are: Ni^{2+} = 12000 cm^{-1}; Co^{2+} = 8200 cm^{-1}; Fe^{2+} = 4600 cm^{-1}; Cr^{3+} = 21700 cm^{-1}; and Ti^{3+} = 5400 cm^{-1}.

5.2 Corundum

The corundum structure is based on a hexagonal closest-packed array of O^{2-} anions in which two-thirds of the octahedral sites are occupied by Al^{3+}. Each Al^{3+} is surrounded by a trigonally distorted octahedron of O^{2-} ions and is not centrally located in the [AlO_6] octahedron. Thus, Al-O distances are 197 pm to three oxygens in one plane and 186 pm to three oxygens in the other plane. Most of the cations substituting for Al^{3+} (octahedral ionic radius 53 pm) in transition metal-bearing Al_2O_3 have significantly larger radii (e.g., Ti^{3+}, 67 pm; V^{3+}, 64 pm; Cr^{3+}, 61.5 pm; Fe^{3+}, 65.5 pm). As a result, the positions of absorption bands in crystal field spectra of transition metal-doped Al_2O_3 are expected to occur at higher energies than in hexahydrated complex ions (Table 1). In addition, the occurrence of cations in non-centrosymmetric distorted octahedral sites (point group C_3) in the trigonal corundum leads to dichroism and hence polarization dependencies of positions and intensities of absorption bands in polarized visible region spectra (McClure, 1962; Burns and Burns, 1984).

The optical spectra of natural and synthetic transition metal-doped corundums have been extensively studied both theoretically and experimentally, and data from measurements at ambient pressures and temperatures are summarized in Table 3. The Δ_o and derived CFSE data are only approximate values because the ground-state splittings, δ, of the t_{2g} orbital group (see Fig. 5b) are unknown but probably are smaller than 1000 cm^{-1}.

High pressure spectral measurements of transition metal-doped Al_2O_3, including several studies of ruby (Stephens and Drickamer, 1961; Goto et al., 1979) show pressure-induced shifts of the crystal field bands to higher energies. Results summarized in Figure 7c indicate that increases of the crystal field splitting parameters for several cations in Al_2O_3 are in remarkable close agreement with trends predicted from the $\Delta \propto 1/R^5$ relationship (Eqn. 9). The effect of temperature on the crystal field spectra of ruby, on the other hand, is to shift band maxima to lower energies (McClure, 1962; Parkin and Burns, 1980). Two frames of reference for CFSE values of Cr^{3+} in Al_2O_3 are 23850 cm^{-1} at 460 kbar and 19200 cm^{-1} at 900°C, compared to 21780 cm^{-1} at ambient pressures and temperatures.

5.3 Garnets

The garnet structure hosts transition metal cations in sites having three different coordination symmetries: 8-fold (triangular dodecahedral), 6-fold (octahedral) and 4-fold (tetrahedral). In silicate garnets, apart from tetrahedral Fe^{3+} ions in titaniferous melanites and schorlomites, the predominant site occupancies involve divalent cations in the larger 8-coordinated and trivalent cations in the smaller 6-coordinated sites (Meagher, 1980).

The octahedral site in the garnet structure is only slightly distorted and is more regular than the 6-coordinated sites found in olivine and pyroxenes. Its point group symmetry is D_{3h}, so that all six metal-oxygen distances are identical. They range from Al^{3+}-O distances of 188.6 pm and 192.4 pm in pyrope and grossular, respectively, to Fe^{3+}-O = 202.4 pm in andradite, Cr^{3+}-O = 198.5 pm in uvarovite, V^{3+}-O = 198.8 pm in goldmanite, to (Al^{3+},Cr^{3+})-O = 190.5 pm in chrome-pyrope. The crystal field parameters for a variety of garnets containing trivalent transition metal ions summarized in Table 4 reflect these variations of metal-oxygen distances. The effect of pressure on crystal field spectra of synthetic uvarovite is shown in Figure 8, and parameters derived from the spectra are summarized in Table 5. The B Racah and site compressibility data are discussed later (Sections 6 and 7).

The 8-coordinated triangular dodecahedral site approximates a distorted cube (point group symmetry D_2), in which there are two sets of four identical metal-oxygen distances. In almandine these Fe^{2+}-O distances are 220.0 pm and 237.8 pm (mean: 230.0 pm), while in pyrope they are 219.6 pm and 234.2 pm (mean: 226.9 pm).

The absorption spectra of almandine garnets such as that shown in Figure 9a, contain three bands near 4400, 5800 and 7600 cm^{-1} corresponding to transitions to the three upper levels of t_{2g} orbital energy levels. The energy level diagram shown in Figure 9b for an almandine with 70 percent $Fe_3Al_2Si_3O_{12}$ (Burns, 1970a, p. 85) utilizes the energy separation of 1100 cm^{-1} for split lower level e_g orbitals derived from temperature variations of almandine Mössbauer spectra (Huggins, 1975). The derived CFSE, estimated from $(0.6\Delta_c + 0.5\delta)$ is 3780 cm^{-1} for Fe^{2+} in this almandine. The spectrum of a pyrope with 20 percent $Fe_3Al_2Si_3O_{10}$, having absorption bands at 4540, 6060 and 7930 cm^{-1} (Burns, 1970a, p. 84) yields larger values, Δ_c = 5680 cm^{-1} and CFSE = 3910 cm^{-1} for Fe^{2+}, which are consistent with the smaller metal-oxygen distances in pyrope. Over the composition range Alm_1-Alm_{100}, therefore, corresponding ranges of Δ_c and CFSE are 5750-5350 cm^{-1} and 3960-3700 cm^{-1}, respectively. Smith and Langer (1983) measured the effects of pressure on the 7600 cm^{-1} and 5800 cm^{-1} bands of a synthetic almandine. By 100 kbar the bands had shifted to 8200 and 6700 cm^{-1}. Estimates of Δ_c and CFSE for Fe^{2+} in almandine were 6350 cm^{-1} and 4270 cm^{-1}, respectively, at 100 kbar (Smith and Langer, 1983). The almandine (Alm_{70}) spectra measured at 400°C (Parkin and Burns, 1980) indicate temperature-induced shifts of the absorption bands to 7620, 5850 and 4150 cm^{-1}, indicating Δ_c = 5370 cm^{-1} and CFSE = 3720 cm^{-1}.

5.4 Olivine

Because the olivine structure contains two contrasting 6-coordinated sites which should give distinguishable crystal field spectra, the electronic spectra of several transition metal-bearing olivines have been studied extensively in attempts to detect preferred site occupancies and to determine differences of CFSE. Most measurements have been made on Fe^{2+} olivines of the fayalite-forsterite and fayalite-tephroite series (Burns, 1970b; Hazen et al., 1977), but some spectral data also exist for Cr-, Ni-, and Co-bearing olivines (G.R. Rossman, unpublished data). Typical polarized absorption spectra of fayalite and synthetic Ni_2SiO_4 are illustrated in Figure 10.

The two 6-coordinated sites of olivine are both distorted from octahedral symmetry, the centrosymmetric M1 site (point group C_i), having smaller average cation-oxygen distances than the non-centrosymmetric M2 site (point group C_s). Pertinent metal-oxygen distances for transition metal-bearing olivines are summarized by Brown (1980). Polarized absorp-

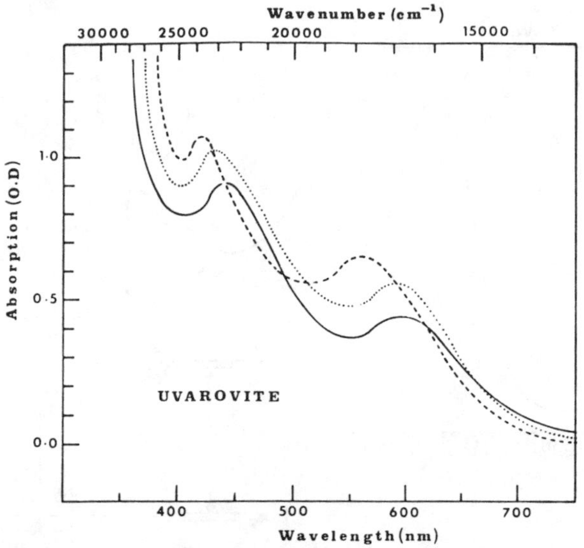

Figure 8. Pressure variations of the crystal field spectra of Cr^{3+} in synthetic uvarovite (from Abu-Eid, 1976) ———— 1 atmos.; .... 9.5 GPa; - - - - - - 17.7 GPa.

Table 4. Crystal Field Parameters for Trivalent Transition Metal Ions in Garnet Structures

Mineral	M^{3+}-O Distance (pm)	$\Delta_o(cm^{-1})$	CFSE(cm^{-1})	Reference
		V^{3+}-garnets		
goldmanite	198.8	17100	13680	a
grossular	192.4	17670	14140	a
pyrope	188.6	18500	14800	a
		Cr^{3+}-garnets		
uvarovite	198.5	16667	20000	b
uvarovite	198.5	16260	19510	c
grossular	192.4	16500	19800	c
andradite	202.4	16100	19320	c
pyrope	188.6	17790	21350	c
pyrope	188.6	17400	20880	a
		Mn^{3+}-garnets		
synthetic Mn-grossular	-	15000	10890	d
		Fe^{3+}-garnets		
andradite	202.4	12600	0	e

a K. Schmetzer and J. Otteman, N. Jahrb. Mineral. Abh. 136, 146 (1979)
b R.M. Abu-Eid and R.G. Burns, Amer. Mineral. 61, 391 (1976)
c G. Amthauer, N. Jahrb. Mineral. Abh. 126, 158 (1976)
d K.R. Frendrup and K. Langer, N. Jahrb. Mineral. Mh. no. 6, 245 (1981)
e R.K. Moore and W.B. White, Canad. Mineral. 11, 791 (1972)

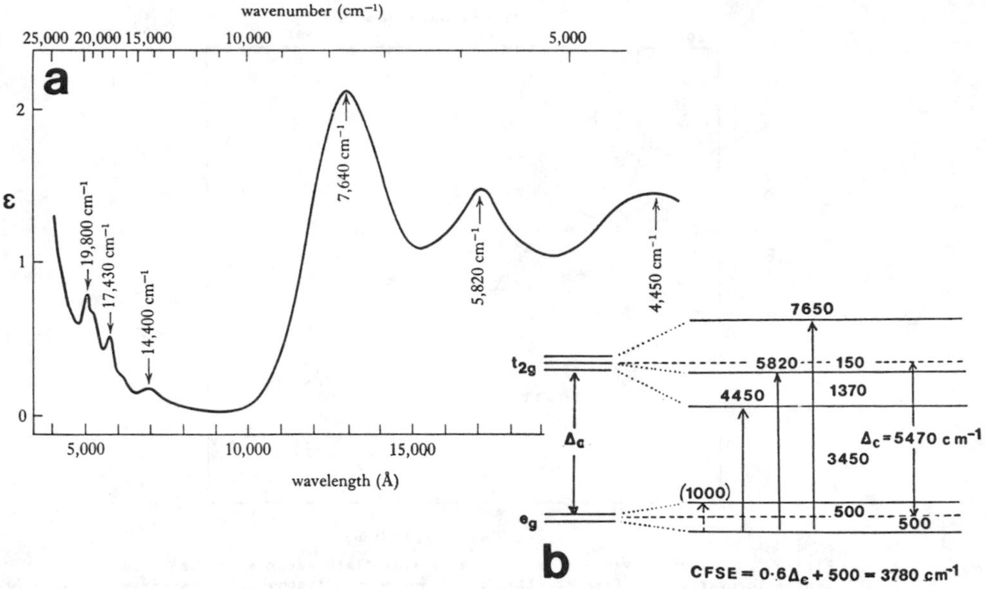

Figure 9. Crystal field spectrum and energy levels of Fe^{2+} in almandine garnet: (a) spectrum of $Alm_{70}Pyr_{30}$ (modified from Burns, 1970a, p. 85); (b) energy level diagram of 8-coordinate Fe^{2+} in alamandine.

Table 5. Spectral Data and Derived Parameters for Cr^{3+} in Synthetic Uvarovite at High Pressures

Parameter	Pressure, GPa 0.0001	4.2	9.5	17.7	19.7	Reference
ν_1 (cm^{-1})	16667	16810	17007	17668	17820	a
ν_2 (cm^{-1})	22727	22846	23041	23697	23840	a
B (cm^{-1})	589	585	582	577	575	b
β_{35}	0.642	0.637	0.634	0.629	0.624	c
ionicity change (%)	-	-0.8	-1.1	-2.1	-2.4	d
$\delta\Delta$ (cm^{-1})	-	143	340	1001	1153	e
K (GPa)	-	427	449	491	474	f
$\beta \times 10^3$ (GPa^{-1})	-	2.34	2.23	2.04	2.11	g
K_0 (GPa)	-	431	447	472	453	h

a Modified from: R.M. Abu-Eid and R.G. Burns, Amer. Mineral. 61, 391 (1976); Note: $\nu_1 \equiv \Delta$
b Racah parameter B calculated from Equation 17
c $\beta_{35} = B/B_0$, where $B_0 = 918$ cm^{-1} for field-free Cr^{3+}
d ionicity change = B/B (at 0.0001 GPa)
e $\delta\Delta = \Delta - \Delta$ (at 0.0001 GPa)
f site bulk modulus calculated from Equation 18
g site compressibility calculated from Equation 18
h zero pressure polyhedral bulk modulus calculated from Equation 20

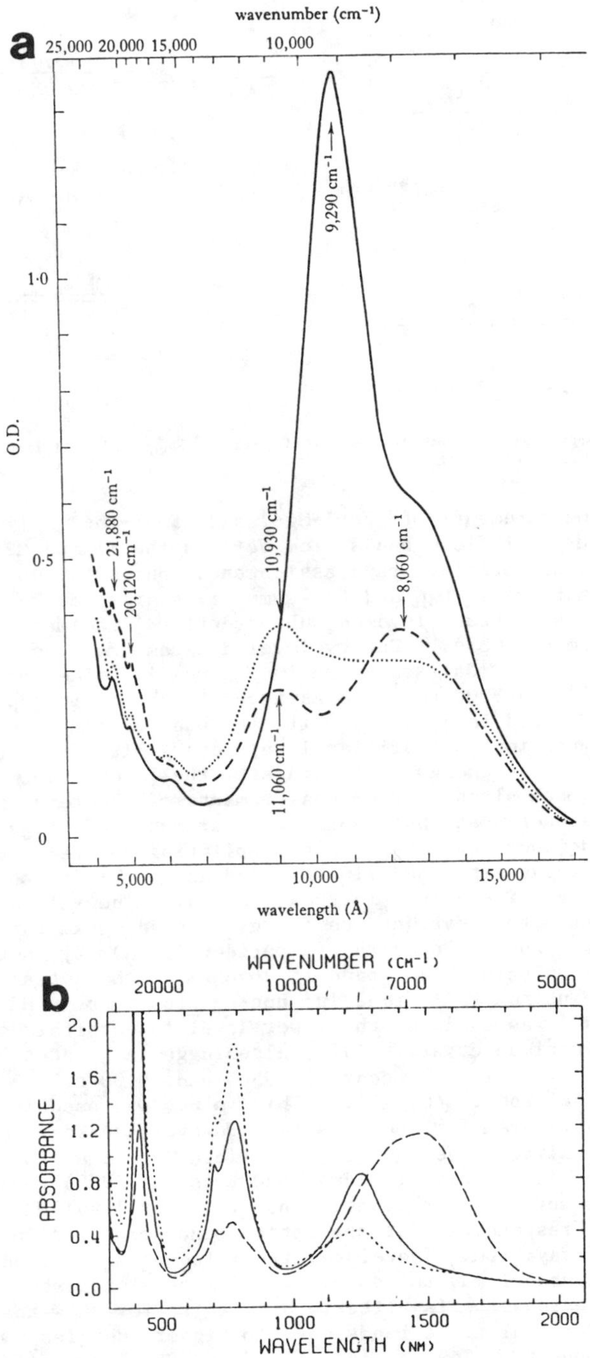

Figure 10. Polarized absorption spectra of Fe^{2+} and Ni^{2+} olivines: (a) fayalite, Fa_{96}, from Rockport, Massachusetts; thickness 80 μm (from Burns, 1970a, p. 81); (b) synthetic α-Ni_2SiO_4; thickness 35 μm (from Rossman et al., 1981). [Optic orientation: α = b; β = c; γ = a; ………………… α-spectra; – – – – – – – – β-spectra; ———— γ-spectra]

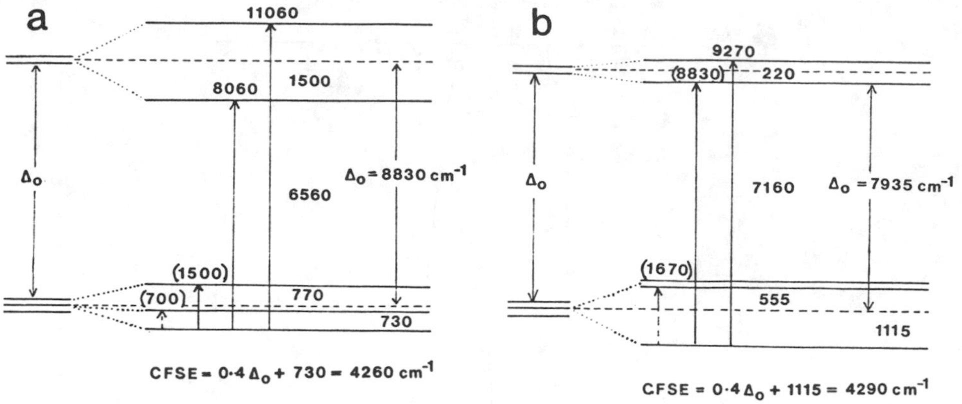

Figure 11. Energy level diagrams for Fe^{2+} in fayalite, Fa_{96} (modified from Burns, 1970a, p. 84): (a) M1 site; (b) M2 site.

tion spectral measurements of Fe^{2+}-Mg^{2+} and Fe^{2+}-Mn^{2+} olivines (Burns, 1970b) enabled crystal field bands from Fe^{2+} in the M1 and M2 sites to be identified. In his original band assignments, Burns (1970b) modelled the M1 and M2 sites to have D_{4h} and C_{3v} symmetries, respectively, but lower site symmetries have been invoked subsequently (Runciman et al., 1973; Goldman and Rossman, 1985). Energy level diagrams derived from the spectra of fayalite (Fig. 10a) are shown in Figure 11. The absorption bands around 11000 cm^{-1} and 8000 cm^{-1} are assigned to the split upper levels of e_g orbitals of M1 site Fe^{2+} ions, while the band centered near 9300 cm^{-1} represents a transition to a similar level of M2 site Fe^{2+} ions. Uncertainties exist for the energy splittings of the lower level t_{2g} orbitals of Fe^{2+} ions in both sites. From measurements of the temperature dependence of the Mössbauer quadrupole splitting parameter of fayalite, Huggins (1976) estimated lower level t_{2g} orbital splittings of Fe^{2+} to be 620 cm^{-1} and 1400 cm^{-1} for one site, and 710 cm^{-1} and 1500 cm^{-1} for the other site, but large uncertainties exist in these values. The values of 700 and 1500 cm^{-1} were used to construct the energy level diagram for the M1 site. The polarized absorption spectra of peridot (Fa_{10}) by Runciman et al. (1973) indicated an absorption band at 1668 cm^{-1}, the intensity of which suggests an origin from Fe^{2+} ions in non-centrosymmetric M2 sites. The value of 1670 cm^{-1} was used as the lower level t_{2g} orbital splitting for the M2 site. Runciman et al. (1973) also suggested that transitions to the upper level e_g orbitals occur at 9540 and 9100 cm^{-1}, indicating a small splitting of about 440 cm^{-1}. This value was used in constructing the energy level diagram for the M2 site. However, analogies between isostructural triphyllite ($LiFePO_4$) crystal field spectra (which originate from M2 site Fe^{2+} ions only) (Goldman and Rossman, 1985) and γ-polarized fayalite spectra suggest that transitions to upper level M2 site Fe^{2+} e_g orbitals may be responsible for absorption bands centered near 9300 cm^{-1} and 8000 cm^{-1} of fayalite. Therefore, the value of Δ_0 derived for M2 site Fe^{2+} ions in Figure 11 may be too high. Compositional variations of the olivine spectra across the fayalite-forsterite series show that peak maxima of all Fe^{2+} crystal field bands move to higher energies with decreasing Fe_2SiO_4 component. The ranges of Δ_0 and CFSE of Fe^{2+} ions in the composition range Fa_{10} to Fa_{100} are estimated to be

M1 site: Δ_o = 9670-8830cm^{-1}; CFSE = 4600-4250cm^{-1},

M2 site: Δ_o = 8210-7930cm^{-1}; CFSE = 4400-4280cm^{-1}.

Several high pressure spectral measurements have been made of Mg-Fe^{2+} olivines. The dominant band due to M2 site Fe^{2+} ions shows a pronounced pressure-induced shift of about 12-16 cm^{-1}/kbar to higher energies for fayalite and forsteritic olivines (Shankland et al., 1974; Smith and Langer, 1982). Bands attributed to M1 site Fe^{2+} ions show smaller shifts at elevated pressures (Mao and Bell, 1972; Abu-Eid, 1976). These trends have been interpreted as indicating less distortion of the olivine M1 site at high pressures. Smith and Langer (1982) concluded from the different pressure-induced shifts of absorption bands that the CFSE of M2 site Fe^{2+} might increase at a faster rate with pressure than the M1 site Fe^{2+} value, thereby influencing the partitioning of Fe^{2+} between the two sites. However, without information about lower level splittings of the t_{2g} orbitals in both sites at high pressures, CFSE's of Fe^{2+} cannot be estimated accurately from high pressure crystal field spectra of olivine.

Crystal field spectra of Ni^{2+}-bearing olivines include those on Ni_2SiO_4 shown in Figure 10b and Ni^{2+}-doped forsterites (Burns, 1970a, p. 157; Wood, 1974; Rossman et al., 1981). The diffuse reflectance spectra of powdered nickel-bearing olivines (Wood, 1974) lack the resolution of polarized absorption spectra of single crystals (Rossman et al., 1981). Nevertheless, using a manual curve-resolver, Wood (1974) fitted the spectrum of $Ni_{0.1}Mg_{1.9}SiO_4$ and estimated the CFSE of Ni^{2+} ions in the two sites to be 27.3 kcal (9550cm^{-1}) and 25.7 kcal. (8990cm^{-1}) per gram atom for the M1 and M2 sites, respectively. The higher CFSE for the M1 site appeared to support the evidence for the enrichment of Ni^{2+} ions in this site of the olivine structure (Rajamani et al., 1975). The polarized absorption spectra of single crystals of Ni_2SiO_4 (Fig. 10b) and $(Mg,Ni)_2SiO_4$ show broad bands centered at approximately 6750-6970 cm^{-1} and 8020-7860 cm^{-1} (Rossman et al., 1981), from which better estimates of the CFSE's of Ni^{2+} in the olivine M1 and M2 sites might be 9620-9430 cm^{-1} and 8100-8360 cm^{-1}, respectively.

5.5 Silicate spinels

Although only a few transition metal-bearing silicate spinels (γ-phases) have been synthesized, spectral data for Fe_2SiO_4 and Ni_2SiO_4 have been obtained and employed in discussions of the olivine→spinel transition described later. The transition metal ions in these silicate spinels are accommodated in the 6-coordinated B sites, which show only small trigonal distortion (point group C_{3v}) from octahedral symmetry. The crystal field spectra are further simplified by showing no polarization dependences for the cubic spinel phase.

The crystal field spectrum of Fe_2SiO_4 spinel illustrated in Figure 12a contains a broad, slightly asymmetric band centered around 11430 cm^{-1} (Mao and Bell, 1972; Burns and Sung, 1978), leading to the crystal field parameters for Fe^{2+} of Δ_o = 10760 cm^{-1} and CFSE = 4970 cm^{-1}. These values are significantly higher than the values estimated for Fe^{2+} in the olivine M1 and M2 sites. The spectra of Ni_2SiO_4 illustrated in Figure 12b contain an intense band at 9150 cm^{-1} at 1 bar with a prominent shoulder at about 8000 cm^{-1}. At 120 kbar, these two features shifted to 10100 cm^{-1} and

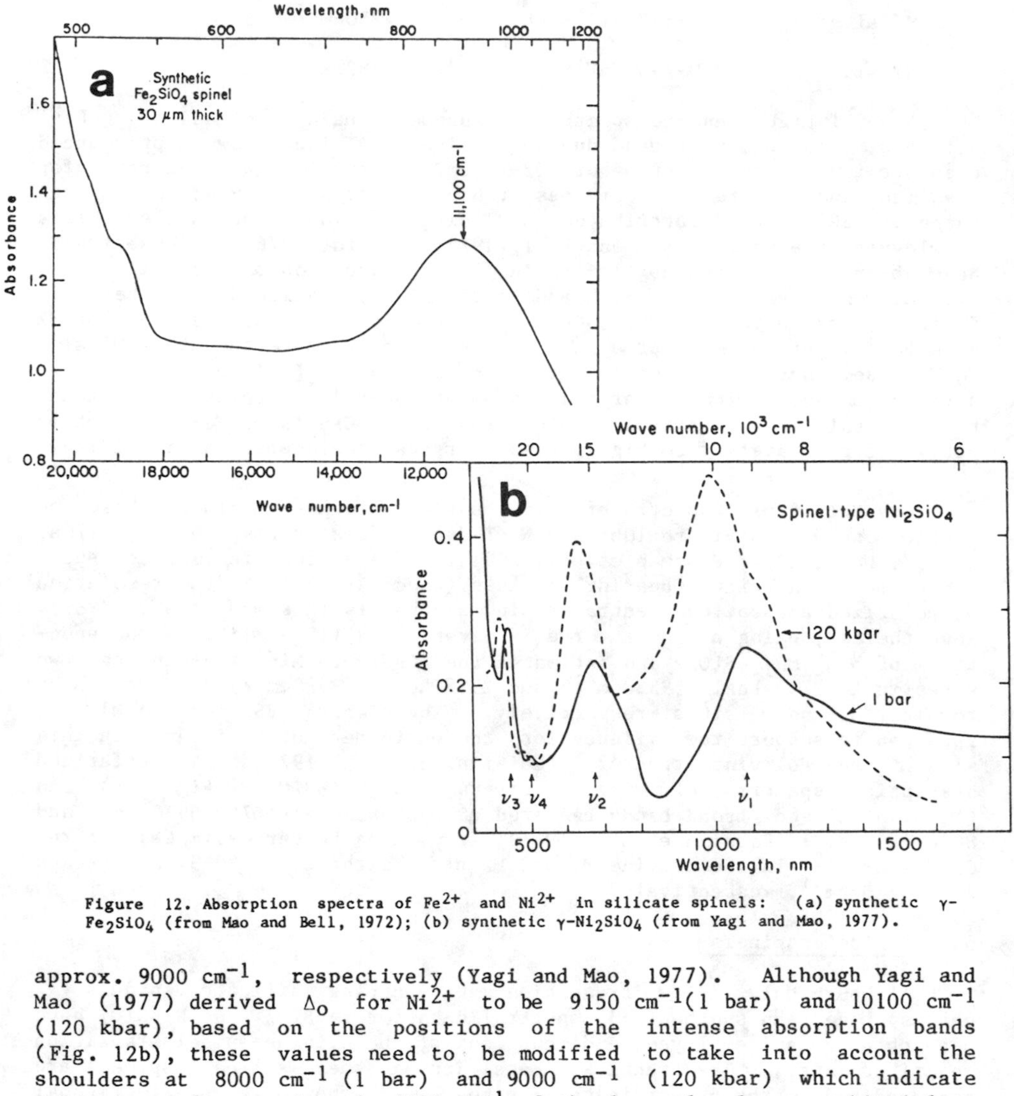

Figure 12. Absorption spectra of Fe^{2+} and Ni^{2+} in silicate spinels: (a) synthetic γ-Fe_2SiO_4 (from Mao and Bell, 1972); (b) synthetic γ-Ni_2SiO_4 (from Yagi and Mao, 1977).

approx. 9000 cm^{-1}, respectively (Yagi and Mao, 1977). Although Yagi and Mao (1977) derived Δ_0 for Ni^{2+} to be 9150 cm^{-1}(1 bar) and 10100 cm^{-1} (120 kbar), based on the positions of the intense absorption bands (Fig. 12b), these values need to be modified to take into account the shoulders at 8000 cm^{-1} (1 bar) and 9000 cm^{-1} (120 kbar) which indicate splittings of approximately 1100 cm^{-1} of the lower level t_{2g} orbital levels. The corrected crystal field splitting parameters for Ni^{2+} in Ni_2SiO_4 spinel for a range of pressures are

$$\Delta_0 = 8400\ cm^{-1} - 9400\ cm^{-1} \quad (1\ bar - 120\ kbar),$$

$$CFSE = 10080\ cm^{-1} - 11280\ cm^{-1} \quad (1\ bar - 120\ kbar).$$

Note that in both cases, the CFSE of Fe^{2+} and Ni^{2+} in the silicate spinel (γ-phase) are higher than the values for both the M1 and M2 sites of the olivine (α-phase) structure. This factor is particularly relevant to olivine→spinel transformations discussed in Section 10.1.

5.6 Pyroxenes

The pyroxene structure, like olivine, has two distinctive cation sites which contrast markedly with respect to oxygen ligand-type, metal-oxygen distances, and distortion. These factors, in turn, profoundly influence cation site occupancies, with the result that pyroxenes have been studied extensively by spectroscopic techniques (Rossman, 1980) to determine intracrystalline cation ordering.

In orthopyroxenes, the 6-coordinated M2 site is a very distorted polyhedron (point group C_1, but approximately C_{2v}) with a range of metal-oxygen distances, the longest being to two oxygen atoms bridging $[SiO_4]$ tetrahedra in the pyroxene single chains. The oxygen polyhedron surrounding the M1 site is only slightly distorted from octahedral symmetry, the range of metal-oxygen distances being due to the M1 cation not being located at the center of the octahedron. Ranges of Fe^{2+}-oxygen distances in orthoferrosilite, for example, are: M1 site: 204-217 pm (mean = 213.5 pm); M2 site: 204-252 pm (mean = 222.8 pm). In clinopyroxenes, Ca^{2+} or Na^+ ions occupy the M2 positions, and the 6-coordinated M1 site is a somewhat more distorted octahedron. In hedenbergite, the M1 site Fe^{2+}-oxygen distances range from 208 to 216 pm (average = 213.0 pm). The structural data compiled by Cameron and Papike (1980) for transition metal-bearing pyroxenes show that the average M1-oxygen distance depends on the ionic radius of the cation in the M1 site.

The crystal field spectra of Fe^{2+}-bearing pyroxenes have been studied extensively and the data are summarized elsewhere (Hazen et al., 1978; Rossman, 1980). Several assignments of the spectra have been proposed, particularly for M2 site Fe^{2+} ions, and the results of Goldman and Rossman (1977) for the lower level splittings in a bronzite Fs_{15} are utilized in the energy level diagrams for orthoferrosilite Fs_{86} shown in Figure 13. The approximate values of Δ_o and CFSE calculated for Fe^{2+} in the two sites of orthopyroxenes in the composition range Fs_{15}-Fs_{86} are

M1 site: Δ_o = 9460 - 9150 cm^{-1}; CFSE = 4280 - 4160 cm^{-1},

M2 site: Δ_o = 7265 - 6900 cm^{-1}; CFSE = 3805 - 3660 cm^{-1}.

Corresponding data for Fe^{2+} in the calcic clinopyroxene M1 site along the diopside-hedenbergite join are approximately Δ_o = 8850 cm^{-1}; CFSE = 4000 cm^{-1}. In Table 6 are summarized crystal field parameters derived from the optical spectra of transition metal-bearing pyroxenes.

6. COVALENT BONDING AND RACAH PARAMETERS FROM OPTICAL SPECTRA

Visible-region spectra not only are the sources of CFSE data for transition metal ions, but also provide a measure of covalency bonding differences among host structures and enable estimates to be made of changes of covalent bond character at high pressures. Covalent bonding interactions are expressed semi-empirically in the ligand field model of transition metal spectroscopy in terms of Racah B and C parameters, which in field-free (gaseous) cations are a measure of interelectron repulsions. In transition metal-bearing crystals, however, they become a measure of

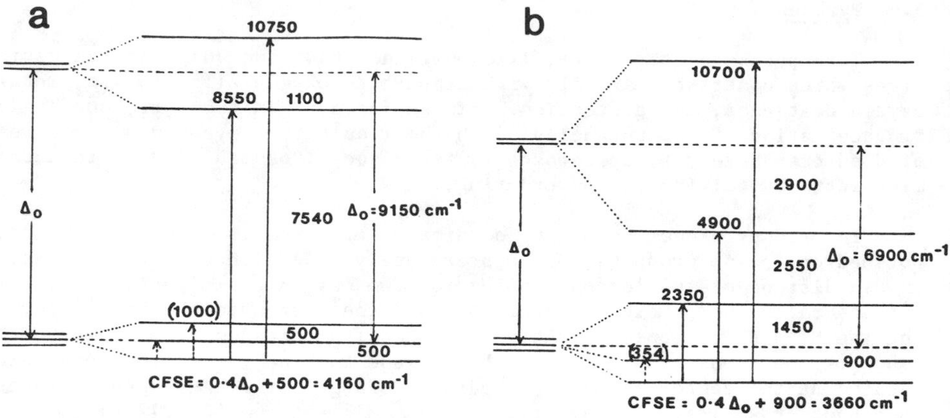

Figure 13. Energy level diagrams for Fe^{2+} ions in orthoferrosilite, Fs_{86} (modified from Burns 1970a, p. 91): (a) M1 site; (b) M2 site.

Table 6. Crystal Field Parameters for Transition Metal-Bearing Pyroxenes

Mineral or Phase	Δ_0 (cm^{-1})	CFSE	Site Occupancy	References
enstatite-orthoferrosilite	9460-9150	4280-4160	Fe^{2+}/M1	a,b
enstatite-orthoferrosilite	7265-6900	3805-3660	Fe^{2+}/M2	a-d
Ni^{2+}-enstatite	6900	8280	Ni^{2+}/M1	e
	or 5500	6600	Ni^{2+}/M2	a,f
Co^{2+}-enstatite	8370	6700	Co^{2+}/M1	a,e
Cf^{3+}-enstatite	15380	18460	Cr^{3+}/M1	a,d
hedenbergite	8850	4000	Fe^{2+}/M1	a,b,d
Ni^{2+}-diopside	8400	10800	Ni^{2+}/M1	a,e
Co^{2+}-diopside	8020	6420	Co^{2+}/M1	a,e
Cr^{3+}-diopside	15500	18600	Cr^{3+}/M1	a,g
Cr^{3+}-jadeite	15600	18720	Cr^{3+}/M1	e
ureyite	15270	18320	Cr^{3+}/M1	h
Ti^{3+}-diopside	18470	7900	Ti^{3+}/M1	i,j
$NaTiSi_2O_6$	18770	8000	Ti^{3+}/M1	j,k
V^{3+}-diopside	16700	13360	V^{3+}/M1	l

a This work; R.G. Burns, Mineralogical Applications of Crystal Field Theory, Cambridge Univ. Press (1970)
b W.A. Runciman, D. Sengupta and M. Marshall, Amer. Mineral. 58, 444 (1973)
c D.S. Goldman and G.R. Rossman, Amer. Mineral. 62, 151 (1977)
d G.R. Rossman, Ch. 3 in: Pyroxenes, C.T. Prewitt, ed., Rev. Mineral. 7, 93 (1980)
e W.B. White, G.J. McCarthy and D.E. Scheetz, Amer. Mineral., 56, 72 (1971)
f G.R. Rossman, R.D. Shannon and R.K. Waring, J. Solid State Chem. 39, 277 (1981)
g H.K. Mao, P.M. Bell and J.S. Dickey, Jr., Ann. Rept. Geophys. Lab., Yearbook 71, 538 (1972)
h G.R. Rossman, pers. comm. (1985)
i E. Dowty and J.R. Clark, Amer. Mineral. 58, 230 (1973)
j R.G. Burns and F.E. Huggins, Amer. Mineral. 58, 955 (1973)
k C.T. Prewitt, R.D. Shannon and W.B. White, Contrib. Mineral. Petrol. 35, 77 (1972)
l K. Schmetzer, N. Jahrb. Mineral. Abh. 144 73 (1982)

bond covalency and provide indications of deviations from the purely electrostatic crystal field model. The C/B ratio is usually close to 4, but usually the Racah B parameter is taken as the measure of changes of covalent bonding. The degree of covalency of a metal-ligand bond is expressed as the covalency parameter β, also known as the nephelauxitic (Greek: "cloud-expanding") ratio, which is the ratio of the Racah B parameter in the compound (mineral) to that of the free cation, B_0 (Jorgensen, 1964). The value of B is proportional to the average reciprocal radius of the partly-filled 3d shell of a transition metal ion, which increases when electrons are shared with surrounding ligands. Therefore, B for transition metal compounds is always smaller than B_0 for the free cation. The nephelauxitic ratio is related to two parameters: one, a central field covalency parameter (designated β_{35}) which arises from the screening of the nuclear charge of the cation by the ligand; and the other, a symmetry-restricted covalency parameter (designated β_{33} or β_{55}) which arises from delocalization of d electrons onto the ligands (Reinen, 1969; Konig, 1971). Here, we focus on the central field covalency parameter, β_{35}, since it is influenced by all 3d orbitals. To evaluate B, and hence β_{35}, for a transition metal-bearing mineral, the energies of at least two absorption bands in a spectrum are required. This eliminates cations with $3d^1$ and $3d^9$ configurations (e.g., Ti^{3+}, Cu^{2+}) because they give only one absorption band in a regular octahedral (or tetrahedral) site from which the crystal field splitting parameter (Δ_0) only can be estimated. The remaining cations have several electronic transitions (spin-allowed and spin-forbidden), which are related to both parameters Δ and B. The most convenient of these cations are those with $3d^3$ and $3d^8$ configurations, because Δ and B can be calculated from two relatively intense, low energy spin-allowed transitions. We shall confine our attention to Cr^{3+}.

For Cr^{3+} in an octahedral site, the lowest energy spin-allowed transition produces an absorption band with frequency ν_1 which is equivalent to Δ_0, from which the CFSE of Cr^{3+} may be estimated directly. Such data are summarized in Section 5. The second spin-allowed transition gives rise to absorption band ν_2, the energy of which is expressed by

$$\nu_2 = 1/2(15B + 30\Delta) - 1/2[(15B - 10\Delta)^2 + 12B\Delta]^{1/2} \qquad (16)$$

(Konig, 1971), from which the B Racah parameter may be evaluated:

$$B = [(2\nu_1 - \nu_2)(\nu_2 - \nu_1)]/(27\nu_1 - 15\nu_2). \qquad (17)$$

Spectral data for several Cr^{3+}-bearing minerals discussed in Section 5 are summarized in Table 7. The Racah B parameters were computed from Equation 17 and the nephelauxitic parameters, β_{35}, were calculated using the B_0 value for Cr^{3+} of 918 cm^{-1}. Pressure variations of the B Racah parameter for uvarovite are also listed in Table 5. For phases containing low concentrations of Cr^{3+}, there is a trend towards smalaler Racah B parameters with decreasing metal-oxygen distances in the host structures (Abu-Eid and Burns, 1976; Amthauer, 1976; Schmetzer, 1982). This trend is also consistent with the results for uvarovite at high pressures, which indicate a 2.5% decrease of Racah B parameter over a 20 GPa pressure range, consistent with compression of the Cr^{3+}-oxygen bond. A similar decrease occurs for ruby, the value of B for Cr^{3+} in Al_2O_3 changing from 630 cm^{-1} to 610 mc^{-1} over the pressure range 10 GPa (Drickamer and Frank, 1973).

Table 7. Spectral Data for Cr^{3+} in Oxides and Silicates

Mineral	ν_1 cm^{-1}	ν_2 cm^{-1}	CFSE cm^{-1}	B cm^{-1}	Mean M-O Distance pm	Reference
periclase	16200	22700	19440	650.6	211	a
ruby	18450(‖) 18000(⊥)	25200(‖) 24400(⊥)	21780	618.7	191	b
spinel	18520	24900	22220	612.1	190	c
pyrope	17790	24150	21350	615.6	189	d
grossular	16500	22290	19800	645.4	193	d
forsterite	16900	23500	20280	654.9	212	e
enstatite	15380	22220	18460	712.7	208	f
diopside	15500	22000	18600	650.8	208	g
ureyite	15600	22000	18720	645.6	200	h
uvarovite	16667	22727	20000	589	199	i

a W. Low, Phys. Rev. 105, 801 (1957)
b D.S. McClure, J. Chem. Phys. 36, 2757 (1962). Note: The band frequencies correspond to light polarized parallel and perpendicular to the c axis.
c D.T. Sviridov et al., Opt. Spektrosk. 35, 102 (1973)
d G. Amthauer, N. Jahrb. Mineral. Mh. no. 6, 245 (1981)
e B.E. Scheetz and W.B. White, Contrib. Mineral. Petrol. 37, 221 (1972)
f G.R. Rossman, Ch. 3 in: Pyroxenes, C.T. Prewitt, ed., Rev. Mineral. 7, 93 (1980)
g H.K. Mao, P.M. Bell and J.S. Dickey, Jr., Ann. Rept. Geophys. Lab., Yearbook 71, 538 (1972)
h W.B. White, G.J. McCarthy and R.E. Scheetz, Amer. Mineral. 56, 72 (1971); G.R. Rossman (pers. comm., 1985)
i See Table 5

Racah B parameters have also been calculated for Mn^{2+} and Fe^{3+} in several minerals (Manning, 1970; Keester and White, 1968). High pressure spectral measurements of Mn^{2+}-bearing garnets (Smith and Langer, 1983) again show a decrease in B values with increasing pressure, suggesting a relative decrease in ionicity of Mn^{2+}-O bonds of about 0.86% over 11.2 GPa. These relatively small decreases of Racah parameters of Mn^{2+}- and Cr^{3+}-bearing minerals suggest that their covalent bonding characters do not change dramatically over pressure ranges applicable to the Earth's mantle.

7. SITE BULK MODULI FROM OPTICAL SPECTRA

Equations 12-14 which were discussed in Section 3 and may be summarized as follows:

$$K = \frac{1}{\beta} = -\frac{VdP}{dV} = \frac{5\Delta dP}{3\,d\Delta} \tag{18}$$

show the relationship between the pressure variation of crystal field splitting (Δ), site compressibility (β) and bulk modulus or incompressibility (K) of the coordination polyhedron about a transition metal ion.

The bulk modulus of a coordination site at zero (atmospheric) pressure, K_0, may also be estimated from the first-order Birch-Murnaghan equation of state,

$$P = 3K_0\ [(V_0/V)^{7/3} - (V_0/V)^{5/3}]/2, \qquad (19)$$

which may be rewritten using Equation 11:

$$P = 3K_0\ [(\Delta/\Delta_0)^{7/5} - (\Delta/\Delta_0)]/2. \qquad (20)$$

These equations, which are based on an ionic point-charge model, neglect changes of geometry of the coordination site with increasing pressure. They are strictly applicable to mineral structures containing regular octahedral (or cubic or tetrahedral) sites showing negligible distortion, for which Δ may be determined directly from absorption spectra. Equations 18 and 20 are more applicable, therefore, to cations in octahedral sites in the periclase, corundum, spinel and garnet structures compared to the more distorted 6-coordinated sites of olivine and pyroxene. Furthermore, the bulk modulus of a crystal depends on the bulk moduli of component coordination polyhedra and the manner in which these polyhedra are linked. Structures such as periclse, spinel and garnet with extensive edge-sharing between coordination polyhedra in three dimensions should have large bulk moduli similar to those of constituent cation sites (Hazen and Finger, 1979; Hazen, this volume).

The crystal field spectral data for Cr^{3+} in synthetic uvarovite summarized in Table 5 contain values of site compressibilities β and site bulk moduli K at different pressures. The zero pressure bulk modulus of the $[CrO_6]$ octahedron in this garnet structure, K_0 = 450 GPa, may be compared with the value 220 GPa estimated from high pressure crystal structural data for the $[AlO_6]$ octahedron in pyrope and grossular (Hazen and Finger, 1979). In Table 8 are compared polyhedral bulk moduli values for other transition metal-bearing minerals for which crystal field spectral and crystal structural data are available at high pressures. Given the assumption of high ionicity inherent in Equation 12, there is remarkable consistency between the two sets of data.

8. SPIN-PAIRING TRANSITONS

The energy level diagrams developed in Section 5 are based on energy differences, either obtained experimentally from measurements of optical spectra, or determined indirectly from corroborative spectral measurements. However, inherent in the estimates of CFSE's derived from these energy level diagrams are assumptions that each transition metal cation coordinated to oxygen ligands has a high-spin configuration; that is, the 3d electrons are assumed to conform with Hund's rules and to occupy singly as many 3d orbitals as possible in the (unexcited) ground state. Such assumptions of high-spin states are vindicated by magnetic susceptibility and interatomic distance data; low-spin states for cations in oxide structures are known only for Co^{3+} and Ni^{3+}. Octahedrally-coordinated transition metal ions with four, five, six, and seven 3d electrons have two alternative electronic configurations corresponding to high-spin and low-spin states, and the possibility exists that spin-pairing transitions can

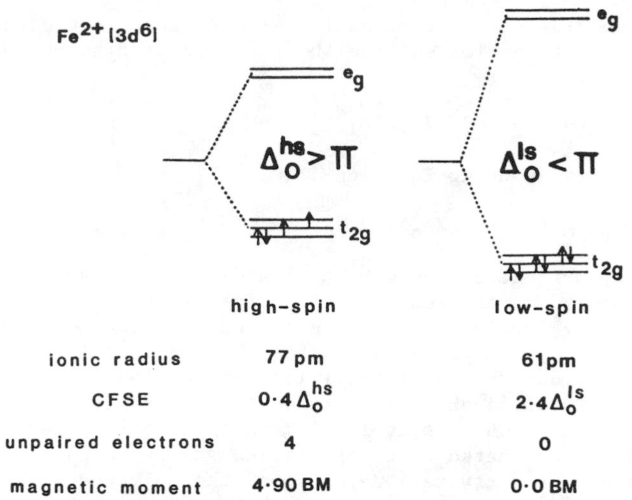

Figure 14. Electronic configurations and physical properties of octahedrally coordinated Fe^{2+} ions in (a) high-spin and (b) low-spin configurations.

Table 8. Polyhedral Bulk Moduli Data for Transition Metal-Bearing Minerals

Mineral	Cation	K_o (spectral) (GPa)	Ko (crystal structure) (GPa)	Reference
periclase	Mg^{2+}	-	$[MgO_6]$ in MgO = 161	a
	Fe^{2+}	148	$[FeO_6]$ in FeO = 153	a,b
	Co^{2+}	139	$[CoO_6]$ in CoO = 185	a,c
	Cr^{3+}	133	-	c
corundum	Al^{3+}	-	$[AlO_6]$ in Al_2O_3 = 240	a
	Cr^{3+}	303	$[CrO_6]$ in Cr_2O_3 = 250	a,c
	V^{3+}	303	$[VO_6]$ in V_2O_3 = 180	a,c
pyrope	Mg^{2+}	-	$[MgO_8]$ = 130	a
almandine	Fe^{2+}	123.5	-	d
grossular	Al^{3+}	-	$[AlO_6]$ = 220	a
uvarovite	Cr^{3+}	453	-	e
γ-Ni_2SiO_4	Ni^{2+}	157	$[NiO_6]$ = 150	a,f

a R.M. Hazen and L.W. Finger, J. Geophys. Res. 84, 6723 (1979)
b T.J. Shankland, A.G. Duba and A. Woronow, J. Geophys. Res. 79, 3273 (1974); Section 5.1
c H.G. Drickamer and C.W. Frank, Electronic Transitions and the High Pressure Chemistry and Physics of Solids. Chapman and Hall, Ltd., (1973); Section 5.1 and Figure 7
d G. Smith and K. Langer, N. Jahrb. Miner. Mh. 12, 541 (1983)
e R.M. Abu-Eid (1976); see Table 5
f T. Yagi and H.K. Mao (1979); but see Section 5.5

be induced in them by increased pressure. Indeed, such pressure-induced changes of spin state have been demonstrated in Co^{3+} in corundum-type Co_2O_3 (Chenavas et al., 1971) and in Fe^{2+} replacing Mn^{2+} in pyrite-type MnS_2 (Drickamer and Frank, 1973). However, shock-induced spin-pairing suggested for Fe^{3+} in Fe_2O_3 (e.g., Syono et al., 1971; Goto et al., 1982) is debatable (Syono et al., 1984).

Properties of transition metal cations with dual electronic configurations are summarized in Figure 14, which applies to octahedrally coordinated Fe^{2+}. Large differences between high-spin and low-spin states exist for the magnetic susceptibility, ionic radius (molar volume) and CFSE, which are expected to profoundly influence geophysical and geochemical properties of the Lower Mantle (Strens, 1976; Gaffney and Anderson, 1973; Burns, 1976).

The criteria determining a spin-pairing transition may be expressed by

$$\Delta = A\,\Pi \,, \qquad (21)$$

where Δ is the crystal field splitting, Π is the spin-pairing energy, and the numerical coefficient A is governed by the change of metal-ligand interatomic distance, R, between the low-spin and high-spin states. Attempts have been made to evaluate these terms (Ohnishi, 1978; Goto et al., 1982; Ohnishi and Sugano, 1981) using the $\Delta \propto R^{-5}$ law (Eqn. 9), the Birch-Murnaghan equation of state (Eqn. 19), and pressure-variations of Racah B parameter data which affect the choice of integer n in the $B \propto R^n$ dependence. Such calculations indicate that pressure-induced high-spin to low-spin transitions might occur in Fe^{2+} in FeO at about 25-40 GPa, and at 70-130 GPa for MnO, CoO and Fe_2O_3 (Ohnishi, 1978). Careful experimental measurements are required to provide more accurate data for calculating these spin-pairing transitions, and to demonstrate unambiguously that high-spin to low-spin transitions occur in minerals relevant to the Lower Mantle.

9. STABILIZATION OF UNUSUAL CATION VALENCIES

Low symmetry environments have the ability to stabilize unusual valencies, particularly when the cation involved is subject to the Jahn-Teller effect. Such cations include 6-coordinated Cr^{2+} and Mn^{3+}, both of which have four 3d electrons and electronic configurations $(t_{2g})^3(e_g)^1$. The fourth 3d electron has the potential of stabilizing the cation located in certain very distorted sites because large splittings are induced of upper level e_g orbitals, thereby lowering the energy of the singly occupied e_g orbital. The existence and stability of Mn^{3+} in the very distorted 6-coordinated site in the epidote structure is readily shown by the optical spectra of piemontites (Burns, 1970a, p. 55), from which CFSE's were estimated. Divalent chromium might also be predicted to be stabilized in distorted sites in the olivine and pyroxene structures.

Olivines in basalts from the Moon and in diamond inclusions from kimberlites contain significant amounts of chromium, leading to the suggestion (Burns, 1975) that Cr^{2+} ions occur in these olivines from the Moon and the Upper Mantle. The Cr^{2+} ion is predicted to be stabilized in the olivine M1 site, and such a site occupancy is consistent with optical spectral measurements of chrome forsterite (Burns, 1975).

Divalent chromium has also been suggested to occur in blue chrome diopsides, and again its presence is indicated by crystal field spectral measurements (Mao et al., 1972; Burns, 1975). The CFSE of Cr^{2+}, which may be accommodated in the very distorted pyroxene M2 site, is approximately 7900 cm^{-1}, compared to ≈ 5400 cm^{-1} if the site were not distorted. Failure to recognize Cr^{2+} has led to untenable hypotheses about the crystal chemistry of chromium in the diopside structure. For example, certain features in the optical spectra of synthetic blue chrome diopsides were assigned (Ikeda and Yagi, 1982) to low spin Cr^{3+} in tetrahedral sites in the pyroxene structure. Although dual electronic configurations are possible for Cr^{3+}, $3d^3$, in tetrahedral coordination, it is extremely unlikely that the energy separation between lower level e orbitals and higher level t_2 orbitals (Fig. 4c) is sufficiently large to induce spin-pairing in Cr^{3+}. The discussion in Section 8 suggests that the very small tetrahedral crystal field splitting parameter, Δ_t, of transition metal cations (Eqn. 3) prevents the existence of low-spin states in tetrahedral sites in oxide structures, let alone Cr^{3+}, which has exceptionally high CFSE in octahedral sites (Table 1). Therefore, Cr^{2+} in the pyroxene M2 site is the more preferable explanation of the color and optical spectra of blue diopsides.

10. INFLUENCE OF CFSE ON HIGH PRESSURE PHASE TRANSITIONS

Some of the consequences of transition metal cations existing in low-spin states in the Earth's interior are described in Section 8. However, the presence of transition metal ions in mineral structures may modify phase equilibria at high pressures as a result of increased CFSE acquired by certain cations in dense phases. The additional electronic stabilization can influence both the depth in the mantle at which a phase transition occurs and the distribution coefficents of transition metals in coexisting dense phases of the Lower Mantle.

10.1 The olivine→spinel transition

Experimental phase equilibrium studies have confirmed deductions from seismic velocity data that below 400 km, olivine and pyroxene, abundant in the Upper Mantle, are transformed to more dense spinel and garnet phases. Syono et al. (1971) first suggested that CFSE influences pressures of olivine→spinel transformations in silicates and germanates containing transition metal ions. They proposed that a linear relationship exists between the transition pressure, P_t, and r_R/r_M, the ratio of ionic radii of divalent cations (R = Mg, Mn, Fe, Co, Ni, Zn) and tetrahedral ions (M = Si, Ge), as illustrated in Figure 15. The basis of their proposition is as follows. At equilibrium in an olivine→spinel transition,

$$\Delta G = \Delta E + P_t\Delta V - T\Delta S, \qquad (22)$$

where ΔG, ΔE, ΔV, and ΔS are differences of free energy, internal energy, molar volume, and entropy, respectively, between olivine and spinel at the

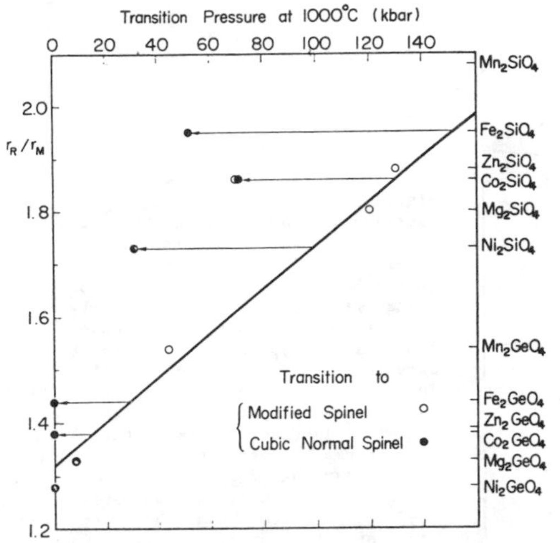

Figure 15. Transition pressures at 1000°C for various olivines to β-phase (modified spinel) or γ-phase (spinel) as a function of ionic radius ratio: divalent cation (R) to Si^{4+} or Ge^{4+} (M) (from Syono et al., 1971). Note that cations acquiring excess CFSE in spinel over olivine (e.g., Fe^{2+}, Co^{2+}, Ni^{2+}) deviate from a linear trend.

transition pressure, P_t. Therefore,

$$P_t = (T\Delta S - \Delta E/\Delta V). \qquad (23)$$

Syono et al. (1971) incorrectly assumed that both ΔV and ΔS for the olivine→spinel transition are each approximately independent of composition, so that transition pressures P_t are directly proportional to ΔE. However, while the three quantities ΔE, TΔS and $P_t\Delta V$ may make approximately equal contributions to ΔG in Equation 22, the values of ΔS and ΔV can vary by factors of 2 to 5 (A. Navrotsky, pers. comm.; Burns and Sung, 1978). Syono et al. further assumed that ΔE is linearly proportional to r_R/r_M to conform with the apparent linear relationship between P_t and r_R/r_M for many silicate and germanate spinels. Syono et al. noted, however, that the transition metal ions acquiring CFSE (Fe^{2+}, Co^{2+}, Ni^{2+}) deviate significantly from this linear trend. For example, the transition pressures at 1000°C for Fe_2SiO_4 and Ni_2SiO_4 are predicted from the P_t versus r_R/r_M plot (Fig. 15) to be about 150 kbar and 100 kbar, respectively, but are in fact measured experimentally to be about 60 kbar and 30 kbar, respectively. Since CFSE contributes to internal energy, ΔE, Syono et al. suggested that the excess CFSE in the spinel structure over that of olivine tends to lower the transition pressures of olivine→spinel transitions in Fe_2SiO_4 and Ni_2SiO_4.

Attempts to quantify the magnitude of the CFSE on transition pressures of olivine→spinel transitions have been made using absorption spectral data for these phases. Mao and Bell (1972) estimated the CFSE of Fe^{2+} in fayalite and Fe_2SiO_4 spinel from spectra of these phases measured at ambient pressures and temperature, and estimated that CFSE lowers P_t for Fe_2SiO_4 by about 98 kbar. Yagi and Mao (1977) using 120 kbar spectral data for Ni_2SiO_4 spinel (Fig. 12b) calculated that the CFSE difference for Ni^{2+} in the olivine and spinel phases lowers P_t by 140 kbar which, they noted, is in the right direction but appears too large in comparison with

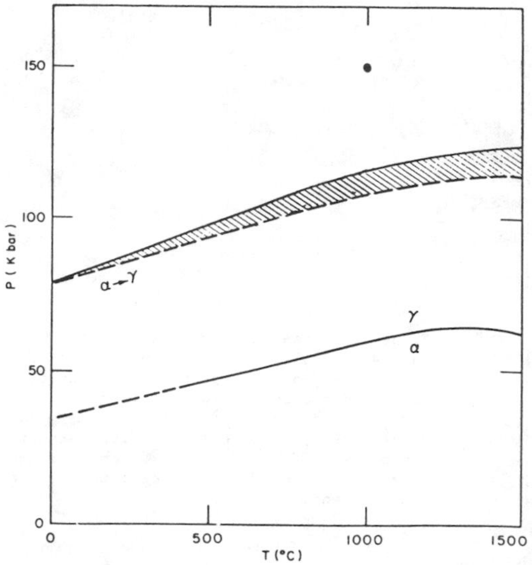

Figure 16. Semi-quantitative phase equilibrium diagram showing the effect of crystal field stabilization on the olivine→spinel transition in Fe_2SiO_4 (after Burns and Sung, 1978). The bottom solid line is the experimentally determined phase boundary. The upper solid line is a hypothetical phase boundary assuming no crystal field stabilization in Fe^{2+} on the olivine→spinel transition. The curves are depicted as bands broadening with increasing temperature because the thermal expansion of fayalite is larger than that for Fe_2SiO_4 spinel. The black dot is the transition pressure for Fe_2SiO_4 predicted from the r_R/r_M ratio plot of Syono et al. (1971) (see Fig. 15).

other olivine→spinel transitions. Yagi and Mao suggested that in order to define more accurately the role of CFSE on phase transitions, further information is needed on pressure dependencies of CFSE and compressibilities of the phases, as well as energy levels of cations in distorted octahedral sites. Such factors were taken into account by Burns and Sung (1978) who critically examined effects of crystal field stabiliation on the olivine→spinel transition in the system Mg_2SiO_4-Fe_2SiO_4.

Burns and Sung (1978) calculated free energy changes, ΔG_{CFS}, due to differences of Fe^{2+} crystal field splittings between the spinel and olivine structures from

$$\Delta G_{CFS} = \Delta CFSE - T\Delta S_{CFS} \quad (24)$$

as functions of P and T using crystal field spectral data for these phases at high pressures and temperatures, where $\Delta CFSE$ and ΔS_{CFS} are differences of crystal field stabilization enthalpies and electronic configurational entropies of Fe_2SiO_4 spinel and fayalite at the transition temperature and pressure. The results of these calculations indicated that ΔG_{CFS} is always negative, showing that crystal field stabilization of Fe^{2+} promotes the olivine→spinel transition in Fe_2SiO_4, and expands the stability field of spinel at the expense of olivine in the system Mg_2SiO_4-Fe_2SiO_4. Because of crystal field effects, the transition pressures for the olivine→spinel transitions in Fe_2SiO_4 is lowered by about 50 kbar at 1000°C (Fig. 16). Since olivines of the Upper Mantle contain approximately 10 mole % Fe_2SiO_4, their transition pressures may be decreased by about 5 kbar due to the presence of Fe^{2+} in the crystal structures. This means that the depth of the olivine→spinel transition in a typical Upper Mantle forsteritic olivine is 15 km shallower than it would be if iron were absent from the minerals.

10.2 Partitioning of iron in post-spinel phases of the Lower Mantle

Although Upper Mantle olivine is transformed into the more dense isochemical spinel phase and delineates the onset of the Transition Zone in the Earth's interior at 400 km, the $(Mg,Fe)_2SiO_4$ stoichiometry appears to be unstable in the Lower Mantle relative to denser oxide structure-types. One post-spinel transformation that has been studied recently is the spinel to periclase plus perovskite transition

$$\underset{\text{spinel}}{\gamma(Mg,Fe)_2SiO_4} \rightarrow \underset{\text{magnesiowüstite}}{(Fe,Mg)O} + \underset{\text{perovskite}}{(Mg,Fe)SiO_3} \qquad (25)$$

which is believed to occur below 650 km. Such a disproportionation reaction raises the possiblity that each breakdown product has a different Fe/Mg ratio.

In investigations of phase relations in the system $MgO\text{-}FeO\text{-}SiO_2$ at high P and T, compositions $(Mg_{1-x}Fe_x)_2SiO_4$ yielded magnesiowüstites with higher Fe/Mg ratios than coexisting perovskites (Yagi et al., 1979; Bell et al., 1979). For example,

$$(Mg_{0.85}\ Fe_{0.15})_2SiO_4 \rightarrow (Mg_{0.74}\ Fe_{0.26})O + (Mg_{0.96}\ Fe_{0.04})SiO_3. \qquad (26)$$

The strong partitioning of iron into magnesiowüstite is the result of higher CFSE of Fe^{2+} in the periclase structure. Various estimates have been made of the CFSE of Fe^{2+} in dense oxide structures modelled as potential mantle mineral phases (Gaffney, 1972; Burns, 1976). All estimates indicate that octahedrally coordinated Fe^{2+} (in periclase, for example) has a considerably higher CFSE than Fe^{2+} ions in 8- to 12-coordinated sites of the perovskite structure. Yagi et al. (1979) calculated the excess CFSE of Fe^{2+} in periclase-type FeO over perovskite-type $FeSiO_3$ may be as high as 12.5 kcal $mole^{-1}$. The large extra CFSE factor thus favors the concentration of iron in magnesiowustite and the depletion of iron in a coexisting perovskite phase in the Lower Mantle.

There are two factors that may complicate this simple model of iron fractionation between Lower Mantle phases. The first is a possible phase change of MgO from the NaCl to CsCl structure-type, which would produce a coordination site change from octahedral to cubic. According to Equation 3, the smaller cubic crystal field splitting might reduce the CFSE of Fe^{2+} ions in 8-coordinated sites when the (Mg,Fe)O phase transforms to the CsCl structure. The second complication is the possibility of a spin-pairing transition in Fe^{2+} ions discussed in Section 9. The smaller ionic radius of low-spin Fe^{2+} could induce iron to enter octahedral sites in the perovskite structure, leading to iron enrichment in the $(Mg,Fe)SiO_3$ phase of the deep Lower Mantle.

11. SUMMARY

This chapter demonstrates how environments about transition metal cations in minerals influence 3d orbital energy levels, giving rise to absorption bands in the visible region. Conversely, by understanding the origin of such optical spectra, information may be obtained about environments surrounding transition metal cations and how they influence geochemical and geophysical properties of these elements. The crystal field model provides the simplest interpretation of optical spectra and enables semi-quantitative 3d orbital energy level diagrams to be constructed for cations in a number of common oxide and silicate minerals. Although these lack the rigor of more quantitative molecular orbital energy level diagrams currently being computed for a variety of coordination clusters containing transition metals, the effects of covalent bonding do not seriously affect important parameters calculated from the point-charge crystal field model. The crystal field spectra are relatively straightforward for transition metal cations occurring in regular or slightly distorted coordination sites in cubic phases, so that reliable crystal field stabilization energies may be estimated for cubic sites in garnets and octahedral sites in periclase, garnet and silicate spinel structures. The pleochroism or polarization-dependent crystal field spectra of non-cubic minerals, such as corundum, olivine and pyroxenes, complicate estimates of CFSE's of transition metal cations located in very distorted octahedral sites in these mineral structures. Nevertheless, trends in CFSE values, particularly those obtained from spectra at elevated pressures, enable geophysical parameters such as site compressibilities and polyhedral bulk moduli of transition metal-bearing minerals to be estimated. Differences of CFSE form the basis for evaluating pressures at which various transitions occur, including spin-pairing in Fe^{2+} in the Lower Mantle. They also enable partitioning of iron cations in post-spinel phases in the Lower Mantle to be deduced.

12. WORKED PROBLEMS

Problem 1.

"Rediscover" crystal field theory from the following thermodynamic and spectroscopic data for divalent cations of the first transition series [heats of hydration (kJ. mole^{-1}); octahedral crystal field splittings for hexahydrated cations (cm^{-1})]; Ca [2464; 0]; Sc [-; -]; Ti [2732; -]; V [2778; 12600]; Cr [2795; 13900]; Mn [2736; 7800]; Fe [2845; 10400]; Co [2916; 9300]; Ni [2996; 8500]; Cu [3000; 12600]; Zn [2933; 0].

Solution: The heats of hydration data are plotted in Figure 1a. Note the characteristic double-humped curve. The crystal field splitting data are those contained in Table 1. Taking Ni^{2+}, for example, its CFSE = $1.2\Delta_0$ = $1.2 \times (8500)cm^{-1}$ = 10200 cm^{-1}. Since 10000 cm^{-1} = 28.6 k.cal = 119·7 kJ, the CFSE of Ni^{2+} is 122 kJ (g. ion)$^{-1}$. Therefore, the "corrected" heat of hydration is (2996-122) or 2874 kJ. mole^{-1}.

Problem 2.

Estimate the CFSE's of divalent Mn, Fe, Co, Ni, Cu, and Zn in their corresponding sulfides from the lattice energy data summarized below. Comment on the usefulness of this approach. Lattice energies (kJ. mole^{-1}) are: CaS, 3046; MnS, 3364; FeS, 3494; CoS, 3577; NiS, 3669; CuS, 3753; ZnS, 3623.

Solution: Comparable data are plotted in Figure 1b. Again, note the double-humped curve, and the linear trend connecting the lattice energies for CaS, MnS and ZnS which acquire zero CFSE. The lattice energy of NiS, for example, lies approx. 150 kJ above the line connecting the MnS and ZnS values. Therefore the CFSE of Ni^{2+} in NiS is about 150 kJ $(gm. ion)^{-1}$.

Many of these sulfides are opaque, so that spectroscopically-determined Δ's, and hence CFSE's, are difficult to obtain. Also note that the sulfides exhibit four different structure-types: rocksalt (CaS, MnS), zinc blende (MnS, ZnS), nickel arsenide (FeS, CoS, NiS), and covellite (CuS). The fact that the linear and double-humped trends exist for the lattice energies of these sulfides is attributed to the small contribution (<5%) that CFSE's make to the lattice energies. It is the relative values of CFSE's between adjacent elements that is important in crystal field theory.

Problem 3.

The polarized absorption spectra at different pressures of a titanian fassaite from the Allende meteorite are shown in Figure 17.

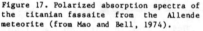

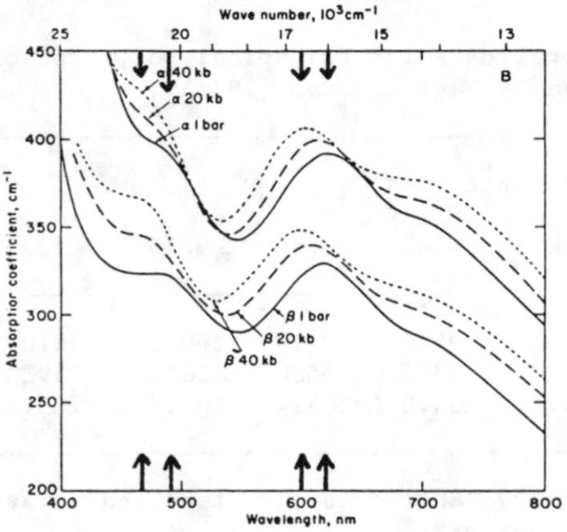

Figure 17. Polarized absorption spectra of the titanian fassaite from the Allende meteorite (from Mao and Bell, 1974).

(i) Estimate the upper level splittings of e_g orbital energy levels at 1 atmosphere and 40 kbar.

(ii) Assuming lower level splittings of the t_{2g} orbitals at each pressure to be about 1000 cm^{-1}, (compare Fig. 13a in the text), construct an energy level diagram for Ti^{3+} ions in the distorted octahedral M1 site.

(iii) Calculate the Δ_o and CFSE of Ti^{3+} in the clinopyroxene structure at 1 atmosphere and 40 kbar.

(iv) Estimate the site incompressibility of the $[TiO_6]$ octahedron (Eqn. 20). Comment on the reliability of this polyhedral bulk modulus value.

Solution: The locations of the peak maxima are denoted by arrows in Figure 17. (Note that the broad band centered near 700 nm is assigned to the $Ti^{3+} \rightarrow Ti^{4+}$ intervalence charge transfer transition and is not relevant to the problem.) The peaks due to crystal field transitions in Ti^{3+} are at 16000 and 20390 cm^{-1} (1 atmos.) and at 16670 and 21190 cm^{-1} (40 kbar or 4.0 GPa).

(i) The upper level splittings of e_g orbitals are: at 1 atmos. (20390 - 16000) = 4390 cm^{-1}; and at 4.0 GPa, 4520 cm^{-1}.

(ii) Follow the procedure shown in Figure 13a.

(iii) At 1 atmos, the components of Δ_o are: one-half 4390 = 2195 cm^{-1}; and (16000 - 500) = 15500 cm^{-1}. Therefore, Δ_o = 17695 cm^{-1}. The CFSE of Ti^{3+} is 0.4 Δ_o or about 7080 cm^{-1}. Similarly, at 4.0 GPa the values are: Δ_o = (2260 + 16170) = 18430 cm^{-1}; CFSE = 7370 cm^{-1}.

(iv) Using Equation 20,

$$K_o = 2(4.0 \text{ GPa})/3[(18430/17695)^{7/5} - 18430/17695)] = 156 \text{ GPa}.$$

Problem 4.

The following crystal field spectral data for the spinel polymorph of Ni_2SiO_4 at high pressures were reported by Yagi and Mao (1977):

Parameter	Pressure, GPa							
	0.0001	1.46	2.10	3.01	4.47	6.86	9.70	12.1
ν_1 (cm^{-1})	9150	9240	9260	9350	9540	9710	10000	10100
ν_2 (cm^{-1})	14780	14930	15040	15170	15310	15580	15820	15920
R (pm)	206.3	205.9	205.6	205.4	205.0	204.4	203.7	203.2

(i) Using the layout scheme suggested by Table 5 in the text, calculate for Ni^{2+} at each pressure the values of:

(a) Δ_o; (b) CFSE; (c) B; (d) β_{35}, assuming B_o = 915cm^{-1}; (e) K; (f) β; and (g) K_o.

(ii) Plot the changes of each of these parameters as a function of the change of the Ni^{2+}-O distance R. How does (Δ_p/Δ_o) versus (R_p/R_o) variation compare with the inverse-fifth law dependence (Eqn. 10)?

(iii) The text (Section 5.5) suggests that values of Δ_o at 12.0 GPa and 0.0001 GPa should be 9400 cm^{-1} and 8400 cm^{-1}, respectively. How do these revised values affect estimates of CFSE, B, K, β, and K_o at 12.1 GPa?

(iv) Predict the partitioning of Ni^{2+} if γ-Ni_2SiO_4 were to disproportionate to periclase plus perovskite structure-types at high pressures.

(v) Following the procedure described by Burns and Sung (1978), calculate the influence of CFSE on the transition pressure of the olivine→spinel transformation in Ni_2SiO_4 at 1000°C.

Solution: (i) Note that typical spectra are illustrated in Figure 12b. Using the data at 12.1 GPa:

(a) $\Delta_o = \nu_1 = 10100\ cm^{-1}$.

(b) CFSE of Ni^{2+} ($3d^8$) = $1.2\ \Delta_o = 12120\ cm^{-1}$.

(c) Using Equation 17,

$$B = [(2 \times 10100 - 15920)(15920 - 10100)]/(27 \times 10100 - 15 \times 15920) = 735\ cm^{-1}$$

(d) $\beta_{35} = B/B_o = 735/915 = 0.8$. The value of B at 0.0001 GPa is 782 cm^{-1}. Therefore, the percentage decrease of ionicity at 12.1 GPa is 6%.

(e) Using Equation 18 and the value of Δ at 0.0001 GPa,

$$K = 5 \times 9150 \times 12.1/3 \times (10100 - 9150) = 194\ GPa.$$

(f) Using Equation 18, $\beta = 1/194 = 5.1 \times 10^{-3}\ GPa^{-1}$.

(g) Using Equation 20,

$$K_o = 2 \times 12.1/3[(10100/9150)^{7/5} - (10100/9150)] = 181\ GPa.$$

(ii) Changes of Δ_o and B with pressure are plotted in Yagi and Mao (1977, p. 508). The ratio (Δ_p/Δ_0) varies as $(R_p/R_0)^{-6.5}$.

(iii) The revised value of K_o, for example, is 158 GPa.

(iv) The CFSE of Ni^{2+} in γ-Ni_2SiO_4 at 10.0 GPa is about 12000 cm^{-1}. In MgO, the CFSE is also about 12000 cm^{-1} (see Section 5.1). In the 12-coordinated site of perovskite, the magnitude of Δ_d is approximately one-half the octahedral crystal field splitting (see Eqn. 3). Since the CFSE of Ni^{2+} in a dodecahedral site is $0.8\ \Delta_d$ (compare Fig. 4b), or $[0.8(0.5 \times 12000)] = 4800\ cm^{-1}$; this low value suggests that Ni^{2+} would be strongly enriched in the periclase phase.

ACKNOWLEDGMENTS

I thank George Rossman, Gordon Smith, and Kurt Langer for providing unpublished spectral data. Many concepts developed in the chapter arose from discussions over the years with Frank Huggins, Rateb Abu-Eid, Chien-Min Sung, Kathleen Parkin, David Sherman, Don Goldman, Gordon Smith, and George Rossman. I appreciate the super-human efforts of Francis Doughty for getting the manuscript into shape. This research has been supported by grants from NASA (grant no. NSG-7604) and NSF (grant no. EAR83-13585).

REFERENCES

Abu-Eid, R.M. (1976) Absorption spectra of transition metal-bearing minerals at high pressures. In: The Physics and Chemistry of Minerals and Rocks. R.G.J. Strens, ed., J. Wiley, New York, p. 641-675.

__________ and Burns, R.G. (1976) The effect of pressure on the degree of covalency of the cation-oxygen bond in minerals. Amer. Mineral. 61, 391-397.

Amthauer, G. (1976) Crystal chemistry and color of chromium-bearing garnet. N. Jahrb. Mineral. Abh. 126, 158-186.

Ballhausen, C.J. (1966) Introduction to Ligand Field Theory. McGraw-Hill, New York, 298 pp.

Bell, P.M., Yagi, T. and Mao, H.K. (1979) Iron-magnesium distribution coefficients between spinel [$(Mg,Fe)_2SiO_4$], magnesiowüstite [(Mg,Fe)O], and perovskite [$(Mg,Fe)SiO_3$]. Ann. Rept. Geophys. Lab., Yearbook 78, 618-621.

Brown, G.E., Jr. (1980) Olivines and Silicate Spinels, Chapter 11 in: Orthosilicates. P.H. Ribbe, ed., Rev. Mineral. 5, 275-381.

Burdett, J.K. (1978) A new look at structure and bonding in transition metal complexes. Adv. Inorg. Chem. Radiochem. 21, 113-146.

Burns, R.G. (1970a) Mineralogical Applications of Crystal Field Theory. Cambridge University Press, Cambridge, England, 224 pp.

__________ (1970b) Crystal field spectra and evidence of cation ordering in olivine minerals. Amer. Mineral. 55, 1608-1632.

__________ (1975) On the occurrence and stability of divalent chromium in olivines included in diamonds. Contrib. Mineral. Petrol. 51, 213-221.

__________ (1976) Partitioning of transition metals in mineral structures of the Mantle. In: The Physics and Chemistry of Minerals and Rocks. R.G.J. Strens, ed., J. Wiley, New York, p. 556-572.

__________ (1982) Electronic spectra of minerals at high pressures: how the Mantle excites electrons. In: High-Pressure Researches in Geoscience. W. Schreyer, ed., E. Schweizerbart'sche Verlagsbuchhandlung, Stuttgart, p. 223-246.

__________ (1985) Electronic spectra of minerals. Chapter 3 in: Chemical Bonding and Spectroscopy in Mineral Chemistry. F.J. Berry and D.J. Vaughan, eds., Chapman and Hall, Ltd., London, p. 63-101.

__________ and Burns, V.M. (1984) Optical and Mössbauer spectra of transition-metal-doped corundum and periclase. In: Structure and Properties of MgO and Al_2O_3 Ceramics. W.D. Kingery, ed., Amer. Ceram. Soc., Columbus, Ohio, Adv. Ceram. 10, 46-61.

__________ and Sung, C-M. (1978) The effect of crystal field stabilization on the olivine→spinel transition in the system Mg_2SiO_4-Fe_2SiO_4. Phys. Chem. Minerals 2, 349-364.

Cameron, M. and Papike, J.J. (1980) Crystal chemistry of silicate pyroxenes. Chapter 2 in: Pyroxenes. C.T. Prewitt, ed., Rev. Mineral. 7, 5-92.

Chenavas, J., Joubert, J.C., and Marezio, M. (1971) Low-spin — high-spin state transition in high pressure cobalt sesquioxide. Solid State Comm. 9, 1057-1060.

Cotton, F.A. (1971) Chemical Applications of Group Theory, 2nd ed. Wiley-Interscience, New York, 386 pp.

Drickamer, H.G. and Frank, C.W. (1973) Electronic Transitions and the High Pressure Chemistry and Physics of Solids. Chapman and Hall, London, 220 pp.

Figgis, B.N. (1966) Introduction to Ligand Fields. Interscience Publ., New York, Chapter 5.

Gaffney, E.S. (1972) Crystal-field effects in mantle minerals. Phys. Earth Planet. Inter. 6, 385-390.

__________ and Anderson, D.L. (1973) Effect of low-spin Fe^{2+} on the composition of the lower mantle. J. Geophys. Res. 78, 7005-7014.

George, P. and McClure, D.S. (1959) The effect of inner orbital splitting on the thermodynamic properties of transition metal compounds and coordination complexes. Progr. Inorg. Chem. 1, 382-463.

Goldman, D.S. and Rossman, G.R. (1977) The spectra of iron in orthopyroxene revisited: the splitting of the ground state. Amer. Mineral. 62, 151-157.

__________ and __________ (1985) The effect of site distribution on the electronic properties of ferrous iron. Phys. Chem. Minerals (in preparation).

Goto, T., Ahrens, T.J. and Rossman, G.R. (1979) Absorption spectra of Cr^{3+} in Al_2O_3 under shock compression. Phys. Chem. Minerals 4, 253-263.

__________, __________ , __________ and Syono, Y. (1980) Absorption spectrum of shock-compressed Fe^{2+}-bearing MgO and the radiative conductivity of the lower mantle. Phys. Earth Planet. Inter. 22, 277-288.

__________ Sato, J. and Syono, Y. (1982) Shock-induced spin-pairing transition in Fe_2O_3 due to the pressure effect on the crystal field. In: High Pressure Research in Geophysics. S. Akimoto and M.H. Manghnani, eds., D. Reidel Publ. Co., Dordrecht, p. 595-609.

Hazen, R.M. and Finger, L.W. (1979) Bulk modulus-volume relationships for cation-anion polyhedra. J. Geophys. Res. 84, 6723-6728.

__________ , Mao, H.K. and Bell, P.M. (1977) Effects of compositional variation on absorption spectra of lunar olivines. Proc. 8th Lunar Sci. Conf., Suppl. 8, Geochim. Cosmochim. Acta 1, 1081-190.

__________ , Bell, P.M. and Mao, H.K. (1978) Effects of compositional variation on absorption spectra of lunar pyroxenes. Proc. 9th Lunar Planet. Sci. Conf., 2919-2934.

Henderson, P. (1982) Inorganic Geochemistry. Pergamon Press, Oxford, 353 pp.
Huggins, F.E. (1975) The 3d levels of ferrous ions in silicate garnets. Amer. Mineral. 60, 316-319.
________ (1976) Mössbauer studies of iron minerals under pressures of up to 200 kilobars. In: The Physics and Chemistry of Minerals and Rocks. R.G.J. Strens, ed., J. Wiley, New York, p. 613-640.
Ikeda, K. and Yagi, K. (1982) Crystal field spectra for blue and green diopsides synthesized in the join $CaMgSi_2O_6$-$CaCrAlSiO_6$. Contrib. Mineral. Petrol. 81, 113-118.
Jorgensen, C.K. (1962) The nephelauxitic series. Progr. Inorg. Chem. 4, 73.
Keester, K.L. and White, W.B. (1968) Crystal-field spectra and chemical bonding in manganese minerals. Mineral. Mag. 1966 IMA vol., 22-35.
Konig, E. (1971) The nephelauxitic effect: calculation and accuracy of interatomic repulsion parameters in cubic high-spin d^2, d^3, d^7, and d^8 systems. Structure and Bonding 9, 175-211.
Lever, A.P.B. (1985) Inorganic Electronic Spectroscopy, 2nd ed. Elsevier Amsterdam.
Manning, P.G. (1969) Racah parameters and their relationship to lengths and covalencies of Mn^{2+} and Fe^{3+} oxygen bonds in silicates. Canadian Mineral. 10, 677-688.
McClure, D.S. (1962) Optical spectra of transition metal ions in corundum. J. Chem. Phys. 36, 2757-2779.
Mao, H.K. and Bell, P.M. (1972) Interpretation of the pressure effect on the optical absorption bands of natural fayalite to 20 kb. Crystal field stabilization of the olivine→spinel transition. Ann. Rept. Geophys. Lab., Yearbook 71, 524-528.
________ and ________ (1974) Crystal-field effects of trivalent titanium in fassaite from the Pueble de Allende meteorite. Ann. Rept. Geophys. Lab., Yearbook., 73, 488-492.
________ , ________ and Dickey, J.S., Jr. (1972) Comparison of the crystal field spectra of natural and synthetic chrome diopside. Ann. Rept. Geophys. Lab., Yearbook 71, 538-541.
________ , ________ and Yagi, T. (1982) Iron-magnesium fractionation model for the Earth. In: High-Pressure Research. S. Akimoto and M.H. Manghnani, eds., D. Reidel Publ. Co., p. 319-325.
Marfunin, A.S. (1979) Physics of Minerals and Inorganic Materials. Springer-Verlag, New York, 340 pp.
Meagher, E.P. (1980) Silicate garnets. Chapter 2 in: Orthosilicates. P.H. Ribbe, ed., Rev. Mineral. 5, 25-66.
Ohnishi, S. (1978) A theory of the pressure-induced high-spin to low-spin transition of transition-metal oxides. Phys. Earth Planet. Inter. 17, 130-139.
________ and Sugano, S. (1981) Strain interaction effects on the high-spin - low-spin transition of transition-metal compounds. J. Phys. C. 14, 39-55.
Orgel, L.E. (1952) The effects of crystal fields on the properties of transition metal ions. J. Chem. Soc. 4756-4761.
Parkin, K.M. and Burns, R.G. (1980) High temperature crystal field spectra of transition metal-bearing minerals: relevance to remote-sensed spectra of planetary surfaces. Proc. 11th Lunar Planet. Sci. Conf., Suppl. 12, Geochim. Cosmochim. Acta 1, 731-755.
Rajamani, V., Brown, G.E. and Prewitt, C.T. (1975) Cation ordering in Ni-Mg olivine. Amer. Mineral. 60, 292-299.
Reinen, D. (1969) Ligand field spectroscopy and chemical bonding in Cr^{3+}-containing oxidic solids. Structure and Bonding 6, 30-51.
Rossman, G.R. (1980) Pyroxene Spectroscopy. Chapter 3 In: Pyroxenes. C.T. Prewitt, ed., Rev. Mineral. 7, 93-116.
________ , Shannon, R.D. and Waring, R.K. (1981) Origin of the yellow color of complex nickel oxides. J. Solid State Chem. 39, 277.
Runciman, W.A., Sengupta, D. and Gourley, J.T. (1973) The polarized spectra of iron in silicates. II Olivine. Amer. Mineral. 58, 466-470.
Schmetzer, K. (1982) Absorption spectroscopy and color of V^{3+}-bearing natural oxides and silicates - a contribution to the crystal chemistry of vanadium. N. Jahrb. Mineral. Abh. 144, 73-106.
Shankland, T.J., Duba, A.G. and Woronow, A. (1974) Pressure shifts of optical absorption bands in iron-bearing garnet, spinel, olivine, pyroxene and periclase. J. Geophys. Res. 79, 3273-3282.
Smith, G.H. and Langer, K. (1982) Single crystal spectra of olivines in the range 40,000-5,000 cm^{-1} at pressures up to 200 kbars. Amer. Mineral. 67, 343-348.
________ and ________ (1983) High pressure spectra up to 120 kbars of the synthetic garnet end members spessartine and almandine. N. Jahrb. Mineral Mh., 541-555.
Stephens, D.R. and Drickamer, H.G. (1961) Effect of pressure on the spectrum of ruby. J. Chem. Phys. 35, 427-429.
Strens, R.G.J. (1976) Behavior of iron compounds at high pressure, and the stability of Fe_2O in planetary mantles. In: The Physics and Chemistry of Minerals and Rocks. R.G.J. Strens, ed., J. Wiley, New York, p. 545-554.
Sugano, S. and Ohnishi, S. (1984) Electron theory of transition-metal compounds under high pressure. In: Material Science of the Earth's Interior. I. Sunagawa, ed., Terra Scientific Publ. Co., Tokyo, p. 173-189.

Syono, Y., Ito, A., and Morimoto, S. (1984) Mössbauer study of the high pressure phase of Fe_2O_3. Solid State Comm. 50, 97-100.

__________ , Tokonami, M. and Matsui, Y. (1971) Crystal field effects on the olivine→spinel transformation. Phys. Earth Planet. Inter. 4, 347-352.

Wood, B.J. (1974) Crystal field spectrum of Ni^{2+} in olivine. Amer. Mineral. 59, 244-248.

__________ (1981) Crystal field electronic effects on the thermodynamic properties of Fe^{2+} minerals. In: Thermodynamics of Minerals and Melts. R.C. Newton, A. Navrotsky and B.J. Wood, eds., Springer-Verlag, New York, Adv. Phys. Geochem. 1, 63-84.

Yagi, T. and Mao, H.K. (1977) Crystal-field spectra of the spinel polymorphs of Ni_2SiO_4 at high pressures. Ann. Rept. Geophys. Lab., Yearbook 76, 505-508.

__________ , Bell, P.M., and Mao, H.K. (1979) Phase relations in the system MgO-FeO-SiO_2 between 150 and 700 kbar at 1000°C. Ann. Rept. Geophys. Lab., Yearbook 78, 614-618.

Chapter 9. Robert M. Hazen

COMPARATIVE CRYSTAL CHEMISTRY and the POLYHEDRAL APPROACH

1. INTRODUCTION

The macroscopic behavior of crystals is a consequence of atomic-scale interactions. Compression and thermal expansion, two of the more obvious examples of macroscopic mineral properties, occur because of the variations in interatomic distances that result from changing pressure and temperature. During the past fifteen years more than 80 crystallographic studies have been published on the variation of mineral structures with pressure and temperature (Hazen and Finger, 1982). These studies in comparative crystal chemistry reveal systematic structural behavior that has led to prediction of mineral structures under conditions deep within the earth. Comparative crystal chemistry, furthermore, provides important constraints on models of interatomic bonding.

The first publications in comparative crystal chemistry were often weighty affairs, filled with massive tables and complex analysis. A single classic paper on high-temperature crystal chemistry of six pyroxenes (Cameron et al., 1973) incorporates 26 separate structure refinements. Eighteen tables with more than 3000 separate cell parameters, bond distances and angles, thermal parameters, and other data (not to mention tens of thousands of unpublished structure factors) were required for this one paper. Given the plethora of nonambient structural parameters, it became obvious that a simple conceptual framework was needed if any general relationships of comparative crystal chemistry were to emerge.

The first attempts to identify simplifying relationships in the compression and expansion of mineral structures were based on the principles of crystal chemistry developed by Linus Pauling (1960). Pauling recognized the dominance of nearest-neighbor interactions in determining crystal structure and properties. The identification of cation polyhedra as fundamental structural units, and the consequent adaptation of associated bonding parameters such as cation and anion formal (i.e., integral) charge, cation-anion distance, and coordination number, helped Pauling to reduce the complexities of inorganic structures to a set of "rules" of great simplicity and predictive applications.

Pauling's rules relate to structural topology. With the help of his polyhedral approach, similar relationships can be identified that relate to variations of structure with pressure and temperature. It is now possible to predict the high-pressure and high-temperature structural variations of most minerals (Bish and Burnham, 1980; Hazen and Finger, 1982). The objectives of this paper are to review these advances in comparative crystal chemistry and, in addition, to demonstrate how the polyhedral approach can be applied to other crystal properties in order to elucidate the relationships between microscopic and macroscopic behavior.

2. COMPARATIVE CRYSTAL CHEMISTRY

This section is adapted from "Comparative Crystal Chemistry" by R.M. Hazen and L.W. Finger (Wiley, New York, 1982).

2.1 Nonambient crystallography

Comparative crystal chemistry is an empirical science. Precise structural parameters at high pressure and high temperature are necessary to resolve the subtle shifts of atomic positions that are manifest in compression and thermal expansion. It is important, therefore, to understand the capabilities and limitations of crystal heaters and pressurizers that are used in crystallographic research.

High-pressure crystallography. Diamond is remarkable, not only for its hardness, but also because of its transparency to many ranges of electromagnetic radiation. These properties have led to the present widespread use of diamonds in generating high-pressure environments for physical experimentation. The opposed-diamond-anvil configuration is incorporated into all high-pressure, single-crystal devices (Fig. 1). More than a dozen designs of diamond cells have been used in single-crystal research, but the simplest and most widely employed is based on the screw-tightened, miniature cell of Merrill and Bassett (1974). The design, construction, and use of this cell, as modified by Hazen and Finger (1977a; Fig. 2), are detailed by Hazen and Finger (1982).

All high-pressure cells for single-crystal x-ray diffraction have somewhat restricted access to reciprocal space. In the standard miniature device, for example, only about a third of reciprocal space to 60°2Θ is accessible. High- pressure data sets are further limited by the reduced intensity of an x-ray beam that must pass through diamonds and beryllium components, and by the increased background scattered radiation. High-pressure structure refinements, therefore, are usually of lower precision than corresponding structure determinations under room conditions.

The maximum pressure obtainable with the Merrill and Bassett pressure cell is about 65 kbar. Other cells have been employed in single-crystal studies to pressures in excess of 200 kbar (Hazen et al., 1981); however, studies at pressures above 65 kbar are complicated by the requirements for thin crystals that reduce scattering intensities and more massive diamond supports that can further restrict access to reciprocal space.

High-temperature crystallography. Given the fundamental nature of high-temperature research, particularly in the earth sciences, it is not surprising that heaters were applied to x-ray cameras as early as the 1920's, shortly after the development of powder x-ray diffraction as a useful identification technique.

Radiative single-crystal heaters are the most versatile and widely used type today. First applied to single-crystal x-ray studies by Foit and Peacor (1967), these heaters have been adopted and modified by many workers. The most common design now in use is based on the model devel-

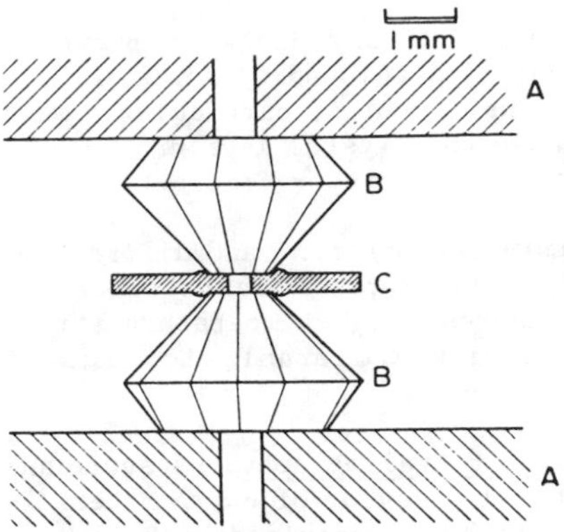

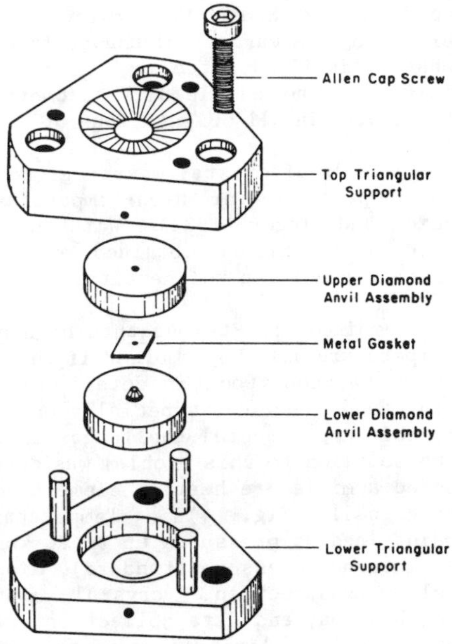

Figure 1 (above). Opposed-diamond-anvil configuration with a metal foil gasket. (A) Diamond supports, (B) diamond anvils, (C) gasket. From Hazen and Finger (1982).

Figure 2 (right). Exploded view of the miniature diamond-anvil cell of Merrill and Bassett (1974), as modified by Hazen and Finger (1977a). Load is applied by three screws. From Hazen and Finger (1982).

Figure 3 (below). Radiative heater for single-crystal x-ray diffraction as designed by Y. Ohashi. From Hazen and Finger (1982).

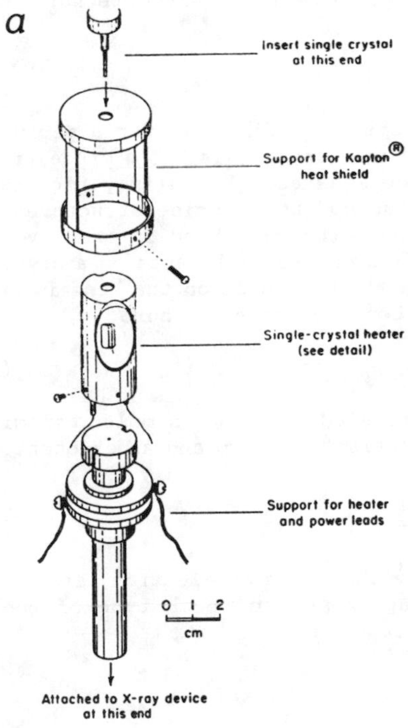

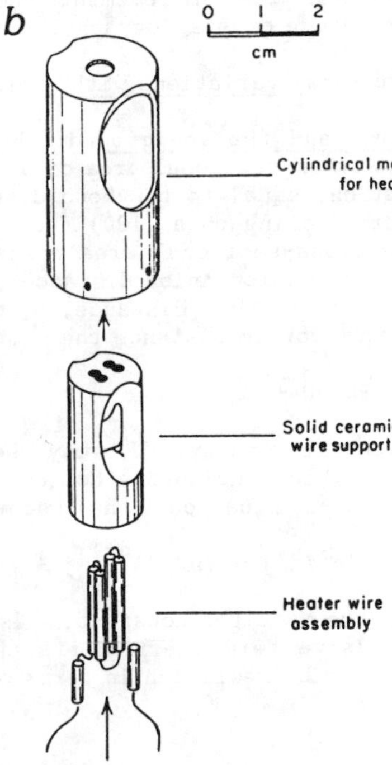

oped at the State University of New York at Stony Brook by Brown et al.(1973). A variety of the resistance heater, developed by Dr. Yoshikazu Ohashi, is illustrated in Figure 3. This instrument is particularly well suited to four-circle diffractometry because the crystal is surrounded by the heater in all orientations of the main diffractometer circle.

Details of crystal mounting, temperature calibration, and diffractometer operation for high-temperature crystallography were presented by Hazen and Finger (1982). With these procedures, high-temperature structure data can be obtained with a precision comparable to that of ambient-condition refinements.

Combined high-temperature/high-pressure crystallography. Pressure and temperature must be combined if the structural behavior of earth materials is to be understood in detail. Crystallography at combined high temperature and pressure, especially on oriented single crystals, presents formidable experimental challenges that have only recently been partly met. One solution to this problem was developed by Hazen and Finger (1981), who added a miniature heating element to the Merrill and Bassett (1974) diamond cell (Fig. 4). Temperatures in excess of 400°C have been maintained at pressures up to 20 kbar in studies of fluorite (now used as an internal pressure standard), diopside, and other compounds. Details of cell construction, crystal mounting, pressure and temperature calibration, and data collection are provided by Hazen and Finger (1982). As a result of limitations of the present design, the heated pressure cell has been employed primarily in the determination of phase boundaries of materials with reversible transitions, rather than for complete structure refinements, which are of low precision. Great opportunities exist, therefore, for the improvement and application of high-pressure and high-temperature crystallography.

2.2 Structural variations with pressure

Pressure and the ionic bond. Pressure is defined as force per unit area. The cross-sectional area of a bond in a crystal is, to a crude first approximation, equal to the bond distance squared. (Visualize, for example, a force acting on a (100) face of the sodium chloride structure, in which the cross-sectional area associated with each bond is exactly the square of the cation-anion distance.) If a pressure, P, acts on a crystal with cation-anion bond distance, d, then the net force on the bond is $F_P = Pd^2$. At equilibrium distance the sum of bonding forces is zero:

$$\partial U/\partial D + F_P = 0 . \quad (1)$$

Bond potential energy, U, may be modeled with a simple binomial expression, which includes a Coulombic attractive term and an exponential repulsive term. Equation 1 thus becomes

$$[Ae^2/d^2] - [nB/d^{(n+1)}] + Pd^2 = 0 , \quad (2)$$

where A is the Madelung constant, e is the charge on an electron, and B and n are repulsive term coefficients characteristic of each type of bond. Equation 2 may be rewritten in terms of pressure:

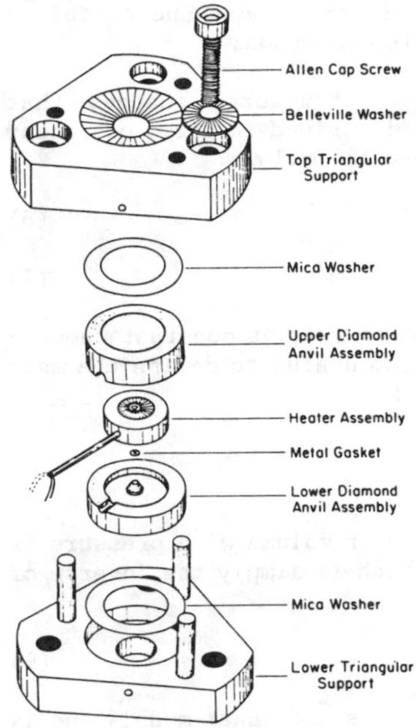

Figure 4. Exploded view of the high-temperature, high-pressure diamond-anvil cell. From Hazen and Finger (1981); reproduced by permission of the American Institute of Physics.

$$P = [nB/d^{(n+3)}] - [Ae^2/d^4] \ . \tag{3}$$

The fractional change of interatomic distance with pressure (i.e., the bond linear compressibility) is given by

$$-\frac{1}{d(\partial P/\partial d)} = \frac{1}{\frac{(n+3)nB}{d^{(n+3)}} - \frac{4Ae^2}{d^4}} \ , \tag{4}$$

which, combined with Equation 3, gives

$$-(1/d)(\partial d/\partial P) = \frac{1/(n+3)}{P + [(n-1)/(n+3)][Ae^2/d^4]} \tag{5}$$

Thus, if the values of bonding parameters n, B, d, and A are known, then bond compressibility may be calculated. These four bonding terms are positive; therefore, compressibility is also always positive. (In a few unusual cases, such as $BiVO_4$, complex many-body interactions lead to pressure expansion of one bond while other bonds compress; such complex interactions cannot be modeled by the simple-pair potential approach of Equations 1 to 5.) In practice Equation 5 is adequate for the prediction of relative, but not absolute, values of bond compressibilities (Hazen, 1975). The two-term bond potential model is too simple for calculating accurate derivatives of the potential function, and the use of d^2 as the area term in the pressure-force calculation is only a first order approximation. More sophisticated bonding models, such as the self-consistent symmetrized augmented plane wave method (Bukowinski, 1980) and the modified electron-gas method (Tossell, 1980), have proved more successful in predicting compression of ionic bonds in simple compounds.

Bond distance variations with pressure. Structural changes that result from application of pressure are best described in terms of linear and volume compressibilities, which are defined as follows:

$$\text{Linear:} \quad \beta_\ell = -1/d(\partial d/\partial P)_{T,X} \,, \tag{6}$$

$$\text{Volume:} \quad \beta_V = -1/V(\partial V/\partial P)_{T,X} \,, \tag{7}$$

where subscripts T and X refer to partial derivatives at constant temperature and composition, respectively. It is also useful to define the mean compressibility between two pressures, P_1 and P_2:

$$\text{Mean:} \quad \bar{\beta}_V = \frac{-2}{(V_1 + V_2)} \times \frac{(V_2 - V_1)}{(P_2 - P_1)} \,. \tag{8}$$

An important parameter that relates the change of volume with pressure is the bulk modulus, K (in units of pressure), which is simply the inverse of compressibility:

$$\text{Bulk modulus:} \quad K = 1/\beta_V \,. \tag{9}$$

Bulk modulus is closely related to Young's modulus and shear moduli and is thus commonly cited in physics and geophysics literature. Compressibility is more widely cited in chemical thermodynamics literature because of its close functional relationship to thermal expansivity (see below).

Data on the variation of structural dimensions come from both three-dimensional structure refinements and unit cell data on simple, constrained structures. The structures of NaCl, CsCl, CaF_2, and cubic ZnS are all fully constrained with no variable atomic coordinates. A knowledge of their unit cell dimensions provides all interatomic distances as well. Other simple structures, including rutile, corundum, hexagonal ZnS, and ZrO_2, also have compressibilities that are closely related to changes in interatomic distances. In addition, complete three-dimensional structure refinements on crystals at high pressure have been reported for more than 40 compounds, primarily oxides and silicates. High-pressure studies on simple and complex crystals are summarized by Hazen and Finger (1982) and provide a wealth of information on the variation of structure with pressure.

In reviewing the compressibility behavior of minerals, it is useful first to examine the magnitude of changes in bond distances in various materials (Fig. 5). The least compressible units in minerals and mineral-like materials are short, rigid ionic bonds from highly charged cations, such as W^{6+} and Si^{4+}, and strong covalent bonds, such as C-C in diamond and B-N in boron nitride. The most compressible units, as much as three orders of magnitude more compressible than W-O, are weak van der Waals bonds in molecular and layer compounds and long cation-anion bonds in alkali halides. Bonds of intermediate distance and strength, such as Mg-O and Ca-F, have intermediate compressibilities.

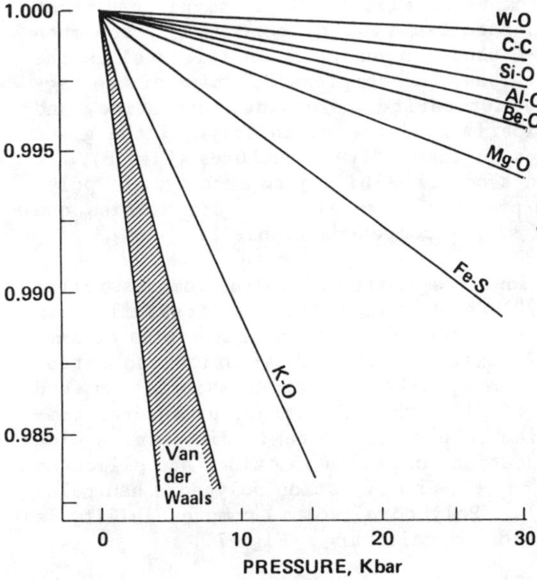

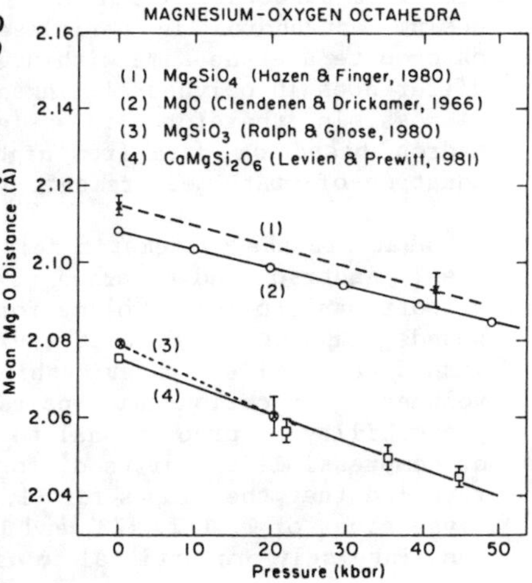

Figure 5 (left). Comparative linear compressibilities of some common bonds in minerals.

Figure 6 (right). Magnesium-oxygen bond distances versus pressure for octahedra in several compounds. After Hazen and Finger (1982).

Table 1. Selected volume compressibilities for polyhedra*

Cation	Anion	Coordination	$\beta_V \times 10^3$ kbar^{-1}	Cation	Anion	Coordination	$\beta_V \times 10^3$ kbar^{-1}
Li^+	F^-	6	1.5	K^+	O^{2-}	12	3.7
Na^+	F^-	6	2.2	Na^+	O^{2-}	8	3.1
K^+	F^-	6	3.4	Mg^{2+}	O^{2-}	6	0.62
Na^+	Cl^-	6	4.2	Ca^{2+}	O^{2-}	6	0.91
Na^+	Br^-	6	5.0	Ba^{2+}	O^{2-}	6	1.4
Na^+	I^-	6	6.2	Be^{2+}	O^{2-}	4	0.40
Cs^+	Cl^-	8	5.5	Al^{3+}	O^{2-}	4	<0.4
Ca^{2+}	F^-	8	1.2	Al^{3+}	O^{2-}	6	0.42
Mg^{2+}	F^-	8	1.0	Fe^{3+}	O^{2-}	6	0.43
Ca^{2+}	S^{2-}	6	2.3	Si^{4+}	O^{2-}	4	<0.4
Zn^{2+}	S^{2-}	4	1.3	Ti^{4+}	O^{2-}	6	0.4
Pb^{2+}	S^{2-}	6	2.1	Re^{6+}	O^{2-}	8	0.15

*After Hazen and Finger (1982).

Given the wide range of bond compressibilities, it is significant that the average compressibility of a specific type of bond (e.g., the mean compressibility of octahedral Mg-O bonds in an oxide or silicate) is the same from structure to structure. Thus, the compressibilities of the magnesium octahedron in periclase, forsterite, diopside, enstatite, and phlogopite are the same within experimental error, in spite of the great differences in polyhedral linkages in these five structures (Fig. 6). It is possible, therefore, to assign a compressibility to each type of polyhedron based on data from high-pressure crystal structure studies and equation-of-state measurements of simple compounds (Table 1).

What are the systematic relationships among polyhedral compressibilities? Anderson and Anderson (1970) demonstrated that compressibility is proportional to molar volume for a number of isostructural series of compounds. Hazen and Prewitt (1977) extended this observation and established a similar relationship for polyhedral compressibilities and volumes. For cation-anion pairs of given formal charge, polyhedral compressibility is proportional to the cube of interatomic distance. A plot of compressibility versus d^3 for cation polyhedra in oxides and silicates revealed that the slopes for +1, +2, +3, and +4 cation polyhedra had relative values of 1, 1/2, 1/3, and 1/4. Polyhedral volume compressibility is thus inversely proportional to cation formal charge (Fig. 7),

$$\beta_V = [0.133 d^3/z_c] \ \ \mathrm{Mbar}^{-1} \ . \qquad (10)$$

Equations similar to 10 were derived for halides, sulfides, and other compounds by Hazen and Finger (1979a), who found that polyhedral compressibilities in all of these compounds could be modeled by a single expression,

$$\beta_{poly} = 0.133 \ [d^3/S^2 z_c z_a] \ \ \mathrm{Mbar}^{-1} \ , \qquad (11)$$

where β_{poly} is the volume compressibility of the cation polyhedron, z_a is the anion formal charge, and S^2 is a scaling factor defined as 0.5 for all oxygen-based polyhedra, and calculated to be 0.75 for halides, 0.40 for chalcogenides (sulfides, selenides, and tellurides), 0.25 for phosphides, arsenides, and antimonides, and 0.20 for carbides. Note that S^2 conforms to the relative ionicity of the different anions in question. Equation 11, which has been called the polyhedral bulk modulus-volume relationship because of its similarity to the equations of Anderson and Anderson (1970), models successfully the ionic bonds in alkali halides, the bonds of intermediate character in silicates, and the covalent bonds in carbides. It is even possible to treat the compressibility of C-C bonds in diamond by selecting +4 and -4 as cation and anion formal charge.

A few exceptions to this empirical relationship provide insight to the nature of bond compression. Some bonds, including V-O in V_2O_3 and Zn-O in ZnO, are significantly more compressible than would have been predicted by the bulk modulus-volume relationship. These bonds are also significantly more covalent than the average cation-oxygen bond, and the use of $S^2 = 0.5$ may be inappropriate in these cases. It thus appears that, for a given cation-anion pair, more covalent bonds may be somewhat more compressible.

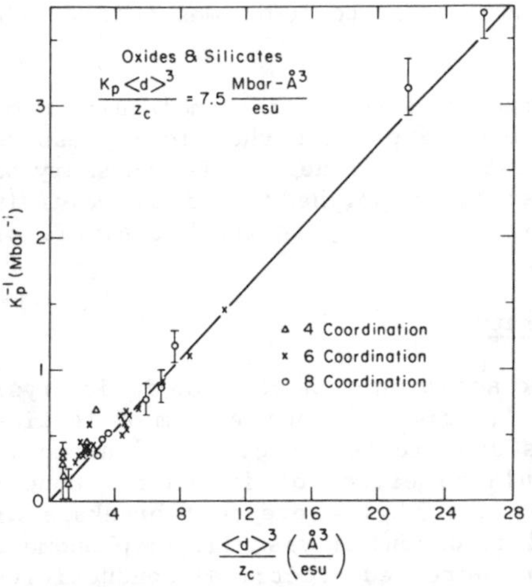

Figure 7. The bulk modulus-volume relationship for polyhedra in oxides and silicates. Triangles represent tetrahedra; crosses, octahedra; and circles, eight- or greater-coordinated sites. Error bars represent one estimated standard deviation in polyhedral bulk moduli for selected polyhedra in silicates. The line is a weighted linear-regression fit of all data, constrained to pass through the origin. From Hazen and Finger (1979a); reproduced by permission of the American Geophysical Union.

Compounds with the cesium chloride structure have cation-anion bonds that are significantly less compressible than predicted by the bulk modulus-volume relationship. The cesium chloride structure, with eight anions at the corners of a unit cube and a cation at the cube's center, is unique in the high degree of face sharing between adjacent cation polyhedra. In cesium chloride-type compounds the shortest cation-cation distances are only 15% longer than the cation-anion distances, in contrast to the 50-75% greater separation in most other structures. It is probable, therefore, that cation-cation repulsion plays a much greater role in cesium chloride compression than in the other structures represented in Figure 7. The bulk modulus-volume relationship, which incorporates only the bonding characteristics of the primary coordination sphere, is not valid for structures with such extensive polyhedral face sharing and strong second-nearest-neighbor interactions.

Other structural changes with pressure. In addition to bond compression, there are at least three ways that a structure of fixed topology can be modified by pressure: polyhedral distortion, interpolyhedral bond-angle bending, and intermolecular compression. It is possible for cation polyhedra to distort without changes in cation-anion distances. In the minerals studied to date, this type of distortion has not been observed, and polyhedral distortion does not appear to be an important compression mechanism in most materials.

Significant compression may result from interpolyhedral bond bending. Corner-linked structures, notably the framework silicates, may be extremely compressible even though the constituent cation-anion bonds are quite rigid. Bond-angle bending, in general, requires less energy than bond compression (Gibbs, 1982) and thus may be the principal cause of volume changes with temperature and pressure. The largest linear compressibilities observed in crystals are those associated with intermolecular compression. In crystals of inert gas, including argon and neon (Finger et al., 1981), compressibilities are as much as three orders of magnitude greater than in silicates. In layer compounds, such as graphite and talc, interlayer compressibility is commonly from 10 to 100 times that within the layers (Hazen and Finger, 1978a).

Given the large magnitude of interpolyhedral and intermolecular compression effects, it is often impossible to predict the high- pressure behavior of a structure from bond-distance data alone. It is necessary to consider both the nature of the constituent polyhedra and the ways in which the polyhedra are linked together in any thorough analysis of high-pressure behavior.

2.3 Structural variations with temperature

Temperature and the ionic bond. The addition of heat to an ionic crystal increases the energy of that crystal, primarily in the form of lattice vibrations or phonons. If ionic bonds are treated as classical harmonic oscillators, then the principal calculated effect of increased temperature is increased vibration amplitude, with the eventual breakage of bonds at high temperature. This model is useful in visualizing phenomena such as melting, site disordering, or increased electrical conductivity at high temperature. The purely harmonic model of atomic vibrations is not adequate to explain many crystal properties, however, and anharmonic vibration terms must be considered in any analysis of the effect of temperature on crystal structure (see an introduction by Kittel, 1971).

An important consequence of anharmonic motion is the change of equilibrium bond distance with temperature, i.e., thermal expansion. A useful intuitive approach to understanding thermal expansion stems from the fact that the potential diagram of an ionic bond, with potential energy plotted as a function of interionic separation (Fig. 8), is asymmetric about the minimum potential. Thus, as the potential energy of the system increases with the addition of heat, the equilibrium separation changes in proportion to the asymmetry. There is no constraint on the skew of the asymmetry, and thermal expansion of a bond may be positive or negative.

Dimensional changes of a crystal may be represented by the coefficient of thermal expansion, α, which is exactly analogous to compressibility:

$$\text{Linear:}\quad \alpha_{\ell} = 1/d(\partial d/\partial T)_{P,X}\ , \tag{12}$$

$$\text{Volume:}\quad \alpha_{V} = 1/V(\partial V/\partial T)_{P,X}\ , \tag{13}$$

where subscripts P and X denote partials at constant pressure and composition, respectively. Another useful measure of temperature variations is the mean coefficient of expansion between two temperatures, T_1 and T_2:

$$\text{Mean}\ \alpha_{T_1,T_2} = 2/(d_1+d_2)[(d_2-d_1)/(T_2-T_1)] = \alpha_{(T_1+T_2)/2}\ . \tag{14}$$

The latter parameter is commonly reported in studies of high-temperature structural variations. For want of a more satisfactory theoretically based relationship, most volume-thermal expansion data for crystals are presented by a simple second-order polynomial:

$$V = V_0 + aT + bT^2\ . \tag{15}$$

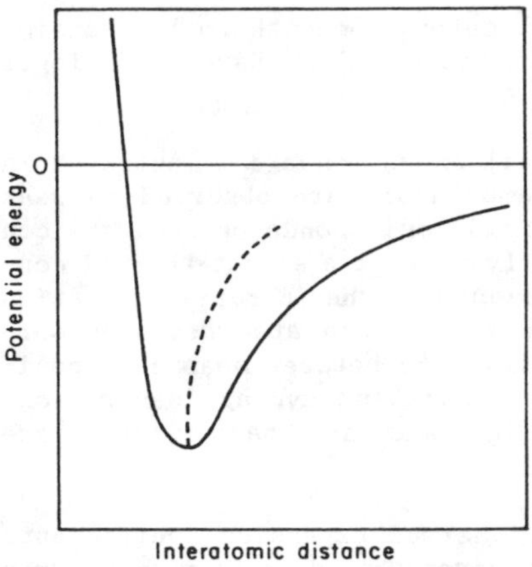

Figure 8. Equilibrium interatomic distance of a diatomic oscillator increases as a function of temperature because of the asymmetry of the potential function. After Hazen and Finger (1982).

Much of the description of structural variations with temperature depends on thermal expansion coefficients of volume or linear structural elements. It is important, therefore, to recognize the limits in the accuracy of reported coefficients. Crystal expansion may be determined by a number of different techniques, on crystals or on powders. Internal precision of these techniques is often better than 1%, yet different studies on the same material may differ by 10%. A variety of factors may contribute to these differences. Impurities, crystalline defects, and sample preparation may influence the results. But even in well-crystallized, pure compounds such as periclase and corundum there exists a considerable range of published values for thermal expansion. Until such time as techniques and sample descriptions become more standardized the accuracy of any given study must be conservatively taken as ±5%, even though reported precision may be much smaller.

Errors in bond-distance expansion coefficients must be significantly greater than 5% because of the unknown effects of correlated thermal motion. A bond distance from an x-ray structure refinement is the separation of average atomic positions, which is generally less than the average separation of the atoms (because of vibrations off the line of bonding -- Busing and Levy, 1964). Thus the thermal expansion of a bond based on distances between atomic centers may be significantly less than the true thermal expansion of the bond. It is important to remember in the subsequent discussion that the important effects of thermal motions have not been included because of our lack of ability to document the detailed nature of atomic motions. A complete understanding of bond expansion must await this additional information.

<u>Bond distance variations with temperature</u>. Data on the thermal expansion of bonds, like compressibility data, come from both three-dimensional structure refinements and unit cell data on simple, constrained structures. In addition to expansion data on about 100 simple compounds, there are now available high-temperature structure refinements of more than 40 minerals and mineral-like compounds. These refinements provide a wealth of data on the expansion of approximately 130 different

cation-oxygen polyhedra in oxides and silicates. Bond thermal expansion data from both types of studies have been compiled by Hazen and Finger (1982).

The relative magnitudes of bond thermal expansion are illustrated in Figure 9. The greatest changes with temperature are observed in weak cation-anion bonds in alkali halides. Rigid ionic bonds between silicon and oxygen, on the other hand, may actually undergo a slight thermal contraction. This phenomenon may be a consequence of the increased vibration amplitudes of the silicon and oxygen atoms; as each atom deviates more from the line of bonding, the observed distance between mean atom positions may decrease in order to maintain a constant average separation. Bonds with intermediate strength, such as Mg-O and Ca-F, have intermediate thermal expansion.

In spite of the wide range of bond thermal expansion coefficients illustrated in Figure 9, the value for any given type of cation polyhedron appears to be independent of structure. Magnesium-oxygen octahedra thus have the same linear expansion coefficient (about 14×10^{-6} $°C^{-1}$) in periclase, diopside, forsterite, phlogopite, tremolite, and other silicates (Fig. 10). Similarly, all silicon-oxygen tetrahedra show zero or slightly negative expansion (except in special cases where aluminum-silicon disorder takes place at high temperature, thus changing both the composition and size of the tetrahedron). As with polyhedral compressibility, it is thus possible to assign to each type of cation-anion polyhedron a coefficient of volume thermal expansion (Table 2).

Examination of data in Table 2 reveals an intriguing clustering of thermal expansion coefficients. Values of about 50, 14, and 9×10^{-6} $°C^{-1}$ dominate the list. This fact led Hazen and Prewitt (1977) to the observation that bond thermal expansion is inversely proportional to Pauling bond strength, z_c/n, where z_c is cation formal charge and n is the coordination number of the cation polyhedron:

$$\alpha_V = 13 \times 10^{-6} \, [n/z_c] \quad °C^{-1} \; . \tag{16}$$

Equation 16 is applicable to the volume thermal expansion of cation-oxygen polyhedra in a wide variety of oxides and silicates. Thus, for example, all divalent cation-oxygen octahedra have volume expansion coefficients of about 4×10^{-5} $°C^{-1}$. (It is remarkable that measured thermal expansions of the rocksalt-type oxides, including NiO, MgO, CoO, MnO, CdO, SrO, BaO, and CaO, are the same within ±5%, even though the mass and ionic volumes of these divalent cations differ by more than a factor of two.)

Equation 16 may be generalized to polyhedra with anions other than oxygen in the same way that Equation 10 was modified to yield Equation 11:

$$\alpha_V = 13 \times 10^{-6} \, [n/z_c z_a S^2] \quad °C^{-1} \; . \tag{17}$$

The empirical term, S^2, is the same as in Equation 11. It is surprising that this "scaling factor" should have the same relative values for expressions describing the very different phenomena of bond compression and bond thermal expansion. The physical significance of S^2 remains uncertain, yet there does seem to be some generality in its application.

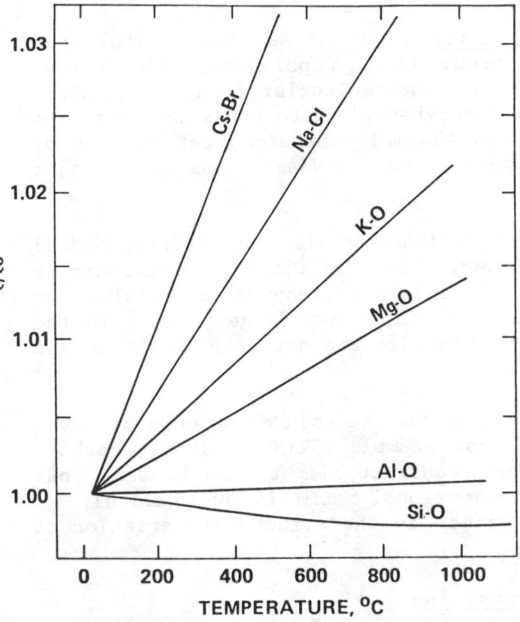

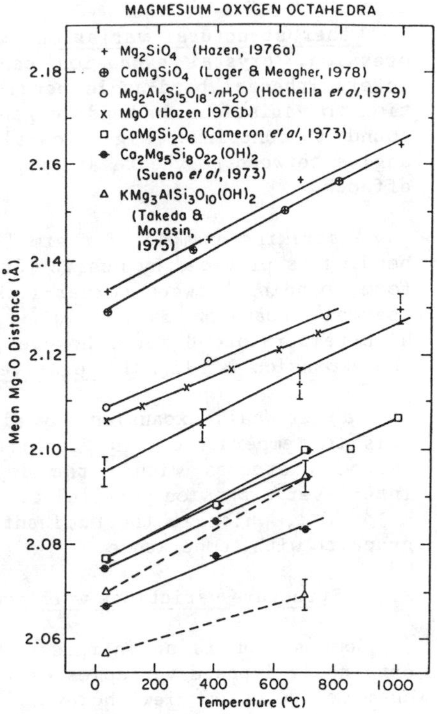

Figure 9 (above). Comparative linear thermal expansivities of some common bonds in minerals.

Figure 10 (right). The variation of average Mg-O bond distances in seven different compounds with magnesium octahedra, plotted versus temperature. From Hazen and Finger (1982).

Table 2. Selected linear thermal expansivities for polyhedra*

Cation	Anion	Coordination	$\bar{\alpha}_\ell \times 10^6$ °C^{-1}	Cation	Anion	Coordination	$\bar{\alpha}_\ell \times 10^6$ °C^{-1}
Na^+	Cl^-	6	51	Na^+	O^{2-}	7	35
K^+	Cl^-	6	46	K^+	O^{2-}	6	21
K^+	Br^-	6	49	Li^+	O^{2-}	6	20
Rb^+	Br^-	6	44	Mg^{2+}	O^{2-}	6	14
Li^+	F^-	6	46	Fe^{2+}	O^{2-}	6	13
Cs^+	Br^-	8	68	Ca^{2+}	O^{2-}	6	15
Ca^{2+}	F^-	8	21	Ba^{2+}	O^{2-}	6	15
Pb^{2+}	S^{2-}	6	22	Be^{2+}	O^{2-}	4	9
Na^{2+}	S^{2-}	4	9	Al^{3+}	O^{2-}	6	9
Be^{3+}	N^{3-}	4	13	Fe^{3+}	O^{2-}	6	9
Ta^{4+}	C^{4-}	6	7	Al^{3+}	O^{2-}	4	1
C^{4+}	C^{4-}	4	3.5	Si^{3+}	O^{2-}	4	<0.1

*After Hazen and Finger (1982).

Other structural variations with temperature. As with crystal compression, crystal expansion can occur through polyhedral distortion, interpolyhedral bond-angle bending, or intermolecular changes, in addition to variations in bond lengths. Polyhedral distortions have not been found to contribute significantly to thermal expansion, but bending of angles between polyhedra and expansion of van der Waals bonds can be major effects.

A striking example of thermal expansion controlled by interpolyhedral bending is provided by quartz (Skinner, 1966). In the low-temperature, α form, bending between corner-linked, silicon tetrahedra is possible and thermal expansion is 7×10^{-3} °C^{-1}, a relatively large value. In the high-temperature β form, however, tetrahedra are not free to rotate and the expansion is slightly negative.

Van der Waals expansion may also control the volume response of crystals to temperature. In graphite, for example (Skinner, 1966), carbon-carbon distances within the layers contract slightly on heating, but interlayer expansion parallel to the hexagonal c axis is approximately 3×10^{-3} °C^{-1}, which is the dominant effect in the structural variation of graphite with temperature.

2.4 Structure variations with composition

Composition is an intrinsic variable, and thus is distinct from the externally imposed variables of temperature and pressure. From a crystal chemical point of view, however, it is instructive to consider continuous structural variations with composition, just as with changes in temperature and pressure. A useful set of definitions is the coefficients of compositional expansion, γ:

$$\text{Linear:} \quad \gamma_\ell = 1/d(\partial d/\partial X)_{T,P} \,, \tag{18}$$

$$\text{Volume:} \quad \gamma_V = 1/V(\partial V/\partial X)_{T,P} \,. \tag{19}$$

Note that Equations 18 and 19 have exactly the same form as those for compressibility (6 and 7) and thermal expansion (12 and 13). A useful convention in applying Equations 18 and 19 is that, if two cations are in solid solution, then X refers to the mol fraction of the larger cation. The coefficient of compositional expansion will thus always be positive. The units of γ are fractional change in volume or distance per mole.

The principal continuous structural variations that occur because of atomic substitution result from differences in ionic radii. Cations of similar size and valence commonly substitute for each other in minerals and similar compounds. In a few instances, such as solid solution between zirconium and hafnium, two cations are so similar that the compositional change has little measurable effect on structure. Generally, however, substituting cations differ by a few percent in size, and thus result in changes in polyhedral volumes.

Cation-anion distances in a disordered polyhedron will vary from unit cell to unit cell, depending on the cation present in each specific location. X-ray diffraction techniques can be used to determine only the mean bond distance, d, averaged over the entire crystal for each symmetrically distinct polyhedron. The net change in mean bond distance, Δd, is given by

$$\Delta d = \Delta X_2(r_1 - r_2) , \tag{20}$$

where X_2 is the fractional change in occupancy of cation 2, and r_1 and r_2 are the ionic radii of the two cations. It is thus possible to calculate the effect of a cation substitution on crystal structure. A particularly useful relationship in this regard is the plot of bond distance versus mole percent aluminum in tetrahedral sites with both Al and Si (Fig. 11). The similarity of electron density of these two cations precludes direct determination of site occupancy by x-ray diffraction. Site occupancies can be deduced, however, from the mean T-O distances.

Equations 18 and 20 may be combined to relate the coefficient of compositional expansion to ionic radii of the substituting cations:

$$\gamma_\ell = 1/d(\partial d/\partial X) = (r_1 - r_2)/d . \tag{21}$$

In this relationship cation 1 may be a single ionic species, or a group of cations with

$$r_i = \sum_{i=1}^{n} f_i r_i/\Sigma f_i , \tag{22}$$

where f_i and r_i are the mole fraction and ionic radii of the ith cation, respectively.

The change in polyhedral volume, Vp, for a given compositional change is also a constant; $(r_1 - r_2)$ is generally small compared to d, and Vp is proportional to d^3. The change in polyhedral volume is thus given by the coefficient of volume compositional expansion:

$$\gamma_V = 3(r_1 - r_2)/d . \tag{23}$$

This simple equation, like the analogous relationships for compressibility (11) and thermal expansion (17), may be used to predict variations of structural parameters with composition.

2.5 Combined P-T-X structure variations

The structural analogy of pressure, temperature, and composition. Hazen (1977) proposed that geometrical aspects of structural variations with pressure, temperature, or composition are analogous in several ways. The fundamental unit of structure for purposes of the analogy is the cation coordination polyhedron. For a given type of cation polyhedron, a given change in pressure, temperature, or composition has a constant effect on polyhedral size, regardless of the way in which the polyhedra are linked. Thus, to a first approximation, polyhedral volume coefficients α_V, β_V, and γ_V are independent of structure.

Polyhedral volume changes with pressure or temperature or composition may be predicted from the simplest of bonding parameters: cation-anion

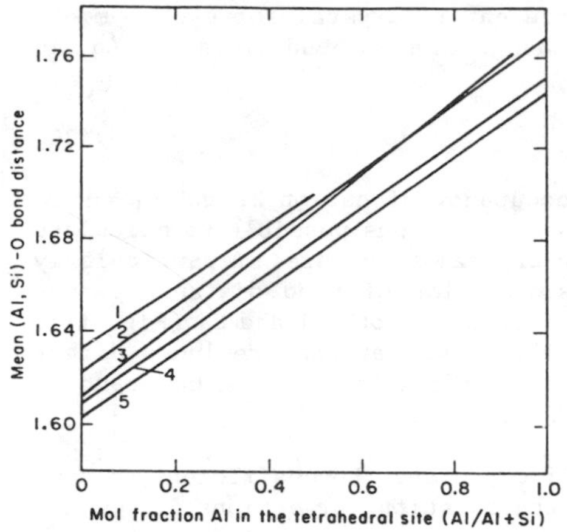

Figure 11. Bond distance versus mole fraction aluminum in several groups of minerals with tetrahedral Al-Si solid solution. From Hazen and Finger (1982).
(1) $CaMgSi_2O_6$-$CaAl_2SiO_6$ (Hazen and Finger, 1977b)
(2) layer silicates (Smith and Bailey, 1963)
(3) micas (Hazen and Burnham, 1973)
(4) feldspars (Smith, 1974)
(5) framework silicates (Smith and Bailey, 1963)

bond distance (d), coordination number (n), formal cation and anion charges (z_c and z_a), and relative ionic radii ($r_1 - r_2$), as given in Equations 11, 17, and 23. In structures with more than one type of cation polyhedron, variations of pressure, temperature, or composition all have the effect of changing the ratios of polyhedral sizes. It is important to remember that the P-T-X analogy extends only to geometrical aspects of a structure. Changes in pressure, temperature, and composition produce very different effects on vibrational properties and electronic structure of solids.

In considering the relationships between structure variations with pressure, temperature, and composition it is useful to represent structural parameters in P-T-X space. All crystalline materials, for example, may be modeled in P-T-X space by surfaces of constant molar volume (isochoric surfaces). One consequence of the structural analogy of pressure, temperature, and composition is that for many substances isochoric surfaces are also surfaces of constant structure (Hazen, 1977). In rocksalt-type oxides such as (Mg,Fe)O, for example, a single unit cell parameter completely defines the structure. Isochoric surfaces are constrained to be isostructural surfaces in P-T-X (Fig. 12).

Isostructural surfaces exist for a large number of more complex compounds. The magnesium-iron silicate spinel, γ-$(Mg,Fe)_2SiO_4$, which is of special interest to geophysicists because of its supposed presence in the earth's mantle, has only two variable structural parameters. The size of the silicon tetrahedron is essentially constant with changes in P, T, and magnesium-iron ratio. The magnesium-iron octahedron, on the other hand, does vary in size; the octahedral volume determines the molar volume and atomic coordinate of the spinel. Isostructural P-T-X surfaces for these silicate spinels, therefore, also correspond to isochoric surfaces. Furthermore, the spinel isochores parallel those of the oxide, (Mg,Fe)O, as illustrated in Figure 12.

Even relatively complex structures such as the alkali feldspars, enstatite, forsterite, and aluminosilicates may have isostructural P-T-X

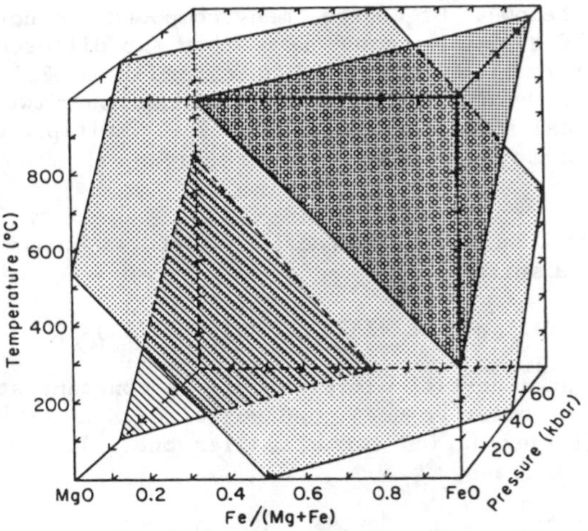

Figure 12. Isostructural surfaces for (Mg,Fe)O in P-T-X space. Three arbitrary isochoric surfaces illustrate the orientation of these planes, which are the same for both magnesium-iron oxides and spinels. From Hazen and Finger (1982).

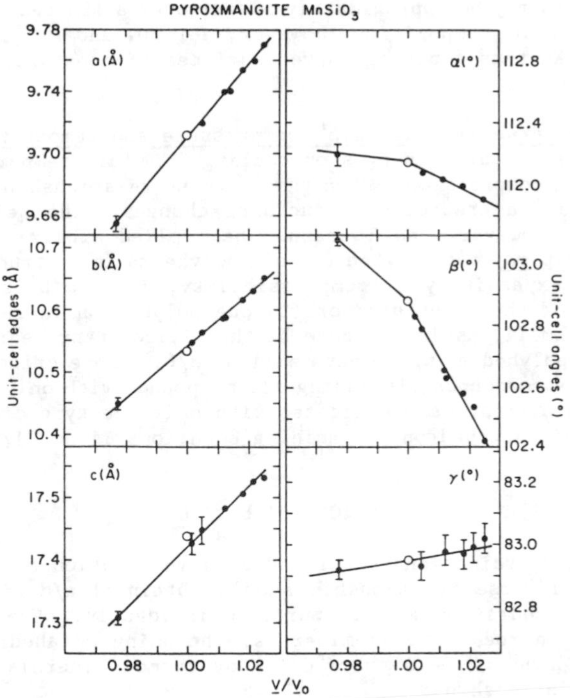

Figure 13. Unit-cell parameters of pyroxmangite versus V/V_0. The continuous variations of these parameters with either temperature ($V/V_0 > 1$) or pressure ($V/V_0 < 1$) is a demonstration of the inverse relationship. Data from Linda Pinckney (pers. comm.), figure from Hazen and Finger (1982).

surfaces (see below). Of course, many compounds do not demonstrate this phenomenon. If a structure has more than two different types of cation polyhedra, for example, then a given change in P, T, or X will commonly not be cancelled by any possible change of the other two variables, unless multiple chemical substitutions are invoked. Multiple compositional variables, of course, increase the dimensions of the P-T-X space under consideration.

All isochoric surfaces have certain features in common. Consider the slopes of such a surface:

$$(\partial P/\partial T)_{\Diamond,X}, \quad (\partial P/\partial X)_{\Diamond,T}, \quad \text{and} \quad (\partial T/\partial X)_{\Diamond,P} , \tag{24}$$

where $\Diamond$ designates partial differentials at constant structure (<u>in addition</u> to constant molar volume), and positive ∂X is defined as the substitution of a larger cation for a smaller one. It follows that for all isostructural surfaces

$$(\partial P/\partial T)_{\Diamond,X} > 0 , \tag{25}$$

$$(\partial P/\partial X)_{\Diamond,T} > 0 , \tag{26}$$

$$(\partial T/\partial X)_{\Diamond,P} < 0 , \tag{27}$$

The surfaces may be approximately planar over a limited range of pressure, temperature, or composition; however, α_V, β_V, and γ_V generally vary with P, T, and X, thus implying curved surfaces of constant volume (Hazen and Finger, 1982).

<u>The "inverse relationship" of pressure and temperature</u>. In numerous compounds structural changes on cooling from high temperature are similar to those on compression. When this inverse relationship obtains it can be very useful in predicting structural changes. The relationship is <u>not</u> universal, however, and it cannot be applied arbitrarily. The inverse relationship <u>may</u> obtain when (1) all polyhedra in a structure have similar ratios of expansivity to compressibility; i.e., α/β is the same for all polyhedra of the structure; or (2) one polyhedron is relatively rigid (α and β are small, as in the case of the silicon tetrahedron) compared with the other polyhedra, which have similar α/β. These criteria are fulfilled by numerous compounds, including all compounds with only one type of polyhedron and a great many silicates with only one type of polyhedron other than the Si tetrahedron. Combining Equations 11 and 17 for polyhedral α and β:

$$(dP/dT)_V \simeq \alpha/\beta = 100n/d^3 \ \text{bar/°C} , \tag{28}$$

where n is coordination number and d is mean cation-anion bond distance. Thus, the inverse relationship should obtain if n/d^3 is similar for all cation polyhedra in a structure. Coincidently, several common cation polyhedra in rock-forming minerals, including octahedral Mg, Fe^{2+}, Al, and Fe^{3+}, have α/β = 65 bar/°C. Many common minerals thus display the inverse relationship.

Conformity with the inverse relationship is best demonstrated by a plot of a variable structural parameter (any distance, angle, volume, or other geometrical aspect of the structure) versus the molar volume or V/V_0. If the inverse relationship is valid, then all structural parameters will show a continuous variation on such a plot (Fig. 13).

Conformity to the inverse relationship implies that, given an atomic topology and a molar volume, the coordinates of all atoms are independent of P, T, and X. Such a circumstance will obtain if nearest-neighbor electrostatic interactions dominate the crystal energy. In cases where the inverse relationship does not obtain it may be assumed that electrostatics do not control structural details.

Structural variations and the prediction of phase equilibria. One of the great challenges of solid state science is the prediction of phase transitions. Although most transformations defy simple explanations, a number of phase transitions appear to be controlled by geometrical limits to specific structural topologies. Principles of comparative crystal chemistry may be used to predict P-T-X transition surfaces in many of these geometrically controlled phase transitions.

In order to facilitate a geometrical approach to phase transformations it is useful to review the topological classification scheme of Buerger (1951, 1972), as elaborated by Megaw (1973). The topology of a structure is defined by Buerger as the set of linkages or bonds between all nearest-neighbor atoms. Reversible or displacive transitions involve no significant alteration of coordination polyhedra or their linkages. The other extreme type of phase transition is the reconstructive transformation in which many bonds are broken and the topology is radically altered. Martensitic or shear transitions are phase changes of intermediate character, in which large portions of the structure are displaced with respect to each other. This topological classification system has proved useful, although in principle all gradations from reversible to reconstructive transitions are possible.

Geometrical limits to structures arise from the misfit of adjacent structural units. In many instances these misfits can be quantified in terms of the absolute or relative sizes of the structural elements, and P-T-X limits may be equated to those conditions at which size limits are violated. Perhaps the first suggestion of a geometrical limit to structure was Pauling's (1939) discussion of the $R^{2+}CO_3$ carbonates. Pauling noted that if the R cation had a radius smaller than about 1.1 A then the calcite form was observed, but for larger cations the aragonite form occurred. He then proposed that the relative stability of the carbonates was a function of cation-to-anion radius ratio, ρ. Similar ratios were proposed for other compounds; for example, compounds of the form $M^{2+}X_2$ assume the rutile structure if $\rho < 0.73$, or the fluorite structure if $\rho >$ 0.73. Pressure, temperature, and composition all influence ionic radius ratios, and thus have analogous effects on the stabilities of simple ionic structures. In these structures an increase in pressure, a decrease in temperature, or substitution of a smaller cation will all have the same effect on the reconstructive transition.

Polyhedral tilting is a common pure displacive phase transition mechanism in ionic compounds (Hazen and Finger, 1979b). Polyhedral tilt transitions may be recognized by five criteria:

1. Structures are composed of corner-linked polyhedra, such as found in framework silicates (quartz, feldspar) or perovskites. The corner-linked framework commonly forms large sites for alkali or alkaline earth cations. Polyhedral tilt transitions occur when polyhedral elements of the framework tilt owing to the changing size of the large cation with changing P, T, or X.

2. The transition is between a high-symmetry or less-distorted form (stable at lower P or higher T), and a low-symmetry or more distorted form (stable at higher P or lower T).

3. The transition is rapid, reversible, and nonquenchable; single crystals are preserved through the transition.

4. Twinning is commonly introduced in the low-symmetry form; the twin law is a symmetry operation of the high-symmetry form that is lost in the transition.

5. The transition is geometrically related to the size of the large site or cavities; therefore, (dP/dT) of the transition equals $(\partial P/\partial T)_V$ of the large site, which equals the ratio α/β for the large site.

Of the characteristics common to all polyhedral tilt transitions, criteria 3 and 4 are generally true of displacive transitions, and criteria 1 and 2 may also obtain for some other types of reversible transitions. The combination of all five criteria, however, appears to be unique to polyhedral tilting (Hazen and Finger, 1982).

Polyhedral tilt transitions can be predicted if geometrical limits are known. Consider the example of the fully disordered alkali feldspars, $(K,Na)AlSi_3O_8$. These feldspars can exist in an untilted form (sanidine and monalbite are the K- and Na-end members, respectively) and a tilted triclinic form (e.g., high albite). Hazen (1976) demonstrated that the transition from monoclinic to triclinic symmetry was geometrically controlled; the monoclinic form can be converted to triclinic by an increase in pressure, a decrease in temperature, or an increase in the smaller sodium cation in the alkali site. Furthermore, the P-T-X polyhedral tilt transition surface (Fig. 14) is also a surface of constant structure.

A number of other phase transitions can be understood in terms of geometrical limits. The martensitic transitions of pyroxene to pyroxenoid (Hazen and Finger, 1982), and the high-temperature stability of trioctahedral micas (Hazen and Wones, 1978) are both amenable to analysis by principles of comparative crystal chemistry. Indeed, in any compound where misfit of adjacent structural elements is energetically unfavorable, such a geometrically controlled transition may occur, and analysis of structural variations with pressure, temperature, and composition may prove enlightening.

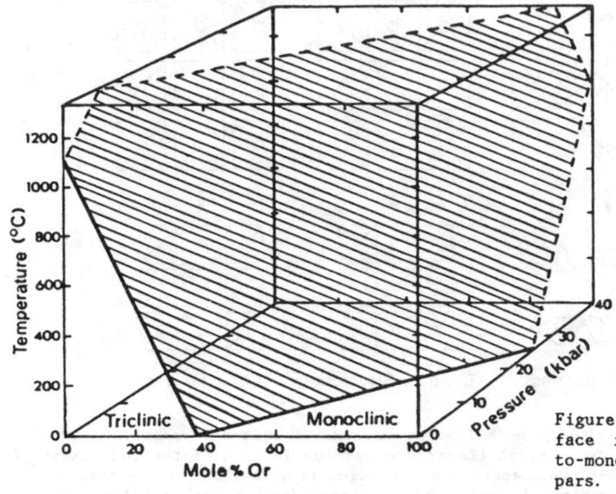

Figure 14. Polyhedral tilt transition surface in P-T-X space for the triclinic-to-monoclinic transition in alkali feldspars. From Hazen (1976); copyright 1976 by the American Association for the Advancement of Science.

3. THE POLYHEDRAL APPROACH

3.1 Polyhedral properties and polyhedral linkages

A significant conclusion of comparative crystal chemistry is that each type of cation polyhedron, which is defined by its distinctive arrangement of a cation and group of anions, possesses a set of well-defined properties. Compressibility, thermal expansivity, and polyhedral volume vary by only a few percent from structure to structure. Polyhedra have characteristic vibration frequencies and amplitudes, and these frequencies and amplitudes vary in characteristic ways with changes in pressure and temperature. It is even possible to assign elastic moduli (shear and compression) and fictive thermochemical properties (heat capacity, enthalpy, and calorimetric entropy) to polyhedra. Thus, in any condensed-matter system, be it solid, liquid, or glass, cation polyhedra may be treated as building blocks that contribute known amounts to the bulk properties of the material.

The objective of the polyhedral approach is first to quantify polyhedral characteristics and then to derive macroscopic material properties from these characteristics and a knowledge of polyhedral linkages. Two cation polyhedra may be bonded by a shared face, a shared edge, a shared corner, or van der Waals forces. The principal difficulty in applying the polyhedral approach is the development of appropriate summation procedures for these different linkage topologies. Consider a two-dimensional analogy of crystal compression (Fig. 15). In each of the four examples a square polygon undergoes a 25% linear compression. In the case of a group of squares the linear compression of the entire array is also 25%. In the other examples, however, the average linear compression is much greater than 25% because of the effects of inter-polygon bending and "intermolecular" (the two-dimensional analog of van der Waals bond) compression. In real crystals, as in these two-dimensional analogs, a knowledge of polyhedral linkages and their effect on bulk properties is essential before any calculations can be performed.

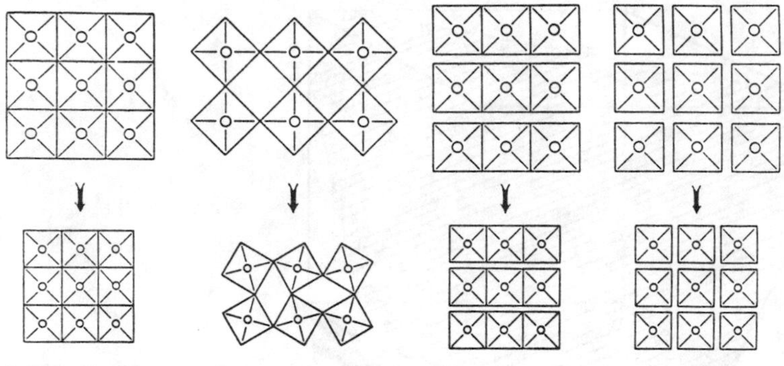

Figure 15. Effects of polyhedral linkage on bulk compression. A square represents a two-dimensional polyhedron, which undergoes a 25% linear compression in each of the four examples. When squares share edges, then the composite compression is also 25%. In the cases of corner-sharing and van der Waals bonding, however, bulk compression is much greater. The layer configuration has very anisotropic compression. After Hazen and Finger (1984).

The polyhedral approach has been applied successfully to the modeling of a number of crystalline properties. Examples of crystal compression, elastic moduli, and thermochemical properties are reviewed below.

3.2 Examples of the polyhedral approach

Calculation of crystal compressibility. Crystal compression occurs by bond shortening (i.e., polyhedral compression), bond-angle bending (i.e., interpolyhedral angle bending), and intermolecular or van der Waals bond compression. Whereas the magnitudes of polyhedral compression are known, force constants associated with bond-angle bending and van der Waals bond compression are not yet well defined. In many instances, therefore, it is not possible to make quantitative predictions of crystal compression. Nevertheless, the magnitudes of the latter two effects are known to be much greater than that of polyhedral compression, so that qualitative aspects of crystal compression can be predicted.

In order to predict structural changes with pressure it is useful first to analyze polyhedral linkages. Imagine each shared edge among the tetrahedral and octahedral units of a structure to be represented by a line in space, and each shared face by a polygon. If this array of lines and polygons forms a continuous three-dimensional network, then interpolyhedral angle bending and intermolecular compression will not occur. The bulk compression of the crystal will be similar to that of the most compressible tetrahedral or octahedral units. In the rocksalt-type oxides, for example, bulk compressibility is precisely that of the constituent polyhedra. Similarly, in the close-packed olivine- and spinel-type orthosilicates the bulk compressibility is close to that of the divalent octahedra. Pyroxenes, garnets, and aluminosilicates are also slightly less compressible than their most compressible polyhedra.

Scheelite-type binary oxides provide a striking example of the close relationship between polyhedral and bulk properties of structures with extensive edge and face sharing (Hazen et al., 1984). All scheelites have the general formula ABO_4, where the A cation is 8-coordinated and the B cation is 4-coordinated. The 8-coordinated polyhedra form a continuous three-dimensional network, and so control the bulk compression. In the mineral scheelite, $CaWO_4$, the bulk compressibility is about 10^{-3} kbar^{-1}, which is typical for divalent calcium coordinated by eight oxygens. In the scheelite form of zirconium germanate, $ZrGeO_4$, compressibility is less than 0.5×10^{-3} kbar^{-1}, because of the relative rigidity of the Zr^{4+} polyhedra. In scheelite-type sodium iodate, on the other hand, the 8-coordinated Na^+ polyhedra are relatively more compressible and yield a bulk compressibility of about 2×10^{-3} kbar^{-1}. Crystals with identical structures, therefore, differ in compressibility by a factor of four because of the differences in large-site ion compressibilities.

In many layer silicates, including micas, chlorites, and serpentines, the network of shared edges and faces extends in only two dimensions, parallel to the layers. Compressibility of these minerals within the layers is comparable to the linear compressibilities of the octahedral-layer cation polyhedra. Compressibility perpendicular to the layer, which is controlled by much weaker vander Waals bonds or long alkali cation-to-oxygen bonds, is many times greater. Extreme anisotropies in bonding, therefore, commonly lead to extreme anisotropies in physical properties.

The most compressible silicate minerals are those in which the network of edge- and face-sharing among octahedra and tetrahedra is completely discontinuous. Framework silicates, including feldspars, zeolites, and quartz, with their three-dimensional arrays of corner-linked tetrahedra are, consequently, often an order of magnitude more compressible than the more densely packed mantle silicates.

In several corner-linked structures of high symmetry, such as cubic perovskite, β-quartz, and the ambient form of ReO_3, interpolyhedral angle bending (i.e., polyhedral tilting) does not occur. These structures, therefore, have very low compressibilities, equal to those of the constituent corner-linked polyhedra. The low-symmetry, tilted modification of each of these structures, however, has an order-of-magnitude <u>greater</u> compressibility.

<u>Calculation of elastic moduli</u>. Au (1984) has adopted the polyhedral approach to the theoretical modeling of elastic moduli of crystalline solids under ambient conditions. Each type of polyhedron, according to Au, may be assigned characteristic shear and compressive moduli, in addition to the characteristic volume compressibility.

There have been numerous attempts in the theoretical modeling of the elastic properties of complex crystal structures of geophysical interest. Most of these approaches rely on conventional lattice-dynamics calculations (Striefler and Barsch, 1972; Weidner and Simmons, 1972) and recently developed structural simulation techniques (e.g., Price and Parker, 1984; Matsui and Busing, 1984). Calculations on complex structures such as olivines and spinels are based on two-body central potentials. The elastic properties of crystalline solids, however, are the manifestation not only of two-body central forces, but also of many-body and noncentral forces within the crystal structure. Conventional lattice calculations, by adopting the atomistic scale as the fundamental modeling unit, are inadequate in addressing the phenomenon of the cooperative interactions in atomic clusters, i.e., the noncentral and many-body forces.

The modified rigid-ion model described by Au (1984), which is an extension of the earlier work by Weidner and Simmons (1972), is based on the coordination polyhedra as the fundamental units in modeling crystal structure. In this polyhedral approach the crystal structure is represented by its constituent coordination polyhedra, which are treated as three-dimensional elastic continua. Elastic moduli, experimentally determined or otherwise assumed, are ascribed to these coordination polyhedra. Finite element analysis is applied to retrieve the complex interatomic force information implicit in the elastic moduli of such a polyhedron. Because the elastic moduli contain information on the nature of static interatomic forces within the crystal structure, the polyhedral approach avoids the difficulties encountered in conventional lattice calculations and provides a basis to model many-body and noncentral forces.

The application of the polyhedral approach by Au (1984) to olivine-type silicates resulted in calculated elastic moduli that differ by only a few percent from observed values. Although the methods outlined by Au are complex, they do lead to moduli of remarkable precision and, furthermore, provide insights about relationships between elastic constants and crystal structure that are not readily apparent with other methods. This approach is now being extended to include extrapolations to mantle conditions and interpolations between end-member compositions. The polyhedral approach, with its future development and extension, will provide a sound phenomenological basis for a better understanding of the physical as well as the thermochemical properties of solids.

Calculation of thermochemical properties. Robinson and Haas (1983) have developed a polyhedral method for the prediction of heat capacity, relative enthalpy, and calorimetric entropy. It has long been known that some thermochemical parameters of minerals can be approximated by summations involving the properties of constituent oxides. Robinson and Haas (1983) reviewed these methods and noted that "The general validity of any of these approaches is rooted in the fact that lattice vibrational modes are the dominant contribution to heat capacity and calorimetric entropy. To the degree that silicate mineral structures can be approximated as oxygen frameworks with vibrating interstitial cations, the overall lattice modes will be largely a function of the modes of the individual cation-oxygen polyhedra. It is this feature which allows the spectroscopist to identify both the coordination and composition of components in mineral phases from their spectral characteristics. This relationship also implies that the standard molar heat capacity and calorimetric entropy of minerals can be estimated by summing, in appropriate proportions, fictive molar heat capacities and calorimetric entropies for the constituent structural groups in minerals." Robinson and Haas (1983) outline a procedure for deriving these fictive polyhedral properties, present heat capacity and entropy coefficients for more than a dozen common polyhedra, and demonstrate that thermochemical properties of many common minerals may be predicted to within ±5%. Not only can their procedure be used to calculate thermochemical properties of minerals that are not suitable for calorimetric measurements but, in addition, the close relationship

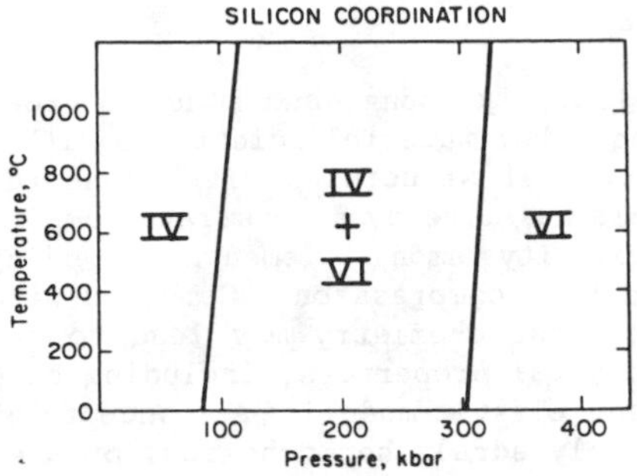

Figure 16. Polyhedral stability fields for Si^{4+}-O^{2-} coordination groups. From Hazen and Finger (1982).

between polyhedral and bulk energetics that is implicit in the Robinson and Haas method provides a clear illustration of the importance of nearest-neighbor bonding groups in mineral behavior.

3.3 Polyhedral phase equilibria

Polyhedra have definite properties — size, shape, coefficients of compressibility and thermal expansion, elastic moduli, and vibrational frequencies. In the extreme extension of the polyhedral approach it is possible to think of the cation polyhedron as also having its own stability field. Beyond certain ranges of pressure, temperature, or composition a given type of polyhedron is not stable. The problem of predicting the stable assemblage of phases at any given P, T, and X, therefore, may be simplified by first identifying the limited number of possible cation polyhedra and then identifying the most stable linkage topologies for that set of polyhedra.

Silicon-oxygen polyhedra provide a classic example of stability fields (Hazen and Finger, 1978b). Under room conditions virtually all silicates have tetrahedrally coordinated Si. Transformations involving a coordination change to octahedral Si occur over a pressure range from about 100 kbar for framework silicates to about 250 kbar for orthosilicates; above 250 kbar virtually all silicates have octahedral Si. It is possible that many Si^{IV}-to-Si^{VI} transitions occur when the average Si-O distance in the tetrahedra reaches a critical minimum distance of about 1.59 Å (Hazen and Finger, 1978b); temperature has little effect on Si-O bond distance, and it also seems to have little effect on the pressure of changes in silicon coordination. These data may be illustrated on a polyhedral phase diagram for silicon (Fig. 16).

Similar diagrams might be constructed for aluminum, magnesium, or other common rock-forming polyhedra, based on the abundant data from phase equilibria and melt-structure studies. Knowledge of polyhedral stability fields not only would facilitate the prediction of stable structures at extreme pressures and temperatures, but would also reveal much about the energetics of bonding in mineral systems.

4. CONCLUSIONS

Comparative crystal chemistry is relatively young, and it holds great promise for providing mineralogists and other material scientists with a detailed picture of the variation of crystal structures with pressure, temperature, and composition. This field has already demonstrated empirical relationships of surprising simplicity among elementary bonding parameters, bond thermal expansion, and bond compression. Combined with the polyhedral approach, comparative crystal chemistry may lead to the prediction of a range of macroscopic crystal properties, including bulk compressibility, thermal expansivity, and elastic moduli. As concepts of geometrical structural limits, fictive polyhedral thermochemical parameters, and polyhedral stability fields are further refined, it is not unreasonable to anticipate a time when phase equilibria may be derived from a knowledge of crystal structure alone. The mineralogist's dream of relating microscopic details to macroscopic behavior may thus be at hand.

5. PROBLEM SETS

Problem 1. Use Equation 11 to predict polyhedral volume compressibilities in the following compounds:

A. Rocksalt-type NaCl, MgO, and PbS (cation-anion distances are 2.81, 2.11, and 2.97 Å, respectively).

B. Pyrope garnet, $Mg_3Al_2Si_4O_{12}$ (cation-anion distances are 2.27, 1.93, and 1.64 Å for Mg-O, Al-O, and Si-O, respectively).

C. Phlogopite mica, $KMg_3AlSi_3O_{10}(OH)_2$ (cation-anion distances are 2.99, 2.06, and 1.65 Å for K-O, Mg-O, and (Al,Si)-O, respectively).

Problem 2. Use Equation 17 to predict polyhedral volume thermal expansivities for the same compounds as in Problem 1. Consider potassium to be 6-coordinated in phlogopite.

Problem 3. In phlogopite the tetrahedral (Al,Si), octahedral (Mg), and interlayer (K) layers comprise ~45%, ~20%, and ~35%, respectively, of the total mica layer thickness. Given these values and the polyhedral data derived above:

A. Predict the compressibility and the thermal expansion of phlogopite both parallel and perpendicular to the layers.

B. Calculate its bulk compression and expansion from the linear values.

(Hint: Compression or expansion within the layers is only as great as the least compressible or expansible polyhedral layer; expansion perpendicular to the layers is the sum of individual layer expansion).

Problem 4. If phlogopite stability is limited by the absolute size of the octahedral Mg layer, then what is the approximate slope, dP/dT, of the stability line?

Problem 5. Consider the distorted perovskite form of $MgSiO_3$, which is presumed to be the dominant silicate in the earth's mantle. The average length of eight Mg-O bonds is 2.21; the average length of six Si-O bonds is 1.79; the density is 4.098 gm/cc. Predict the density of this phase at 50 kbar and 1000°C (equal to a depth of about 100 km below the average continent); at 150 kbar and 1500°C (about 300 km); at 300 kbar and 1900°C (about 600 km), and at 600 kbar and 2300°C (about 1500 km). Plot these values on the earth's density-depth profile provided at the right. (Hints: Assume that this perovskite is 30% Si-octahedra by volume and 70% Mg-polyhedra by volume, and polyhedra fill all space.

Use the following simplified equation for calculating density:

$$\rho_{P,T} = \rho_o[1 - \alpha_v T + \beta_v P] \quad .)$$

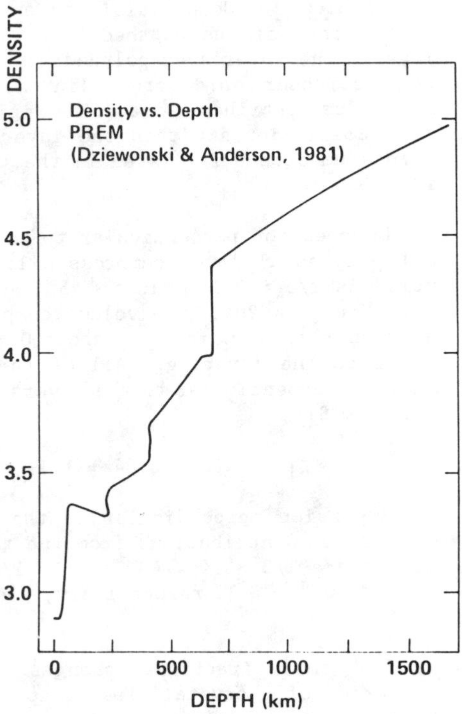

6. SOLUTIONS TO PROBLEMS

Problems 1 and 2. These problems may be solved by simple substitution into Equations 11 and 17, as shown in the table below. The observed values of these polyhedral compressibilities and expansivities are given in the table as well.

Type	Bond	d(A)	S^2	Z_c	Z_a	n	β^*_{calc}	β^*_{obs}	α^{**}_{calc}	α^{**}_{obs}
Rocksalt	Na-Cl	2.81	0.75	1	1	6	3.93	4.17	104	153
	Mg-O	2.11	0.50	2	2	6	0.62	0.62	39	39
	Pb-S	2.97	0.40	2	2	6	2.18	2.08	49	66
Pyrope	Mg-O	2.27	0.50	2	2	8	0.78	0.8(1)	52	39
	Al-O	1.93	0.50	3	2	6	0.32	0.4(1)	26	21
	Si-O	1.64	0.50	4	2	4	0.15	0.3(1)	13	6
Phlogopite	K	2.99	0.50	1	2	6	3.56	3.7(6)	72	63
	Mg-O	2.06	0.50	2	2	6	0.58	0.6(1)	39	43
	Al,Si_3-O	1.65	0.50	3.75	2	4	0.16	0.3(2)	14	0

$*$ × $Mbar^{-1}$ $\quad$ $**$ × 10^{-6} $°C^{-1}$

Problem 3. Compressibility and thermal expansion within mica layers equals that of the octahedral Mg layer, which is the least compressible layer. The interlayer polyhedra vary more with pressure and temperature; the tetrahedra are corner linked, and thus free to rotate even though individual tetrahedra are quite rigid. Thus linear compression and expansion coefficients within the layers are 1/3 the volume coefficients for magnesium octahedra listed in the table above (i.e., 0.19 $Mbar^{-1}$ and 13 × 10^{-6} $°C^{-1}$).

Compression perpendicular to the layers equals the sum of the individual polyhedral layer compressibilities. the tetrahedral layer contribution is 1/3 × 0.16 $Mbar^{-1}$ × 45%, and the octahedral contribution is 1/3 × 0.58 $Mbar^{-1}$ × 20%. the volume compression of the potassium interlayer is 3.56 $Mbar^{-1}$, of which only about 0.39 $Mbar^{-1}$ is possible in the plane parallel to the layering. All of the remaining compression, 1.27 $Mbar^{-1}$, must be perpendicular to the layers. Thus the interlayer contributes 3.17 × 35% to $\beta\perp$:

$$\beta\perp = 0.02 + 0.04 + 1.11 = 1.17 \text{ Mbar}^{-1} .$$

Expansion perpendicular to the layers may be derived in exactly the same way. Contributions from the tetrahedral, octahedral, and potassium layers are 1/3 × 14 × 45% = 2.1, 1/3 × 39 × 20% = 2.6, and 46 × 35% = 16.1 (all × 10^{-6} $°C^{-1}$), respectively, yielding a total expansion of 20.8 × 10^{-6} $°C^{-1}$.

For small fractional changes in volume, the volume compression and expansion of a crystal closely approximates the sum of linear compressibilities or expansivities:

$$(1 - \Delta a)(1 - \Delta b)(1 - \Delta c) \simeq 1 - \Delta a - \Delta b - \Delta c ,$$

when Δa, Δb, and Δc are all small. For mica, therefore,

$$\beta_V \simeq 0.19 + 0.19 + 1.17 = 1.55 \text{ Mbar}^{-1} .$$

$$\alpha_V \simeq 13 + 13 + 21 = 47 \times 10^{-6}\ °C^{-1} .$$

The observed values for phlogopite compression and thermal expansion are 1.7 $Mbar^{-1}$ and 33 × 10^{-6} $°C^{-1}$.

Problem 4. The dP/dT of any transition that is controlled by polyhedral size is simply α/β for that polyhedron. In the case of octahedral magnesium in phlogopite:

$$dP/dT = \alpha_{Mg-O}/\beta_{Mg-O} = 13 \times 10^{-6}\ °C^{-1}/0.21 \text{ Mbar}^{-1} = 62 \text{ bar}/°C .$$

Problem 5. In the perovskite structure the octahedra are corner linked and the 8-coordinated magnesium cations share faces. Polyhedra fill all space so that the total compression and expansion is thus the sum of polyhedral volume changes. Polyhedral volume compression and expansion are as follows:

Bond	d	n	$S^2 z_c z_a$	β_{calc} ($Mbar^{-1}$)	α_{calc} ($\times 10^6\ °C^{-1}$)
Mg-O	2.21	8	2	0.72	52
Si-O	1.79	6	4	0.19	19

for the bulk mineral, therefore:

$$\alpha_V \times 10^6 \simeq 0.3 \times 19 + 0.7 \times 52 = 42\ °C^{-1}$$

$$\beta_V \simeq 0.3 \times 0.19 + 0.7 \times 0.72 = 0.504\ Mbar^{-1}$$

The equation for density (see above) can thus be solved for the several sets of pressure and temperature. In spite of the obvious simplifications in terms of perovskite composition, effects of pressure on thermal expansion, and uncertainties about the earth's P-T profile, the calculated perovskite equation of state is similar to that of the lower mantle.

Note that the density of magnesium perovskite actually decreases with depth through the first 200 kilometers. This phenomena is generally true for close-packed oxides and silicates of the upper mantle because the thermal expansion term is greater than the compression term in this region. Note also that measured density of the earth from seismic studies shows a decrease in this region. Only below 200 kilometers does the compression of minerals become the dominant term in volume changes.

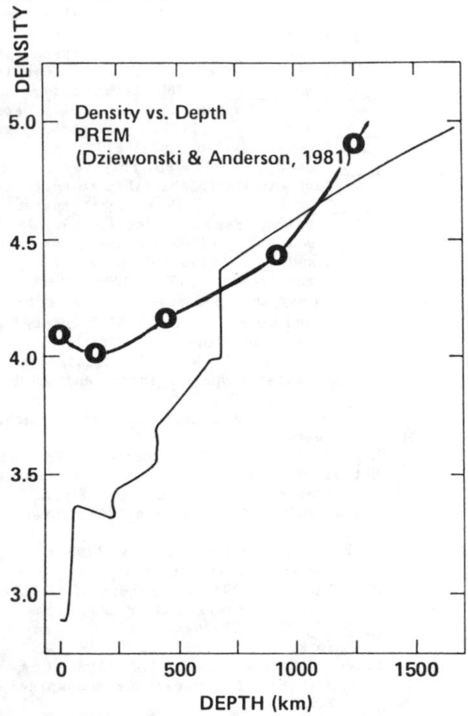

REFERENCES

Anderson, D.L. and Anderson, O.L. (1970) The bulk modulus-volume relationship for oxides. J. Geophys. Res. 75, 3494-3500.

Au, A.Y. (1984) "Theoretical Modeling of the Elastic Properties of Mantle Silicates." Ph.D. thesis, State Univ. New York, Stony Brook, New York.

Bish, D.L. and Burnham, C.W. (1980) Structure energetics of cation-ording in orthopyroxene and Ca-clinopyroxenes. Geol. Soc. Am. Abstr. Progr. 12, 388.

Brown, G.E., Sueno, S. and Prewitt, C.T. (1973) A new single crystal heater for the precession camera and four-circle diffractometer. Am. Mineral. 58, 698-704.

Buerger, M.J. (1951) Crystallographic aspects of phase equilibria. In: R. Smoluchowski, J.E. Mayer, and W.A. Weyl, eds., "Phase Transformations in Solids," Wiley, New York, pp. 183-209.

______ (1972) Phase transformations. Sov. Phys. Crystallogr. 16, 959-968.

Bukowinski, M.S.T. (1980) Effect of pressure on bonding in MgO. J. Geophys. Res. 85, 285-292.

Busing, W.R. and Levy, H.A. (1964) The effect of thermal motion on the estimation of bond lengths from diffraction measurements. Acta Crystallogr. 17, 142-146.

Cameron, M., Sueno, S., Prewitt, C.T. and Papike, J.J. (1973) High-temperature crystal chemistry of acmite, diopside, hedenbergite, jadeite, spodumene, and ureyite. Am. Mineral. 58, 594-618.

Finger, L.W., Hazen, R.M., Zou, G., Mao, H.K. and Bell, P.M. (1981) Structure and compression of crystalline argon and neon at high pressure and room temperature. Appl. Phys. Lett. 39, 892-894.

Foit, F.F. and Peacor, D.R. (1967) A high-temperature furnace for a single-crystal diffractometer. J. Sci. Instrum. 44, 183-185.

Gibbs, G.V. (1982) Molecules as models for bonding in silicates. Am. Mineral. 67, 421-450.

Hazen, R.M. (1975) "Effects of Temperature and Pressure on the Crystal Physics of Olivine." Ph.D. thesis, Harvard Univ., Cambridge, Massachusetts.

______ (1976) Sanidine: predicted and observed monoclinic-to-triclinic reversible phase transformations at high pressure. Science 194, 105-107.

______ (1977) Temperature, pressure, and composition: structurally analogous variables. Phys. Chem. Minerals 1, 83-94.

______ and Burnham, C.W. (1973) The crystal structures of one-layer phlogopite and annite. Am. Mineral. 58, 889-900.

______ and Finger, L.W. (1977a) Modifications in high-pressure, single-crystal diamond-cell techniques. Carnegie Inst. Washington Yearbook 76, 655-656.

______ and ______ (1977b) Crystal structure and compositional variation of Angra dos Reis fassaite. Earth Planet. Sci. Lett. 35, 357-362.

______ and ______ (1978a) The crystal structures and compressibilities of layer minerals. I. SnS_2, berndtite, and II, phlogopite and chlorite. Am. Mineral. 63, 289-296.

______ and ______ (1978b) Crystal chemistry of silicon-oxygen bonds at high pressure: implications for the earth's mantle mineralogy. Science 201, 1122-1123.

______ and ______ (1979a) Bulk modulus-volume relationship for cation-anion polyhedra. J. Geophys. Res. 84, 6723-6728.

______ and ______ (1979b) Polyhedral tilting: a common type of pure displacive phase transition and its relationship to analcite at high pressure. Phase Transitions 1, 1-22.

______ and ______ (1981) High-temperature diamond-anvil pressure cell for single-crystal studies. Rev. Sci. Instrum. 52, 75-79.

______ and ______ (1982) "Comparative Crystal Chemistry." Wiley, New York.

______ and ______ (1984) Comparative crystal chemistry. Am. Sci. 72, 143-150.

______ and Prewitt, C.T. (1977) Effects of temperature and pressure on interatomic distances in oxygen-based minerals. Am. Mineral. 62, 309-315.

______ and Wones, D.R. (1978) Predicted and observed compositional limits of trioctahedral micas. Am. Mineral. 63, 885-892.

______, Finger, L.W. and Mariathasan, J.W.E. (1984) High-pressure crystal chemistry of scheelite-type tungstates and molybdates. J. Phys. Chem. Solids (in press).

______, Mao, H.K., Finger, L.W. and Bell, P.M. (1981) Irreversible unit cell volume changes of wüstite single crystals quenched from high pressure. Carnegie Inst. Washington Yearbook 80, 274-277.

Kittel, C. (1971) "Introduction to Solid State Physics," 4th ed. Wiley, New York.

Matsui, M. and Busing, W.R. (1984) Computational modeling of the structure and elastic constants of olivine and spinel forms of Mg_2SiO_4. Phys. Chem. Minerals 11, 11-59.

Megaw, H.D. (1973) "Crystal Structures: A Working Approach." W.B. Saunders, Philadelphia, PA.

Merrill, L. and Bassett, W.A. (1974) Miniature diamond anvil pressure cell for single- crystal x-ray diffraction studies. Rev. Sci. Instrum. 45, 290-294.

Pauling, L. (1939) "The Nature of the Chemical Bond and the Structure of Molecules and Crystals." Cornell Univ. Press, Ithaca, New York.

______ (1960) "The Nature of the Chemical Bond," 3rd ed. Cornell Univ. Press, Ithaca, New York.

Price, G.D. and Parker, S.C. (1984) Computer simulations of the structural and physical properties of the olivine and spinel polymorphs of Mg_2SiO_4. Phys. Chem. Minerals 10, 209-216.

Robinson, G.P. and Haas, J.L., Jr. (1983) Heat capacity, relative enthalpy, and calorimetric entropy of silicate minerals: an empirical method of prediction. Am. Mineral. 68, 541-553.

Skinner, B.J. (1966) Thermal expansion. In: S.P. Clark, Jr., ed., "Handbook of Physical Constants." Geol. Soc. Am. Mem. 97, 75-96.

Smith, J.V. (1974) "Feldspar Minerals," Vol. 1 and 2. Springer-Verlag, New York.

______ and Bailey, S.W. (1963) Second review of Al-O and Si-O tetrahedral distances. Acta Crystallogr. 16, 801-811.

Striefler, M.E. and Barsch, G.R. (1972) Lattice dynamics at zero wave vector and elastic constants of spinel in the rigid-ion approximation. J. Phys. Chem. Solids 33, 2229-2250.

Tossell, J.A. (1980) Theoretical study of structures, stabilities, and phase transitions in some metal dihalide and dioxide polymorphs. J. Geophys. Res. 85, 6456-6460.

Weidner, D.J. and Simmons, G. (1972) Elastic properties of alpha-quartz and the alkali halides based on an interatomic force model. J. Geophys. Res. 77, 826-847.

Chapter 10. Charles W. Burnham

MINERAL STRUCTURE ENERGETICS and MODELING USING the IONIC APPROACH

From its inception the ionic bonding model has provided a productive framework for understanding the structures of minerals, particularly silicates, oxides, and carbonates. Within the past fifteen years several new computational methods have been developed that permit quantitative modeling of ionic structures, with respect to both their configurations and their cohesive energies. These methods permit one to determine an atomic structure that either best fits an assumed set of interatomic distances, or that has minimum ionic cohesive energy. This chapter will describe these methods, give some examples of their application, point out their strengths and weaknesses, and suggest the direction of future developments.

1. MODELING WITH THE IONIC APPROXIMATION

Impetus to model mineral structures comes from the desire to rationalize and understand observed structures and their properties and to predict structural behavior. Mineralogists and crystal chemists want to understand why certain coordinations are observed for various cations, under what circumstances these coordinations will change, and what controls the distortions of individual cation coordination polyhedra. What structural attributes, for example, control the relative stabilities of the aluminosilicate polymorphs -- andalusite, sillimanite, and kyanite -- as a function of temperature and pressure, or determine the favored stackings of "I-beam" units in pyroxenes and amphiboles or of sheet units in micas and other layer silicates? Are there geometric and topologic factors that limit the ability of a particular coordination polyhedron or an entire structure to undergo cation substitutions? If so, how important are these in controlling the limits of isomorphism? Likewise, what structural properties influence and control the lattice dynamical effects observed in vibrational spectra and inferred from thermodynamic data?

Of course one would like to be able to predict the behavior of crystal structures with changes in temperature, pressure, and composition. The need to develop modeling procedures with such capabilities is most important to simulate conditions unobtainable in the laboratory, such as extremes of temperature and pressure, or compositional variation in a particular structure that is immune to synthesis. A crucial desire is to contrast by modeling the relative stabilities of real versus hypothetical polymorphs as conditions change to predict temperatures and pressures of phase transformations. These capabilities may ultimately prove critical to detailed elucidation of mantle mineralogy (see also Navrotsky, this volume).

For many years the mainstays of qualitative structural understanding and prediction with the ionic model have been Pauling's rules for ionic structures (see e.g., Pauling, 1960), and the notion of ionic radius (see e.g., Shannon, 1976). Pauling's second rule of local electrostatic charge balance provides a basis to judge the adherence of a structure to the principles of ionic bonding. For those many mineral structures in which the anions do not conform exactly to electrostatic charge balance, Baur (1971a) has shown that the departure from charge saturation can be used to predict cation-anion interatomic distances, with overbonded oxygens participating in longer than normal distances and underbonded ones in shorter than normal distances. Various schemes (Brown and Shannon, 1973) have been developed that relate Pauling bond strength to bond length, thus adding some predictive capacity to the basic Pauling procedure where cation-anion bond strength is defined simply as cation charge divided by coordination number. These extensions of Pauling's second rule provide a means of predicting bond lengths semiquantitatively and of assessing coordination number in complex cases where large cations are surrounded by many anions ranging uniformly from normal distances to unreasonably long distances.

The distortions of coordination polyhedra observed in structures that achieve local electrostatic charge balance -- olivine, for example -- remain largely unexplained by empirical methods, except in those cases where sharing of polyhedral edges or faces are known to affect polyhedral geometry. The varying extent of polyhedral edge sharing in the structures of the TiO_2 polymorphs has been noted by Evans (1966) as perhaps rationalizing their relative stabilities (incorrectly, it turns out); and Kamb (1968) suggested that the expected changes in the lengths of shared versus unshared octahedral edges in spinel as a function of pressure might be used to predict the olivine - spinel transformation.

The limited predictive capacity of these qualitative methods is now quite constraining. Development of comprehensive quantitative methods to model ionic structures, as we shall see here, permits us to overcome those constraints and has already significantly enhanced our ability to understand and predict structural behavior. Extensions of these methods can now be used to predict phonon spectra, including their dependence on both temperature and pressure. As outlined by Kieffer (this volume), the spectral properties also provide a determination of the thermal contribution to thermodynamic quantities.

2. THERMODYNAMIC SETTING OF IONIC COHESIVE ENERGY

The ultimate model of a crystal structure would describe exactly the interactions between all atoms and from them derive the atomic arrangement and its dynamical characteristics having the minimum free energy for the temperature and pressure of interest. Procedures described here fulfill just part of this objective; they consider only the static contributions to the free energy. Given the Gibbs free energy as

$$G = H - TS \tag{1}$$

and enthalpy as

$$H = U + PV, \tag{2}$$

the static cohesive energy, often incorrectly termed "lattice energy", comprises the bulk of the internal energy, U. The static cohesive energy corresponds to the work necessary to separate the constituent atoms in a crystalline structure to infinity, thus it is related to interactions embodied in the bonding. The total internal energy contains, in addition, the vibrational energy:

$$U_{tot} = U_{coh} + U_{vib}. \tag{3}$$

Additional contributions from defects and surfaces are ignored. At low pressures the PV term is small relative to U, thus enthalpy is essentially equal to internal energy.

The entropy contains two contributions:

$$S_{tot} = S_{config} + S_{vib}. \tag{4}$$

For ordered structures S_{config} is zero, and the entire entropy is vibrational. In modeling ordered structures with the static lattice approximation, cohesive energies, U_{coh}, can be compared under the assumption that the vibrational characteristics, described by the lattice dynamics and quantified in heat capacity, are similar for similarly bonded structures. This assumption can be tested by an explicit calculation of the vibrational spectrum from the model. Comparison of existing spectroscopic data can also aid in assessing the relative contributions of vibrational effects, as discussed in the chapters by Kieffer and McMillan.

With changes in pressure the PV terms assume importance and can be added easily to calculated values of U_{coh} to approximate the static portion of H. As temperature is increased the main effects on free energy are vibrational. The anharmonicity of interatomic vibrations leads to thermal expansion of both interatomic distances and cell dimensions, which can be determined from a quasiharmonic or anharmonic lattice-dynamical calculation. To the extent that static modeling procedures are based on observed interatomic distances, the static effects of both temperature and pressure are included by virtue of the dependence of calculated cohesive energies on atomic coordinates.

The overall modeling problem for structural studies is divisible into two separable but closely related problems: We wish on the one hand to obtain an atomic configuration that best achieves certain criteria; on the other hand we wish to calculate the U_{coh} for any given atomic configuration. Two structure modeling strategies will be described. One makes use of the <u>D</u>istance-<u>L</u>east-<u>S</u>quares (DLS) procedure, first employed by Meier and Villiger (1969), which uses observed or expected interatomic distances as input data, and adjusts by iterative least-squares all atomic coordinates and unit cell dimensions until an optimum distance structure is obtained, whose energy is then calculated by methods described below. The other uses energy-minimization techniques such as those embodied in the program WMIN (Busing, 1981) that adjust atomic coordinates and cell dimensions until a minimum energy structure is obtained for a given topology and symmetry. With this procedure the energy calculations lead directly to the structure model.

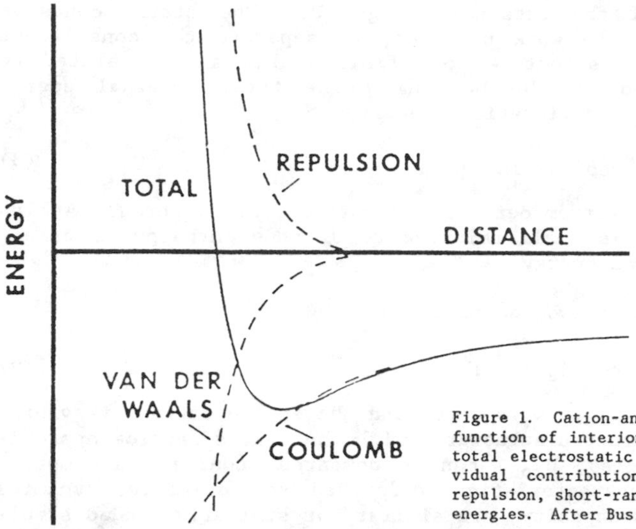

Figure 1. Cation-anion electrostatic energy as a function of interionic separation. Solid line: total electrostatic energy; dashed lines: individual contributions of Coulomb, short-range repulsion, short-range attractive (vander Waals) energies. After Busing (1970).

Calculations of cohesive energies, U_{coh}, are carried out under assumptions of ionic bonding. The model assumes that the interionic forces are adequately described as a combination of electrostatic forces between charged ions obeying Coulomb's law and short-range forces arising from interactions between electron clouds of neighboring ions. The simplest approach ignores contributions to U_{coh} that may arise from crystal field effects, from polarization of anions, or from covalency effects in structures that are only partly ionic. Each of these refinements, however, can be included in extended treatments, some of which are briefly reviewed here.

The reader should be aware of the contrast between the methods described here that apply to infinitely repeating ionic crystals, and ab initio molecular orbital techniques such as those employed by Gibbs and others (cf. Gibbs, 1982) which apply the Hartree-Fock self-consistent field method to atoms in a finite cluster and yield minimum energy configurations of molecular species. It should be emphasized that these approaches are complementary; interaction of the two strategies is proving symbiotic especially in studies of crystals with marked covalent character.

3. COHESIVE ENERGY CALCULATION IN THE IONIC MODEL

The static cohesive energy is most simply formulated in terms of pairwise interactions consisting of a long-range component from Coulomb electrostatic forces between ions and a short-range component arising from orbital interactions between nearest-neighbor ions. With respect to a cation-anion pair interaction, the two components combine to yield the familiar energy well, with its minimum occurring at the observed interionic distance (Fig. 1). Asymmetry of the well leads to anharmonic vibrations. The calculation strategy deals with the terms separately, yielding a net attractive long-range Coulomb term and a net short-range term that is repulsive for nearest-neighbor separations. In this treatment there are no explicit or angle-dependent many-body forces.

3.1. Coulomb electrostatic energy

The Coulomb interionic force is given by

$$f_{ij} = q_i q_j e^2/(r_{ij})^2, \tag{5}$$

where q is the ionic valence, e the electronic charge, and r_{ij} the interionic distance. The energy of interaction between ions i and j is thus

$$U_{ij} = q_i q_j e^2/r_{ij}, \tag{6}$$

which will be negative (attractive) between cations (+q) and anions (-q), and positive (repulsive) between cation pairs or anion pairs. The total Coulomb electrostatic energy is given by

$$U_{cou} = \tfrac{1}{2} \sum_i \sum_j q_i q_j e^2/r_{ij}, \tag{7}$$

where the summation includes all interactions twice and compensation is made with the factor ½. It is easily demonstrated (Boeyens and Gafner, 1969) that the series in (7) converges slowly, primarily because adjacent shells of ions j interacting with ion i have opposite signs.

Ewald (1921) first showed that rapid convergence can be achieved using a Fourier method that sums in reciprocal space. To develop the method further, Bertaut (1952) proceeded from the observation that (7) has Patterson-function-like attributes, and thus it can be manipulated as if it were the self-convolution of a series of charge density functions. Bertaut's formulation for the Coulomb energy of site j, $U_s(j)$ (modified from Ohashi and Burnham, 1972), is:

$$U_s(j) = \{2\pi R^2/V\} q_j \sum_{\mathbf{h}} F(\mathbf{h})\Phi(\alpha) - (2g/R)q_j^2, \tag{8}$$

where the summation is over reciprocal lattice vectors, **h**. F(**h**) is a structure-factor-like expression:

$$F(\mathbf{h}) = \sum_p q_p \exp\{2\pi i(h_1\Delta x_1 + h_2\Delta x_2 + h_3\Delta x_3)\}, \tag{9}$$

where this summation is over the p atoms in the unit cell. Each charge, q, is distributed over a sphere whose radius, R, must be equal to or less than half the shortest interatomic distance. The function $\Phi(\alpha)$ is related to the Fourier transform of a normalized charge distribution function, $\alpha = 2\pi|\mathbf{h}|R$, and g is a constant that also depends on the charge distribution function selected. The second term on the right side of (8) subtracts the potential interaction of atom j with itself. The h_i and Δx_i in (9) are the components of reciprocal lattice vectors **h** and interatomic vectors $\mathbf{x}_{jp}$ respectively. Functions $\Phi(\alpha)$ and constants g for uniform, linear, parabolic, and cubic charge density functions are listed by Jones and Templeton (1956). Templeton and Johnson (1961) derived correction factors that can be applied to compensate for termination errors introduced by summing in (8) only over limited ranges of $|\mathbf{h}|$ and hence α. For a linear charge density function ($\rho(r)$ with $r < R$) of the form $3(R-r)/\pi R^4$, g is 26/35 and $\Phi(\alpha) = 288[\alpha \sin\alpha + 2\cos\alpha - 2]^2/\alpha^{10}$, thus rapid convergence is assured because successive terms in the summation in (8) are proportional to α^{-8}, the inverse 8th power of distance in reciprocal space. Jones and Templeton (1956) asserted that for accuracies between 1% and 0.003%, the linear charge density function is at least as good as any other.

The total Coulomb electrostatic energy per mole is

$$U_{cou(tot)} = \tfrac{1}{2} \sum U_s(i), \tag{10}$$

where the summation includes all atoms in one formula unit, and the factor ½ again compensates for each interaction being included twice.

A number of programs for summing Coulomb electrostatic energies using some variation of the Ewald-Bertaut method are available, including ELEN (Ohashi and Burnham, 1972) and WMIN (Busing, 1981), both of which we have used extensively. Checks of the results obtained using different programs with different algorithms and different charge density functions show complete agreement of fully converged calculations. The method is thus well proven, reliable, and fast.

Some readers are undoubtedly familiar with the Madelung formulation of electrostatic energy, such as given, for example, by Kittel (1971):

$$U_{cou} = A\, q_i q_j\, e^2/r_{ij}, \tag{11}$$

where r_{ij} must be a unique cation-anion distance and A is the structure-dependent Madelung constant. I wish to emphasize that this approach is strictly applicable only to highly symmetric structures that have only one cation and one anion in the asymmetric unit, and only one crystallographically distinct nearest-neighbor cation-anion distance. There can be no true Madelung constant for more complicated structures (i.e. most structures), and the ready availability of programs to carry out full summations renders the Madelung approach obsolete and inappropriate for geologically relevant materials.

3.2. Short-range energy

In contrast to the long-range Coulomb electrostatic energy, computation of short-range energies in the ionic model from first principles is neither straightforward nor simple. Consequently these energies have frequently been ignored in mineral studies; this has often led to incorrect or misleading results. The short-range energy arises from interactions between electrons in orbitals of nearest neighbor cations and anions. Over these distances the energy is repulsive, hence reference is most often made to short-range repulsive energy. For next nearest-neighbor ions, such as O^{2-} - O^{2-}, the magnitude of the short-range energy is much smaller, and may even be attractive. Approximations of these energies have been obtained by a variety of empirical methods, briefly reviewed here.

Early formulations of the short-range energy were exclusively repulsive, with terms proportional to r_{ij}^{-n} and n in the range from 6 to 10 (Kittel, 1971). A more widely employed repulsive expression is that of Born (Born and Huang, 1954):

$$U_{ij(rep)} = \lambda_{ij}\exp(-r_{ij}/\rho_{ij}). \quad (12)$$

Both λ_{ij} and ρ_{ij} are constants for a particular interaction between ions i and j that describe the shape of the repulsive potential as a function of distance r_{ij} (see Fig. 1). The total short-range energy for site i is then

$$U_{i(rep)} = \sum\lambda_{ij}\exp(-r_{ij}/\rho_{ij}), \quad (13)$$

where the summation need extend only over the nearest neighbors, or, in the case of anions, next nearest neighbors as well. While anion-anion interactions lead to small but significant energies, cation-cation interactions may safely be neglected (Bish and Burnham, 1984; Post and Burnham, 1985). The entire repulsive energy per mole is then

$$U_{rep(tot)} = \tfrac{1}{2}\sum U_{i(rep)}, \quad (14)$$

where the summation is over all atoms per formula unit, and the factor ½ compensates for interactions being included twice.

An alternative formulation, due to Gilbert (1968) and employed by several authors including Busing (1970), writes the repulsive energy for an ij interaction as

$$U_{ij(rep)} = (B_i + B_j)\exp[(A_i + A_j - r_{ij})/(B_i + B_j)], \quad (15)$$

where the A_i are radius parameters and the B_i are "softness" parameters assigned to each ion.

A modification of the Born exponential expression writes the short-range interaction energy as

$$U_{ij} = \lambda_{ij}\exp(-r_{ij}/\rho_{ij}) - C_{ij}/r_{ij}^6, \quad (16)$$

where the second term is a dipole-dipole (dispersion, or van der Waals) interaction which is attractive (see Fig. 1). The coefficient C_{ij} may be determined empirically (see Catlow et al., 1982a), or theoretically (Muhlhausen and Gordon, 1981a; Hemley et al., 1985). To be physically meaningful the magnitude of the dispersion terms should be damped at very short distances (Tang and Toennies, 1984).

3.3. Empirical determination of short-range interactions

From compressibility data: If a structure, highly symmetric and usually close-packed, has one kind of coordination polyhedron involving one unique cation-anion distance whose value is directly related to unit

Table 1. Born-type short-range repulsive parameters for Mg-O and Si-O.

Source		λ, kj/mole	ρ, Å
Mg-O			
From MgO bulk compressibility[a]		36018	0.3785
From MgO bulk compressibility[b]		35169	0.3805
Minimum-energy fitting to observed structures[c]		213689	0.2756
Fitting to MEG short-range energies:[d]			
Oxygen shell radius, Å:	0.93	418986	0.2371
	1.01	359196	0.2457
[Fitting range is	1.05	337440	0.2498
1.80-2.35Å only]	1.11	287985	0.2581
Si-O			
From forsterite vibrational spectrum[b]		40468	0.579
From forsterite vibrational spectrum[e]		45741	0.53
Minimum-energy fitting to observed structures[c]		96401	0.3455
Fitting to MEG short-range energies:[d]			
Oxygen shell radius, Å:	0.93	406350	0.2342
	1.01	358150	0.2428
[Fitting range is	1.05	337063	0.2472
1.48-1.80Å only]	1.11	302378	0.2546

a Ohashi and Burnham (1972); MgO bulk modulus = 1.622Mb.
b Bish and Burnham (1984); MgO bulk modulus = 1.61Mb; k_{Si-O} = 3.46 mdyne/Å.
c Catlow et al. (1982b); Mg-O potential also includes dispersion term C_{Mg-O} = 429.4 kj/mole (see eq (16)).
d Post and Burnham (1985); note specified fitting range.
e Lasaga (1980); k_{Si-O} = 4.4 mdyne/Å.

cell volume, then, as shown by Kittel (1971), λ and ρ for that cation-anion pair may be obtained from that structure's bulk modulus. Simple AB compounds with the ZnS, NaCl, or CsCl structures are amenable to this analysis; values obtained for Mg-O in periclase are listed in Table 1. Many cation-anion pairs, such as Al-O and Si-O for example, do not occur in sufficiently simple structures, and one must turn elsewhere to obtain appropriate repulsive parameters.

From vibrational spectra: Frequencies of interatomic stretching and bending vibrational modes are proportional to their force constants. If the frequencies of simple stretching or bending normal modes -- symmetrical Si-O stretching within a tetrahedron for example -- can be identified in the infrared or Raman spectrum of a mineral, then the force constant for that bond can be ascertained under the assumption that the motion is localized (see McMillan, this volume, for some details and cautions). The force constant, k, associated with a local mode such as a bond stretch is the second derivative of potential energy with respect to distance, d^2U/dr^2. For the Si-O bond Lasaga (1980) writes:

$$U_{ij(tot)} = U_{cou} + U_{rep} \tag{17}$$

$$= q_iq_je^2/r_{ij} + \lambda_{ij}\exp(-r_{ij}/\rho_{ij})$$

$$dU/dr = -q_iq_je^2/r_{ij}^2 - \lambda_{ij}/\rho_{ij}\{\exp(-r_{ij}/\rho_{ij})\} \tag{18}$$

$$= 0 \text{ when } r_{ij} = r_{ij}(\text{equil})$$

$$d^2U/dr^2 = 2q_iq_je^2/r^3 + \lambda_{ij}/\rho_{ij}{}^2\{\exp(-r_{ij}/\rho_{ij})\} = k_{ij}. \qquad (19)$$

Knowing r_{ij}(equil) and k_{ij}, one can solve for λ_{ij} and ρ_{ij}. Lasaga (1980) and Bish and Burnham (1984) report repulsive parameters for Si-O based on slightly different k_{Si-O} values; these are compared in Table 1.

Fitting to observed structures: For structures such as those of many silicates in which each cation-anion pair exhibits a range of distances, it is possible to use the observed structures to derive a set of λ_{ij} and ρ_{ij} in the Born formulation, or A_i and B_i in the Gilbert formulation, that best fit the observed structures as minimum energy ones. Catlow and his coworkers have determined Born-type parameters for a number of cation-anion pairs by least-squares fitting, and have used them to study the energetics of defects in ionic materials (Catlow and James, 1982). Further they reported using least-squares analysis with several observed silicate structures (monticellite, sillimanite, sphene, diopside, benitoite, jadeite, and wollastonite), in conjunction with M^{2+}-O repulsive parameters from the oxides and O^{2-}-O^{2-} pair potentials approximated from Hartree-Fock calculations on O^{1-}-O^{1-}, to obtain repulsive parameters for Si-O (Catlow et al., 1982b) (see Table 1).

Busing's (1981) energy minimization program, WMIN, has been used by a number of authors to obtain Gilbert-type repulsive parameters by least-squares fit to observed structures. Busing (1970) himself used Gilbert's (1968) theoretically-derived B for Cl^- to obtain B's for Mg^{2+}, Ca^{2+}, Sr^{2+}, and Ba^{2+} by extrapolating from Gilbert's values for isoelectronic species. He then obtained A's for these species by least-squares. Miyamoto and Takeda (1984) obtained A and B values for Mg, Si, and O by fitting to the Mg_2SiO_4 structure; similar least-squares fittings for A and B values for Ca, Mg, and O using observed structures of Mg_2SiO_4 and γ-Ca_2SiO_4 are reported by Matsui and Busing (1984) (see Table 5 below).

3.4. Theoretical determination of the short-range interactions

In 1972 Gordon and Kim (1972) introduced the electron gas approach as an approximate non-empirical method to evaluate repulsive interactions based on purely ionic theory. After subsequent development by Waldman and Gordon (1979) and Muhlhausen and Gordon (1981a,b), the presently constituted Modified Electron Gas (MEG) theory writes the individual components of the short-range pair potential -- kinetic, exchange, correlation, non-point Coulomb -- as functionals of the electron density of the interacting pair, which is assumed to be simply a superposition of individual ion electron densities, calculated from Hartree-Fock atomic wave functions. The MEG procedure, like all energy calculation procedures, may be implemented for crystal studies assuming pairwise-additive central forces (Cohen and Gordon, 1975, 1976). This procedure essentially ignores many-body contributions to energy, such as those resulting from the overlap of charge density from three or more ions, isotropic expansion or contraction of the spherical ionic charge densities, or anisotropic distortion or polarization of the electron distribution. Muhlhausen and Gordon (1981a) examined the first two of these effects and found the second to be significant. They considered the isotropic component of the electrostatic field at an anion site that is created by the surrounding cations in the crystal. The crystal field is simulated in the calculation of ionic charge density by adjustment of the radius of a shell potential surrounding the anion; self-consistency is achieved when the shell potential matches the particular anion site potential in the crystal. The pair potentials calculated from these shell-stabilized charge densities therefore contain spherical many-body contributions to the energy.

Post and Burnham (1985) have used this refinement of the MEG model to derive short-range repulsive parameters for a variety of common cation-anion pairs in minerals. The MEG calculation using LEMINPI (Muhlhausen and Gordon, 1981a) produces site potentials and short-range energy values as a function of separation distance for each ion pair of concern. Post and Burnham then fit these values to the Born exponential form, and thereby determine λ and ρ for the ion pair over the appropriate range of separation distances. They report correlation coefficients greater than 0.9995 between the MEG pair potentials and the curves calculated with the fitted λ and ρ. Some representative values of λ and ρ are listed in Table 1, where they are compared with empirically determined values; other MEG-based Born-type repulsive parameters are listed by Post and Burnham (1985).

The primary distinction between repulsive energy parameters determined using the theoretical MEG methods and those determined empirically is that the MEG ones represent a purely ionic model, whereas fitting to observed structural data, whether bulk modulus, vibrational spectra, or interatomic distances, implicitly includes non-ideal contributions insofar as they affect the property being observed. It is not possible to evaluate the extent of these contributions, nor to identify them specifically as anisotropic polarizations, covalency effects, or whatever. Depending on the modeling problem, this may or may not be a disadvantage. If the departure of a structure or its properties from those expected in the pure ionic case is of concern, then the MEG potentials are best suited to calculate the expected structure; on the other hand if the goal is to model minimum-energy structures that best fit those observed, then empirically-determined short-range parameters may work best, especially for structures known to have significant departures from complete ionicity. Keep in mind, however, that the empirically-determined parameters include bonding effects whose sources are unidentified.

Further development of MEG methods is now under way. A refinement of the shell-stabilized model in which the ion self energy is evaluated explicitly has improved the precision with which the potentials are calculated for crystals (Hemley and Gordon, 1985). Moreover anisotropic polarization of anion charge density has been included to yield anisotropic interactions appropriate for covalency. Jackson and Gordon (pers. comm.) have obtained very satisfactory models for quartz that represent much improvement over those obtained with the isotropic shell-stabilized models (see below). Efforts are also being made by Gordon and his coworkers to develop methods to handle transition metal ions.

3.5. Static cohesive energy calculations

Static cohesive energies calculated in the manner outlined here have been used primarily in comparisons of similar but slightly different structures to assess their relative stabilities. One must be watchful that such comparisons are realistic, keeping in mind that vibrational energies can contribute several kj/mole at room temperature. Although some vibrational calculations are now being carried out, they have been applied only to simple structures so far. Differences between vibrational energies of similar structures might, depending on circumstances, amount to a few kj/mole at room temperature. For an assessment of these contributions, the reader is again referred to Kieffer's paper in this volume.

If MEG-determined repulsive parameters are used, the total cohesive energy must be further corrected for the fact that artificial shell-stabilized anions do not occur naturally. Post and Burnham (1985) point out that the correction involves addition of the self-energy difference between shell-stabilized and free anions, which is obtained during the Hartree-Fock self-consistent-field calculations. For oxides, the non-existence of O^{2-} requires, in addition, that the difference between the self energies of gas phase O^{1-} and shell stabilized O^{2-} anions be added to the total cohesive energy; Post and Burnham (1985) list those corrections.

Comparisons of calculated total cohesive energies with experimental thermodynamic values show close agreement in the cases of alkali halides and some simple oxides (Busing, 1970; Cohen and Gordon, 1976; Tossell, 1980a; Muhlhausen and Gordon, 1981b). In their study of TiO_2 polymorphs, Post and Burnham (1985) report a cohesive energy for rutile 4.9% lower than that obtained through a Born-Haber cycle calculation, most of the difference probably being accounted for by covalency effects. Their total calculated cohesive energy for quartz of 10924 kj/mole is about 6% less than the experimental value of 11627 kj/mole; this is a surprisingly good result considering the lack of agreement between the MEG minimum energy structure and the observed one. As Post and Burnham (1985) point out, this shows that the covalency effects associated with the Si-O bond in quartz contribute significantly to configuration, particularly Si-O-Si angles, yet contribute little to total cohesive energy. Gibbs' (1982) molecular orbital results on the energy of r_{Si-O} versus Si-O-Si angle bear this out.

Ionic cohesive energies thus comprise most of the internal energy of silicate and oxide minerals. Because the lattice dynamical behavior of similar structures is nearly the same, comparisons of these calculated energies do yield productive insights and do have predictive capabilities, as the examples discussed below will show.

4. STRUCTURE COMPARISONS USING COULOMB ENERGIES ONLY

In some restricted circumstances, comparisons of cohesive energies containing only Coulomb electrostatic terms may prove useful. Giese (1978) and Bish and Giese (1981) exploit one such circumstance by

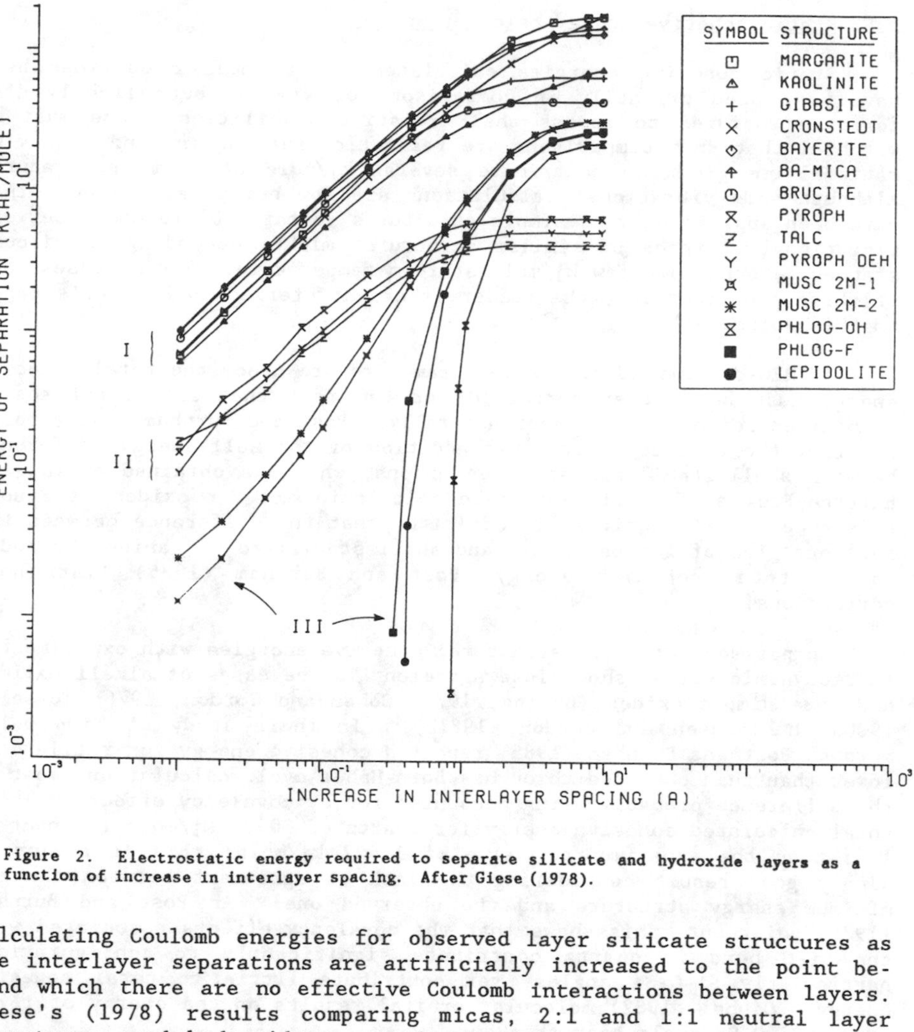

Figure 2. Electrostatic energy required to separate silicate and hydroxide layers as a function of increase in interlayer spacing. After Giese (1978).

calculating Coulomb energies for observed layer silicate structures as the interlayer separations are artificially increased to the point beyond which there are no effective Coulomb interactions between layers. Giese's (1978) results comparing micas, 2:1 and 1:1 neutral layer structures, and hydroxides, reproduced in Figure 2, show clearly the differing nature of interlayer bonding in these structures and provide some surprising but not irrational relationships. The 1:1 layer silicates and hydroxides have relatively strong interlayer interactions due to hydrogen bonding across ther interlayer regions; brittle micas have similarly strong interactions simply due to their layer charges. Dioctahedral and trioctahedral micas differ substantially in their behavior; there are slight repulsions between trioctahedral layers at small separations, and more energy is required to carry out small increases in separation of dioctahedral than trioctahedral layers, presumably a manifestation of the different OH^- orientations in the two structures. Interestingly the neutral 2:1 layers require more energy than normal micas to achieve small separations, but the charges on mica layers result in longer range interlayer interactions. Thus it is more difficult to initiate separation of talc or pyrophyllite layers, but the total energy for complete separation is less than for the corresponding mica layers. (In all mica calculations, the interlayer cations were split evenly and half were associated with each layer as the layer separations were increased.) Bish and Giese (1981) show by the same means that electrostatic interactions between neutral 2:1 and brucite layers in chlorite are substantial; chemical substitutions that lead to formal layer charges obviously increase interlayer interactions, but are not necessary to assure chlorite stability. Giese (1984) has written a comprehensive review covering electrostatic energy calculations on layer silicates.

5. DISTANCE-LEAST-SQUARES STRUCTURE MODELING

The DLS procedure adjusts atomic coordinates and unit cell dimensions of a structure with assumed topology and symmetry until the resultant interatomic distances achieve the best least-squares fit to a set of prescribed input distances. Thus an optimum-distance structure is obtained without any particular assumptions about the nature of the bonding or the interatomic forces. The implicit bonding model is the sum total of those factors that control the observed interatomic distances that are selected as prescriptions for the model structure. The procedure was devised by Meier and Villiger (1969) as a tactic to assist in solving complex highly pseudo-symmetric structures, particularly those of zeolites, which are their major interest. Assumptions of various ordering schemes lead to specific distance prescriptions that yield DLS superstructure models in subgroup symmetries that can then be tested against observed x-ray data. Dollase and Baur (1976) used this technique to great advantage to derive a well-constrained low tridymite superstructure, whose solution might have been totally intractible with x-ray data alone.

5.1. Prescribing distances

In most crystal structures the number of crystallographically independent interatomic distances is greater than the number of variable atom coordinates and cell dimensions, i.e., those not constrained by symmetry to be equal to others or to some specific value. Any modeling problem is, in theory, amenable to DLS treatment if expectation values can be assigned to a number of these distances that exceeds the number of variable structural parameters. Distance prescriptions might come from known distances of the same kind in related structures, from ionic radii tables (or, for that matter, covalent or atomic radii tables depending on bonding), or from compilations of individual bond thermal expansions or compressibilities (e.g. Hazen and Finger, 1982; Hazen, this volume).

Decisions about which distances and how many distances to prescribe depend on the problem. Obviously the greater the degree of overdetermination, the more rapid and sure will be the least-squares convergence. If the structure being modeled is constrained by known cell dimensions, convergence is generally rapid and little difficulty is encountered. If, on the other hand, the unit cell geometry is also to be adjusted -- such as to model known structure types with new compositions, or known structures under temperature and pressure conditions for which no x-ray data exist, or hypothetical structures -- one must carefully select the prescribed distances to assure that the connectivity of the structure is specified. To illustrate this point: Suppose one wanted to model a pyroxene, whose structural modules consist of "I-beams" each containing a strip of M1 octahedra to which two silicate tetrahedral chains are attached, one above and one below (see Fig. 3 or Fig. 8 below). To model the I-beams themselves successfully one will have to prescribe some distances that control I-beam geometry, such as Si-Si, or M1-M1, or the O-O shared octahedral edges; Si-O plus M1-O plus intrapolyhedron O-O distances alone are insufficient. The geometric relations between I-beams also must be specified, most likely using M2-O distances from M2 to O's on different I-beams, or possibly M2-M1 or M2-Si distances between cations on different I-beams, although assumptions about inter-I-beam cation-cation distances are likely to be less good. Full modeling of a tetrahedral framework structure likewise will require that intertetrahedral T-T distances be specified, otherwise only the assumed symmetry controls the framework connectivity, and that is almost always not sufficient. Since these intermodular distances are usually not well constrained, and will frequently be assigned low weights (see next section), convergence is by no means assured. This warning is particularly applicable to modeling of hypothetical structures whose unit cell dimensions are part of the model rather than part of the prescription. It is always more difficult to achieve a "best" arrangement of atoms or coordination polyhedra in a unit cell if the size and shape of the cell are unknown as well.

5.2. Weighting of prescribed distances

The prescribed distances represent expectations of separations of atoms in a state of equilibrium between interatomic forces having no residual internal "stresses." Different interatomic potentials lead to differing degrees of "compliance" among different kinds of distances; these are manifested in a variety of interatomic force constants and hence vibrational frequencies. It was therefore evident at the outset that the prescribed distances had to be weighted according to some scheme that would permit weak bonds to vary more freely than strong bonds. An obvious choice was to use weights equal to classical Pauling bond strengths, but that gave no guidance for O-O and cation-cation weights. With Si-O weight set to 1.0, Meier and Villiger (1969) used O-O weights of 0.3 to 0.5, while Baur (1972) used 0.07, and Bish and Burnham (1984) used 0.14 for certain edge-sharing O-O distances in olivine, 0.07 for other O-O distances less than 3.2 Å, and 0.04 for those greater than 3.2 Å. Bish and Burnham further assigned weights of 0.04 to prescribed cation-cation distances in olivines.

Dempsey and Strens (1976) report using cation-oxygen weights proportional to bond strength with bond strength determined by adjusting the Pauling second-rule value according to (Brown and Shannon, 1973):

$$s = s_0(r_0/r)^n, \tag{20}$$

where s is bond strength in valence units, s_0 is the standard bond strength at standard distance r_o, r is the particular distance being weighted, and n is usually around 5. They weighted O-O distances by "analogy with the repulsion to be expected between two balloons of 'electron gas'..." (Dempsey and Strens, 1976) that led to values near 0.17 valence units, and then they arbitrarily weighted cation-cation interactions at 0.167.

Both Baur (1977) and Bish and Burnham (1984) state that more appropriate weights should be assigned in proportion to cation-anion force constants, where these are known from analysis of infrared or Raman spectra. As an example, the Mg-O and Si-O stretching force constants of forsterite, 0.46 and 3.47 mdyne/Å, obtained by Iishi (1978) lead to $w_{Mg-O} = 0.13$ with $w_{Si-O} = 1.0$. This is less than half the Pauling bond strength value, but it appears to work successfully modeling pyroxenes, as we shall see below. If one were to assign O-O weights in proportion to O-Si-O bending force constants obtained by ab initio molecular orbital methods (Gibbs, 1982), the values would lie near 0.1. Alternatively one might assign values in some proportion to MEG-derived repulsive forces, which have now been reported for a variety of cation-anion, cation-cation, and anion-anion pairs (Post and Burnham, 1985).

Weighting and, in the case of hypothetical structure modeling, the precise selection of distances to be prescribed, are the two most important considerations in developing DLS models. Convergence problems may well be expected if the number of prescribed distances is not at least 1.5 times the number of parameters, perhaps even more if cell dimensions are also to be optimized.

Tests of DLS to simulate known structures repeatedly show convergence with good agreement. Baur (1977) reports reproduction of experimental structures of forsterite, fayalite, $K_2Mg_5Si_{12}O_{30}$, and γ-Co_2SiO_4 to mean deviations between model and observed distances of 0.04 Å or better; Dempsey and Strens (1976) report excellent agreement with natrolite, forsterite, and β-Mg_2SiO_4 structures.

5.3. DLS simulations of isotypic hypothetical structures and of temperature and pressure effects on structures

Baur (1977) comprehensively reviewed DLS modeling that had been reported up to 1977, including those applications in which DLS was used as an aid to solve complex superstructures. Interestingly the first reported use of a DLS kind of procedure (Shoemaker and Shoemaker, 1967) was to develop a trial structure for the alloy $Nb_{48}Ni_{39}Al_{13}$. Because DLS refinements are substantially faster than structure refinements using diffraction-based structure factors, they can save significant time and money during the course of precise refinement of a complex structure.

Early DLS refinements of β-spinels were reported by Baur (1971b). These were carried out with fixed known cell dimensions, and prescribed distances were obtained using estimates based on empirical relations that show the effects of an anion's Pauling charge imbalance on its associated cation-anion distances (Baur, 1971a). The relatively poor agreement of the β-Mg_2SiO_4 DLS model with the experimental structure (mean bond-length deviation of 0.068 Å) was blamed on experimental data, because a later DLS model of β-Co_2SiO_4 agreed much better with the observed structure (mean bond-length deviation of 0.027 Å)(Baur, 1977).

Given the importance of high-pressure phase transformations in MgFe silicates, it is not surprising that DLS modeling was applied early on to analysis of possible hypothetical structures. Baur (1972) simulated three alternatives to olivine, all of which were -- like olivine itself -- hexagonally close-packed and centric. They all satisfied local electrostatic charge balance and had unit cell volumes no larger than that of olivine; none has ever been observed for any A_2BX_4 compound. Relative cohesive energies of these simulated structures were not computed, thus Baur's arguments rationalizing their instability were based on the relative extents of edge sharing (two have more shared edges than olivine), and on relative shared-edge lengths (the mean length of shared edges in the third is longer than the mean of unshared edges). Dempsey and Strens (1976) computed DLS models for several hypothetical phases at high pressures, including $MgSiO_3$ [perovskite] (unobserved in 1976), Mg_2SiO_4 [strontium plumbate], and $NaAlSiO_4$ [calcium ferrite].

If individual bond thermal expansions or isothermal compressibilities are known from high temperature or high pressure structure refinements, from empirical relationships founded on high temperature and pressure structure data (Hazen and Finger, 1982), or from bulk thermal expansion or compressibility data for highly symmetric AX structures, it is a simple matter to simulate a known structure at other temperatures and pressures of interest. Dempsey and Strens (1976) modeled well the anisotropic cell expansion of forsterite and fayalite to 1123K using bond expansion coefficients from Cameron et al. (1973); but their simulated linear cell compressibilities of forsterite to 100 kb were rather poor, because their assumptions about bond compressibilities were based on gross approximations derived from analysis of bulk compressibilities of simple oxides. Khan's (1976) DLS models of the forsterite structure and cell at 300° and 600°C, based on known bond expansions from high temperature structures, agree well with observation.

As more mineral crystal structures are refined at high temperature and high pressure, the observed behavior of more kinds of bonds will become much better known. Full DLS simulations of structures under experimentally inaccessible conditions will then become more realistic. In the meantime there are numerous structures whose cell dimensions are known at a variety of temperatures and pressures; DLS models generated within constrained cells have a very good chance of proving reliable. If, in addition, the cohesive energies of these models are computed, then the insights gained from the models will be further enhanced.

5.4. A hypothetical non-existent structure -- antidiopside

It is well known that the Ca^{2+} in diopside is almost completely constrained to the 8-coordinated M2 site (see e.g., Burnham, 1973). Thus Ca^{2+} and Mg^{2+} never measureably disorder over the M1 and M2 sites, and diopside compositions do not vary significantly toward wollastonite.

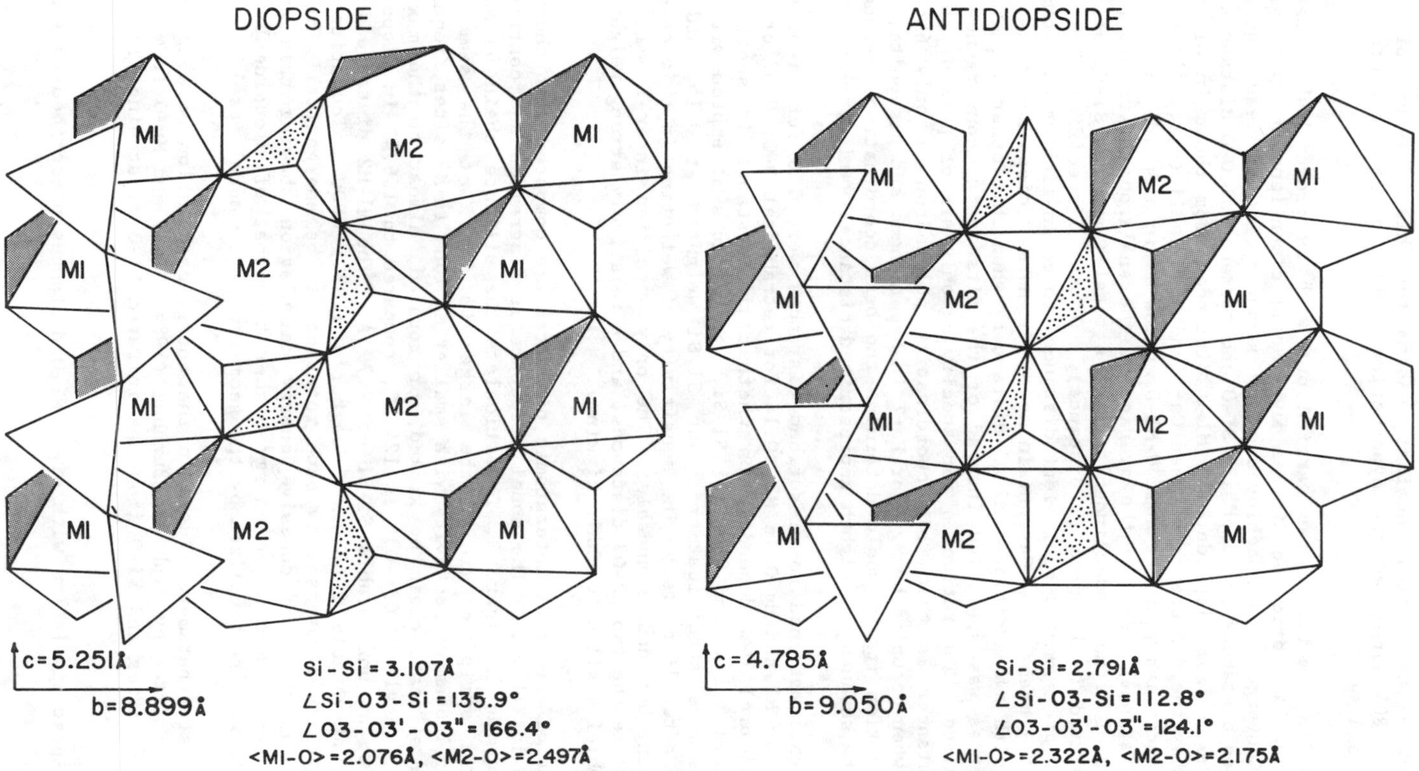

Figure 3. Structures of diopside and DLS model of antidiopside projected onto (100). After Bish and Burnham (1980).

To elucidate the crystal chemical reasons for this behavior, Bish and Burnham (1980) simulated the hypothetical antidiopside structure, with Ca in M1 and Mg in M2.

The DLS simulation was carried out with 29 prescribed distances determining 14 atomic coordinates and 4 cell dimensions in space group C2/c. Convergence was achieved after seven cycles of least-squares. Prescribed distances included all Si-O and tetrahedral O-O distances (10 total) taken from diopside; all M1-O distances from those in diopside expanded by the difference between Ca^{2+} and Mg^{2+} radii (6 total); all M1 O-O distances assuming the same O-M1-O angles as in diopside (12 total); 6 M2-O distances taken as the average of the analagous distances in M2 polyhedra of ortho- and clinoenstatite (6 total); and the Si-Si separation observed in diopside (1 distance). Of the 6 specified M2-O distances, 2 are symmetry-equivalent distances to chain-linking O3 atoms in different I-beams. The Ca-containing M2 in normal diopside is 8-coordinated with two additional M2-O3 distances; when this other set of M2-O3 distances was specified instead of the first pair, convergence was not attained. The rationale for specifying only six of possibly eight M2-O distances is that this coordination polyhedron is only 6-coordinated when occupied by Mg^{2+} or Fe^{2+} in hypersthene and pigeonite, thus the simulation itself would determine the best coordination, depending on how the seventh and eighth unprescribed distances varied.

Weights assigned to prescribed distances were 1.0 for Si-O, 0.13 for Ca-O in M1 and Mg-O in M2, 0.14 for tetrahedral O-O, 0.07 or 0.04 for octahedral O-O depending on whether the distance is shorter or longer than 3.2 Å, and 0.01 for Si-Si. Simulations attempted with the same distance specifications, but Si-Si weighted at 0.15, did not converge after 16 cycles. The sensitivity of weighting in this case is frightening but understandable: The only specified inter-I-beam connections are the two M2-O3 distances, and a relatively strong weight for Si-Si inhibits silicate chain kinking.

Some of the model interatomic distances are compared with those of diopside in Table 2. The model structure is diagrammed and contrasted with diopside in Figure 3. To articulate the silicate chain on an M1 octahedral strip containing the larger Ca ions, at the same time achieving I-beam connectivity with smaller Mg ions in M2 sites, requires that the silicate chain be kinked, or rotated, almost to the maximum possible. Angle O3-O3'-O3" is 124.1°, compared to 166.4° in diopside, leading to an unfavorable Si-Si of 2.79 Å. The M1-M2 distance is an equally unfavorable 2.68 Å. Meagher's (1980) CNDO calculations on silicate chain clusters show such rotations to be energetically highly unfavorable; thus the cohesive energy calculation showing this antidiopside model to have an energy higher than that of diopside by 264 kj/mole is well rationalized by the geometry of the model itself.

Repulsive parameters used in the energy calculation were the same as those reported by Bish and Burnham (1984): Ca-O and Mg-O from oxide compressibilities and Si-O from the infrared Si-O stretching frequency in forsterite.

A repeat simulation with eight M2-O distances prescribed as simply

Table 2. Unit cell and interatomic distances of DLS model of antidiopside, $(Mg_{M2}Ca_{M1}Si_2O_6)$ compared to diopside, $(Ca_{M2}Mg_{M1}Si_2O_6)$.

	Diopside Observed*	Antidiopside DLS Prescribed	Antidiopside DLS Refined
Unit cell			
a,Å	9.746		10.19
b	8.899		9.05
c	5.251		4.78
β,°	105.63		98.43
Si Tetrahedron			
Si-O1(C1)	1.602Å	1.602Å	1.603Å
Si-O2(C1)	1.585	1.585	1.588
Si-O3(C1)	1.664	1.664	1.665
Si-O3(C2)	1.687	1.687	1.687
<O-O> for 6	2.663	all 6	2.667
Si-Si	3.107	3.1[illegible]	2.791
Si-O3-Si	135.9°		112.8°
O3-O3'-O3"	166.4°		124.1°
M1 Polyhedron			
M1-O1(A1,B1)	2.115	2.395	2.376
M1-O1(A2,B2)	2.065	2.345	2.291
M1-O2(C1,D1)	2.050	2.330	2.298
<M1-O>	2.077		2.322
<O-O> for 12	2.710	all 12	3.270
M1-M1			3.210
M2 Polyhedron			
M2-O1(A1,B1)	2.360	2.070	2.121
M2-O2(C2,D2)	2.353	2.011	1.988
M2-O3(C1,D1)	2.561	2.357	2.417
M2-O3(C2,D2)	2.717		(3.529)
<M2-O>	2.498		2.175

* Cameron et al. (1973).

Table 3. Mean M1-O and M2-O interatomic distances, in Å, in intermediate DLS model structures versus those in corresponding end-member structures. After Bish and Burnham (1984).

M1-M2	End-member <M1-O>	Intermediate <M1-O>	Intermediate <M2-O>	End-member <M2-O>
Mg-Fe	2.101	2.104	2.179	2.182
Fe-Mg	2.157	2.154	2.129	2.126
Mg-Co	2.101	2.102	2.139	2.142
Co-Mg	2.119	2.118	2.127	2.126
Ni-Mg	2.076	2.078	2.124	2.126
Mg-Ni	2.101	2.100	2.102	2.102
Fe-Mn	2.157	2.159	2.224	2.227
Mn-Fe	2.185	2.184	2.186	2.182
Mg-Mn	2.101	2.107	2.220	2.227
Mn-Mg	2.185	2.181	2.132	2.126
Mg-Ca	2.101	2.126	2.380	2.392
Ca-Mg	2.346	2.332	2.143	2.126
Mn-Ca	2.185	2.198	2.360	2.392
Ca-Mn	2.346	2.338	2.239	2.227
Fe-Ca	2.157	2.210	2.372	2.392
Ca-Fe	2.346	2.329	2.203	2.182

those for normal diopside reduced by the Ca-Mg ionic radius difference led to convergence after 8 least-squares cycles. While the distance prescription is more straightforward, the model has several highly unfavorable attributes: One M1 O-O distance is 1.98 Å, an M2-Si distance is 2.86 Å, and M1-M1 separation along the octahedral strip is 2.89 Å. The silicate chain is, however, less rotated; O3-O3'-O3" is 149° and Si-Si is 3.02 Å. Nevertheless this model has a cohesive energy 657 kj/mole higher than that of diopside. Davidson et al. (1982) report an exchange free energy, ΔG_{exch}, for the diopside-antidiopside reaction of +160 kj/mole at room temperature and pressure.

5.5. Cation site preferences and exchange energies in olivines

Bish and Burnham (1984) have calculated cohesive energies for ordered and anti-ordered DLS structures of eight binary silicate olivines. Simulations were made with 29 prescribed distances varying 11 atomic coordinates and 3 cell dimensions in the standard olivine space group (Pnma). M1-O and M2-O prescribed distances were taken from appropriate end-member olivine structures. Cation-oxygen weights were in proportion to Pauling bond strengths: $w_{Si-O} = 1.0$ and $w_{M^{2+}-O} = 0.33$. O-O weights were set at 0.14 for edge-shared distances, 0.07 for other O-O less than 3.2 Å and 0.04 for other O-O greater than 3.2 Å. Cation-cation weights were 0.04. Comparison of all the ordered and anti-ordered models reveals clearly the synergistic effect of M1 cations on M2 polyhedra and vice versa. As Table 3 shows, mean M1-O distances are larger or smaller in the binary olivine than in the corresponding end-member depending on whether the M2 cation is larger or smaller than M1, likewise for M2. As the size difference between M1 and M2 increases, so does the departure of M1 and M2 polyhedra from end-member dimensions. This effect is the structural source of non-ideal solid solution behavior and the related volumes, enthalpies, and entropies of mixing.

Cohesive energies of all binary pairs were calculated using M-O repulsive parameters from oxide compressibilities and Si-O parameters from infrared stretching frequencies, as discussed previously. Comparison of the ordered and anti-ordered energies gives a structure exchange energy, which in all eight cases correctly predicts the observed site preferences (Table 4). Large positive exchange energies appear for those cases in which ordering is strong, and small positive values appear in cases where disordering is common. The very slight negative exchange energy for Fe-Mg olivine, favoring Fe^{2+} in M2, would probably revert to a small positive value favoring Fe^{2+} in M1 if crystal field effects were included. These results also show clearly that such predictive success is not achieved if repulsive energies are ignored, or if only individual M1 and M2 site energies of the end-members are compared.

5.6. The future of DLS

Although one might think the clearly more powerful energy minimization techniques, described below, would supplant DLS, that will not happen completely in the near future. DLS can easily model static structures at high temperatures and pressures or both as the wealth of experimental interatomic distance data obtained under such conditions

Table 4. Cohesive energy differences between DLS models of ordered and anti-ordered olivines, $M1M2SiO_4$ (kj/mole). Positive values indicate the cation distribution shown (ordered) has the lower energy; negative values indicate the opposite (anti-ordered) distribution has the lower energy. After Bish and Burnham (1984).

M1	M2	ΔU, Coulomb only	ΔU, Coulomb + repulsive	Observations
Fe	Mg	-20.9	-0.4	Fe usually in M1; very slight ordering; some LFSE from Fe^{2+}
Co	Mg	-11.3	+0.4	Co in M1; partial ordering; LFSE from Co^{2+}
Ni	Mg	+18.0	+9.6	Ni in M1; strong ordering; LFSE from Ni^{2+}
Fe	Mn	+5.4	+28.5	Fe in M1; partial ordering; LFSE from Fe^{2+}
Mg	Mn	+46.4	+46.0	Mg in M1; nearly complete ordering
Mg	Ca	+46.0	+71.5	Mg in M1; complete ordering
Mn	Ca	+69.9	+27.6	Mn in M1; nearly complete ordering
Fe	Ca	-28.0	+43.9	Fe in M1; strong ordering; some LFSE from Fe^{2+}

Table 5. Empirically derived Gilbert-type short range potentials used to calculate minimum-energy models of Mg_2SiO_4 phases.

	Potentials N1[a]	Potentials MB[b]	Potentials P4[c]
A_{Mg}, Å	(0.97)	1.552	1.6297
B_{Mg}, Å	(0.065)	0.0538	0.1969
A_O, Å	1.770	1.791	1.4745
B_O, Å	0.105	0.25	0.0568
A_{Si}, Å	0.608	-	1.0763
B_{Si}, Å	0.0172	-	0.1706
D_{Si-O}, kj/mole	-	-	430.53
q_{Mg}	(+2.0)	(+2.0)	+1.726
q_O	(-2.0)	-1.16	-1.208
q_{Si}	(+4.0)	(+0.64)	(+1.380)

Parameters in parentheses are either fixed or constrained.

a Miyamoto and Takeda (1984).

b Matsui and Busing (1984); q_{Mg} fixed, q_O fitted, q_{Si} constrained for neutrality.

c Price and Parker (1984); D_{Si-O}, q_{Mg} and q_O fitted, q_{Si} constrained for neutrality.

expands. Effects of temperature and pressure must be incorporated into repulsive parameters in order to carry out energy minimizations under such conditions, and these techniques are still very much in the embryonic state. DLS modeling is generally easier to carry out and is certainly faster computationally. Indicators of satisfactory convergence, however, must be carefully monitored.

6. CALCULATING MINIMUM ENERGY CONFIGURATIONS

When the cohesive energy is parameterized in terms of Coulomb electrostatic terms (Eqn. 7) and short-range energy terms (Eqns. 12, 15, or 16), it is, in principle, possible to vary the atomic coordinates and unit cell dimensions until the total cohesive energy achieves a minimum value. The program WMIN, written by Busing (1981), does exactly that. The user has a choice of three minimization algorithms: Newton's method, which calculates first and second derivatives of energy with respect to coordinates; the method of steepest descents, in which only first derivatives are calculated; and the Rosenbrock search method, a vector search procedure that requires no derivatives. Newton's method is fast but works most successfully when the structure is close to its minimum energy configuration, otherwise it may locate local minima, or saddle points, or may not converge at all. The Rosenbrock search procedure is more time consuming, but always locates the real minimum, even from a saddle point (Busing, 1981). A suitable strategy might well involve beginning minimization with the Rosenbrock search procedure and completing it with Newton's method to get final configurations.

WMIN normally uses Gilbert-type repulsions (Eqn. 15), but Burnham and Post (1983) have modified it to use Born-type parameters (Eqn. 12). It can be used to fit either Gilbert-type or Born-type repulsive parameters to known structures, under the assumption they are minimum energy structures.

One can carry out full structure minimizations, or one can carry out partial minimizations varying only a selected set of parameters. Examples of both strategies are discussed below.

Assurances that the structures obtained are in fact minimum energy ones could be derived from the quasiharmonic lattice dynamical stability criterion that all phonon frequencies must be real numbers. If imaginary phonon frequencies are encountered as a function of lattice parameter or atomic position, the structure is unstable with respect to these coordinates. These "soft mode" instabilities may be missed with energy minimization routines that examine only first derivatives. Such a situation could obtain if the structure were confined to a saddle point by any symmetry constraints imposed during minimization.

6.1. Modeling structures using empirically-derived short-range parameters

As representative examples, Miyamoto and Takeda (1984) report minimum energy models of forsterite, β-spinel, and γ-spinel forms of Mg_2SiO_4. Gilbert-type repulsive parameters (Eqn. 15) were obtained by

Table 6. Comparison of observed and minimum-energy forsterite structures.

	Observed[b]	Empirical short-range potentials[a] Gilbert-type[c]	Born-type[d]	Gilbert+Morse[e]	MEG potentials[f]
Unit cell					
a,Å	4.7535(4)	4.799 (+1.0)	5.040 (+6.0)	4.643 (-2.3)	4.874 (+2.5)
b,	10.1943(5)	10.141 (-0.5)	10.187 (-0.1)	10.416 (+2.2)	10.322 (+1.2)
c,	5.9807(4)	5.911 (-1.2)	6.054 (+1.2)	6.124 (+2.4)	5.977 (-0.1)
V,Å^3	289.80(5)	287.67 (-0.7)	310.71 (+7.2)	296.17 (+2.2)	300.55 (+3.7)
Si Tetrahedron					
Si-O1,Å	1.615(3)	1.623 (+0.5)	1.622 (+0.4)	1.608 (-0.4)	1.560 (-3.4)
Si-O2	1.640(3)	1.660 (+1.2)	1.683 (+2.6)	1.645 (+0.3)	1.606 (-2.1)
Si-O3(x2)	1.633(2)	1.634 (+0.1)	1.641 (+0.5)	1.634 (+0.1)	1.586 (-2.9)
<Si-O>	1.630	1.638 (+0.5)	1.647 (+1.0)	1.629 (-0.1)	1.585 (-2.8)
M1 Octahedron					
Mg1-O1(x2),Å	2.083(2)	2.037 (-2.2)	2.100 (-0.8)	2.079 (-0.2)	2.076 (-0.3)
Mg1-O2(x2)	2.074(2)	2.088 (+0.2)	2.186 (+4.9)	2.087 (+0.2)	2.119 (+2.2)
Mg1-O3(x2)	2.145(3)	2.165 (+0.9)	2.230 (+4.0)	2.135 (-0.5)	2.251 (+4.9)
<Mg1-O>	2.101	2.097 (-0.2)	2.172 (+3.4)	2.100 (-0.1)	2.149 (+2.3)
M2 Octahedron					
Mg2-O1,Å	2.166(3)	2.207 (+1.9)	2.205 (+1.8)	2.214 (+2.2)	2.332 (+7.7)
Mg2-O2	2.045(5)	2.042 (-0.2)	2.094 (+2.4)	2.065 (+1.0)	2.085 (+2.0)
Mg2-O3(x2)	2.064(4)	1.971 (-4.5)	2.087 (+1.1)	2.105 (+2.0)	2.035 (-1.4)
Mg2-O3'(x2)	2.208(4)	2.299 (+4.1)	2.330 (+5.5)	2.239 (+1.4)	2.353 (+6.6)
<Mg2-O>	2.126	2.132 (+0.3)	2.189 (+3.0)	2.161 (+1.7)	2.199 (+3.4)
O2-O3 (sh)	2.558(5)	2.539 (-0.7)	2.445 (-4.4)	2.604 (+1.8)	2.426 (-4.8)

a Numbers in parentheses are % deviations from observed values.
b Hazen (1976); esd's are in parentheses.
c Miyamoto and Takeda (1984); model using potentials N1 (Table 5), which yield best agreement for mean bond lengths.
d Potentials from Catlow et al. (1982b); WMIN calculation by Burnham and Post (1983).
e Price and Parker (1984); model using potentials P4 (Table 5), charges varied.
f Post and Burnham (1985); O^{2-} shell radii are: O1=1.03Å, O2=1.05Å, O3=1.08Å.

Table 7. Observed and minimum-energy γ-Mg_2SiO_4 structures.

	Observed[b]	Empirical short-range potentials[a] Gilbert-type[c]	Rigid SiO_4[d]	Gilbert+Morse[e]
a,Å	8.0649(1)	8.042 (-0.3)	8.04 (-0.3)	8.057 (-0.1)
u	0.3685	0.3687	-	0.3690
<Si-O>,Å	1.655(4)	1.653(-0.1)	(1.655)	1.660 (-0.3)
<Mg-O>,Å	2.070(4)	2.063 (-0.3)	2.06 (-0.5)	2.064 (-0.3)

a Numbers in parentheses are % deviations from observed values.
b Sasaki et al. (1982); esd's in parentheses.
c Miyamoto and Takeda (1984); potentials N1.
d Matsui and Busing (1984); potentials MB, rigid SiO_4 tetrahedra.
e Price and Parker (1984); potentials P4, Morse-type term for Si-O.

setting A_{Mg} and B_{Mg} to values fitted by Busing (1970) to $MgCl_2$ and fitting Si and O parameters to the observed forsterite structure. Table 5 lists these parameters and compares them to other empirically-derived ones discussed below; Table 6 compares cell dimensions and interatomic distances of the model forsterite structure obtained this way with several other model structures and the observed one; Table 7 likewise compares three minimum energy γ-spinel structures with the observed one. Miyamoto and Takeda (1984) also modeled both ilmenite and perovskite phases of $MgSiO_3$ using these same repulsive parameters. The model ilmenite structure has unit cell dimensions larger than observed by 1.5%, cell volume larger by 4.3%, $Si\text{-}O_6$ octahedra with mean Si-O larger by 3.1% and mean O-O larger by 2.5%, and $Mg\text{-}O_6$ octahedra with mean Mg-O larger by 0.8% and mean O-O larger by 1.3%. The perovskite model structure has unit cell dimensions larger by 1.5% to 2.9%, and cell volume larger by 6.6%. The mean Si-O distance in the $Si\text{-}O_6$ octahedron is larger by 2.1%, and individual Si-O distances are larger by 1.1% to 3.5%; the longest distance in the model corresponds to the shortest one in the observed structure and vice versa. The mean Mg-O distance in the $[MgO_8]$ square antiprism is long by 3.6%; distance distortions are modeled incorrectly with individual Mg-O distances varying from observed values by +11% to -5%.

Matsui and Busing (1984a) describe a forsterite model parameterized in terms of the three orthorhombic cell dimensions, x and y coordinates for Mg, x and y coordinates for Si, and a rotation parameter of a rigid SiO_4 tetrahedron about the c axis. A_{Mg}, A_O, B_{Mg}, and B_O and the charge on oxygen, q_O, were fitted empirically; q_{Si} was adjusted to maintain neutrality. The model structure has lattice parameters that deviate by up to 4.4% and mean Mg-O distances deviating 0.04 Å from observed values, but the model permits rather good reproduction of elastic constants, the maximum deviation being 21% from the observed value. The same rigid tetrahedron approach was also applied to γ-spinel, yielding a model whose cell parameter differed from that observed (Sasaki et al., 1982) by 0.3%, whose mean Mg-O distance differed by 0.5% (Table 7), and whose calculated elastic constants were in reasonable agreement with observation. It should be stressed here that simulation of elastic constants provides a more severe test of the appropriateness of a particular set of pair potentials than does determining minimum energy atom configurations, since the former depends on second derivatives of potentials while the latter involves only first derivatives.

Price and Parker (1984) report minimum energy models of both α- and γ-Mg_2SiO_4 calculated with Gilbert-type potentials with variable charges empirically fitted to forsterite, to which a supplementary Morse-type potential for Si-O has been added. The objective of including a Morse potential, U_m, is to model better the covalency effects in the Si-O bond; it has the form

$$U_m = D_{ij}\{\exp[-2(r_{ij} - r_{ij}^*)]-2\exp[-\beta_{ij}(r_{ij} - r_{ij}^*)]\}, \qquad (21)$$

where D_{ij} (the well depth at $r_{ij} = r_{ij}^*$), β_{ij}, and r_{ij}^* may be calculated from the vibrational behavior of the diatomic molecule, but, as Price and Parker point out, are better obtained for the crystalline case by fitting to observed structural data. To preserve the capability of varying the charges of Mg and O in the fitting procedure, however, they fitted only D_{Si-O} (Table 5) and set β_{ij} to 1.975 (from spectroscopic data) and r_{ij}^* to 1.63 Å. The lowered charges they obtained from their fitting procedure (Table 5, column 3) compare rather well with those determined for atoms in forsterite by x-ray studies (Fujino et al., 1981): q_{Mg} = 1.75, q_{Si} = 2.1, q_O = -1.40. The potentials thus obtained yield minimum energy structures in better agreement with observation than the other two discussed above. In addition, they yield elastic constants with root-mean-squared deviation from observation of only 15%.

Catlow et al. (1982b) report calculations of minimum energy diopside, wollastonite, rhodonite, and pyroxmangite structures assuming several different divalent cations, using Born-type repulsive parameters including van der Waals terms (Eqn. 16) fitted as described earlier to several silicate structures. They were led to three conclusions based on comparison of the resulting energy values: "(1) The pyroxmangite and rhodonite structures are never ... favored with respect to the diopside and wollastonite structures; (2) pyroxmangite and rhodonite have very similar (cohesive) energies especially for Fe^{2+} and Mn^{2+}; (3) the most energetically favored structure is found to be wollastonite for all ions except Mg^{2+} for which diopside is preferred." The relative energies of the four structure types differed by up to 158 kj/mole for $MgSiO_3$ but only 102 kj/mole for $MnSiO_3$; changing the divalent cation resulted in energy changes up to 720 kj/mole in diopside, 435 kj/mole in wollastonite, 710 kj/mole in rhodonite, and 630 kj/mole in pyroxmangite. Energies of rhodonite and pyroxmangite structures were higher than those of diopside and wollastonite by anywhere from 21 kj/mole for $MgSiO_3$ to 89 kj/mole for $FeSiO_3$. Rhodonite and pyroxmangite energies were reported to differ by 16 kj/mole for $FeSiO_3$ and 8 kj/mole for $MnSiO_3$. For $MgSiO_3$ diopside appears favored over wollastonite by 39 kj/mole; for all other single divalent cation compositions wollastonite appears favored by amounts ranging from 1 kj/mole ($FeSiO_3$) to 63 kj/mole ($SrSiO_3$), increasing as the divalent cation radius increases. Any conclusions drawn from these values are difficult to justify since atomic configurations of the models are not described. Burnham and Post (1983) calculated minimum energy structures for both diopside and forsterite using the empirical potentials Catlow et al. (1982b) reported and found that they departed from reality more than those obtained using empirically fitted Gilbert-type potentials (Tables 6 and 8).

Matsui and Busing (1984b) have followed a similar strategy in modeling diopside as they used on forsterite. They parameterize the silicate chain in terms of a "chain stretching" model in which Si and O3 each have three variable coordinates, but O1 and O2 are constrained to rotate about Si with constant Si-O distances; there are thus 10 variables that describe the chain. To these are added the y coordinates of M1 and M2 plus the four cell parameters. Gilbert-type repulsion parameters for Mg and O were the same as those used on forsterite (Table 5); those for Ca were fitted to the γ-Ca_2SiO_4 structure. The charges (q) on O1 (constrained to = O2), O3, and Si were permitted to vary, with q(Si) constrained to yield net neutrality. Bond angle bending potentials of the form

$$U(\alpha) = (k_\alpha/2)(\alpha - \alpha_o)^2 \qquad (22)$$

Table 8. Comparison of observed and minimum-energy diopside structures.

	Observed[a]	Chain stretching variable charge[b]	Born-type repulsions[c]	MEG repulsions[d]
Unit cell				
a,Å	9.75	9.60 (-1.5)	10.91 (+12)	11.34 (+16)
b	8.90	9.43 (+6.0)	8.67 (-2.6)	11.00 (+24)
c	5.25	5.28 (+0.6)	5.55 (+5.7)	5.41 (+3.0)
β,°	105.6	106.2 (+0.6)	112.6 (+6.6)	118.7 (+12)
V,Å^3	438.6	458.2 (+4.5)	484.9 (+11)	592.3 (+35)
Si Tetrahedron				
Si-O1(C1),Å	1.60	(1.60)	1.63 (+1.9)	1.60 (+0)
Si-O2(C1)	1.59	(1.59)	1.64 (+3.1)	1.60 (+0.6)
Si-O3(C1)	1.66	1.68 (+1.2)	1.70 (+2.4)	1.58 (-4.8)
Si-O3(C2)	1.69	1.67 (-1.2)	1.76 (+4.1)	1.59 (-5.9)
<Si-O>	1.63	1.63 (+0)	1.68 (+3.1)	1.59 (-2.5)
Chain angle:				
O3C1-O3C2-O3C1'	166.4°	e	162.9° (-2.1)	179.5° (+7.9)
Mg (M1) polyhedron				
Mg-O1(A1,B1),Å	2.12	2.24 (+5.7)	2.17 (+2.4)	4.87 (+130)
Mg-O1(A2,B2)	2.06	2.02 (-1.9)	2.06 (+0)	1.96 (-4.9)
Mg-O2(C1,D1)	2.05	2.09 (+2.0)	2.09 (+2.0)	1.96 (-4.4)
<Mg-O>	2.08	2.12 (+3.4)	2.11 (+2.9)	f
Mg-Mg	3.10	3.23 (+4.2)	3.17 (+2.3)	4.98 (+60)
Ca (M2) polyhedron				
Ca-O1(A1,B1),Å	2.36	2.43 (+3.0)	2.31 (-2.1)	2.16 (-8.5)
Ca-O2(C2,D2)	2.35	2.32 (-1.3)	2.83 (+2.0)	2.16 (-8.1)
Ca-O3(C1,D1)	2.56	2.70 (+5.5)	2.53 (-1.2)	3.72 (+45)
Ca-O3(C2,D2)	2.72	2.77 (+1.8)	3.45 (+27)	4.50 (+65)
<Ca-O>	2.50	2.55 (+2.0)	2.78 (+11)	f

Numbers in parentheses are % deviations from observed values.
a Cameron et al. (1973); atom notation from Burnham et al. (1967).
b Matsui and Busing (1984); Gilbert-type potentials (MB, Table 5) with variable charges; $q_{Mg,Ca}$ fixed at +2.0, $q_{O1,O2}$ fitted at -1.26, q_{O3} fitted at 0.905, q_{Si} set to +1.425 to maintain neutrality; see reference for other potential parameters.
c Born-type potentials from Catlow et al. (1982b), WMIN calculation by Burnham and Post (1983).
d Post and Burnham (1985); O^{2-} shell radii: O1,2=1.11Å; O3=0.93Å.
e Value not available.
f Value of no significance.

were included for the Si-O-Si angle and O-Si-O angles, where α is the calculated angle, and α_o (the unconstrained value) and k_α (the force constant) are parameters determined during modeling. In place of Si-O repulsion parameters, they employed a "bond stretching" term for the variable Si-O3 distances of the form

$$U(d) = (k_d/2)(d - d_o)^2, \tag{23}$$

where d is the calculated bond distance, and the unconstrained value, d_o, and the force constant, k_d, are model parameters. This modeling strategy was adopted because it best satisfied the primary objective of calculating elastic constants, at the same time yielding a reasonably good minimum energy structural configuration. Comparison of this model structure with others for diopside (Table 8) shows it to be significantly better than the one based on empirical Born-type repulsive parameters (Catlow et al., 1982b) and far superior to the one based on MEG repulsive parameters (Post and Burnham, 1985). It does not, however, compare as well to the observed structure as does the similarly derived model for forsterite.

It appears that minimum-energy models of forsterite, its polymorphs, and diopside developed using empirically-determined Gilbert-type repulsive parameters are better able to reproduce the known structures than are those calculated with fitted Born-type repulsive parameters. At present the potentials that best model forsterite are those employed by Price and Parker (1984) which contain reduced charges and a Morse-type term to account for covalency of Si-O. As one attempts to employ empirically determined short-range potential parameters to simulate structures different from those used for fitting, it is important to be aware that their suitability may diminish rapidly as differences between the structures increase. Certainly as one proceeds to structures with distances substantially outside the range of fitting, the repulsive parameters are more likely to be inappropriate. Yet, interestingly, the Miyamoto and Takeda (1984) Gilbert-type Si-O potentials fitted to forsterite structures do not perform particularly badly when used to simulate 6-coordinated Si in $MgSiO_3$ ilmenite and perovskite. Their forsterite-fitted Mg-O potentials perform substantially less well, however, on the $[MgO_8]$ square antiprism in perovskite. Although it is not precisely clear whether one can model complex mineral structures better with these procedures than with DLS right now, an important advantage of the energy minimization approach is that the short-range potentials, however characterized, provide the basis for simulation of additional properties such as elastic constants, compressibilities, and dielectric constants.

6.2. Modeling structures using MEG short-range interactions

Minimum energy structures calculated using short-range interactions determined by the MEG procedure outlined earlier provide models illustrating the purest ionic assumptions. Since no empirical information is contained in them, the degree to which observed structures depart from them is a measure of the significance of other effects on bonding, such as anisotropic polarization, covalency, or crystal field effects not included in the model. In this context they are extremely useful when compared with observed configurations.

Models of alkali and alkaline-earth halides and oxides have been shown by several workers to agree well with observation (Cohen and Gordon, 1975, 1976; Tossell, 1980a,b; Muhlhausen and Gordon, 1981a,b), thus demonstrating the general validity of the purely ionic model in those cases. A recent improvement in the calculation procedure includes the volume dependence of the ion self-energies explicitly in the minimizations. Results obtained so far by Hemley and Gordon (1985) and Hemley et al. (1985) indicate improved agreement of the calculated lattice parameters with observation for both alkali halides and alkaline-earth oxides. Muhlhausen and Gordon (1981b) calculated the static equilibrium lattice parameters for both $CaTiO_3$ and $CaSiO_3$ in the cubic perovskite structure. For $CaTiO_3$ the calculated lattice parameter is low by only 0.7% and the calculated static cohesive energy is less than the room temperature experimental value by 4.9%. $CaSiO_3$ transforms to perovskite at pressures above about 160 kbar; the calculated lattice parameter at a simulated pressure of 160 kbar is higher than the high-pressure experimental value by only 1.7%. Hemley et al. have now extended these cal-

Table 9. MEG-based minimum energy structures of rutile and anatase compared with observed structures (after Post and Burnham, 1985).

	Rutile (P4/mnm)		Anatase (I4/amd)	
	Observed	Minimum energy*	Observed	Minimum energy*
<u>Unit cell</u>				
a,Å	4.594	4.491 (-2.2)	3.776	3.689 (-2.3)
c	2.958	3.063 (+3.5)	9.486	10.067 (+6.1)
V,Å^3	62.43	61.78 (-1.0)	135.25	137.00 (+1.3)
<u>Interatomic distances</u>				
Ti-O (x4),Å	1.948	1.961 (+0.7)	1.930	1.909 (-1.1)
Ti-O (x2)	1.980	1.935 (-2.3)	1.973	2.030 (+2.9)
<Ti-O>	1.959	1.952 (-0.4)	1.944	1.949 (+0.3)
O-O (sh)	2.536	2.462 (-2.9)	2.459	2.405 (-2.2)
O^{2-} shell radius:		1.11Å		1.08Å

* Values in parentheses are % deviations from observed.

Table 10. MEG-based minimum energy structure of brookite compared with observed structure (after Post and Burnham, 1985).

	Brookite (Pcab)	
	Observed	Minimum energy*
<u>Unit cell</u>		
a,Å	9.184	9.171 (-0.1)
b	5.447	5.373 (-1.4)
c	5.145	5.224 (+1.5)
V,Å^3	257.38	257.42 (+0)
<u>Interatomic distances</u>		
Ti-O1,Å	1.865	1.904 (+2.1)
Ti-O1'	1.992	1.998 (-0.3)
Ti-O1"	1.994	1.929 (-3.3)
Ti-O2	1.919	1.956 (+1.9)
Ti-O2'	1.946	1.931 (-0.8)
Ti-O2"	2.039	2.017 (-1.1)
<Ti-O>	1.959	1.956 (-0.4)
O1-O1 (sh)	2.485	2.434 (-2.1)
O2-O2 (sh)	2.514	2.462 (-2.1)

O^{2-} shell radii: O1 = 1.11Å, O2 = 1.10Å
* Values in parentheses are % deviations from observed.

Figure 4. TiO_2 POLYMORPHS - CALCULATED COHESIVE ENERGIES

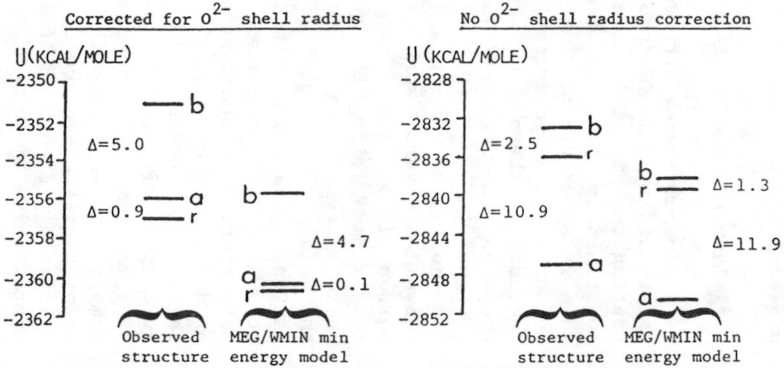

MEG-based cohesive energies of TiO_2 polymorphs. Units are kcal/mole. After Burnham and Post (1983).

r = rutile
b = brookite
a = anatase

culations to include distortions from cubic symmetry, self-energy corrections, and lattice dynamical effects in $MgSiO_3$ as well as $CaSiO_3$ (pers. comm.).

Post and Burnham (1985; see also Burnham and Post, 1983) have used MEG repulsive parameters fitted to the Born formulation (Eqn. 12) to model the TiO_2 polymorphs, quartz, forsterite, and diopside; discussion of these results provides an interesting basis to evaluate the power of the purely ionic approach to model several minerals over a substantial range of ionicity.

Rutile, anatase, and brookite were all modeled in their normal space groups. Shell radii of O^{2-} were adjusted in all cases to match the O^{2-} shell potential to the site potentials; all variable atomic coordinates and cell dimensions were adjusted during the minimizations. Model cell dimensions and interatomic distances reported by Post and Burnham are listed in Tables 9 (rutile, anatase) and 10 (brookite) where they are compared with observation. The agreement is, indeed, quite good; the largest variances are in c of anatase (6.1% too long) and c of rutile (3.5% too long). Mean Ti-O distances are very close; all shared O-O octahedral edges model slightly shorter than observed, suggesting that real Ti-Ti repulsions across the shared edges may be somewhat damped by covalency effects. Distortions of Ti-O octahedra are modeled fairly well: In brookite the longest and shortest of six Ti-O distances match observation but the sequence from shortest to longest does not match; in anatase the distortion sense is correct but the model octahedron is more distorted than the observed one; and in rutile, the magnitude of model distortion is almost correct, but in the wrong sense, so that the apical Ti-O in the model are shorter rather than longer than the other Ti-O distances.

Cohesive energies of the minimum-energy models are displayed in Figure 4, where they are compared with cohesive energies calculated for the observed structures. The relative energies, when corrected for shell radius self energies, show both the observed and mimimum-energy rutile structures to be most stable, and both the observed and minimum-energy brookite structures to be least stable. Note that this is in contrast to the previously accepted qualitative assessment (Evans, 1966) that suggested anatase would be less stable than brookite because the number of shared octahedral edges per octahedron is two in rutile, three in brookite, and four in anatase; the calculated stability sequence is, however, consistent with the natural frequency and abundance of the three polymorphs. If the O^{2-}self-energy corrections are not made, Figure 4 shows the relative stabilities change; the importance of the corrections is self-evident. Comparisons of calculated energy data with two thermodynamic observations are encouraging: The calculated ΔU for anatase $\rightarrow$ rutile is -3.8 kj/mole, while Navrotsky and Kleppa (1967) report ΔH_{298} for the same transformation as -5.4 kj/mole; the absolute cohesive energy calculated for rutile is -9874 kj/mole, while that obtained from a Born-Haber cycle is -10385 kj/mole, a difference of only 5%.

Table 11. MEG-based minimum energy structures of quartz compared with observed structures (after Post and Burnham, 1985).

	Low quartz observed	High quartz observed	Minimum energy structures as function of O2- shell radius 0.93Å	1.01Å	1.06Å
Unit cell					
a,Å	4.91	5.01	4.97	5.05	5.10
c	5.40	5.47	5.52	5.60	5.66
V,Å^3	112.74	118.90	118.00	123.68	127.33
Interatomic distances					
Si-O,Å	1.594	1.609	1.564	1.588	1.603
Si-O'	1.613	1.609	1.564	1.588	1.604
<O-O>	2.618	2.626	2.553	2.592	2.618
O-Si-O angles					
Range,°	108.6-111.4	103.0-114.7	104.4-116.3	104.2-116.5	104.2-116.5
Variance,°	1.4	28.6	29.3	30.8	31.5
Si-O-Si,°	144.6	148.7	162.6	162.8	162.9

Comparison of an MEG-based minimum-energy model for quartz (Post and Burnham, 1985) with the observed structure, Table 11, shows that the model structure matches that of high quartz better than low, even though the modeling was carried out with $P3_121$ constraints, varying four atomic coordinates and two cell dimensions. The model fails to achieve the correct Si-O-Si angle, being some 18° too wide. Nevertheless this is an improvement over the 180° angle reported by Tossell (1980a) for his MEG-based model. Post and Burnham (1985) show that increasing the shell radius on O^{2-} lengthens the Si-O and O-O distances and increases the cell volume while not altering the O-Si-O and Si-O-Si angles. The difficulty modeling the Si-O-Si angle is clearly related to significant covalency effects; Jackson and Gordon (pers. comm.) have largely overcome this inadequacy by employing multiple shells on oxygen to model the anisotropic polarization.

An MEG-based minimum-energy model for forsterite reported by Post and Burnham (1985) is compared with other minimum-energy models constructed with empirical repulsive parameters in Table 6. Shell radii on oxygens in this model were 1.03 Å for O_1, 1.05 Å for O_2, and 1.08 Å for O_3, reflecting the differences in the environments of these crystallographically distinct atoms. While the model Si-O distances are all low, the distortion of the tetrahedron is modeled correctly, thus implying that it has an ionic origin related to the structural configuration itself. Octahedral distortions are modeled correctly at least in part, with differences between minimum and maximum Mg-O distances accentuated. In general, this non-empirical purely ionic minimum energy model is rather respectible, but improvements in the MEG procedure, as have been developed by Hemley et al. (1985) for MgO (see above), will need to be incorporated before such a model can have predictive capacity or be applied more generally to olivines.

Diopside is modeled extremely poorly by MEG-based minimum-energy calculations. The Post and Burnham (1985) results are compared with others calculated using empirical repulsion parameters in Table 8. Cell dimensions are all larger than observed, with cell volume larger by 35%; Si-O distances are all shorter than observed, as they were in both forsterite and quartz, with the mean Si-O shorter by 2.5%; the chain angle O3-O3'-O3" is nearly straight at 179.5°, compared with 166.4° in the observed structure. Certain Mg-O and Ca-O distances are so long in the model that the coordinations of both sites have become four, instead of the observed values of six for Mg and eight for Ca. To explain this behavior, note that in the observed diopside structure the Pauling charge balances, uncorrected for distance variations, are +1.91 on O1 (coordinated to 2Mg's, 1Ca, and 1Si), +1.58 on O2 (coordinated to 1Mg, 1Ca, and 1Si), and +2.5 on O3 (coordinated to 2Ca and 2Si); O1 and O2 are thus underbonded and O3 is overbonded, which qualitatively explains the variations in cation-oxygen distances in the observed structure. Now, if we recalculate using the effective coordinations of Mg and Ca in the MEG-based model of four each, all three oxygens become charge balanced: O1 and O2 both coordinate to 1Mg (+2/4) plus 1Ca (+2/4) plus 1Si (+4/4); O3 coordinates to 2Si (2 x (+4/4)). What this tells us is that the MEG-based model, founded as it is in purely ionic theory, probably represents what the diopside structure would look like if it were completely ionic. This further affirms that Pauling charge balance is a very desireable attribute of purely ionic structures. Substantial corrections to MEG-based potentials to account for non-ionic effects will have to be developed before modeling of complex mineral structures, especially those not satisfying Pauling charge-balance criteria, will be satisfactory.

As a start to achieve such expanded capability, electron gas studies based on a mixed ionic-covalent description of bonding deserve mention. LeSar and Gordon (1982) modeled the structures, compressional effects, and polymorphism of alkali and alkaline-earth hydroxides using hydroxide ions whose charge densities were determined by a molecular-type Hartree-Fock self-consistent field calculation. Tossell (1985) employed this approach to study electron deficient anions and ion pairs in minerals. Charge densities obtained from band-structure calculations (e.g., pseudopotentials) can also be used to calculate crystal properties using electron gas theory, as has been demonstrated by Boyer (1983) for MgO. Such calculated charge densities provide one basis for examining further departures from an ionic description of bonding.

The examples discussed here show that minimum-energy models calculated with the aid of MEG theory can be extremely informative and that the technique has already demonstrated substantial power. When the ability to describe anisotropic polarizations becomes routine and the capacity to handle transition metal ions is developed that power will be substantially enhanced. Already a start has been made to incorporate extensions of the purely ionic model in the electron gas approach; results described here demonstrate that these will be essential to assure a role for MEG theory in future studies of silicates.

7. PREDICTING PHASE TRANSITIONS

The important objective of predicting polymorphic phase transformations with the ionic model is one that has received considerable attention in recent years, particularly as applied to pressure-induced transformations. While Jeanloz discusses this matter in detail elsewhere in this volume, some remarks are appropriate here in the context of structure modeling with the MEG technique.

Cohen and Gordon (1975) studied the NaCl to CsCl transformation in alkali halides. More recently Hemley and Gordon (1985) have calculated the temperature dependence of the transition pressure for NaF and NaCl; they report good agreement with experiment for the sign of the Clapeyron slope at high temperature for NaCl, but their calculated magnitudes differ from the admitedly uncertain experimental numbers.

Prediction of the NaCl to CsCl transition in CaO using the MEG model and its subsequent experimental verification (Jeanloz et al., 1979) served as an early indicator of the potential of the MEG technique. Calculations underestimate ΔV and overestimate the pressure of the transition in both CaO and SrO (Sato and Jeanloz, 1981), thus they underestimate the stability of the CsCl structure relative to the NaCl phase. It is not yet known whether this is an inherent limitation of the ionic model, or whether a more refined treatment of the charge densities will improve the calculations.

Tossell (1980b) predicted on the basis of MEG calculations that SiO_2 would transform from the rutile to the fluorite structure at about 4Mb. Jackson et al. (Hemley, pers. comm.), however, have carried out calculations using more refined charge densities and predict that stishovite will be stable to much higher pressures, and will distort to the $CaCl_2$ structure through a displacive transformation.

Thermally induced displacive transitions in fluoro-perovskites have recently been studied using electron-gas pair potentials in the quasi-harmonic approximation to determine their equations of state (Boyer and Hardy, 1981; Boyer, 1984). These studies relate quasiharmonic phonon instabilities calculated from the model to the temperatures at which transformations from low symmetry to high symmetry structures take place. Similar calculations carried out by Hemley et al. on silicate perovskites indicate that the distorted structures have large stability fields (Hemley, pers. comm.).

While the scope of these studies has so far been limited to rather simple highly symmetric mineral phases and their isotypes, improvements in pair potentials and charge densities combined with enhanced capabilities to calculate phonon spectra will soon make feasible investigation of more complex mineral transformations.

8. APPLICATION OF MODELING TECHNIQUES TO ORDER-DISORDER PROBLEMS

In addition to making a configurational entropy contribution to free energy, substitutional and positional disorder also affect the internal energy by altering slightly the static structural configuration. In many cases the disordering energy has a substantially greater effect on free energy than does the added configurational entropy, although the latter term's influence increases with increasing temperature. Procedures for modeling disordered structures are not yet well developed, and calculation of disordering energies can not yet be done directly.

As we have already seen, ordered and anti-ordered arrangements can be modeled and their relative energies calculated. This energy difference is an exchange energy -- not a disordering energy -- since it represents the energy difference between two ordered states -- not between ordered and disordered states. Giese (1975) brought substantial insight to bear on the issue of disordering energies through his electrostatic energy calculations on columbite, $MnNb_2O_6$. In this mineral's unit cell there are 12 cation sites occupied by $8Nb^{+5}$ and $4Mn^{+2}$, and in the disordered structure the average cation site charge is +4. Although he did not model separate structures in his procedure, he did calculate electrostatic energies for all 495 different possible ordered cation distributions in one unit cell. He also calculated the electrostatic energy for the structure with the average charge of +4 assigned to every site, a procedure which had previously been asserted as equivalent to calculating the energy of the disordered distribution. His results showed the following: Many cation arrangements have nearly the same energy with total energies in the range of -35140 to -40800 kj/mole; 17% of the arrangements have energies within 660 kj/mole of each other; the minimum energy arrangement (-40825 kj/mole) matches the observed ordered columbite arrangement; the energy of the average-charge arrangement is -38500 kj/mole, 650 kj/mole higher than the average of the energies of all 495 arrangements. His points are these: The calculated energy of a cation-disordered structure with average charge assigned to each site is not a correct representation of the disordered cohesive energy; and disordering is likely if there are two or more cation arrangements with nearly the same cohesive energy. Furthermore, as the difference between the cohesive energies of alternative arrangements increases, the higher will be the temperature at which the $T\Delta S$ term will overcome this difference to yield a lower free energy and favor disordering.

8.1. Partial minimizations on structures with complex cation distributions

Several structures that have complex cation distributions, either ordered or disordered, also have additional cations whose positions are affected by these distributions. Examples include feldspars, in which the alkali atom positions respond to changes in tetrahedral Al,Si distributions; amphiboles, whose A-site locations are affected by both tetrahedral and octahedral cation substitutions; and hollandites, whose tunnel cations take up positions that reflect the distributions of cations in octahedral chains surrounding the tunnels. As illustrations of insights to be gained by carrying out partial minimizations, I discuss a recent study of Na atom positions in albite reported by Post and Burnham (1984), and then briefly mention the results of studies on both amphiboles (Docka et al., 1985) and hollandites (Post and Burnham, 1983).

Ribbe et al. (1969), Prewitt et al. (1976), and Winter et al.(1979) have all analysed the large apparent thermal motion of Na in the high albite structure, and have concluded that it is a consequence of

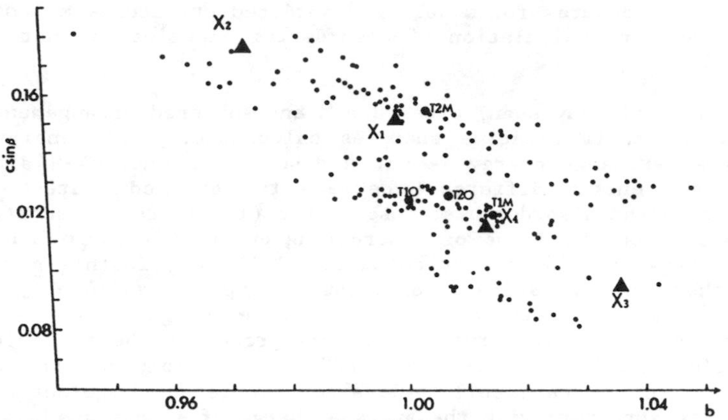

Figure 5. Projection down a of 175 minimum energy Na positions determined for 56 arrangements of 4 Al and 12 Si over 16 tetrahedral positions per unit cell of room-temperature high-albite framework. Triangles and X's are refined Na quarter-atom positions determined by Ribbe et al. (1969) and Prewitt et al. (1976), respectively. Larger solid circles are Na positions corresponding to Al distributed into the indicated site only. After Post and Burnham (1984).

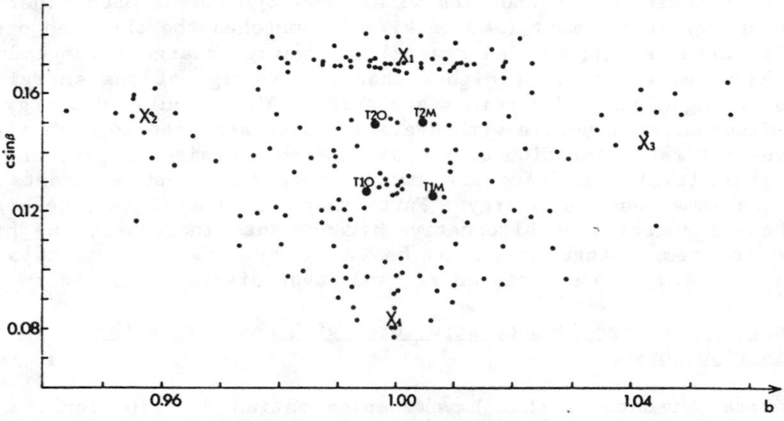

Figure 6. Projection down a of 175 minimum energy Na positions determined for 56 arrangements of 4 Al and 12 Si over 16 tetrahedral positions per unit cell of high albite framework determined at 1090°C by Prewitt et al. (1976). X's are refined Na quarter-atom positions; larger solid circles are Na positions corresponding to Al distributed into the indicated site only. After Post and Burnham (1984).

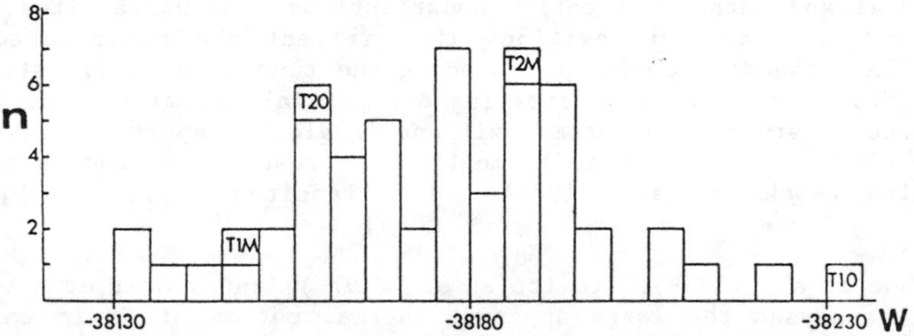

Figure 7. Histogram showing calculated cohesive energies per unit cell (W, kcal/cell) for room temperature high albite after minimizing Na positions. The 56 distinct Al,Si distributions correspond to those included in Figure 5; cases having Al ordered into each of the four distinct tetrahedral sites are indicated. After Post and Burnham (1984).

positional disorder due to disorder of Al and Si in tetrahedral sites, in contrast to the electron-density smearing of Na in low albite which is thought to be due to true anisotropic thermal motion. Both Ribbe et al. (1969) and Prewitt et al. (1976) refined the crystal structures of high albite assuming a quarter-atom model for Na, in which the alkali site is split into four separate sites, each assuming a different position in the alkali cavity and each having its own temperature factor and a statistical occupancy of 25% of an Na atom. Improvements in refinements obtained by this procedure led to the notion implicitly that a "quadripartite" model for high albite, in which there exist four kinds of domains each having its own Na position in the cavity, appeared reasonable. Winter et al. (1979) observed, however, that there were many more local configurations of Al and Si that were possible, and that the quarter-atom refinements did not rule out completely random distributions with each local Al,Si configuration resulting in a unique Na position within the cavity.

Brown and Fenn (1979) carried out some electrostatic energy calculations to address this problem; in addition to Coulomb terms they employed short-range repulsive terms only for Na-O and K-O interactions estimated from bulk compressibility data for NaF and KF. Using the fixed high albite framework reported by Prewitt et al. (1976), redistributing the Al and Si formal charges among tetrahedra, and assuming the Na to occupy each of the quarter-atom positions in turn, their calculations showed that the most favored of the quarter-atom Na positions does depend on Al,Si distribution. Interestingly, however, the single Na position appeared energetically favored for Al in T2O and T2m. Comparing energies for a given Na position with different Al,Si distributions showed that the lowest energy was always obtained with Al in T1O, regardless of Na position.

Post and Burnham (1984) employed MEG-derived repulsive parameters in partial energy minimizations of the feldspar structure, using the observed high albite oxygen framework that was kept fixed, and minimizing the Na position for all ordered distributions of 12Si and 4Al over the 16 tetrahedral sites in one unit cell that do not violate the Al-avoidance principle. The resulting Na positions are displayed in Figure 5, along with the quarter-atom Na positions determined by Ribbe et al. (1969) and Prewitt et al. (1976). Notice the good correspondence, supporting the notion that the quarter-atoms are simply an approximation to a real Na electron density in high albite that averages all these positions in response to the existence of many local Al,Si distributions. Notice also that the quarter-atom positions do not correspond to those that would be expected if Al occupied only T1O, T1m, T2O, and T2m sites in disordered high albite. Figure 6 shows the distribution of Na positions calculated using the high albite 1090°C framework configuration reported by Prewitt et al. (1976) along with the four quarter-atom Na positions determined in that high temperature refinement. Notice again that the quarter-atom positions are attempting to mimic an electron density distribution that corresponds to the superposition of all these Na positions, rather than simply those calculated for Al in T1O, T1m, T2O, and T2m. Interestingly the quarter-atoms around which the density of calculated positions is high consistently have lower temperature factors in the reported structure refinements than do those around which the density of Na positions is lower. Post and Burnham (1984) argue that these calculations demonstrate a total lack of a domain structure in high albite. The histogram of calculated energies for various Al,Si distributions, Figure 7, shows that many distributions have similar energies and that the configuration with Al in T1O has the lowest energy. Both the tendency to disorder and the Al site preference in low albite are thus rationalized.

Docka et al. (1985) have carried out partial minimizations on amphibole structures to investigate the effect of tetrahedral and octahedral cation substitutions and distributions, of F^- for OH^- substitutions, of $[Fe^{3+} + O^{2-}]$ for $[Fe^{2+} + OH^-]$ substitutions, and of K for Na A-site substitutions on the minimum-energy position of the A-site. Their calculations show that the A-site does respond to these substitutions, with migration of the minimum off the conventional 2/m location to positions either on the 2-fold or the mirror if the local cation charge distribution is symmetrical, or to the general position if it is not. A-site displacements away from the 2/m site ranged up to 0.8 Å, and were generally larger with Na rather than K as the A-site occupant. These partial minimizations were carried out using lowered symmetry and doubled cells to permit modeling of rational local configurations, and using a fixed array of oxygen and tetrahedral and octahedral cation positions taken from the tremolite structure; only the A-site coordinates were varied during energy minimizations.

Post and Burnham (1983) showed by minimizing tunnel cation positions in hypothetical Ti,Al and Ti,Mg hollandites, $[A_x(Ti,Al)_8O_{16}]$ and $[A_x(Ti_{7.33}Mg_{.67})O_{16}]$ with A = Ba^{2+}, K^+, or Na^+, that the minimum-energy positions were significantly affected by octahedral cation distributions. Their results show that these positions could be displaced by up to 1 Å as the octahedral cation distribution was altered.

8.2. Short-range ordering energies

In contrast to long-range ordering energetics and calculation of cation site preferences, analysis of short-range ordering preferences attempts to quantify the energetics of next-nearest neighbor interactions to assess the liklihood of various local configurations of cations in adjacent or close polyhedra. Such calculations involve full or partial minimizations of models with substantially reduced symmetry and multiple cells. As an example of this application, I discuss work reported by Cohen and Burnham (1985) on the energetics of short-range ordering in Ca-Tschermak's (CaTs) pyroxene $[CaAl(AlSi)O_6]$, fassaite $[Ca_2(MgAl)(AlSi)_2O_{12}]$, and omphacite $[(NaCa)(MgAl)Si_4O_{12}]$. Energies of various cation arrangements were calculated using MEG-derived repulsive parameters, a fixed observed CaTs structure, and DLS-determined structures (based on fixed observed cell dimensions) for fassaite and omphacite. To get a suitable number of distinct cation arrangements, structures with doubled c-axes were used. Calculations were carried out for 35 cation arrangements in CaTs, 49 in fassaite, and 32 in omphacite.

Short-range ordering parameters were defined for several schemes. In CaTs, for example, the short-range ordering parameter involving local

Table 12a. Short-range ordering energies in aluminous clinopyroxenes (after Cohen and Burnham, 1985).

	Cation pair	Separation distance,Å	Pairs/ 6 Oxy	Reaction	SRO energy kj/6 O	Esd
CaTs	T1A-T2A (A)	3.125	2	AlAl+SiSi-2AlSi	-73	5
	T1A-T2C (B)	3.421	1	AlAl+SiSi-2AlSi	-24	4
	T1A-T1B (C)	3.984	1	AlAl+SiSi-2AlSi	-9	4
Fassaite	M1-T pairs	*	6	MgAl+AlSi-MgSi+AlAl	-187	7
	M1-M1(1)	3.08	1	MgMg+AlAl-2MgAl	-38	2
Omphacite	M1-M1(1)	3.08	1	MgMg+AlAl-2MgAl	-36	3
	M2-M2(1)	4.38	1	NaNa+CaCa-2NaCa	-10	3
	M1(1)-M2	3.19	2	NaMg+CaAl-NaAl+CaMg	-60	5
	M1-M2	3.42	1	NaMg+CaAl-NaAl+CaMg	-30	4

* Three M1-T pairs at 3.24, 3.26, and 3.44Å have statistically indistinguishable ordering energies.

Table 12b. Cohesive energies of four possible long-range tetrahedral ordering schemes for CaTs (after Cohen and Burnham, 1985).

Space group	Pairs ordered favorably	Cohesive energy, kj/6 O mole
$P2_1/n$	A, B, C	-32909
C2	A, B	-32903
$C\bar{1}$	A, C	-32884
P2/n	A only	-32856

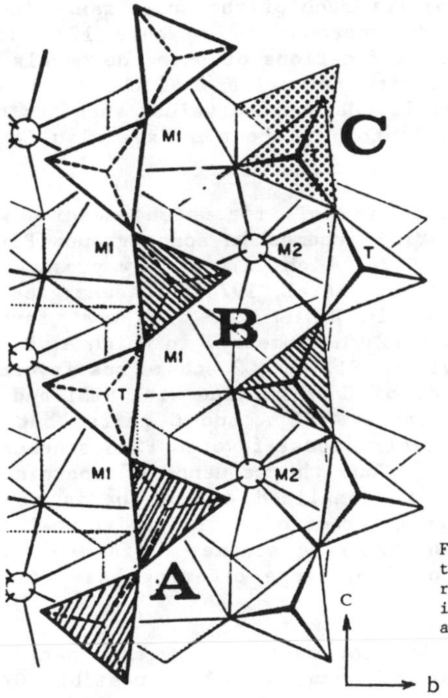

Figure 8. Three distinct next-nearest neighbor tetrahedral pairs in C2/c clinopyroxene. A pairs related by a c-glide, B pairs related by an inversion center; C pairs related by a two-fold axis. After Cohen and Burnham (1985).

distributions of tetrahedral Al and Si is:

$$s(i) = 2P(i,AlSi) - P(i,AlAl) - P(i,SiSi) \tag{24}$$

where the P's are probabilities of AlSi, AlAl, and SiSi pairs occurring in the ith kind of tetrahedral pair; with equal numbers of tetrahedral Al's and Si's, s(i) can range from -1 (complete segregation of Al and Si), through 0 (complete short-range disorder), to +1 (complete short-range order, i.e., only AlSi pairs). There are three tetrahedral pairs in CaTs in which the interaction energies are likely to be significant: A pairs are adjacent tetrahedra in the same silicate chain related by the c-glide; B pairs are in adjacent chains in the same (100) plane related by inversion; and C pairs are on opposite sides of an M2 polyhedron in different I-beams and are related by a 2-fold axis. The T-T separations are 3.13 Å for the A pairs, 3.42 Å for the B pairs, and 3.98 Å for the C pairs; these pairs are illustrated in Figure 8. In addition to the T-T pairs, in fassaite there may be short-range ordering involving three M1-T interactions at 3.24, 3.26, and 3.44 Å, and Mg,Al short-range ordering through an M1-M1 interaction at 3.08 Å. In omphacite there may be short-range ordering among adjacent M1 octahedra, among adjacent M2 polyhedra, in an M1-M2 pair at 3.19 Å, and in another M1-M2 pair at 3.42 Å.

The short-range order parameters characterizing each cation distribution for which an energy calculation was made were determined, after which multiple regressions were computed of calculated energies versus short-range order parameters. The short-range ordering energies reported by Cohen and Burnham (1985) for each of the short-range interactions in the three pyroxenes are reproduced in Table 12a; their analysis of errors introduced by simplifications of both the models and the calculation procedures combined with comparisons of the calculated short-range energies with experimentally determined values suggests that the numbers may be too large by a factor of about 5 (see also Cohen, 1984).

Four of the cation distributions in CaTs for which energies were calculated correspond to Al,Si ordering schemes in space groups $P2_1/n$, C2, $C\bar{1}$, and P2/n, all subgroups of C2/c which CaTs is now presumed to exhibit (Grove and Burnham, 1974; Okamura et al. 1974); the energies for these four ordered states are listed in Table 12b. The most favored long-range ordering scheme would have $P2_1/n$ symmetry, in which A, B, and C pairs would all order favorably with AlSi; the C2 scheme has favorably ordered A and B pairs, but two kinds of C pairs, one with SiSi and one with AlAl; the $C\bar{1}$ scheme has favorable AlSi A and C pairs, but unfavorable B pairs; and the energetically least-favored P2/n scheme has favorable ordering only of A pairs. Thus the sequence of energies of these ordering schemes is readily rationalized on the basis of the number of favorable tetrahedral pair interactions. The additional fact that the cohesive energies of these ordering schemes differ so little easily explains why no evidence for long-range ordering has yet been observed in CaTs.

The short-range ordering energies for fassaite suggest that long-range ordering might be observed in subgroups of C2/c, possibly C2 or

P2. The fact that observed long-range ordering of omphacite shows more ordering of MgAl on M1-M1 pairs than of NaCa on M2-M2 pairs (Curtis et al., 1975) is consistent with the short-range ordering energies shown in Table 12a.

9. THE FUTURE

Several developments in both the static and the dynamical aspects of structure modeling are likely to occur during the next decade. Procedures that account for anisotropic polarizations when calculating MEG repulsive parameters will become available for general use, and treatments of non-closed-shell ions by electron gas methods will be developed. Extension of lattice dynamical calculations with the electron-gas ionic model to more complex crystal structures is now underway, thus predictions of thermal expansivity of complex structures are nearly within reach. There is continuing refinement of methods used to incorporate many-body contributions in calculations of phonon spectra and elastic constants, and combined ionic-covalent methods will be further extended to include the results of molecular-orbital and band-structure techniques. As these developments materialize, mineralogists will finally be able to calculate mineral free energies at temperatures and pressures in the earth's mantle. Meanwhile DLS methods and existing minimization techniques will find expanded applications that will continue to build our quantitative understanding of mineral crystal chemistry.

ACKNOWLEDGMENTS

I am grateful for years of fruitful collaboration with Drs. David Bish and Jeffrey Post, both of whom have been serious partners in efforts to provide meaning to structure modeling and realism to calculations of cohesive energies. A number of former and present students have also contributed significantly to these efforts: Dr. Yoshikazu Ohashi, who got me started down this road in the early 1970's, Page Chamberlain, Ronald Cohen, Fred Allen, Janet Docka, Dr. Linda Pinckney; I appreciate their enthusiasm and fresh intuition. The perspectives of Prof. Gerry Gibbs have always been important to me. I especially thank Dr. Russell Hemley for his close scrutiny of the draft, and for his generous willingness, on short notice, to suggest some significant changes and additions. The volume editors, Dr. Susan Kieffer and Prof. Alexandra Navrotsky, also made very helpful suggestions. Research reported here carried out by my collaborators and me has been generously supported by NSF Grant EAR 7920095.

REFERENCES

Baur, W.H. (1971a) The prediction of bond length variations in silicon-oxygen bonds. Amer. Mineral. 56, 1573-1599.

________ (1971b) Geometric refinement of the crystal structure of β-Mg_2SiO_4. Nature Phys. Sci. 233, 135-137.

________ (1972) Computer simulated crystal structures of observed and hypothetical Mg_2SiO_4 polymorphs of low and high density. Amer. Mineral. 57, 709-731.

________ (1977) Computer simulation of crystal structures. Phys. Chem. Minerals 2, 3-20.

Bertaut, F. (1952) L'energie electrostatique de reseaux ioniques. J. Phys. Radium 13, 499-505.

Bish, D.L. and Burnham, C.W. (1980) Structure energetics of cation-ordering in orthopyroxene and Ca-clinopyroxenes. Geol. Soc. Amer. Abstr. Progr. 12, 388.

________ and ________ (1984) Structure energy calculations on optimum distance model structures: application to the silicate olivines. Amer. Mineral. 69, 1102-1109.

________ and Giese, R.F. (1981) Interlayer bonding in IIb chlorite. Amer. Mineral. 66, 1216-1220.

Boeyens, J.C.A. and Gafner, G. (1969) Direct summation of Madelung energies. Acta Crystallogr. A25, 411-414.

Born, M. and Huang, K. (1954) Dynamical Theory of Crystal Lattices. Oxford University Press, London, p. 19-37.

Boyer, L.L. (1983) Bonding and equation of state for MgO. Phys. Rev. B27, 1271-1275.

________ (1984) Parameter-free equation of state calculations for $CsCaF_3$. J. Phys. C17, 1825-1832.

________ and Hardy, J.R. (1981) Theoretical study of the structural phase transition in $RbCaF_3$. Phys. Rev. B24, 2577-2591.

Brown, G.E. and Fenn, P.M. (1979) Structure energies of the alkali feldspars. Phys. Chem. Minerals 4, 83-100.

Brown, I.D. and Shannon, R.D. (1973) Empirical bond-strength-bond-length curves for oxides. Acta Crystallogr. A29, 266-282.

Burnham, C.W. (1973) Order-disorder relationships in some rock-forming silicate minerals. Ann. Rev. Earth Planet. Sci. 1, 313-338.

________, Clark, J.R., Papike, J.J., and Prewitt, C.T. (1967) A proposed crystallographic nomenclature for clinopyroxene structures. Z. Kristallogr. 125, 109-119.

________ and Post, J.E. (1983) Modified electron gas (MEG) calculations: A panacea for mineral structure energetics and modeling? Geol. Soc. Amer. Abstr. Progr. 15, 537.

Busing, W.R. (1970) An interpretation of the structures of alkaline earth chlorides in terms of interionic forces. Trans. Amer. Crystallogr. Assoc. 6, 57-72.

________ (1981) WMIN, a computer program to model molecules and crystals in terms of potential energy functions. U.S. Nat'l Technical Info. Service, ORNL-5747.

Cameron, M., Sueno, S., Prewitt, C.T., and Papike, J.J. (1973) High-temperature crystal chemistry of acmite, diopside, hedenbergite, jadeite, spodumene, and ureyite. Amer. Mineral. 58, 594-618.

Catlow, C.R.A. and James, R. (1982) Disorder in TiO_{2-x}. Proc. Royal Soc. London, A384, 157-173.

________, Dixon, M., and Mackrodt, W.C. (1982a) Interionic potentials in ionic solids. In Catlow, C.R.A., and Mackrodt, W.C., Eds., Computer Simulation of Solids, Springer-Verlag, Berlin, p. 115-161.

________, Thomas, J.M., Parker, S.C., and Jefferson, D.A. (1982b) Simulating silicate structures and the structural chemistry of pyroxenoids. Nature 295, 658-662.

Cohen, A.J. and Gordon, R.G. (1975) Theory of the lattice energy, equilibrium structure, elastic constants, and pressure-induced phase transitions in alkali-halide crystals. Phys. Rev. B12, 3228-3241.

________ and ________ (1976) Modified electron gas study of the stability, elastic properties, and high pressure behavior of MgO and CaO crystals. Phys. Rev. B14, 4593-4605.

Cohen, R.E. (1984) Statistical mechanics of aluminous pyroxenes: effects of short-range order on thermodynamic properties. Geol. Soc. Amer. Abstr. Progr. 16, 474.

________ and Burnham, C.W. (1985) Energetics of ordering in aluminous pyroxenes. Amer. Mineral. 70 (in press).

Curtis, L., Gittins, J., Kocman, V., Rucklidge, J.C., Hawthorne, F.C., and Ferguson, R.B. (1975) Two crystal structure refinements of a P2/n titantian ferro-omphacite. Can. Mineral. 13, 62-67.

Davidson, P.M., Grover, J., and Lindsley, D.H. (1982) $(Ca,Mg)_2Si_2O_6$ clinopyroxenes: A solution model based on nonconvergent site-disorder. Contrib. Mineral. Petrol. 80, 88-102

Dempsey, M.J. and Strens, R.G.J. (1976) Modelling crystal structures. In Strens, R.G.J., Ed., Physics and Chemistry of Minerals and Rocks, Wiley, London, p. 443-458.

Docka, J.E., Post, J.E., Bish, D.L., and Burnham, C.W. (1985) Positional disorder in the A site of C2/m clinoamphibole: model energy calculations and probability studies. (In preparation).
Dollase, W.A. and Baur, W.H. (1976) The superstructure of meteoritic low tridymite solved by computer simulation. Amer. Mineral. 61, 971-978.
Evans, R.D. (1966) An Introduction to Crystal Chemistry, 2nd ed. Cambridge University Press, New York.
Ewald, P.P. (1921) The calculation of optical and electrostatic lattice potentials. Ann. Phys. (Leipzig) 64, 253-287.
Fujino, K., Sasaki, S., Takeuchi, Y., and Sadanaga, R. (1981) X-ray determination of electron distribution in forsterite, fayalite and tephroite. Acta Crystallogr. B37, 513-518.
Gibbs, G.V. (1982) Molecules as models for bonding in silicates. Amer. Mineral. 67, 421-450.
Giese, R.F. Jr. (1975) Electrostatic energy of columbite/ixiolite. Nature 256, 31-32.
_______ (1978) The electrostatic interlayer forces of layer structure minerals. Clays and Clay Minerals 26, 51-57.
_______ (1984) Electrostatic energy models of micas. In Bailey, S.W., Ed., Micas. Reviews in Mineralogy 13, 105-144.
Gilbert, T.L. (1968) Soft-sphere model for closed-shell atoms and ions. J. Chem. Phys. 49, 2640-2642.
Gordon, R.G. and Kim, Y.S. (1972) Theory for the forces between closed-shell atoms and molecules. J. Chem. Phys. 56, 3122-3133
Grove, T.L. and Burnham, C.W. (1974) Al-Si disorder in calcium tschermak's pyroxene, $CaAl_2SiO_6$. Trans. Amer. Geophys. Union 55, 1202.
Hazen, R.M. (1976) Effects of temperature and pressure on the crystal structure of forsterite. Amer. Mineral. 61, 1280-1293.
_______ and Finger, L.W. (1982) Comparative Crystal Chemistry. Wiley, New York.
Hemley, R.J. and Gordon, R.G. (1985) Theoretical study of NaF and NaCl at high pressures and temperatures. J. Geophys. Res. (in press.)
_______, Jackson, M., and Gordon, R.G. (1985) First-principles theory for the equations of state of minerals to high pressures and temperatures. Geophys. Res. Ltrs. (submitted).
Iishi, K. (1978) Lattice dynamics of forsterite. Amer. Mineral. 63, 1198-1208.
Jeanloz, R., Ahrens, T.J., Mao, H.K., and Bell, P.M. (1979) B1-B2 transition in calcium oxide from shock-wave and diamond-cell experiments. Science 206, 829-830.
Jones, R.E. and Templeton, D.H. (1956) Optimum atomic shape for Bertaut series. J. Chem. Phys. 25, 1062-1063.
Kamb, B. (1968) Structural basis of the olivine-spinel stability. Amer. Mineral. 53, 1439-1455.
Khan, A.A. (1976) Computer simulation of the thermal expansion behavior of some non-cubic crystals: forsterite, anhydrite and scheelite. Acta Crystallogr. A32, 11-16.
Kittel, C. (1971) Introduction to Solid State Physics, 4th ed. John Wiley and Sons, New York, NY.
Lasaga, A.C. (1980) Defect calculations in silicates: olivine. Amer. Mineral. 65, 1237-1248.
LeSar, R. and Gordon, R.G. (1982) Electron-gas model for molecular crystals: Application to the alkali and alkaline-earth hydroxides. Phys. Rev. B25, 7221-7237.
Matsui, M. and Busing, W.R. (1984a) Computational modeling of the structure and elastic constants of the olivine and the spinel forms of Mg_2SiO_4. Phys. Chem. Minerals 11, 55-59.
_______ and _______ (1984b) Calculation of the elastic constants and high-pressure properties of diopside, $CaMgSi_2O_6$. Amer. Mineral. 69, 1090-1095.
Meagher, E.P. (1980) Stereochemistry and energies of single two-repeat silicate chains. Amer. Mineral. 65, 746-755.
Meier, W.M. and Villiger, H. (1969) Die Methode der Abstandsvergeinerung zur Bestimmung der Atomkoordinaten idealisierter Gerustrukturen. Z. Kristallogr. 129, 411-423.
Miyamoto, M. and Takeda, H. (1984) An attempt to simulate high pressure structures of Mg-silicates by an energy minimization method. Amer. Mineral. 69, 711-718.
Muhlhausen, C. and Gordon, R.G. (1981a) Electron-gas theory of ionic crystals, including many-body effects. Phys. Rev. B23, 900-923.
_______ and _______ (1981b) Density-functional theory for the energy of crystals: test of the ionic model. Phys. Rev. B24, 2147-2160.
Navrotsky, A. and Kleppa, O.J. (1967) Enthalpy of the anatase-rutile transformation. J. Amer. Ceram. Soc. 50, 626.
Ohashi, Y. and Burnham, C.W. (1972) Electrostatic and repulsive energies of the M1 and M2 cation sites in pyroxenes. J. Geophys. Res. 77, 5761-5766.
Okamura, F.P., Ghose, S., and Ohashi, H. (1974) Structure and crystal chemistry of calcium tschermak's pyroxene, $CaAlAlSiO_6$. Amer. Mineral. 59, 549-557.

Pauling, L. (1960) The Nature of the Chemical Bond, 3rd ed. Cornell Univ. Press, Ithica, NY.
Post, J.E. and Burnham, C.W. (1983) Modeling tunnel cation displacements in hollandites using structure energy calculations. Geol. Soc. Amer. Abstr. Progr. 15, 663.
________ and ________ (1984) Disordering in high albite: insights from electrostatic energy minimizations. Geol. Soc. Amer. Abstr. Progr. 16, 625.
________ and ________ (1985) Ionic modeling of mineral structures and energies in the electron gas approximation: TiO_2 polymorphs, quartz, forsterite, diopside. Amer. Mineral. (submitted).
Prewitt, C.T., Sueno, S., and Papike, J.J. (1976) The crystal structures of high albite and monalbite at high temperatures. Amer. Mineral. 61, 1213-1225.
Price, G.D. and Parker, S.C. (1984) Computer simulation of the structural and physical properties of the olivine and spinel polymorphs of Mg_2SiO_4. Phys. Chem. Minerals 10, 209-216.
Ribbe, P.H., Megaw, H.D., and Taylor, W.H. (1969) The albite structures. Acta Crystallogr. B25, 1503-1518.
Sasaki, S., Prewitt, C.T., Sato, Y., and Ito, E. (1982) Single-crystal x-ray study of γ-Mg_2SiO_4. J. Geophys. Res. 87, 7829-7832.
Sato, Y. and Jeanloz, R. (1981) Phase transition in SrO. J. Geophys. Res. 86, 11773-11778.
Shannon, R.D. (1976) Revised effective ionic radii and systematic studies of interatomic distances in halides and chalcogenides. Acta Crystallogr. A32, 751-767.
Shoemaker, C.B. and Shoemaker, D.P. (1967) The crystal structure of the M phase, Nb-Ni-Al. Acta Crystallogr. 23, 231-238.
Tang, K.T. and Toennies, J.P. (1984) An improved simple model for the van der Waals potential based on universal damping functions for the dispersion coefficients. J. Chem. Phys. 80, 3726-3741.
Templeton, D.H. and Johnson, Q.C. (1961) Computation of Madelung sum and crystal energies. In Pepinsky, R., and Robertson, J.M., Ed., Computing Methods and the Phase Problem in X-ray Crystal Analysis. Pergamon, New York, p. 150-153.
Tossell, J.A. (1980a) Calculation of bond distances and heats of formation for BeO, MgO, SiO_2, TiO_2, FeO, and ZnO using the ionic model. Amer. Mineral. 65, 163-1737.
________ (1980b) Theoretical study of structures, stabilities, and phase transitions in some metal dihalide and dioxide polymorphs. J. Geophys. Res. 85, 6456-6460.
________ (1985) Ab initio SCF MO and modified electron gas studies of electron deficient anions and ion pairs in mineral structures. Physica (in press).
Waldman, M. and Gordon, R.G. (1979) Scaled electron gas approximation for intermolecular forces. J. Chem. Phys. 71, 1325-1329.
Winter, J.K., Okamura, F.P., and Ghose, S. (1979) A high-temperature structural study of high albite, monalbite, and the analbite -> monalbite phase transition. Amer. Mineral. 64, 409-423.

Chapter 11. Raymond Jeanloz

THERMODYNAMICS of PHASE TRANSITIONS

1. INTRODUCTION

Most minerals transform to new phases when subjected to changes in pressure or temperature within the range that is spanned inside the Earth. The geological importance of phase transformations is that they can play a central role in the thermal and chemical evolution of a planet. For example, the formation of melt is thought to be the fundamental process by which the planetary interior differentiates chemically. Also, due to the potentially high mobility of the liquid, partial melting can significantly enhance the outward transfer of heat, hence the long-term cooling of the planetary body (BVSP, 1981).

Solid-state transformations may be tectonically important, as well. For example, the basalt-eclogite and olivine-spinel transitions are expected to increase markedly the driving force for subduction of slabs (Ahrens and Schubert, 1975; Schubert, et al., 1975). As this is one of the dominant forces in the plate-tectonic cycling of the crust and underlying mantle which make up the lithosphere, the occurrence and kinetics of these transformations are likely to play an important role in the geological evolution of the Earth. In contrast, the basalt-eclogite transition is not expected to occur within Venus due to the high surface temperature that is present, and this may explain why subduction and Earth-like plate tectonics are not observed on this planet (Anderson, 1981). Furthermore, it is well known that the geophysically observed structure of the Earth's interior is largely determined by phase transitions (both solid-solid and solid-liquid) which are therefore central to any understanding of the present state and geological history of our planet (Ringwood, 1975).

Aside from the geological motivation, there is substantial crystal-chemical interest in studying phase transformations because of the insights these provide into what determines the stability of crystal structures: under given conditions, why does a particular structure exist, and what determines its chemical and physical properties?

On the one hand, it is possible to associate specific types of bonding with particular crystal structures. A case in point is the relation between covalency and tetrahedral coordination (Harrison, 1980). A mineralogical illustration is that tetrahedrally coordinated SiO_2 (e.g., quartz) is highly covalent, but is expected to become significantly more ionic upon transformation to the octahedrally coordinated high-pressure phase stishovite. If bonding character remains constant, on the other hand, it is possible to understand the different thermodynamic properties of two phases in terms of changes in the atomic packing geometry across the transformation. Hence, the relative entropies of different metallic and ionic structures can be explained to a large degree in terms of coordination geometries (Zener, 1947; Jeanloz, 1982).

The purpose of this chapter is to present a conceptually simple overview of equilibrium phase transitions in minerals. Thus, it is the latter, crystal chemical aspect of mineral transformations that is emphasized, and the geological implications are relegated to a few examples. In many cases, bonding character does not change significantly across a phase transition and, because bonding is fundamentally a quantum mechanical problem, changes in bond character are ignored. Rather, it is the influence of atomic packing geometry on the thermodynamic properties of mineral structures which is specifically examined. As the relationship between mineral structure and properties is well illustrated by way of discontinuous (first order) transformations, continuous (higher order) transitions are not addressed because these involve a level of dynamics well beyond the scope of this discussion (e.g., Stanley, 1971; Goodstein, 1975; Carpenter, this volume).[1] Nevertheless, a useful (but admittedly simplistic) picture emerges which allows phase diagrams to be interpreted in terms of physical properties that are determined at the atomic scale.

The approach presented here is simplified, but its main purpose is to shed light on broad trends observed in phase transitions that are common among minerals. The main emphasis is on understanding the general physical processes that determine the thermodynamic changes across phase transitions, and only rough estimates are made of the magnitudes involved. To obtain quantitative results for specific cases, more refined and less general arguments are necessary. This difference in emphasis between the present chapter and others in this volume (e.g., those by Kieffer or Navrotsky) explains some of the apparent differences in the statements made and the equations used.

2. ENERGETICS OF MINERAL TRANSFORMATION

Starting with a solid, either solid-solid or solid-liquid transformations can occur with increasing temperature, but only solid-solid transitions generally occur with increasing pressure (the exception of melting with a volume decrease is not common). The former are driven by the entropy change at the transition, ΔS_{tr}, whereas the latter are driven by the volume change, ΔV_{tr}. That is, the high-temperature phase has a

[1] Order-disorder transitions that are often of mineralogical and occasionally of geophysical interest can, to a large degree, be successfully modeled in terms of "static" statistical mechanics, which is in the same spirit as the discussion given here (Christian, 1975; Jackson et al., 1974; Navrotsky, 1977; Putnis and McConnell, 1980).

larger entropy and the high-pressure phase has a smaller volume than the starting phase. In general, both the volume change and the entropy change can come into play, and the pressure (P)-temperature (T) slope of the coexistence boundary for two phases is given by the Clapeyron equation:

$$\frac{dP}{dT} = \frac{\Delta S_{tr}}{\Delta V_{tr}} \quad . \tag{1}$$

In actuality, one or the other, ΔS_{tr} or ΔV_{tr}, often dominates, so that transitions can be thought of as either due to high pressure ($\Delta S_{tr} \sim 0$) or high temperature ($\Delta V_{tr} \sim 0$), as is illustrated in Figure 1. In fact, the idea that phase boundaries have a significant but finite slope is largely a bias due to the low pressures at which most experiments are conducted ($P \lesssim 0.1\, K_0$, as described below): when the pressure axis is properly scaled, the phase boundaries essentially parallel either the pressure or temperature axis, as illustrated schematically in the figure. This conclusion is more quantitatively substantiated below.

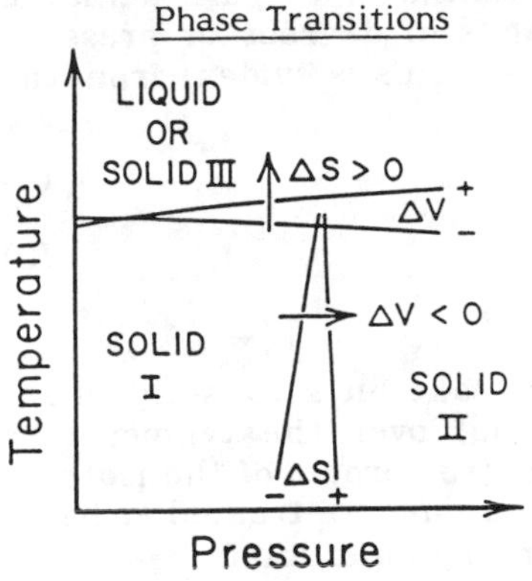

Figure 1. Schematic pressure-temperature phase diagram illustrating the difference between pressure-induced transitions, which are driven by $-\Delta V_{tr}$ (Solid I → Solid II), and temperature-induced transitions, which are driven by ΔS_{tr} (Solid I → Liquid or Solid III). Both positive and negative Clapeyron slopes are possible for the phase boundaries at either type of transition, depending on the signs (+ or −) of the entropy and volume changes, respectively (Equation 1).

Separating the contributions from the volume change and the entropy change is also useful for considering phase transitions at a microscopic level. Specifically, a volume decrease (pressure-induced transition) involves packing the atoms more tightly in the high-pressure phase. Entropy, however, is determined by the vibrational frequencies (hence, interatomic bond strengths and distances) and also by the atomic packing geometry. For example, a major contribution to the entropy increase on melting is due to the fact that the melt is disordered relative to the order that characterizes the crystalline solid. Thus, given two structures (not necessarily with crystalline order), the changes in thermal properties, including entropy, can be understood fairly simply.

To give an idea of the energetics involved in phase transitions, consider first a pressure-induced transformation (see also Navrotsky, this volume). The internal energy change is

$$\Delta U_{tr} \sim - P_{tr} \Delta V_{tr} \quad , \tag{2}$$

assuming a negligible entropy change. Typical transformation pressures and volume changes for silicates are of the order of 10 GPa (10^2 kbar) and 0.5 cm^3/atomol (i.e., several percent relative change in volume; note that 1 cm^3 = 1 J/MPa).[2] Thus, energy changes of about 5 to 10 kJ/atomol, equivalent to about 0.1 eV or 10^3 K temperature, are involved in high-pressure transformations.[3] Similarly, the energy on melting,

$$\Delta U_{tr} \sim T_{tr} \Delta S_{tr} \tag{3}$$

(assuming negligible volume change), is determined by the entropy change, which is roughly found to be of the order of the gas constant (R = 8.314 J/K·atomol). As melting temperatures are in the range of 10^3 K, the energy change is again found to be around 10 kJ/atomol, or 0.1 eV.

Energy changes of several kJ/atomol (tenths of eV) are achieved with either temperature changes of thousands of degrees or pressure changes of tens of GPa (hundreds of kilobars). This is evident from the differential for the internal energy

$$\Delta U(T,V) = \int C_V dT - \int P dV \tag{4}$$

$$\cong 3R\Delta T + \frac{(\Delta P)^2 V}{K_T} \quad , \tag{5}$$

in which the specific heat, C_V, is assumed to take on a constant, high-temperature value of 3R (Dulong-Petit value) over the temperature interval ΔT, and the isothermal bulk modulus (reciprocal of the isothermal compressibility, $K_T = - V(\partial P/\partial V)_T$) and volume are treated as being constant over the pressure interval ΔP in deriving Equation 5.

Substituting typical values for silicates ($V \sim 5$ cm^3/atomol, $K_T \sim 100$ GPa) shows that ΔU changes by kJ/atomol if either $\Delta T \sim 10^3$ K or $\Delta P \sim$ 10 GPa. Pressures and temperatures of this magnitude occur already in the upper mantle, that is, in the outermost region of the Earth. In this sense, the conditions of the deep planetary interior represent a significant thermodynamic perturbation of minerals, which is comparable to that associated with major phase transformations.

---

[2] Throughout this paper, "atomol" is used as an abbreviation for "mole of atoms". If one does not distinguish between the different atoms making up a compound, "gram-atom" is an equivalent term.

[3] One electron volt (eV; implicitly, per atom) is equal to 96.5 kJ/atomol, and corresponds to 11605 K (energy E = kT with k being Boltzmann's constant) or 8066 cm^{-1} (energy E = h $c_l \bar{\nu}$, with h and c_l being Planck's constant and the velocity of light; spectroscopists refer to the frequency $\bar{\nu}$ as wavenumber).

3. SOLID-SOLID TRANSFORMATIONS: VIBRATIONAL ENTROPY

All solid-solid transformations can be thought of as being induced by pressure at 0 K. In Figure 1, this is schematically evident for the I-II transition, as well as the I-III transition with negative Clapeyron slope (the 0 K transition pressure would be very high). The I-III transition with positive Clapeyron slope may only be accessible at a (hypothetical) negative pressure at 0 K. In all cases, however, the entropies of the phases and ΔS_{tr} vanish at 0 K, thus leading to a vanishing Clapeyron slope (Eqn. 1).

Evidently, the Clapeyron slope at higher temperatures is determined by the temperature dependencies of ΔS_{tr} and ΔV_{tr}. The volume change is unlikely to be very dependent on temperature because it is determined by ΔV_{tr} at 0 K and by the relative magnitudes of the thermal expansion coefficients $\left[\alpha = (1/V)(\partial V/\partial T)_P\right]$ of the two phases. As $\alpha \sim 1$ to 5×10^{-5} K^{-1} at zero pressure and high temperature, a 1000 K change in temperature causes only a few percent change in volume. The differential volume change across the transition is much smaller yet, typically amounting to a 20 percent (or less) change in $\Delta V_{tr}/V$ over 1000 K. Also, α decreases with increasing pressure or decreasing temperature (α vanishes at 0 K), so the effect of temperature on ΔV_{tr} can be ignored for this general discussion (some caution is required, however, in case ΔV_{tr} is very small or α changes by a large amount across the transition).

Consequently, the Clapeyron slope of solid–solid phase transitions is mainly determined by the entropy change across the transition, which reflects the differences in the lattice–vibrational spectra of the phases. Instead of considering the full vibrational spectrum, however, we simplify the analysis by describing the solid in terms of a single, harmonic vibrational frequency, ν, which is an average of all the frequencies actually present in the spectrum. In terms of this Einstein model, the entropy of phase I (denoted by subscript 1) at temperature T is (e.g., Slater, 1939; Wallace, 1972)

$$\frac{S_1}{3R} = \left(\frac{\Theta_1}{T}\right)\left[\exp\left(\frac{\Theta_1}{T}\right) - 1\right]^{-1} - \ln\left[1 - \exp\left(\frac{-\Theta_1}{T}\right)\right] \quad , \qquad (6)$$

with the characteristic temperature $\Theta = h\nu/k$ (h is Planck's constant). Note that according to Equation 6, S increases with increasing T/Θ, and therefore with either increasing T or decreasing Θ. For reference, the average frequency observed in a typical vibrational spectrum of a mineral is $\bar{\nu} \cong 500$ cm^{-1}, which corresponds to $\Theta = 720$ K.[4]

[4] Note that the Debye temperature, Θ_D, is related to the characteristic temperature of this model by $\Theta_D = \sqrt{5/3}\,\Theta$ (Wallace, 1972). The difference between Θ and Θ_D is due to the different ways in which the frequencies are averaged in the two models, and results in $\Theta_D = 930$ K in the present case. The frequencies ν and $\bar{\nu}$ are in Hertz (cycles per second) and cm^{-1}, respectively: $\nu = c_l \bar{\nu}$.

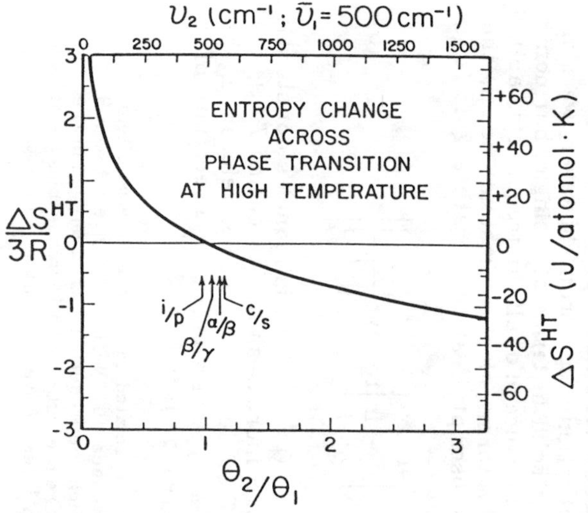

Figure 2 (left). The high-temperature entropy change across a pressure-induced transition (ΔS_{tr} in J K^{-1} $atomol^{-1}$ or $\Delta S_{tr}/3R$ on a $atomol^{-1}$ basis) is shown as a function of the ratio of characteristic temperatures, Θ_2/Θ_1, as given in Equation 7. The average vibrational frequency for the high-pressure phase is given on the upper scale, assuming an average vibrational frequency $\bar{\nu}_1 = k\Theta_1/hc_l = 500\ cm^{-1}$ (see footnote 3 and text). Note that the entropy change is negative (positive Clapeyron slope) for an increase in vibrational frequency and positive (negative Clapeyron slope) for a decrease in vibrational frequency across the transition. Based on experimentally determined entropies of transformation, typical values of Θ_2/Θ_1 are shown for high-pressure transitions in silicates: olivine (α) → β-phase (β) → γ-spinel (γ) Mg_2SiO_4, coesite (c) → stishovite (s) SiO_2 and ilmenite (i) → perovskite (p) $MgSiO_3$ (Data sources: Ito and Yamada, 1982; Jeanloz and Thompson, 1983; Akaogi et al., 1984; Akaogi and Navrotsky, 1984).

Figure 3. Entropy change across a pressure-induced transition, ΔS_{tr}, shown as a function of normalized temperature, T/Θ_1, for several values of Θ_2/Θ_1 (see Equation 6). (a) The entropy change is shown as in Figure 2 with Θ_1 = 720 K (equivalent to $\bar{\nu}_1 = 500\ cm^{-1}$) being assumed for the upper scale. (b) The entropy change is shown normalized to its high-temperature value, ΔS_{tr}^{HT} (see Figure 2), in order to illustrate the similarity with a specific heat curve (C_V versus T/Θ).

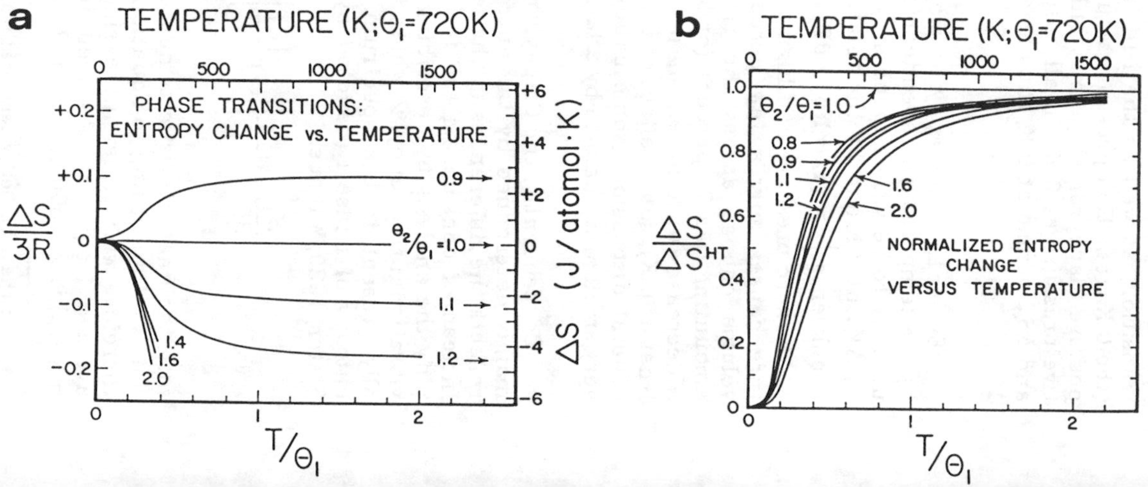

Equation 6 can be differentiated to show that $(1/S)\ (\partial S/\partial T)_V \sim 10^{-3}\ K^{-1}$ at elevated temperatures ($T \gtrsim \Theta$), which is about two orders of magnitude greater an effect of temperature on entropy than was found on volume. At low temperatures, the relative change of entropy with temperature is greater yet (exceeding $10^{-2}\ K^{-1}$ below 100 K), so the assumption that ΔV_{tr} is essentially constant in comparison with ΔS_{tr} is a safe one over a wide temperature range.

It is worth detailing, at this point, the assumptions inherent in the present use of Equation 6. Most important is the fact that only harmonic, vibrational contributions to the entropy are considered. Electronic, magnetic and order-disorder contributions are not included, for example. Also, intrinsic anharmonicity of the lattice vibrations is ignored, although the volume dependence of Θ that will be taken into account in the subsequent discussion (see Eqn. 12) allows for the most significant nonharmonic effects to be included (this emulation of the true lattice vibrations by a volume-dependent harmonic model is called the quasiharmonic approximation). Finally, details of the temperature-dependent thermodynamic properties are not expected to be quantitatively reproduced, particularly at low temperatures. This is because, in contrast to Kieffer's model (Kieffer, this volume), the whole vibrational spectrum is being approximated by a single, average frequency in the present analysis. Nevertheless, thermodynamic properties become insensitive to the details of the vibrational spectrum at high temperatures ($T \gtrsim \Theta$), and despite its crudeness, the present model contains many of the trends observed for phase transitions.

In terms of the present model, ΔS_{tr} is determined by the temperature at the transformation and by the ratio of the characteristic temperatures of the two phases involved, Θ_2/Θ_1 (the ratio of Debye temperatures would be equivalent, at this level of approximation). What, then, are typical values of these parameters for high-pressure mineral transformations?

By expanding Equation 6 for large values of T/Θ, the entropy change at high temperatures is given by

$$\frac{\Delta S_{tr}^{HT}}{3R} = \ln\left(\frac{\Theta_1}{\Theta_2}\right) \tag{7}$$

to first order, with ΔS being $S_2 - S_1$. This relationship can be used to derive the change in average vibrational frequency (or characteristic temperature) across a high-pressure mineral transformation. A few specific examples are shown in Figure 2, and a more complete description of extracting average vibrational frequencies from high-temperature entropy data for crystals is given by Wallace (1972). Note that whether a change of coordination occurs ($MgSiO_3$ ilmenite → perovskite; SiO_2 coesite → stishovite) or does not occur (Mg_2SiO_4 olivine → β—phase → γ—spinel), the average frequency typically changes by less than 10 percent.

Intuitively, an increase in Θ (or average vibrational frequency) can be associated with an increase in the tightness of bonding and therefore a decrease in entropy, as would be found under compression. For example, simulating the interatomic bond by a one-dimensional harmonic oscillator with force constant (bond strength) k and mass m (average atomic mass), the vibrational frequency is (e.g., Slater, 1939; Moore, 1972)

$$\nu = \sqrt{\frac{k}{m}} \quad . \tag{8}$$

As the average atomic mass remains constant across a transition from one polymorph to another, the change in Θ or ν can be directly correlated with a change in effective bond strength. Thus, a negative Clapeyron slope (positive ΔS_{tr}) is associated with a decrease in Θ and effective bond strength k across the transition.

The temperature dependence of the Clapeyron slope (i.e., of ΔS_{tr} in the present model) is given by Equation 6. Starting from zero at 0 K, the Clapeyron slope increases in magnitude to the high-temperature limiting value which is determined by the ratio Θ_2/Θ_1 (Fig. 3). The maximum temperature dependence of the phase boundary (maximum absolute value of ΔS_{tr}) is essentially achieved by 700 K ($T/\Theta \cong 1$) for typical mineral transformations. Above this temperature, a constant Clapeyron slope is expected.

At temperatures below Θ, the curvature in the phase-transition boundary, or the temperature dependence of ΔS_{tr}, reflects the quantized nature of the lattice vibrations in the crystal structures. In fact, the normalized curves of entropy change resemble typical specific heat curves (Fig. 3b) because the entropy is just a function of a single parameter (the nondimensional temperature $\tau = T/\Theta$) in this model. Thus, for a small perturbation in Θ (e.g., $\delta\Theta/\Theta \lesssim 10$ percent), the corresponding change in entropy at a given temperature is

$$\delta S \cong -C_V \frac{\delta\Theta}{\Theta} \quad , \tag{9}$$

because

$$\left(\frac{\partial S}{\partial \Theta}\right)_T = \left(\frac{\partial \tau}{\partial \Theta}\right)_T \frac{dS}{d\tau} = \left(\frac{\partial \tau}{\partial \Theta}\right)_T \left(\frac{\partial T}{\partial \tau}\right)_V \left(\frac{\partial S}{\partial T}\right)_V \tag{10a}$$

$$= \left(\frac{-T}{\Theta^2}\right)(\Theta)\frac{C_V}{T} = -\frac{C_V}{\Theta} \quad . \tag{10b}$$

To summarize, the Clapeyron slope for a solid-solid transition is expected to increase from zero at 0 K to a constant limiting value at high temperature. The ratio of the average vibrational frequencies or effective bond strengths of the two structures can be immediately derived from the slope of the straight-line phase-transition boundary at elevated temperatures. Also, the slope of the phase-transition boundary is expected to vary with temperature for $T < \Theta$, unless $\Theta_1 = \Theta_2$ (i.e., $\Delta S_{tr} = 0$ throughout). In principle, the absolute values of Θ_1 and Θ_2 (not just the ratio) could be determined from the curvature of the phase boundary in this low-temperature regime (Fig. 3). In practice, however, sluggish kinetics are likely to mask the curvature. For this reason, Clapeyron slopes are usually well determined only at high temperatures, which explains why observed P-T phase boundaries are commonly straight lines for solid-state transformations.

The present model is limited in that only the harmonic vibrational contribution to the entropy is considered. Thus, deviations from harmonicity (e.g., a force constant k which depends on interatomic distance) and order-disorder phenomena (including defect reactions) would generally lead to curvature in the phase boundary at high temperatures, as would a variation of ΔV_{tr} with temperature. Given the precision with which most phase diagrams are known, however, these effects are small.

4. GEOMETRICAL ENTROPY AND MELTING

Most solid—solid transformations can be thought of as being driven by the volume change. As is evident from Figure 2, the entropy change is relatively small and can be understood fairly simply in terms of a shift in lattice vibrational frequencies across the transition. For a temperature-induced transition such as melting, the entropy change must be looked at in greater detail, however. It is clear, for example, that the disorder of the liquid relative to the crystal makes up a significant part of the entropy change.

One convenient way to examine the entropy change of transition is to separate it into two components:

$$\Delta S_{tr} = \int_{V_1}^{V_2} \left(\frac{\partial S}{\partial V}\right)_T dV + \Delta S_{geom} \quad . \tag{11}$$

The first term reflects the fact that entropy depends on volume, so one part of ΔS_{tr} is due to the fact that the volume changes across the transformation. This does not make up all of the observed ΔS_{tr}, however, and the second term arises from the fact that the microscopic geometrical configuration of the atoms change at the phase transition.

Loss of crystalline order upon melting, changes of coordination and even subtle changes of atomic packing across solid-state transformations are all examples of the contributions to this geometrical entropy change. As formally defined in Equation 11, ΔS_{geom} is that part of the transformation entropy that cannot be explained by the volume change ΔV_{tr}.

The change of entropy with volume can be partially understood in terms of the effect of compression on bonding forces and vibrational frequencies. Thus, just as a crystal becomes less compressible under increasing pressure, the effective "spring constant" k (Eqn. 8) would be expected to increase and the bond become increasingly stiff with pressure. The average vibrational frequency therefore increases with pressure (Eqn. 8), and the Grüneisen parameter

$$\gamma = -\left(\frac{\partial \ln \nu}{\partial \ln V}\right)_T = -\left(\frac{\partial \ln \Theta}{\partial \ln V}\right)_T \tag{12}$$

is positive. As the vibrational entropy is taken to be a function of T and Θ only (Eqn. 6), the effect of compression on the entropy of a crystal is simply

$$\left(\frac{\partial S}{\partial V}\right)_T = \left(\frac{\partial S}{\partial \Theta}\right)_T \left(\frac{\partial \Theta}{\partial V}\right)_T \tag{13a}$$

$$= -\frac{C_V}{\Theta}\left(\frac{\partial \Theta}{\partial V}\right)_T \tag{13b}$$

$$= \frac{\gamma C_V}{V} \quad , \tag{13c}$$

where Equations 10 and 12 are used to derive lines 13b and 13c.

As noted above, the entropy is expected to decrease with increasing pressure (decreasing volume), which is what is predicted from Equation 13 because γ, C_V and V are all positive. It is clear, therefore, that at least part of the entropy change on transformation can be attributed to the effect of the volume change on the vibrational frequencies. The magnitude of this effect can be estimated by noting that the Maxwell relation,

$$\left(\frac{\partial S}{\partial V}\right)_T = \left(\frac{\partial P}{\partial T}\right)_V = \alpha K_T \quad , \tag{14}$$

combines with Equation 13c to yield the relative changes of frequency and volume from thermodynamic data (α, K_T, C_V, V):

$$\gamma = \frac{\alpha K_T V}{C_V} \sim 1 \text{ to } 2 \quad . \tag{15}$$

Thus, for a typical volume decrease across a solid-solid transition of 3 to 5 percent, the vibrational frequency would be expected to increase by 3 to 10 percent, which is in accord with much of the existing data. As shown in Figure 2, the observed entropy change for several transitions implies a shift of average vibrational frequency of this magnitude.

The trouble with this line of argument is that it can only explain transitions with positive Clapeyron slopes (ΔS_{tr} and ΔV_{tr} of the same sign according to Eqn. 14). In actuality, the geometrical entropy change can dominate over the first term in Equation 11 and lead to a transition entropy that is of opposite sign of the volume change. Furthermore, temperature-driven transitions such as melting typically exhibit a large entropy change, whereas the volume change can be quite small. Thus, it is worth estimating the effects of disordering and the corresponding magnitude of ΔS_{geom}. To reiterate, the geometrical entropy change, as defined here, is simply that part of ΔS_{tr} which is not associated with the volume change upon transformation.

For any transformation, the part of ΔS_{tr} due to the volume change ΔV_{tr} can be approximately determined by evaluating the volume dependence of the entropy in one phase or the other (Fig. 4):

$$\Delta S)_T \cong \gamma C_V \frac{\Delta V_{tr}}{V} \quad . \tag{16}$$

The remaining entropy change at constant volume, $\Delta S)_{V,T}$, gives the geometrical entropy

$$\Delta S_{geom} \cong \Delta S)_{V,T} = \Delta S_{tr} - \gamma C_V \frac{\Delta V_{tr}}{V} \tag{17}$$

which can take on positive or negative values regardless of the sign (or magnitude) of ΔV_{tr}. Although $\Delta S)_T$ is of the same sign as ΔV_{tr}, the geometrical contribution allows ΔS_{tr} to be unrelated in magnitude or sign to $\Delta S)_T$ or ΔV_{tr}. In fact, ΔS_{geom} and $\Delta S)_T$ typically cancel each other in high-pressure solid-state transformations involving a coordination increase. This is caused by the effect of changing the atomic packing geometry on the lattice vibrations, and is discussed at greater length below. In contrast, both terms $\Delta S)_T$ and ΔS_{geom} are positive upon melting (for the usual case of a positive Clapeyron slope) and ΔS_{tr} is enhanced.

To get an idea of the magnitude of the geometrical entropy change, consider the disordering associated with melting. Entropies of fusion for minerals are listed in Table 1 (normalized to R and on an atomol^{-1} basis), and the geometrical contribution has been estimated from Equation 17 with crystal values being used for C_V and γ. Because of uncertainties in the high-temperature data, as well as the approximations used in this analysis, ΔS_{geom} is probably known to no better than 10 to

Table 1. ENTROPY OF FUSION FOR MINERALS

Mineral (Composition)	T_m (K)	$\frac{\Delta V_m}{V}$	$\frac{\Delta S_m}{R}$	$\frac{\Delta S_{geom}}{R}$
Forsterite (Mg_2SiO_4)	1860	0.04	0.6-1.2 (estimated)	0.5-1.1
Fayalite (Fe_2SiO_4)	1490	0.09	1.03	0.7
Protoenstatite ($MgSiO_3$)	1840	0.15	1.01	0.4
Diopside ($CaMgSi_2O_6$)	1665	0.19	1.00	0.2
Pseudowollastonite ($CaSiO_3$)	1817	0.04	0.76	0.6
Albite ($NaAlSi_3O_8$)	1373	0.09	0.42	0.2
Anorthite ($CaAl_2Si_2O_8$)	1830	0.05	0.69	0.5
Sanidine ($KAlSi_3O_8$)	1473	0.09	0.36	0.1
Cristobalite (SiO_2)	1999	0.09	0.18	0
Halite (NaCl)	1074	0.27	1.58	0.3
Sylvite (KCl)	1043	0.21	1.52	0.6

Sources of data: Jackson (1977), Stebbins et al. (1984), Stebbins (pers. comm., 1984), Wolf and Jeanloz (1984).

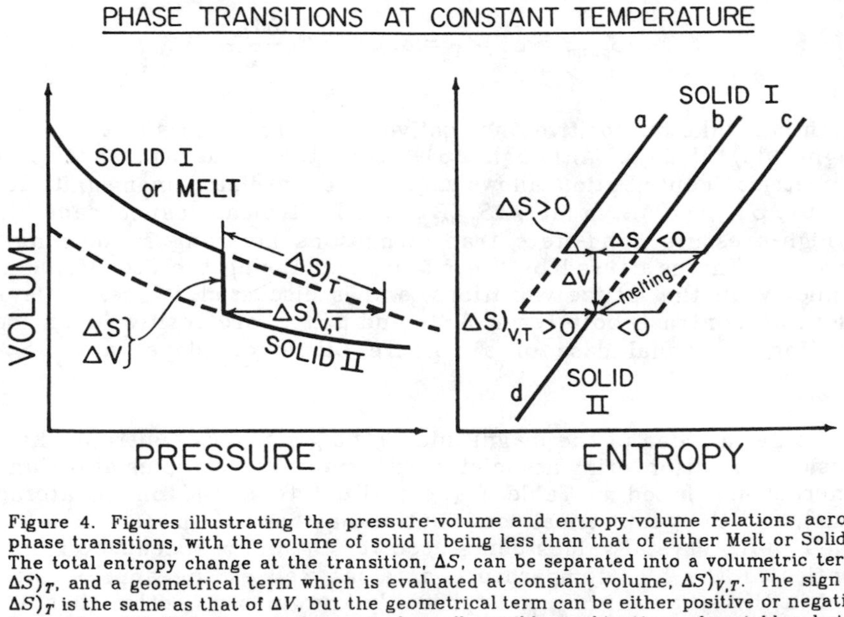

Figure 4. Figures illustrating the pressure-volume and entropy-volume relations across phase transitions, with the volume of solid II being less than that of either Melt or Solid I. The total entropy change at the transition, ΔS, can be separated into a volumetric term, $\Delta S)_T$, and a geometrical term which is evaluated at constant volume, $\Delta S)_{V,T}$. The sign of $\Delta S)_T$ is the same as that of ΔV, but the geometrical term can be either positive or negative for a given ΔV. This diagram does not show all possible combinations of variables, but is used to illustrate the separation of ΔS_{tr} as given in Equations 11 and 17.

50 percent in most cases. Still, the results are broadly consistent with expectations in that highly polymerized melts such as $NaAlSi_3O_8$ or SiO_2 are characterized by small values of ΔS_{geom} (relatively minor disordering on melting), whereas more dissociated melts (e.g., Mg_2SiO_4 and Fe_2SiO_4) are associated with large ΔS_{geom}. In the case of SiO_2, the melt is so well ordered (polymerized) that essentially the full entropy of fusion can be accounted for by the volume change on melting. Clearly, there are additional phenomena, other than the effect of volume on entropy, that could be specifically taken into account before evaluating ΔS_{geom}. For example, changes of polyhedral coordination on melting or contributions from order-disorder reactions in either solid or melt might be considered (e.g., Waseda, 1980; Stebbins et al., 1984). In the present approach, however, these nonvolumetric effects are all lumped into ΔS_{geom}.

For relatively dissociated silicate melts, the disordering entropy on fusion is roughly 0.5 R to 1 R (4 to 8 J atomol^{-1}K^{-1}). A similar analysis for metals (Oriani, 1951) also demonstrates that values in this range are appropriate for nonpolymerized (poorly structured) melts. Furthermore, Stishov (1969) has shown from analyses of shock-wave data that at high pressures a limiting value of the entropy of fusion, $\Delta S_{tr} \sim 1$ R, is achieved and can be interpreted as a disordering entropy (see also Tallon, 1982).

This range of values for the geometrical entropy change on melting makes sense based on simple models of the entropy of liquids. For example, if the liquid is thought of as an assemblage of N atoms (small spheres) and $N\omega$ holes ($\omega \ll 1$), the disordering entropy associated with the ways in which these can rearrange themselves on a lattice of $N(1+\omega)$ points (representing the liquid) is

$$\Delta S_{Disorder} \cong N\mathrm{k}[(1+\omega)\ln(1+\omega) - \omega\ln\omega] \tag{18}$$

(Slater, 1939). Equation 18 is simply based on Boltzmann's statistical mechanical definition of entropy as being determined by the number of configurations W accessible to a system:

$$S = \mathrm{k}\ln W \ , \tag{19}$$

with k being Boltzmann's constant.

The ratio of holes to atoms, ω, can be related to the ratio of the packing fractions of the liquid and solid. The packing fraction, ξ, is defined as the sum of the atomic volumes divided by the volume within which they are contained (the volume of the system), and

$$\xi_{liq} / \xi_{sol} = \frac{1}{1+\omega} \ . \tag{20}$$

For comparison, static packing fractions for closest packed (ordered), random closest packed and random loosely packed arrays of uniform spheres are 0.74, ~ 0.64 and ~ 0.60, respectively (Bernal, 1959, 1964).

The geometrical entropy change is shown as a function of ω in Figure 5, and the corresponding liquid packing fraction is indicated for a closest packed solid. It is worth noting, however, that most minerals are not closest packed: for silicates and oxides $0.31 \lesssim \xi_{sol} \lesssim 0.72$ (Fairbairn, 1943). Also, dynamic packing fractions for liquids are lower than the static values derived by Bernal. Both computer simulations and data analyses based on liquid-state theory yield packing fractions in the range $0.40 \lesssim \xi_{liq} \lesssim 0.50$ for monatomic liquids near the melting point (e.g., Stroud and Ashcroft, 1972; Shimoji, 1977).

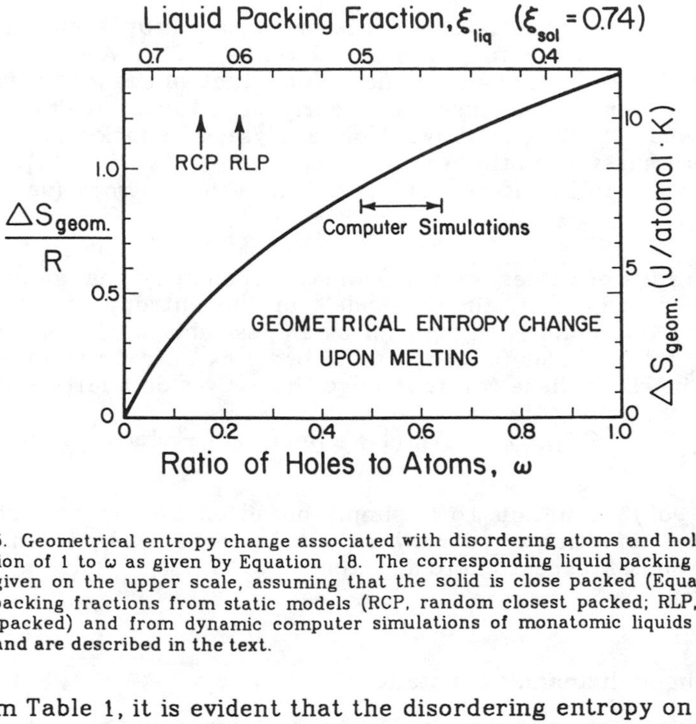

Figure 5. Geometrical entropy change associated with disordering atoms and holes in the proportion of 1 to ω as given by Equation 18. The corresponding liquid packing fraction, ξ_{liq}, is given on the upper scale, assuming that the solid is close packed (Equation 20). Liquid packing fractions from static models (RCP, random closest packed; RLP, random loosely packed) and from dynamic computer simulations of monatomic liquids are also shown, and are described in the text.

From Table 1, it is evident that the disordering entropy on fusion of minerals corresponds to ω ranging from about 0.6 for dissociated liquids to values less than 0.05 for polymerized liquids. If one uses Equation 20 and the known packing fractions of the minerals (Fairbairn, 1943), the mineral melts are in the range $0.42 \lesssim \xi_{liq} \lesssim 0.56$. This is a very crude estimate (e.g., the liquid is treated as an array of uniform spheres and holes), but it is clearly in the expected range. Thus, aside from any contribution due to the volume change on melting, the entropy of fusion is at least equal to ΔS_{geom}, which is roughly 0.5 R to 1 R for nonpolymerized melts.

5. A MODEL PHASE DIAGRAM

Unlike the effect of temperature, pressure has a relatively large effect on ΔV_{tr}. Typical values of K_T for minerals are of the order of 100 GPa (1 Mbar), so a 1 percent change in volume already occurs with a pressure variation of 1 GPa or so. Thus, pressure affects the Clapeyron slope of melting and other temperature-induced transitions both directly, by changing ΔV_{tr}, and indirectly through the change of $\Delta S)_T$ and hence ΔS_{tr} (Eqns. 16 and 17).

A pressure-volume relationship or equation of state is required in order to evaluate the change of ΔV_{tr} with pressure. The Eulerian finite-strain formalism provides the most satisfactory equation of state, as has been empirically shown in a number of studies (Birch, 1952, 1977, 1978; Jackson and Niesler, 1982; Knittle and Jeanloz, 1984; Heinz and Jeanloz, 1984a; Knittle et al., 1985b). Truncating to second order for simplicity, the normalized pressure is

$$P/K_0 = \frac{3}{2}\left[\left(\frac{V}{V_0}\right)^{-5/3} - \left(\frac{V}{V_0}\right)^{-7/3}\right] , \tag{21}$$

with subscript zero indicating zero pressure values. Equation 21 is used here under isothermal conditions, so K_0 is understood to be the isothermal bulk modulus at zero pressure.

From Equation 21, the volume of phase 1 relative to its zero-pressure value, V_1/V_{01}, is given as a function of P/K_{01}. Similarly, for phase 2, V_2/V_{02} is given as a function of P/K_{02}. Starting with a volume change at zero pressure

$$\left.\frac{\Delta V_{tr}}{V_1}\right)_{P=0} = \frac{V_{02} - V_{01}}{V_{01}} , \tag{22}$$

the high pressure value of ΔV_{tr} is given by

$$\frac{\Delta V_{tr}}{V_1} = \left.\frac{\Delta V_{tr}}{V_1}\right)_{P=0} + \frac{(V_2/V_{02}) - (V_1/V_{01})}{(V_1/V_{01})} + \cdots , \tag{23}$$

with higher order terms being negligible if $\Delta V_{tr}/V_1 \ll 1$, as is typically the case. The second term on the right side of Equation 23, and thus the change in $\Delta V_{tr}/V_1$ with pressure (i.e., with increasing P/K_{01}), can

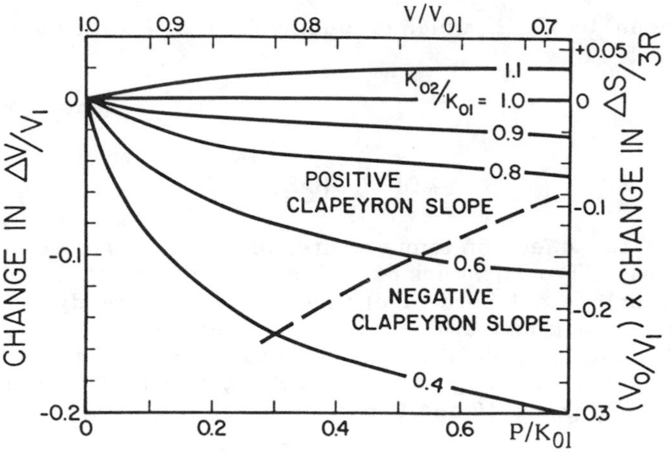

Figure 6. The effect of pressure (normalized to the bulk modulus, K_{01}) on the relative volume change across a temperature-induced transition such as melting is shown as a function of the ratio of bulk moduli, K_{02}/K_{01}. This result is based on Equations 21 to 23, and is used to model the effect of pressure on melting transitions with a positive volume change at zero pressure. The volume compression of the solid is shown on the upper scale and the change in the volumetric part of the entropy of fusion is given by the right-hand scale (see text).

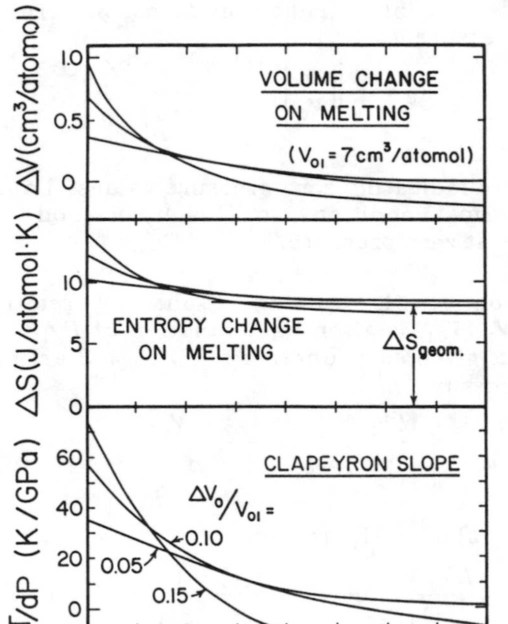

Figure 7. Model of melting transitions as a function of pressure for three values of $\Delta V_{tr}/V_1$ at zero pressure: 5, 10 and 15 percent. The volume change (upper panel), entropy change (middle panel) and Clapeyron slope (lower panel) are shown as a function of normalized pressure, P/K_{01}. Further details are given in the text.

therefore be evaluated as a function of K_{02}/K_{01} by way of Equation 21. The result is shown in Figure 6.

Corresponding to a change in $\Delta V/V$, the change in ΔS_{tr} is given by Equation 16 (the geometrical entropy change is assumed to be unaffected by pressure). In Figure 6, typical values of $\gamma \sim 1.5$ and $C_V \sim 3$ R have been assigned and there is a slight correction of roughly V_{01}/V_1 to the change in ΔS_{tr}. This last point is minor, but arises from the empirical fact that neither γ nor V are typically constant with pressure, but the ratio γ/V is (also, $C_V \sim$ constant at high temperatures in Equation 16).

The ratio K_{02}/K_{01} can be roughly correlated with the volume change $\Delta V_{tr}/V_1$ (at zero pressure) by way of velocity-density systematics (Appendix A). The result is that

$$-\frac{\Delta K/K}{\Delta V/V} \sim 4 \quad , \tag{24}$$

so the bulk modulus is expected to decrease by 20, 40 and 60 percent (K_{02}/K_{01} = 0.8, 0.6 and 0.4) for transitions with zero-pressure volume increases $\Delta V_{tr}/V_1)_{P=0}$ of 5, 10 and 15 percent respectively. Taking the last of these cases as an example, $\Delta V_{tr}/V_1$ decreases by 15 percent and hence vanishes by the pressure $P/K_{01} = 0.3$ (Fig. 6, $K_{02}/K_{01} = 0.4$). Consequently, the Clapeyron slope dT/dP changes sign above this pressure, which can be taken to be roughly 15 to 30 GPa for typical values of K_{01} (~ 50 to 100 GPa for minerals). The change-over from positive to negative Clapeyron slopes is indicated for a range of K_{02}/K_{01} (or $\Delta V_{tr}/V_1$) in Figure 6. It must be emphasized, however, that these results may not be quantitatively accurate because of the approximations that have been made (extrapolation of Equation 21 to high P/K and use of velocity-density systematics). Nevertheless, the qualitative trends are well illustrated.

Applying these results to melting, the three specific values $\Delta V_{tr}/V_1$ = 0.05, 0.10 and 0.15 are considered for the relative volume change on fusion at zero pressure. These are typical of the range observed for minerals. The volume change on melting, the entropy of fusion and the resulting Clapeyron slope are shown as functions of P/K_{01} in Figure 7. Note that a constant value ΔS_{geom} = 1 R has been included in ΔS_{tr}, based on the discussion above. Decreasing the geometrical entropy change increases the Clapeyron slope dT/dP at a given pressure. The same effect results when translating from P/K_{01} to pressure if a large value is assigned to the bulk modulus.

This analysis of melting curves can be combined with the earlier discussion of pressure-induced (solid-solid) transitions to form a generalized P-T phase diagram (Fig. 8). By integrating the Clapeyron slopes given in Figure 7, the shift in the melting temperature (T_m) is given as a function of the non-dimensional pressure P/K_{01}. For specificity, results for a zero pressure bulk modulus and melting temperature K_{01} = 100 GPa and T_m^0 = 1500 K are shown in Figure 8. Similarly, the shift in the transition pressure (P_{tr}) for a solid—solid transition is given as a function of non-dimensional temperature T/Θ (see Fig. 3) for a typical range of Θ_2/Θ_1 (Fig. 2).

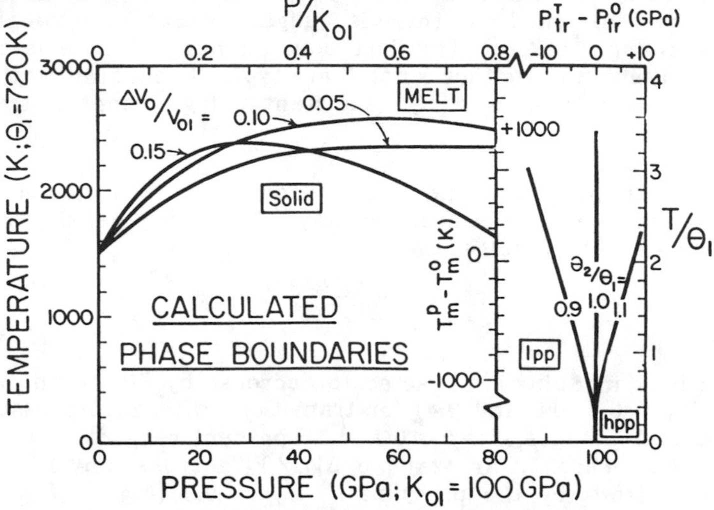

Figure 8. A model phase diagram summarizing the calculated phase boundaries between solid and melt (left-hand portion), and between low- (lpp) and high-pressure (hpp) phases (right-hand portion). For specificity, zero-pressure values for the melting point (T_m^0 = 1500 K) and solid bulk modulus (K_{01} = 100 GPa) have been assigned, as have values for the characteristic temperature of the low-pressure phase (Θ_1 = 720 K) and the zero-Kelvin transition pressure between the solid phases (P_{tr}^0 = 100 GPa). These values are typical for silicates and oxide minerals, as are the ranges shown for the volume change on melting ($\Delta V_0/V_{01} \cong$ 5 to 15 percent) and the ratio of characteristic temperatures ($0.9 \lesssim \Theta_2/\Theta_1 \lesssim 1.1$). Note the similarity of this model phase diagram to the qualitative version shown in Figure 1.

Note that from Equations 1 and 9 the change in transition pressure with temperature is

$$dP_{tr} \cong \frac{C_V}{\Delta V_{tr}} \frac{\Delta\Theta}{\Theta} dT \quad , \tag{25}$$

with Θ, ΔV_{tr} and $\Delta\Theta$ being considered independent of temperature at the present level of approximation. Integration yields the pressure at the solid-solid phase transition

$$P_{tr}^T = P_{tr}^0 + \frac{\Delta\Theta/\Theta}{\Delta V_{tr}} U(T/\Theta) \quad , \tag{26}$$

in which P_{tr}^0 is the 0 K transition pressure and $U(T/\Theta)$ is the internal energy at non-dimensional temperature T/Θ (U at T = 0 is set to zero for convenience). The functional form of a solid-solid $P-T$ boundary is therefore predicted to be the same as that for the internal energy in the present model.[5] As mentioned previously, however, only the high-temperature, straight-line segment of the U(T) curve is likely to be experimentally observed due to kinetic considerations.

The main point to note is that the slope of the melting curve is essentially determined by the equations of state of the coexisting phases, whereas the solid-state transformation boundary mostly reflects the shift in the vibrational spectrum across the transition. The former roughly parallels the pressure axis and the latter parallels the temperature axis, in accord with the conclusion that these transitions are mainly driven by ΔS_{tr} and ΔV_{tr}, respectively. This generalization tends to fail most at low pressure and high temperature; that is, the conditions at which most experimental work has been done to date. In a sense, high-pressure data are required in order to get a less biased view of the main thermodynamic factors that control phase transitions.

Only fusion curves with $\Delta V_{tr} > 0$ at zero pressure have been considered in the discussion so far. In the present model, a negative dT/dP Clapeyron slope ($\Delta V_{tr} < 0$) at zero pressure would flatten and eventually become positive at elevated pressures because K_{02}/K_{01} would be greater than 1.0 (Fig. 6). Obviously, $\Delta V_0/V_{01} = 0$ would lead to a zero Clapeyron slope with this approach, so a relatively pressure-independent fusion curve is generally expected over a broad pressure range.

Due to the simplistic approximations in this model (in particular, the extrapolation of the equation of state), little credence should be given to the pressure at which the fusion curve is predicted to go through a maximum. In fact, changes in melt structure and properties due to compression may preclude the occurrence of a maximum. Nevertheless, existing data for silicates strongly support the conclusion that the bulk modulus of the liquid is much less than that of the solid and hence the fusion curves of minerals flatten out at high pressure as the molar volume of the liquid approaches that of the solid ($K_{02}/K_{01} \sim 0.4$ to 0.8, the range of values used above, is appropriate: Rigden et al., 1984; Rivers, 1985).

This last point is only qualitatively demonstrated here, but it is likely to have important implications for the chemical differentiation, history of geological or tectonic activity, surface petrology, and long-term thermal evolution of terrestrial (rock or Earth-like) planets (Walker et al., 1978; Stolper et al., 1981; BVSP, 1981). The reason for this is that partial melting is thought to be central to the process of planetary

[5] If the temperature dependence of volume is included, Equation 26 must be modified, but it turns out that the functional form remains the same, to first order, because $\alpha(T)$ approximately follows a $C_V(T)$ form: $\partial\alpha/\partial T \sim \partial C_V/\partial T$ (see Eqn. 15; Suzuki, 1977, and Suzuki et al., 1979). Thus, from Equation 12 Θ varies proportionately to the volume with temperature and V varies as $U(T)$, with the result that the temperature dependence of P_{tr} is still like $U(T)$.

differentiation, and the melt which emerges from the planetary interior determines the near-surface (crustal) composition. Also, considerable amount of heat transfer out of the planet is associated with this melt extraction: a case in point is the mid-ocean ridge system on Earth which accounts for the bulk of the global heat flux. With increasing pressure, the molar volumes of melt and solid approach each other, and hence the density difference between these phases vanishes. As it is the lower density of the melt which causes it to rise relative to the solid (i.e., relative to the source region for the partial melt), melts can only be extracted from relatively low-pressure regions of planets. Thus, melts would tend to emerge out of the bulk only of small planets (e.g., 65 percent of the moon) or from the shallow depths of large planets (e.g., a few hundred kilometers depth or 5 percent of the Earth).

In fact, when the solid and melt volumes approach each other, as they are predicted to do at high pressures, the density difference is dominated by fractionation effects (the compositional difference between liquid and coexisting solid in a compositionally complex system that is partially molten). For example, melts from the Earth's upper mantle are enriched in iron relative to the source region. If this fractionation persists with depth, all that might be required for melts to be trapped within the deep Earth would be that the molar volume of the liquid decrease to within 1 or 2 percent of that of the solid. Below this depth, melts would tend to sink, due to their relatively high density, and partial melting could serve as an impediment to cooling of the planet rather than an enhancement, as is the case at shallower depths.

6. COORDINATION CHANGES

Just as the entropy of fusion can be separated into geometrical and volumetric contributions, the same can be done for solid-solid transformations. Such an analysis is helpful for going beyond a phenomenological level of description, and attempting instead to predict the magnitude and sign of Clapeyron slopes. The essence is in modeling the change in average vibrational frequency, or Θ, across a transition. Clearly, the volumetric contribution can be understood in terms of the effect of compression on the vibrational frequency spectrum through the Grüneisen parameter (Eqn. 12). What is less obvious is how to understand the effect of changing the atomic packing geometry on the vibrational frequencies.

In order to concentrate on this last point, it is useful to consider changes in coordination across high-pressure transformations. Not only are coordination increases common across pressure-induced transformations, but they also involve a drastic change in microscopic

B1 (NaCl-type) Structure

$$z = 8$$
$$n_1 = 6$$
$$n_2 = 12$$

$$r_{ca} = \frac{\sqrt{4}}{4}\, a$$
$$r_{aa} = r_{cc} = a$$

B2 (CsCl-type) Structure

$$z = 2$$
$$n_1 = 8$$
$$n_2 = 6$$

$$r_{ca} = \frac{\sqrt{12}}{4}\, a = 1.09\, r_{ca}^{B1} \left(\frac{V(B2)}{V(B1)} \right)^{\frac{1}{3}}$$
$$r_{aa} = r_{cc} = a$$

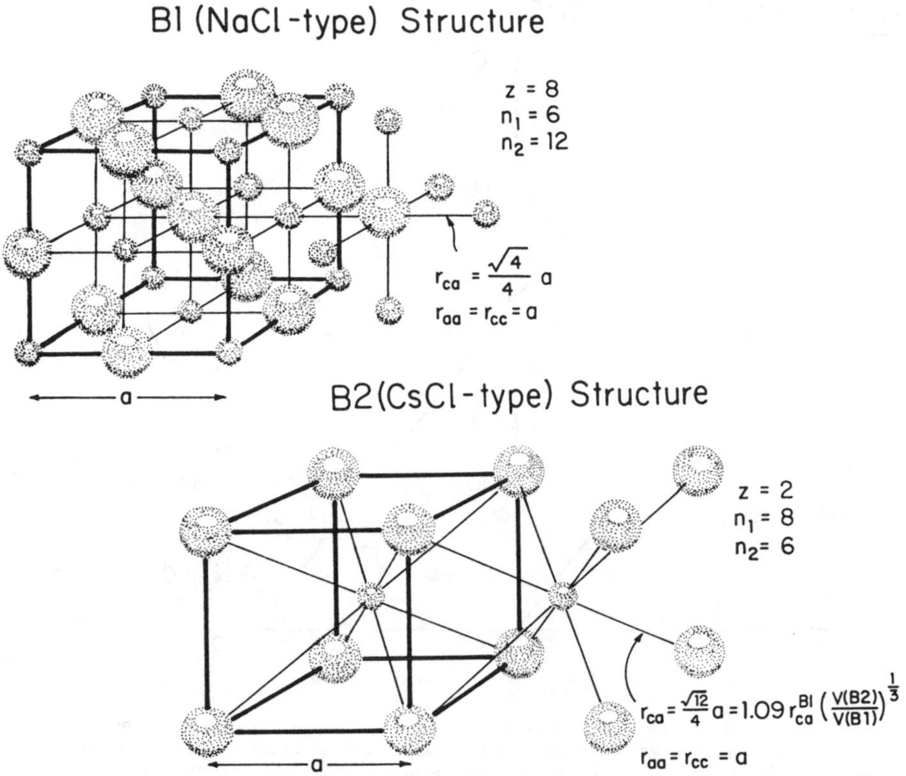

Figure 9. NaCl-type (B1) and CsCl-type (B2) crystal structures, showing the differences in first- (n_1) and second-neighbor (n_2) coordinations. The unit cells are shown with identical volumes (a^3) to illustrate the increased first-neighbor bond lengths (cation-anion distance r_{ca}) in the high-coordination phase. For a real transformation the lattice parameter a (or second-neighbor distances between cations, r_{cc}, and anions, r_{aa}) must decrease from the B1 to the B2 phase. Note that the number of molecules per unit cell also decreases from 4 to 1.

configuration. An example for which a considerable amount of experimental data is available is the B1 (NaCl structure) to B2 (CsCl structure) transition which is observed in many alkali halides and alkaline earth monoxides at elevated pressures (Fig. 9). What is important to recognize about this and other transformations involving coordination increases is that were it not for the volume decrease across the transition, the first-neighbor interatomic distance increases by a large amount (over 9 percent for the B1-B2 transition). In fact, the first-neighbor distance increases for all known B1-B2 transitions, but by smaller amounts for transitions with larger volume decreases. Thermodynamically, the

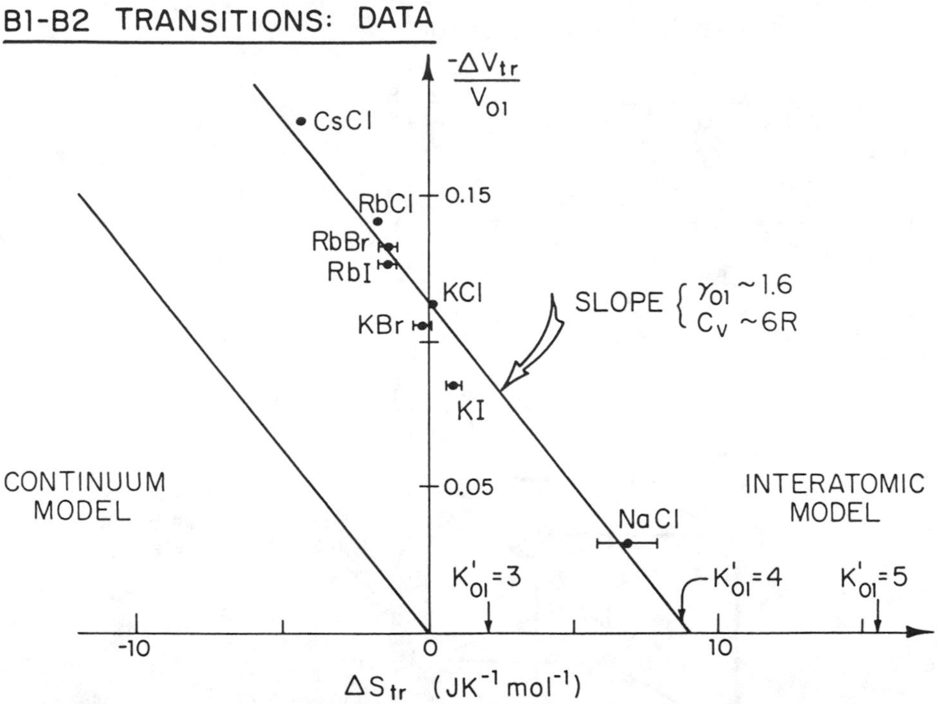

Figure 10. Correlation between the entropy change and volume change at B1-B2 transitions in alkali halides. The continuum model, in which only the volumetric contribution to ΔS_{tr} is considered, reproduces the slope of the observed trend (with the indicated values of Grüneisen parameter, γ, and specific heat, C_V), but incorrectly predicts an intercept of zero entropy change. The intercepts (geometrical entropy change) given by a simple interatomic model (see Appendix B) are correlated with the pressure derivative of the bulk modulus, K'_{01} (arrows in figure) and are consistent with the data, as shown. From Jeanloz (1982), in which original data sources and further discussion can be found.

volume must decrease across a pressure-induced transition, and this is accomplished by second- (and more distant) neighbor distances decreasing. These considerations must be taken into account in order to understand the change in Θ across a solid-state transformation.

If only the volumetric part of ΔS_{tr} is considered, all pressure-induced transformations would have positive Clapeyron slopes with the entropy change directly proportional to the volume change (Eqn. 16). That this is not observed is well illustrated by the Clapeyron slopes of the alkali-halide transformations (Fig. 10). The entropy change is indeed correlated with the volume change, but ΔS_{tr} does not approach zero as ΔV_{tr} becomes small (Bassett et al., 1968). Rather, there is a geometrical entropy change of about 4.5 J K^{-1} $atomol^{-1}$ which makes ΔS_{tr} considerably larger (more positive) than would be expected if only the volume change were taken into account. What is remarkable about this value of ΔS_{geom} is that it is comparable to that found for melting, and yet no disordering is involved.

In order to understand the origin of the change in vibrational entropy due to a change of configuration (as before, defined as being at constant volume), it is necessary to calculate the vibrational frequencies for both structures. A simple lattice dynamical approach has been presented elsewhere (Jeanloz, 1982), and is summarized in Appendix B, but the essence can be understood in terms of what Jeanloz and Roufosse (1982) call a molecular-based model. In this approach, one concentrates on the first-neighbor distance (r) and assumes that it determines the vibrational frequencies. Thus, from Equation 12

$$\frac{\partial \ln \nu}{\partial \ln r} = \frac{\partial \ln \Theta}{\partial \ln r} = -3\gamma \quad , \tag{27}$$

because r is proportional to $V^{1/3}$. In addition, the Grüneisen parameter is observed to decrease with compression such that

$$q \equiv \frac{\partial \ln \gamma}{\partial \ln V} = \frac{1}{3}\frac{\partial \ln \gamma}{\partial \ln r} \tag{28}$$

is positive (for NaCl, $q \cong 1.5$ according to the data of Boehler, 1981, for example).

Aside from the change in interatomic distance, the geometrical entropy change must include the effect of changing coordination. As shown in Appendix B, the frequency is proportional to the square root of the coordination number. This can be qualitatively understood from Equation 8 for the one-dimensional oscillator in that the effective force constant, k, is a sum over the "spring constants" for each bond (six and eight in the case of the B1 and B2 structures, respectively). Combining this result with Equations 27 and 28 yields, upon integration

$$\frac{\Theta(\mathrm{B2})}{\Theta(\mathrm{B1})} = \left[\frac{8}{6}\right]^{1/2} \exp\left[\left[\frac{\gamma(\mathrm{B1})}{q}\right]\left[1 - \left[\frac{r(\mathrm{B2})}{r(\mathrm{B1})}\right]^{3q}\right]\right] \quad . \tag{29}$$

The ratio of first-neighbor bond lengths is determined purely from the geometry of the structures and depends on the volume change at the transition (Fig. 9):

$$\frac{r(\mathrm{B2})}{r(\mathrm{B1})} = \left[\frac{27}{16}\right]^{1/6}\left[1 + \frac{\Delta V_{tr}}{V_1}\right]^{1/3} \quad . \tag{30}$$

To determine the geometrical entropy change at this molecular level, ΔV_{tr} is set to zero so that $r(\mathrm{B2}) = 1.091\ r(\mathrm{B1})$ and $\Theta(\mathrm{B2}) = 0.692\ \Theta(\mathrm{B1})$, according to Equation 29 and with typical values $\gamma(\mathrm{B1}) = 1.6$ and $q = 1.5$. the corresponding geometrical entropy is (Eqn. 7) $\Delta S_{geom} = 9.2$ J K^{-1} atomol^{-1}, which is of the correct sign but much larger than that observed (Fig. 10). By concentrating only on the changes in nearest-neighbor distances and coordination number, a qualitative understanding of how the vibrational frequencies are changed with changing structure (atomic configuration) emerges, but the geometrical effect is overestimated. This is to be compared with the value that is predicted from a purely volumetric or continuum approach, $\Delta S_{geom} = 0$ (Eqn. 16, with $\Delta V_{tr} = 0$).

In general, both the continuum and molecular aspects must be combined in that the former is appropriate for long-range or long-wavelength interactions, whereas the latter is appropriate for short-wavelength interactions. A proper combination of interactions at all length-scales is inherent to lattice dynamical calculations, for example. In the present case, suffice it to say that if the geometrical entropy change is estimated as the average of those derived at the molecular and continuum scales, the resulting value is $\Delta S_{geom} = 4.6$ J K^{-1} atomol^{-1}, which agrees well with the data.

Adding the volume term (Eqn. 16 with $\gamma = 1.6$ and $C_V = 3R$ on a atomol^{-1} basis) gives the relation

$$\Delta S_{tr}^{B1-B2} = 4.6 + 4.8\frac{\Delta V_{tr}}{V_{01}} \text{ JK}^{-1} \text{ atomol}^{-1} \quad , \tag{31}$$

which is in excellent agreement with observations (Fig. 10). Note also that $\Delta S_{tr} = 0$ when $\Delta V_{tr} / V_1 \cong -0.12$. Contrary to what would be predicted if one only considers the molecular scale of interactions, this is not the volume change at which r remains unchanged (from Eqn. 30, $r(B2) = r(B1)$ for $\Delta V_{tr} / V_1 = -0.23$). Rather, at this volume change the molecular and continuum models predict values that are similar but of opposite sign for ΔS; hence, ΔS_{tr} estimated as the average of the values from these two models is zero for $\Delta V_{tr} / V_1 \cong -0.12$, in accord with the data. The main conclusion is that all of the entropy changes (or Clapeyron slopes) for alkali halides, including the trend shown in Figure 10, can be understood in terms of the combined molecular and continuum models presented here.

To some extent, the success of such simple models for entropy changes results from the fact that the average vibrational frequencies tend not to change much across solid-solid, pressure-induced transitions (Fig. 2). For many common minerals, which are characterized by complicated crystal structures, low-pressure transformations often involve a repacking of atomic polyhedra with no change in primary coordination (e.g., quartz—coesite; olivine—β-phase—γ-spinel). As the primary bond distances remain unaffected, ΔS_{tr} is mainly given by the volume contribution, $\gamma C_V \Delta V_{tr} / V$, which is therefore small ($\Delta V_{tr} / V_1 < -0.1$ typically). A large volume decrease is typically associated with an increase in coordination so that the volumetric and geometrical terms in ΔS_{tr} tend to cancel each other out. At very high pressures, ΔV_{tr} can be relatively small, even with a coordination change, so the geometrical contribution can win out and ΔS_{tr} becomes positive (negative Clapeyron slope), although probably not large. Examples of this are provided by the NaCl B1-B2 transformation and probably by the ilmenite-perovskite transformation in $MgSiO_3$. The transition pressures are about 30 and 20 GPa, respectively, and in both cases the coordination increases from sixfold to eightfold (for $MgSiO_3$, both Mg and Si are in sixfold coordination in ilmentite; Mg becomes pseudoeightfold coordinated whereas Si remains sixfold coordinated in the orthorhombically distorted perovskite structure that is observed: Yagi et al., 1978). As noted by Navrotsky (1980) and Jeanloz (1982), one petrological consequence of these crystal-chemical generalizations is that phase transitions deep inside the Earth are more likely to have negative Clapeyron slopes than is the common experience at low pressures. Overall, the dP/dT slopes are expected to be small, and neither positive nor negative phase boundaries can probably be considered more "normal" (see Fig. 2); the existing data suggest that the former are more common only because most experiments to date have been biased towards low pressures.

The geometrical entropy is only one of many properties that can be examined at a microscopic level. With an increase in bond length due to a coordination increase, the average vibrational frequency has been shown to decrease even though the volume decreases by a small amount. Similarly, the bulk modulus does not increase with a coordination change as much as would be expected from the continuum systematics because of the counteracting influence of increased first-neighbor distances. This effect has been experimentally demonstrated in a number of cases (Sato and Jeanloz, 1981; Jeanloz, 1982; Heinz and Jeanloz, 1984b). Also, it helps to explain why the changes in properties across the 670 km discontinuity, which defines the boundary between the Earth's upper and lower mantle, are not consistent with the continuum velocity-density systematics (Appendix A; Jeanloz and Thompson, 1983). That is, because (Mg,Fe)SiO_3 perovskite is formed at about this depth, there is a significant increase in coordination, and the corresponding increase in bulk modulus is smaller than is predicted from the volume change alone.

Anharmonic properties such as the Grüneisen parameter, thermal expansion coefficient and lattice thermal resistivity (reciprocal of the conductivity) normally decrease under compression; that is, with decreasing bond length under pressure (e.g., q in Eqn. 28 is positive). Therefore, the increased bond length across a coordination change would be expected to enhance these properties. As a result, high-pressure phases with high coordination have higher entropy, are more anharmonic and are therefore more like high-temperature phases than would be predicted from their high density (small volume) alone. Roufosse and Jeanloz (1983) used this idea to explain why the lattice thermal conductivity typically decreases across B1-B2 transitions, even though compression normally increases the conductivity. This effect could be of interest for studies of the Earth's internal dynamics because the driving forces for convection (heat transfer by tectonic movements) are enhanced wherever the thermal conductivity is diminished.

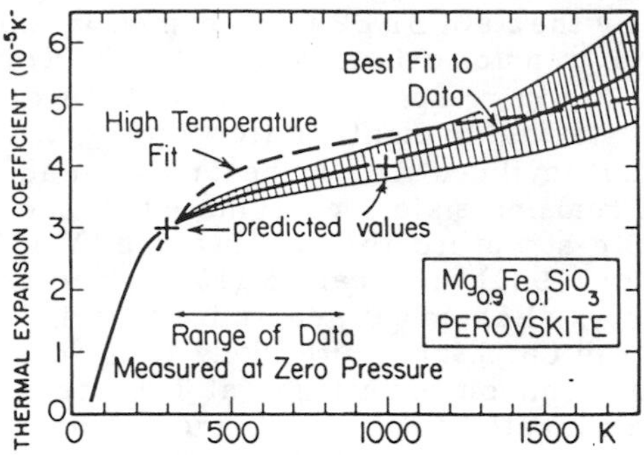

Figure 11. Volumetric thermal expansion coefficient, α, of the high-pressure perovskite phase of $Mg_{0.9}Fe_{0.1}SiO_3$ as measured by Knittle et al. (1985a). Data were collected by X-ray diffraction at zero pressure and between 290 K and 900 K. Two fits to the data are shown: a fit to the high-temperature data alone (dashed) and a fit to all of the data (solid, with shading indicating the estimated uncertainty). In either case, a relatively large value of $\alpha \sim 3.5$ to 4.5×10^{-5} K^{-1} is required at high temperatures. Values predicted by Jeanloz and Roufosse (1982; crosses) are shown for comparison.

Similarly, Jeanloz and Roufosse (1982) predicted that because of high coordination, the thermal expansion coefficient of (Mg,Fe)SiO_3 perovskite is large: comparable to that of the low-pressure phases olivine or magnesiowüstite, and much larger than that of the β -phase and γ -spinel high-pressure (but relatively low coordination) phases. As noted by Jeanloz and Thompson (1983) and Jackson (1983), the predicted value of $\alpha > 2.5 \times 10^{-5}$ K^{-1} at high temperatures is incompatible with a model of uniform composition (pyrolitic or peridotitic) throughout the Earth's mantle. That is, seismically observed densities for the lower mantle are too high to be explained with a conventional upper mantle composition if perovskite has such a large expansivity. Instead, an enrichment of silica or iron relative to pyrolite (for example) would be required for the lower mantle.

Recent measurements (Knittle et al., 1985a) confirm the prediction that α for silicate perovskite is close to that of magnesiowüstite (Fig. 11). This result supports the conclusion that the mantle is stratified, and therefore the upper and lower mantle do not intermix significantly but are separately convecting regions. Based on a microscopic model, which has now been experimentally confirmed, one can conclude that the lower mantle, the largest single region in the Earth, is of different composition and intrinsically denser (by a few percent) than the upper mantle. The exact composition of the lower mantle cannot be uniquely determined. Nevertheless, the evidence for mantle stratification has important consequences for the long-term evolution of the planet because the time scale for cooling a layered mantle is about ten times that for an unstratified mantle (Richter and McKenzie, 1981). Coupled with the idea that melts cannot emerge from the deep interior of a large planet, it becomes evident that the Earth is likely to maintain its internal heat, and hence its tectonic activity or geological vigor, over much longer time scales than is possible for small planets. Is it any surprise, then, that the Moon, Mars and Mercury are geologically dead at present compared to the Earth?

7. PROBLEM SET

Problem 1.

What are the relative effects of pressure and temperature on the average vibrational frequency and hence the high-temperature entropy? Compare the magnitudes involved.

Answer

In our model, the average vibrational frequency, or Θ, depends on volume alone. Thus,

$$\left(\frac{\partial \ln \nu}{\partial T}\right)_P = \left(\frac{\partial \ln \Theta}{\partial T}\right)_P \frac{d\ln\Theta}{d\ln V}\left(\frac{\partial \ln V}{\partial T}\right)_P \tag{1.1}$$

or

$$\left(\frac{\partial \ln \Theta}{\partial T}\right)_P = -\alpha\gamma \ , \tag{1.2}$$

and

$$\left(\frac{\partial \ln \nu}{\partial P}\right)_T = \left(\frac{\partial \ln \Theta}{\partial P}\right)_T = \frac{d\ln\Theta}{d\ln V}\left(\frac{\partial \ln V}{\partial P}\right)_T \tag{1.3}$$

or

$$\left(\frac{\partial \ln \Theta}{\partial P}\right)_T = -\frac{\gamma}{K_T} \ . \tag{1.4}$$

With typical values of the Grüneisen parameter, coefficient of thermal expansion and isothermal bulk modulus ($\gamma \cong 1.5$, $\alpha \cong 2\text{x}10^{-5}\text{K}^{-1}$, $K_T \cong$ 100 GPa), a 1 percent change in average frequency requires temperature and pressure changes of 333 K and 667 MPa (6.7 kbar). Because α is typically so small, the effect of pressure on Θ is very large compared to that of temperature.

The entropy is a function of the single parameter $\tau = T/\Theta$, so either from the above equations or from standard thermodynamic relations

$$dS = \left(\frac{\partial S}{\partial T}\right)_V dT + \left(\frac{\partial S}{\partial V}\right)_T \left(\frac{\partial V}{\partial P}\right)_T dP \tag{1.5}$$

$$= \frac{C_V}{T} dT - \frac{\gamma C_V}{K_T} dP \ . \tag{1.6}$$

Again, using typical high-temperature values for the thermodynamic properties (as above, and C_V = 3 R atomol^{-1}, T = 1000 K), a 333 K temperature change and a 667 MPa pressure change, either of which would be sufficient to change Θ by one percent, affect the entropy by 1 R and −0.03 R, respectively. The entropy is much more sensitive to temperature than pressure, whereas pressure affects the vibrational frequencies more than temperature. Increasing pressure decreases the entropy because Θ increases and therefore τ decreases, but increasing temperature increases S directly both via $\tau = T/\Theta$ and by decreasing Θ (cf. Eqn. 6).

Problem 2.

One unrealistic aspect of the model of fusion presented in this chapter is that it ignores liquid-structure effects such as those associated with coordination change upon melting or the effect of polymerization within the liquid. For example, the first-neighbor coordination decreases from six to about five (average) upon melting of NaCl-structure halides, and SiO_4 tetrahedra act as individual anionic units even in the most dissociated silicate melts, such as for Mg_2SiO_4 (Y. Waseda, 1980, The Structure of Non-Crystalline Materials, McGraw-Hill, New York, 326 pp.). Describe how estimates of the geometrical entropy change (Eqn. 17) and how modeling of the disordering entropy (Eqn. 18) upon melting are affected by these effects of coordination change and melt polymerization. Give numerical results for the two examples mentioned above: NaCl and Mg_2SiO_4.

Answer

The NaCl liquid is five-coordinated on average, which is modeled as an equal mixture of four- and six- coordinated molecules. The geometrical entropy change for the six- to four-coordination change in halides can be emulated by that of the B3 (ZnS structure) → B1 (NaCl structure) transition in the solid. With V(B3) = V(B1), the nearest-neighbor distances are related by r(B3) = 0.8660 r(B1) and the geometrical entropy due to coordination change is given by the "interatomic model" (Eqn. B7) as $\Delta S_{tr}^{HT} = -0.45$ R for $r_0/\rho = 9$, $K'_0 = 4.6$ (this geometrical entropy change varies from −0.41 R to −0.56 R as r_0/ρ and K'_0 vary from 8 and 4.1 to 12 and 5.6). Were it not for the disordering upon melting, the coordination decrease from 6 to 5 associated with the volume increase on melting would result in a ***lowering*** of entropy by about 0.23 R (0.21 R to 0.28 R for the range of r_0/ρ just given). Hence, the entropy change due to disordering is about 0.23 R larger than the values listed in Table 1 for the alkali halides, and would amount to about 0.7 (±0.2) R.

We are considering only a very minor form of polymerization or molecular aggregation in the melt. Instead of a molecule of Mg_2SiO_4 becoming dissociated to seven atoms (as in the monatomic treatment of the disordering entropy), it is more plausibly treated as dissociating to 2 Mg^{2+} ions and one $(SiO_4)^{4-}$ ion. Thus, any monatomic-based estimate for the disordering entropy (e.g., ~ 1 R from Eqn. 18 and Fig. 5) should be reduced by a multiplicative factor of 3/7 (~ 0.43 R).

These two estimates of liquid-structural contributions to thermodynamic properties illustrate the fact that current uncertainties in melt structure translate to uncertainties of a few tenths of R (a few J atomol^{-1} K^{-1}) in modeling the entropy of fusion.

Problem 3.

Model the high-temperature specific heat and entropy of a glass using the simplistic approach of this chapter.

Answer

A glass becomes distinct from the liquid upon cooling through the glass transition temperature, T_g, which is bounded by the melting temperature T_m at infinite cooling rate and a critical temperature T_0 for a hypothetical, infinitely slow cooling rate without crystallization: $T_0 \lesssim T_g \lesssim T_m$ (see, R. Zallen, 1983, The Physics of Amorphous Solids, Wiley, New York, 304 pp.). At T_0, which is never reached in practice, the entropy and volume of the supercooled liquid would be identical to those of the crystal, whereas S and V of the liquid are actually observed to be greater than for the crystal (barring a few exceptions for volume).

Upon cooling to $T_m > T \gtrsim T_g$, the supercooled liquid is fundamentally no different from the liquid above T_m. Its volume is larger than that of the crystal (usually), as is its entropy (due both to the larger volume and to the disordering entropy). The specific heat is often difficult to examine in detail because of the strong influence of a first-order transition on this property (in principle, it becomes unbounded at the transition temperature). Nevertheless, the specific heat is expected to be approximately equal to the high-temperature value for the crystal (Dulong-Petit value of 3 R per atomol). If disordering in the liquid increases with temperature, a plausible situation which has so far been ignored, the specific heat should be somewhat larger in the liquid than in the crystal, and this is often observed.

At T_g, the disordering contribution to the entropy is lost or "frozen out". Hence, entropy abruptly decreases by this amount (~ 0.5 to 1 R atomol^{-1}) and there is a corresponding spike in the specific heat (usually small) as the liquid is cooled through T_g. According to our simple analysis, the remaining entropy difference between glass and crystal at high temperature is simply due to the volume difference and amounts to (see Eqns. 13 and 17):

$$S_{glass} - S_{crystal} \cong \gamma C_V \left(\frac{V_{glass}}{V_{crystal}} - 1 \right) \qquad 3.1$$

with $\gamma C_V \sim 4.5$ R atomol^{-1}. As expected, the entropy difference in Equation 3.1 vanishes as T_g approaches T_0 and the volume of the glass approaches that of the solid. Below T_g, of course, any contribution to the specific heat due to the temperature dependence of the disordering entropy (in the liquid) must vanish. Thus, a specific heat close to that of the crystal is expected for the glass.

The main thermodynamic difference between crystal and glass is that due to the volume difference. There may be associated structural differences, such as the T coordination changes discussed in Problem 2, but these presumably vanish as T_g approaches T_0. The effect of disordering on the average vibrational frequency of the glass, and hence on the high-temperature thermal properties, is not expected to be large according to the present approach. Observations bear this out (see Zallen's book mentioned above).

Problem 4.

In the Debye model, the characteristic frequency is proportional to an average acoustic velocity (Appendix A). Show that this is also the case for a simple potential model of ionic crystals at zero pressure. Use the Born-Mayer model with first-neighbor repulsions only (Appendix B).

Answer

The energy of the crystal structure and its derivatives are given as functions of the nearest-neighbor distance, r, in the Born-Mayer model:

$$\Phi(r) = -\frac{\alpha_M Z^2 e^2}{r} + n\lambda \exp(-r/\rho) \quad , \qquad 4.1$$

$$\Phi'(r) = \frac{\alpha_M Z^2 e^2}{r^2} - n\frac{\lambda}{\rho} \exp(-r/\rho) \quad , \qquad 4.2$$

$$\Phi''(r) = -\frac{2\alpha_M Z^2 e^2}{r^3} + n\frac{\lambda}{\rho^2} \exp(-r/\rho) \quad , \qquad 4.3$$

with α_M, Z, e and n being the Madelung constant, average valence on each atom, charge of the electron, and nearest-neighbor coordination number; λ and ρ are empirical parameters defining the repulsive potential (see M. Born and K. Huang, 1954, Dynamical Theory of Crystal Lattices, Oxford, London, 420 pp., for further details).

The pressure (P) and bulk modulus (K) are given by appropriate derivatives of Φ:

$$P = -\frac{1}{3hr^2}\Phi'(r) = \frac{1}{3hr^2}\left[\frac{-\alpha_M Z^2 e^2}{r^2} + n\frac{\lambda}{\rho}\exp(-r/\rho)\right] \qquad 4.4$$

and

$$K \equiv -\left(\frac{\partial P}{\partial \ln V}\right)_T = \frac{1}{9hr}\Phi''(r) + \frac{2}{3}P \qquad 4.5$$

$$= \frac{1}{9hr}\left[\frac{-2\alpha_M Z^2 e^2}{r^3} + n\frac{\lambda}{\rho^2}\exp(-r/\rho)\right] + \frac{2}{3}P \quad , \qquad 4.6$$

with $h = V/r^3$. Note that for central potentials a term that is directly proportional to P invariably appears in the equations for elastic moduli, as in 4.5 and 4.6. Also, the zero-pressure (subscript zero) equilibrium conditions by which λ and ρ are derived from the observed r_0 (or V_0) and K_0 drop out of Equations 4.4 and 4.6 right away when P is set to zero:

$$\lambda = \frac{\alpha_M Z^2 e^2}{n r_0^2}\rho \exp(r_0/\rho) \qquad 4.7$$

and

$$\rho = \left(\frac{9V_0 K_0}{\alpha_M Z^2 e^2} - \frac{2}{r_0}\right)^{-1} \qquad 4.8$$

Blackman's sum rule for the mean-squared frequency is written here as

$$3s <\Omega^2> = \overline{\mu^{-1}}\, \nabla^2\Phi(r) \quad , \qquad 4.9$$

with s being the number of atoms per formula unit and Ω being in units of s^{-1} (see Appendix B). Because Φ depends only on r, $\nabla^2 = d^2/dr^2 + (2/r)\, d/dr$, and the Coulomb ($1/r$-dependent) term vanishes in Equation 4.9. Thus, at zero pressure:

$$3s <\Omega_0^2> = -\frac{2\alpha_M Z^2 e^2}{r_0^2}\left(\frac{2}{\rho_0} - \frac{1}{\rho}\right)\overline{\mu^{-1}} \qquad 4.10$$

$$= \frac{18}{r_0^2} K_0 V_0 \overline{\mu^{-1}} \quad , \qquad 4.11$$

with the mean inverse mass being given by the masses of the cation and anion, M_c and M_a:

$$\overline{\mu^{-1}} = \frac{1}{2}\left(\frac{1}{M_c} + \frac{1}{M_a}\right) \quad . \qquad 4.12$$

But density, D, is proportional to $(V_0\overline{\mu^{-1}})^{-1}$, so the root-mean-square frequency is proportional to the bulk sound velocity $V_\varphi = \sqrt{K/D}$ as in the Debye model (Appendix A):

$$<\Omega_0^2>^{1/2} = AV_\varphi \propto \nu_D \qquad 4.13$$

with A proportional to $\sqrt{6/sr_0^2}$.

This result was first shown by M. Blackman in 1942 (Proc. Roy. Soc. Lond. A181, 58).

Problem 5.

Empirical interatomic potentials are typically constrained by the volume and bulk modulus of a crystal at zero pressure (Appendix B).

Show that observations on a high-pressure phase transition (transition pressure and volumes of the coexisting phases) can also be used to constrain a simple interatomic potential model, and can therefore be used to model the properties of both phases.

Answer

Under simplifying assumptions, the interatomic potential for nearest neighbor distance r is:

$$\Phi = -\frac{\alpha_M Z^2 e^2}{r} + \sum \varphi(\alpha_{ij} r) \quad , \qquad 5.1$$

with α_M, Z, e, α_{ij} and φ being the Madelung constant, average valence, charge on the electron, ratio of jth coordination-shell radius around atom i to the distance r, and a repulsive pair potential that includes the intrinsic quantum effects (see W.A. Harrison, 1980, Electronic Structure and the Properties of Solids, Freeman, San Francisco, 582 pp.). In principle, α_M, Z and φ can depend on angular strains, internal strains and many-body effects, and these are implicitly included (or ignored). Lattice sums such as in Equation 5.1 are taken in such a way as to avoid double counting.

Equilibrium at the transition pressure (indicated by subscript t) between phases 1 and 2 (subscripts 1, 2) requires

$$P_t(V_{t1} - V_{t2}) + \frac{\alpha_{M2} Z_2^2 e^2 h_2^{1/3}}{V_{t2}^{1/3}} - \frac{\alpha_{M1} Z_1^2 e^2 h_1^{1/3}}{V_{t1}^{1/3}} + \sum \varphi_1(r_{t1}) = \sum \varphi_2(r_{t2}) \qquad 5.2$$

and

$$\left(\frac{V_{t2}}{h_2}\right)^{1/3} \left\{ \left(\frac{V_{t2}}{V_{t1}}\right) \left[\frac{\alpha_{M1} Z_1^2 e^2 h_1}{V_{t1}} + \left(\frac{h_1}{V_{t1}}\right)^{1/3} \sum \alpha_{ij1} \varphi'_1(r_{t1}) \right] - \frac{\alpha_{M2} Z_2^2 e^2 h_2}{V_{t2}} \right\} = \sum \alpha_{ij2} \varphi'_2(r_{t2}) \quad , \qquad 5.3$$

with $h = V/r^3$ being purely a geometrical factor, $\varphi' \equiv d\varphi/dr$, and P and V are pressure and volume. Equations 5.2 and 5.3 assume that P_t, V_{t1} and V_{t2} can be determined for the zero-Kelvin static lattice. As noted in the text, thermal corrections to these variables from room temperature are small and, in any case, can be explicitly derived once φ is known.

A two-parameter potential model can be constrained from Equations 5.2 and 5.3. For example, the Born-Mayer potential between nearest-neighbors only,

$$\varphi = \lambda \exp(-r/\rho) \quad , \qquad 5.4$$

yields

$$\sum \varphi(r_t) = n\lambda \exp(-r_t/\rho) \qquad 5.5$$

and $$\sum \alpha_{ij} \varphi'(r_t) = -n\frac{\lambda}{\rho}\exp(-r_t/\rho) \qquad 5.6$$

with n being the nearest-neighbor coordination number. Substituting Equations 5.5 and 5.6 into 5.2 and 5.3 yields two nonlinear equations in which the only unknowns are λ and ρ, given that the transition pressure and volumes, P_t, V_{t1} and V_{t2}, between two (known) structures has been determined independently.

As illustrated in Problem 4, one can determine the vibrational frequencies and hence the first-order thermal corrections to the transition pressures and volumes once φ is known. In this way, even high temperature values of P_t and V_t are useful constraints on the potential.

Further constraints on Φ, such as the zero-pressure volume and bulk modulus (or pressure derivative) can be used to refine the potential model. For example, more distant-neighbor interactions can be included or φ can be extended to include structure-dependent effects, bond-angle dependencies, or other many body effects. In this regard, Heinz and Jeanloz (1984, Phys. Rev. B 30, 6045) have illustrated how measurements of the equations of state cf two polymorphs can provide further information on interatomic potential models.

APPENDIX A

The Continuum Approach: Debye Model, Velocity-Density Systematics and Entropy-Volume Correlation

In this approach, the phase is treated as an isotropic continuum and there is no attempt to build into the model the actual structure of the crystal or liquid. Probably the most important example is the Debye model for the specific heat (e.g., Slater, 1939; Kieffer, this volume), from which one can derive the velocity-density systematics common in geophysics (Anderson, 1967; Shankland, 1972, 1977). Also, entropy-volume correlations such as those used by Helgeson, et al. (1978) can be explained. Such correlations obviously fail wherever the continuum approach fails.

Lattice vibrational waves of frequence ν and wavelength λ through a crystal at velocity.

$$c = \lambda\nu \quad . \qquad (A1)$$

What Debye recognized is that no wavelengths shorter than interatomic distances can exist in a crystal lattice. Thus, one can take the minimum wavelength as being determined by an average interatomic distance, r. For example,

$$\lambda^3_{\min} = \frac{4}{3}\pi r^3 \quad . \qquad (A2)$$

For simplicity, r is just the edge-length of a box that contains one atom, on average. That is,

$$r^3 = \frac{V}{N} \quad , \tag{A3}$$

with V being volume per atomol and N being Avogadro's number. Therefore, the maximum frequency, which is called the Debye frequency, is, from Equations A1 to A3, simply

$$\nu_D = \frac{c}{\lambda_{\min}} = \frac{c}{(\frac{4}{3}\pi V/N)^{1/3}} \quad . \tag{A4}$$

The corresponding Debye temperature is simply $\Theta_D = h\,\nu_D/k$ (h and k are Planck's constant and Boltzmann's constant, respectively).

Both compressional and shear waves can propagate through the continuum, in general. These travel at acoustic velocities V_P and V_S, so c is an average over these values. Brillouin (1946) has suggested, for example, that for the Debye model a good average is

$$c = \left\{\frac{2}{V_S^3} + \frac{1}{V_P^3}\right\}^{-1/3} \tag{A5}$$

because there are two types of shear wave (horizontal particle motion and vertical particle motion) and one longitudinal wave for a given direction. Note that these are all related to the bulk modulus (K) and density (D) by way of Poisson's ratio (σ):

$$V_P = \left[\frac{3K(1-\sigma)}{D(1+\sigma)}\right]^{1/2} \tag{A6}$$

$$V_S = \left[\frac{3K(1-2\sigma)}{D(2+2\sigma)}\right]^{1/2} \tag{A7}$$

and

$$D = \overline{m}/V \quad , \tag{A8}$$

with $\overline{m}$ being the mean atomic weight (V is still volume per atomol). Note that for a liquid, $\sigma = 0.5$, $V_S = 0$ and $V_P = (K/D)^{1/2} = V_\varphi$, which is the bulk sound velocity.

Taking logarithms and differentiating equation A4 yields

$$\frac{d\ln c}{d\ln V} = -\frac{d\ln c}{d\ln D} = \frac{1}{3} - \gamma \quad , \tag{A9}$$

where

$$\gamma = -\frac{d\ln \nu_D}{d\ln V} \quad . \tag{A10}$$

Obviously, if σ is constant, then the relative changes in compressional or bulk velocities are also given by Equation A9:

$$\frac{d\ln V_P}{d\ln D} = \frac{d\ln V_\varphi}{d\ln D} = \gamma - \frac{1}{3} \quad . \tag{A11}$$

This is the velocity-density correlation that was first empirically found by Birch (1961).

Empirically, the velocity-density correlation is found to satisfy data for minerals and rocks with the logarithmic derivative of Equation A11 between 1 and 2 (Table A1).

Table A1.

EMPIRICAL RESULTS FOR VELOCITY-DENSITY SYSTEMATICS

	$\frac{d\ln V_\varphi}{d\ln D}$
Ultrasonic data on rocks (Birch's law):	1 to 2
Seismic equation of state:	~ 1.5
Effect of pressure, $\frac{1}{2}(K' - 1)$:	~ 1.5
Effect of temperature, $\frac{1}{2}(\delta - 1)$:	~ 1 to 2.5
670 km discontinuity (PREM):	0.36

This is in accord with expectations because γ is typically between 1 and 2. Also, the effects of either pressure or temperature can be examined. For the bulk modulus these are

$$K'_T = \left.\frac{\partial K}{\partial P}\right)_T = -\left(\frac{\partial \ln K}{\partial \ln V}\right)_T \tag{A12a}$$

and

$$\delta_T = \frac{-1}{\alpha K}\left(\frac{\partial K}{\partial T}\right)_P = \frac{-K}{\alpha}\left(\frac{\partial \alpha}{\partial P}\right)_T = -\left(\frac{\partial \ln K}{\partial \ln V}\right)_P \quad . \tag{A12b}$$

Again, the available data suggest that the logarithmic derivative of bulk sound velocity with density is about 1 to 2 for most minerals (Birch, 1952, 1968). In contrast, the velocity change at the 670 km discontinuity is much smaller than would be expected for the density increase (Table A1; Dziewonski and Anderson, 1982). Two possible reasons for this, that the bulk composition of the rock changes and that coordination changes at this depth, are discussed in the text. Both factors seem to be involved.

As the entropy at high temperatures is proportional to the logarithm of the average vibrational frequency (Eqns. 6 and 7),

$$dS^{HT} = 3R\gamma d\ln V \quad , \tag{A13}$$

according to Equation 12. Applying the first part of Equation 28 and integrating yields the entropy volume correlation, with V_0 being a reference volume:

$$S(V) = S(V_0) + 3R\frac{\gamma(V_0)}{q}\left[\left(\frac{V}{V_0}\right)^q - 1\right] \quad . \qquad \text{(A14)}$$

As $0 \lesssim q \lesssim$, the entropy is essentially linearly dependent on volume in this model. As before, these results can only be considered reliable to the extent that the continuum model is.

APPENDIX B

Interatomic Model for Entropy Change with Coordination Change

The bonding forces between atoms i and j (distance r_{ij}) are represented by a central pair potential, φ_{ij} (r_{ij}), which can be determined theoretically or constrained empirically. The mean squared frequency of the lattice is then given by Blackman's sum rule (Blackman, 1942):

$$<\nu^2> = \frac{1}{(2\pi)^2 3s}\sum_{i,j}{}' \frac{1}{M_i}\nabla^2\varphi_{ij}(r_{ij}) \qquad \text{(B1)}$$

with s and M_i being the number of atoms per formula unit and the mass of atom i, respectively, and the prime on the summation means that the "self-term" $i = j = 0$ is excluded. Note that any Coulombic ($1/r$ dependence) term in φ_{ij} cancels in the Laplacian of Equation B1. Thus, for an ionic substance the mean squared frequency is determined solely by the repulsive part of the interatomic potential.

As an example, consider a Born-Mayer model with repulsion between first neighbors (distance r) of the form (e.g., Born and Huang, 1954):

$$\varphi = \lambda \exp(-r/\rho) \quad . \qquad \text{(B2)}$$

In this ionic model, the crystal is held together by Coulomb forces (which can be ignored), and λ and ρ are empirical parameters of the potential which are determined from the zero-pressure equilibrium conditions. Thus, knowing the interatomic distance (r_0) and pressure derivative of the bulk modulus (K'_0) at zero pressure, the potential parameters are given by

$$\lambda = \frac{\alpha_M(ze)^2\rho}{nr_0^2}\exp(r_0/\rho) \qquad \text{(B3)}$$

and

$$K'_0 = \frac{(r_0/\rho)^2 + 3(r_0/\rho) - 12}{3(r_0/\rho) - 6} \quad , \qquad \text{(B4)}$$

with α_M, ze and n being the Madelung constant for the crystal structure, the valence electron charge and the first-neighbor coordination number, respectively. Substituting into Equation B1, the mean-square frequency is

$$<\nu^2> = \frac{n\lambda}{(2\pi)^2 3s\rho}\left[\frac{1}{\rho} - \frac{2}{r}\right]\left(\frac{M_a + M_c}{M_a M_c}\right) \exp(-r/\rho) \quad . \qquad \text{(B5)}$$

Subscripts a, c and 0 have been used to indicate the anion, cation and zero-pressure conditions in these equations.

Assuming that the characteristic temperature can be given by the root mean-square frequency,

$$\Theta = \frac{h}{k}(<\nu^2>)^{1/2} \quad , \qquad \text{(B6)}$$

the high-temperature harmonic contribution to the entropy change is derived from Equation 7:

$$\Delta S_{tr}^{HT} = \frac{3}{2}R\left(\frac{r_2 - r_1}{\rho}\right)\ln\left[\left(\frac{n_1}{n_2}\right)^{1/2}\frac{\left(\frac{1}{\rho} - \frac{2}{r_1}\right)}{\left(\frac{1}{\rho} - \frac{2}{r_2}\right)}\right] \qquad \text{(B7)}$$

on a atomol^{-1} basis (subscripts 1 and 2 indicate the two phases). Note that s and the parameter λ have dropped out of Equation B7, and the ratio Θ_1/Θ_2 is proportional to the square root of the ratio of coordination numbers.

For B1-B2 transformations, the geometrical part of the entropy change (ΔS_{tr} as $\Delta V_{tr} \to 0$) can be directly estimated because $r_2 = (27/16)^{1/6} r_1 \cong 1.091\ r_1$ in this case (Eqn. 30). Substituting also $n_1 = 6$, $n_2 = 8$, Equation B7 reduces to

$$\Delta S_{geom}^{B1-B2} = 1.14\left(\frac{r_1}{\rho}\right) - 12.47\ln\left[\frac{1.33(r_1/\rho) - 2.44}{(r_1/\rho) - 2}\right] \qquad \text{(B8)}$$

in J K^{-1} atomol^{-1}. As r/ρ is determined by K' (Eqn. B4), the geometrical entropy change can be related to the pressure derivative of the bulk modulus. The result is shown in Figure 10 (Interatomic Model) and is in fairly good agreement with the data. Further details are given in Jeanloz (1982) and Jeanloz and Roufosse (1982).

ACKNOWLEDGEMENTS

Work supported by the A.P. Sloan Foundation, Exxon Research and Engineering Company, and the National Science Foundation. I thank E. Knittle, P. Lichtner and J. Stebbins, as well as the Short Course organizers, A. Navrotsky and S.W. Kieffer, for thoughtful comments which have considerably improved the presentation.

REFERENCES

Ahrens, T.J. and Schubert, G. (1975) Gabbro-eclogite reaction rate and its geophysical significance. Rev. Geophys. Space Phys. 13, 383-400.

Akaogi, M. and Navrotsky, A. (1984) The quartz-coesite-stishovite transformations: New calorimetric measurements and calculation of phase diagrams. Phys. Earth Planet. Int. (in press)

________ Ross, N.L., McMillan, P. and Navrotsky, A. (1984) The Mg_2SiO_4 polymorphs (olivine, modified spinel and spinel)—thermodynamic properties from oxide melt solution calorimetry, phase relations, and models of lattice vibrations. Am. Mineral. 69, 499-512.

Anderson, D.L. (1967) A seismic equation of state. Geophys. J. R. Ast. Soc. 13, 9-30.

________(1981) Plate tectonics on Venus. Geophys. Res. Lett. 8, 309-311.

BVSP [Basaltic Volcanism Studies Project] (1981) Basaltic Volcanism on the Terrestrial Planets. Pergamon, New York, 1286 pp.

Bassett, W.A., Takahashi, T., Mao, H.K. and Weaver, J.S. (1968) Pressure-induced phase transformation in NaCl. J. Appl. Phys. 39, 319-325.

Bernal, J.D. (1959) A geometrical approach to the structure of liquids. Nature 183, 141-147.

________(1964) The structure of liquids. Proc. Roy. Soc. London A280, 299-322.

Birch, F. (1952) Elasticity and constitution of the Earth's interior. J. Geophys. Res. 57, 227-286.

________(1961) The velocity of compressional waves in rocks to 10 kilobars, Part 2. J. Geophys. Res. 66, 2199-2224.

________(1968) Thermal expansion at high pressures. J. Geophys. Res. 73, 817-819.

________(1977) Isotherms of the rare gas solids. J. Phys. Chem. Solids 38, 175-177.

________(1978) Finite strain isotherm and velocities for single-crystal and polycrystalline NaCl at high pressures and 300° K. J. Geophys. Res. 83, 1257-1268.

Blackman, M. (1942) On the relation of Debye theory and the lattice theory of specific heats. Proc. R. Soc. London A181, 58-67.

Boehler, R. (1981) Adiabats $(\partial T/\partial P)_s$ and Grüneisen parameter of NaCl up to 50 kilobars and 800°C. J. Geophys. Res. 86, 7159-7162.

Born, M. and Huang, K. (1954) Dynamical Theory of Crystal Lattices. Oxford, London, 420 pp.

Brillouin, L. (1946) Wave Propagation in Periodic Structures. McGraw-Hill, New York, 247 pp.

Christian, J.W. (1975) The Theory of Transformations in Metals and Alloys, Part I. Pergamon, New York, 586 pp.

Dziewonski, A.M. and Anderson, D.L. (1982) Preliminary reference Earth model. Phys. Earth Planet. Int. 10, 12-48.

Fairbairn, H.W. (1943) Packing in ionic minerals. Bull. Geol. Soc. Am. 54, 1305-1374.

Goodstein, D.L. (1975) States of Matter. Prentice-Hall, Englewood Cliffs, New Jersey, 500 pp.

Harrison, W.A. (1980) Electronic Structure and the Properties of Solids. Freeman, San Francisco, 582 pp.

Heinz, D.L. and Jeanloz, R (1984a) The equation of state of the gold calibration standard. J. Appl. Phys. 55, 885-893.

________ and Jeanloz, R. (1984b) Compression of the B2 high-pressure phase of NaCl. Phys. Rev. B30, 6045-6050.

Helgeson, H.C., Delaney, J.M. and Nesbitt, N.W. (1978) Summary and critique of the thermodynamic properties of rock-forming minerals. Am. J. Sci. 278A, 1-229.

Ito, E. and Yamada, H. (1982) Stability relations of silicate spinels, ilmenites and perovskites. In: High-Pressure Research in Geophysics, S. Akimoto and M.H. Manghnani, eds. Center Acad. Pub., Tokyo, 405-419.

Jackson, I. (1977) Melting of some alkaline-earth and transition-metal fluorides and alkali fluoberyllates at elevated pressures: A search for melting systematics. Phys. Earth Planet. Int. 14, 143-164.

________ (1983) Some geophysical constraints on the chemical composition of the Earth's lower mantle. Earth Planet. Sci. Lett. 62, 91-103.

———Liebermann, R.C. and Ringwood, A.E. (1974) Disproportionation of spinels to mixed oxides: Significance of cation configuration and implications for the mantle. Earth Planet. Sci. Lett. 24, 203-208.

———and Niesler, H. (1982) The elasticity of periclase to 3 GPa and some geohysical implications. In: High-Pressure Research in Geophysics, S. Akimoto and M.H. Manghnani, eds. Center Acad. Pub., Tokyo, 93-113.

Jeanloz, R. (1982) Effect of coordination change on thermodynamic properties. In: High-Pressure Research in Geophysics, S. Akimoto and M.H. Manghnani, eds. Center Acad. Pub., Tokyo, 479-498.

———and Roufosse, M.C. (1982) Anharmonic properties: Ionic model of the effects of compression and coordination change. J. Geophys. Res. 87, 10763-10772.

———and Thompson, A.B. (1983) Phase transitions and mantle discontinuities. Rev. Geophys. Space Phys. 21, 51-74.

Knittle, E. and Jeanloz, R. (1984) Structural and bonding changes in cesium iodide at high pressures. Science 223, 53-56.

———Jeanloz, R. and Smith, G. (1985a) The thermal expansion of silicate perovskite and stratification of the Earth's mantle (submitted).

———Rudy, A. and Jeanloz, R. (1985b) High-pressure phase transition in CsBr. Phys. Rev. B31, 588-590.

Moore, W.J. (1972) Physical Chemistry. Prentice-Hall, Englewood Cliffs, New Jersey, 977 pp.

Navrotsky, A. (1977) Calculation of effect of cation disorder on silicate spinel phase boundaries. Earth Planet. Sci. Lett. 33, 437-442.

———(1980) Lower mantle phase transitions may generally have negative pressure-temperature slopes. Geophys. Res. Lett. 7, 709-711.

Oriani, R.A. (1951) The entropies of melting of metals. J. Chem. Phys. 19, 93-97.

Putnis, A. and McConnell, J.D.C. (1980) Principles of Mineral Behavior. Blackwell, Oxford, 257 pp.

Richter, F.M. and McKenzie, D.P. (1981) On some consequences and possible causes of layered mantle convection. J. Geophys. Res. 86, 6133-6142.

Rigden, S.M., Ahrens, T.J. and Stolper, E.M. (1984) Densities of liquid silicates at high pressures. Science 226, 1071-1074.

Ringwood, A.E. (1975) Composition and Petrology of the Earth's Mantle. McGraw-Hill, New York, 618 pp.

Rivers, M.L. (1985) Ph.D. Thesis, Univ. California, Berkeley, CA.

Roufosse, M.C. and Jeanloz, R. (1983) Thermal conductivity of minerals at high pressure: The effect of phase transitions. J. Geophys. Res. 88, 7399-7406.

Sato, Y. and Jeanloz R. (1981) Phase transition in SrO. J. Geophys. Res. 86, 11773-11778.

Schubert, G., Yuen, D.A. and Turcotte, D.L. (1975) Role of phase transitions in a dynamic mantle. Geophys. J. Roy. Astron. Soc. 42, 705-735.

Shankland, T.J. (1972) Velocity-density systematics: Derivation from Debye theory and the effect of ionic size. J. Geophys. Res. 77, 3750-3758.

——— (1977) Elastic properties, chemical composition and crystal structure of minerals. Geophys. Surveys 3, 60-100.

Shimoji, M. (1977) Liquid Metals Academic, New York, 391 pp.

Slater, J.C. (1939) Introduction to Chemical Physics. Dover, New York, 521 pp.

Stanley, H.E. (1971) Introduction to Phase Transitions and Critical Phenomena. Oxford, 308 pp.

Stebbins, J.F., Carmichael, I.S.E. and Moret, L.K. (1984) Heat capacities and entropies of silicate liquids and glasses. Contrib. Mineral. Petrol. 86, 131-148.

Stishov, S.M. (1969) Melting at high pressures. Sov. Phys. Usp. 11, 816-830.

Stolper, E., Walker, D., Hager, B.H. and Hays, J.F. (1981) Melt segregation from partially molten source regions: The importance of melt density and source region size. J. Geophys. Res. 86, 6261-6271.

Stroud, D. and Ashcroft, N.W. (1972) Theory of melting of simple metals: Application to Na. Phys. Rev. B5, 371-383.

Suzuki, I (1975) Thermal expansion of periclase and olivine and their anharmonic properties. J. Phys. Earth 23, 145-149.

———Okajima, S. and Seya, K. (1979) Thermal expansion of single-crystal manganosite. J. Phys. Earth 27, 63-69.

Tallon, J.L. (1982) The fundamental entropy change in transitions from the liquid state. Phys. Lett. 87A, 361-364.

Walker, D., Stolper, E.M. and Hays, J.F. (1978) A numerical treatment of melt/solid segregation: Size of the eucrite parent body and stability of the terrestrial low-velocity zone. J. Geophys. Res. 83, 6005-6013.

Wallace, D.C. (1972) Thermodynamics of Crystals. Wiley, New York, 484 pp.

Waseda, Y. (1980) The Structure of Non-Crystalline Materials. McGraw-Hill, New York, 326 pp.

Wolf, G.H. and Jeanloz, R. (1984) Lindemann melting law: Anharmonic correction and test of its validity for minerals. J. Geophys. Res. 89, 7821-7835.

Yagi, T., Mao, H.K. and Bell, P.M. (1978) Structure and crystal chemistry of perovskite-type $MgSiO_3$. Phys. Chem. Minerals 3, 97-110.

Zener, C. (1947) Contributions to the theory of beta-phase alloys. Phys. Rev. 71, 846-851.